Plant cell structure and metabolism
Second edition

J. L. Hall B.Sc., D. Phil., A.R.C.S.
Professor of Biology,
University of Southampton

T. J. Flowers B.Sc., M.Sc., Ph.D.
Lecturer in the School of Biological
Sciences, University of Sussex

R. M. Roberts M.A., D. Phil.
Professor in Biochemistry,
J. Hillis Miller Health Center,
University of Florida

Longman
London and New York

Longman Group Limited
Longman House
Burnt Mill, Harlow, Essex, UK

Published in the United States of America
by Longman Inc., New York

First published 1974
Reprinted with minor corrections, 1976
Second edition 1982

British Library Cataloguing in Publication Data

Hall, J. L.
    Plant cell structure and metabolism.—2nd ed.
    1. Plant cells and tissues
    I. Title II. Flowers, T. J. III. Roberts, R. M.
    581.87'61   QK725

    ISBN 0-582-44408-X

Printed in Singapore by Singapore National Printers

Contents

Preface to the first edition

This book is written for students in plant physiology and biochemistry; it considers plant metabolism in terms of cell structure rather than under the broader headings of lipids, carbohydrates, nitrogen metabolism, etc. In recent years a great deal has been learnt of the relationship between structure and function in plant cells and of the biochemical properties of the various organelles and other cellular inclusions. It is now known that what we see by microscopy is not only the result of past biochemical activity but itself represents ordered units of metabolic activity within the cell. Thus it is now possible to approach the study of plant metabolism in a different way. For example, textbooks written five or less years ago contained little or no reference to the function of microbodies, Golgi bodies and lysosomes in plant cells but so much progress has now been made that we feel that each of these structures warrants a separate chapter. We now know more of the relationships between cellular structures: between nucleus and cytoplasm; between chloroplast, mitochondria, and microbodies; between cell wall, Golgi bodies and microtubules; and of the ribosomes associated with some organelles. However, this is not a book entirely about plants since considerably more is known of certain subjects, e.g. protein synthesis, in relation to microbial or animal systems. In these instances, we have discussed the most well described system and indicated the extent of our knowledge in plants.

The considerable diversity in the structure of courses in British and American Universities and Colleges makes it difficult to state very precisely the level at which the text is aimed. We assume some knowledge of chemistry but otherwise the text may be suitable for various levels. The first two chapters are essentially introductory. Chapter 1 provides an outline of cell structure and familiarizes the reader with some of the techniques used in cell science. The second chapter gives a sufficient background of cell chemistry for the book to be read by the non-specialist and without the aid of other biochemical texts. Membranes, which form the main basis for cellular compartmentation, are then discussed in some detail. The remaining chapters are concerned with the structure and biochemical properties of the soluble phase of the cell and of the major cellular organelles and are arranged so that related topics are grouped. The metabolic and structural relationships that exist between the various organelles mean that the text contains numerous cross references. Furthermore, since each chapter may often be read as a separate entity, material may be summarized in one chapter which is dealt with more fully elsewhere.

The literature cited is not intended to cover each subject fully. However, since we feel that it is worthwhile for students to consult original publications whenever possible, we quote a number of important papers in each chapter; these papers appear in journals found in most libraries. The 'Further Reading' lists are designed to give a sample of up-to-date books and reviews which treat certain important topics in more detail than is possible in this text. References included in the 'Further Reading' lists are not cited again.

We would like to thank a number of physiologists and biochemists who have read and criticized various chapters in this book. These are: Dr C. A. Allen, Dr M. Baig, Dr J. Barber, Dr R. J. Ellis, Dr G. E. Hobson, Dr Rachel M. Leech, Dr R. J. Mans, Dr R. J. Miller, Dr T. W. O'Brien, Dr R. Sexton, Dr R. G. Stanley, Dr P. A. Whittaker. Any errors of fact or interpretation that remain are entirely of our own making. We are also grateful to Dr H. Aldrich, Dr E. H. Newcomb, Dr B. E. Juniper, Dr Rachel M. Leech, Professor R. D. Preston, Dr J. Pickett-Heaps, Dr H. H. Mollenhauer and many others who kindly provided micrographs and other illustrative material which is such an important part of the text. Our thanks are also due to Mr Robert Welham and Dr Gillian Harbinson for their help in editing the manuscript, Miss Carol Davie and Mrs Margaret Ward who prepared many of the figures and Mrs Sylvia Wilkinson and Mrs Phyllis Mann who did much of the typing.

J. L. Hall
T. J. Flowers
R. M. Roberts

Preface to the second edition

The aims and general format of this second edition remain very much as with the first, although a considerable proportion of the material has been rewritten to bring the text up-to-date and several new topics have been added. The chapter on membranes has been expanded to cover new developments and includes a general section on transport processes which relates to many other chapters in the book. We have also included a section on nitrogen metabolism which has been placed in the chapter on chloroplasts since a number of important enzymes of nitrogen metabolism are located in this organelle. In addition, there is an extra chapter on isolated protoplasts since these structures are now widely used in plant cell biology and are referred to in various parts of the book.

Once again we should like to thank a number of physiologists and biochemists who have read and criticized various chapters in the book. These are Professor D. A. Baker, Dr J. A. Bryant, Dr J. E. Kay, Dr L. J. Ludwig, Dr A. L. Moore, Dr G. R. Stewart, Dr A. R. D. Taylor, Dr I. Vasil, Dr M. Wallis and Dr A. R. Yeo. Any errors of fact or interpretation that remain are entirely of our own making. We are also grateful to numerous researchers who provided micrographs and other illustrations which are such a valuable part of the text. We would also like to thank Mrs Jill Davey and Mrs Nicola Ford who did much of the typing.

J. L. H.
T. J. F.
R. M. R.

Abbreviations frequently used in the text

AMP, ADP, ATP	adenosine 5′-mono-, di- and triphosphate
ATPase	adenosine triphosphatase
Co-A	coenzyme A
DCMU	3-(3,4-dichlorophenyl 1)-1,1 dimethylurea
DNA	deoxyribonucleic acid
E'_0	standard oxidation reduction potential
EDTA	ethylenediaminetetracetic acid
EMP pathway	Embden-Meyeroff-Parnas or Glycolytic pathway
ER	endoplasmic reticulum
Fd	ferredoxin
FAD	flavin adenine dinucleotide
FMN	flavin mononucleotide
FP	flavoprotein
GTP	guanosine triphosphate
K_m	Michaelis constant
NAD	nicotinamide adenine dinucleotide
NAD^+	oxidized NAD
NADH	reduced NAD
NADP	nicotinamide adenine dinucleotide phosphate
$NADP^+$	oxidized NADP
NADPH	reduced NADP
P_i	inorganic phosphate
PP_i	inorganic pyrophosphate
PEP	phosphoenolpyruvate
PPP	pentose phosphate pathway
RNA	ribonucleic acid
mRNA	messenger RNA
rRNA	ribosomal RNA
tRNA	transfer RNA
S	Svedberg unit
TCA cycle	tricarboxylic acid cycle (Krebs' cycle, citric acid cycle)
tris	2 amino-2 hydroxymethylpropane, 1-3 diol
UDPG	uridine diphosphoglucose

Abbreviated titles of journals cited in the text

Adv. Bot. Res.	Advances in Botanical Research.
Adv. Carbohyd. Chem.	Advances in Carbohydrate Chemistry.
Adv. Enzymol.	Advances in Enzymology.
Amer. J. Bot.	American Journal of Botany.
Ann. Bot.	Annals of Botany.
Ann. Inst. natl. Rech. agron.	Annales de l'Institut national de la Recherche agronomique.
Ann. Rev. Biochem.	Annual Review of Biochemistry.
Ann. Rev. Microbiol.	Annual Review of Microbiology.
Ann. Rev. Physiol.	Annual Review of Physiology.
Ann. Rev. Plant Physiol.	Annual Review of Plant Physiology.
Arch. Biochem. Biophys.	Archives of Biochemistry and Biophysics.
Bact. Rev.	Bacteriological Reviews.
Biochem. Soc. Trans.	Biochemical Society Transactions.
Biochemistry	Biochemistry.
Biochem. Biophys. Res. Commun.	Biochemical and Biophysical Research Communications.
Biochem. J.	Biochemical Journal.
Biochem. Soc. Symp.	Biochemical Society Symposia.
Biochem. Z.	Biochemische Zeitschrift.
Biochim. Biophys. Acta.	Biochimica et Biophysica Acta.
Biol. Rev.	Biological Reviews.
Brit. med. Bull.	British Medical Bulletin.
Can. J. Bot.	Canadian Journal of Botany.
Carls. Res. Commun.	Carlsberg Research Communications.
Circulation	Circulation.
Expl. Cell Res.	Experimental Cell Research.
Eur. J. Biochem.	European Journal of Biochemistry.
FEBS Letters	Federation of European Biochemical Societies Letters.
Int. Rev. Cytol.	International Review of Cytology.
Int. Rev. expl. Path.	International Review of Experimental Pathology.
J. Amer. Chem. Soc.	Journal of the American Chemical Society.
J. Am. Oil Chem. Soc.	Journal of the American Oil Chemists Society.
J. biol. Chem.	Journal of Biological Chemistry.
J. Cell Biol.	Journal of Cell Biology.

J. cell. comp. Physiol.	Journal of Cellular and Comparative Physiology.
J. Cell Sci.	Journal of Cell Science.
J. exp. Bot.	Journal of Experimental Botany.
J. gen. Physiol.	Journal of General Physiology.
J. Lipid Res.	Journal of Lipid Research.
J. Microscopie	Journal de Microscopie.
J. molec. Biol.	Journal of Molecular Biology.
J. Roy. micr. Soc.	Journal of the Royal Microscopical Society.
J. theoret. Biol.	Journal of Theoretical Biology.
J. Ultrastruct. Res.	Journal of Ultrastructural Research.
Life Sci.	Life Sciences.
Molec. gen. Genet.	Molecular and general Genetics.
Nature, Lond.	Nature.
Nature New Biol.	Nature New Biology.
New Phytol.	New Phytologist.
Phil. Trans. Roy. Soc. Ser. B	Philosophical Transactions of the Royal Society Series B.
Physiol. Plant.	Physiologia Plantarum.
Physiol. Rev.	Physiological Reviews.
Phytochemistry	Phytochemistry.
Plant Physiol.	Plant Physiology.
Plant Cell Physiol.	Plant and Cell Physiology.
Planta	Planta.
Proc. natl. Acad. Sci. U.S.A.	Proceedings of the National Academy of Sciences of the United States of America.
Proc. N.Y. Acad. Sci.	Proceedings of the New York Academy of Sciences.
Proc. Roy. Soc.	Proceedings of the Royal Society.
Protoplasma	Protoplasma.
Protoplasmatologia	Protoplasmatologia.
Quart. Rev. Biophys.	Quarterly Review of Biophysics.
Science, N.Y.	Science.
Sci. Amer.	Scientific American.
Smithson. misc. Collns.	Smithsonian Miscellaneous Collections.
Sub-cell. Biochem.	Sub-cellular Biochemistry.
Symp. Soc. exp. Biol.	Symposia of the Society for Experimental Biology.
Symp. Quant. Biol.	Symposia of Quantitative Biology.
Trans. Am. Soc. Artificial Internal Organs	Transactions of the American Society of Artificial Internal Organs.
Z. Naturforschg.	Zeitschrift für Naturforschung.
Z. Pflanzenphysiol.	Zeitschrift für Pflanzenphysiologie.

1 Introduction to cell science

Cells were first described by Robert Hooke in 1665 when he used his simple microscope to examine sections of cork. He used the term 'cell' to describe the small walled units which could be seen in plant tissues; the biological significance of this discovery was, however, not appreciated until some considerable time later. The wall itself was considered to be the plant tissue while the existence and importance of the cell contents were not realized. However, as microscopes and preparative techniques improved during the next 100 years or so, it became possible to examine the contents of cells. Between 1831–33 Robert Brown described a small, usually spherical, body which he called the nucleus, as a regular feature of plant cells and, at about the same time (1846), Mohl gave the name *protoplasm* to the thin mucilaginous layer found inside the cell wall of living plant cells. Later the term *cytoplasm* was used to denote the whole living material except for the nucleus. Still later, about 1880, Schimper and Meyer are generally credited with the discovery of the plastids, a heterogeneous group of cell organelles that includes the chloroplasts. Thus by about 100 years ago the largest structural elements of the plant cell (wall, cytoplasm-containing nucleus and plastids, central vacuole) had been described. These discoveries resulted from developments in microscope construction and improvements in preparative and staining techniques; in later sections of this book we shall see further examples of the correlation between our knowledge of cell biology and progress in the techniques for studying cells.

At the same time the discoveries of cell structure were being made, the important generalization known as the cell theory was emerging. This is generally associated with the names of Schleiden and Schwann who, in 1838–39, were the first to bring together the ideas and discoveries of the time into a generalized theory which stated that cells containing nuclei are the fundamental units of structure for both plants and animals. Twenty years later, in 1859, Virchow propounded the next important generalization: that cells arise only from the pre-existing cells by division. By 1866 Haeckel had realized that the nucleus was responsible for the storage and transmission of hereditary characters.

Since the formulation of these ideas, our knowledge of cells has advanced quite rapidly and has been derived from two basically different approaches. On one hand, there has been the microscopic examination of intact cells which for a long time relied on developments in light microscopy but was greatly stimulated in the 1950s by the introduction of the electron microscope into biological research. Closely associated with microscopy has been the development of techniques such as enzyme cytochemistry and autoradiography that allow the visual localization of certain molecules in intact cells and tissues. In contrast to the microscopical approach, biochemists have attempted to investigate the chemistry of cells by the techniques which involve disruption of cell structure, the separation of cell parts by centrifugation, and the examination of the chemical activity of the separated parts. For a long time these two disciplines, microscopy and biochemistry, remained apart but in the last thirty years or so they have been merged in an attempt to relate structure and function in living cells.

The results of the application of these methods to a wide variety of plants, animals and microorganisms has led to a new and interesting expression of the ideas of the cell theory by Lwoff (1962). He writes that when living organisms are considered at the cellular level, one sees unity. There is unity of plan since each cell possesses a nucleus embedded in cytoplasm; unity of function since the metabolism is basically the same for all cells; and unity of composition since all cells are composed of the same basic macromolecules, which in turn are composed of the same small molecules. Many examples of this unity will be found in the course of the book.

The plant cell

The cell is the fundamental unit of living things, in terms of both structure and function. In simple terms, the plant cell can be divided into two main regions: the cell wall and the protoplast. The latter consists of the nucleus, the cytoplasm, which contains a variety of membrane-bound organelles, and frequently, a large vacuole containing an aqueous medium of salts and various organic molecules acting as a reservoir for the metabolic activity of the cell. However, although the great majority of plant cells conform to this general picture, there is considerable variation in their size, shape and structure and this is normally closely related to the function of the particular cell type.

Size

At whatever level we study cells we must attempt to express the size of objects under examination in some standard unit so that we can compare one cell or cell part with another, and one person can relate his observations with anothers. Cells are generally small so that the units chosen to express cell size are subdivisions of a metre. The most widely used are:

millimetre	(mm)	10^{-3} m
micrometre	(μm = μ = micron)	10^{-6} m
nanometre	(nm = mμ)	10^{-9} m
picometre	(pm = $\mu\mu$)	10^{-12} m

There is considerable variation in the literature in the units and abbreviations used by various workers but the abbreviations mm, μm, nm and pm have now been adopted as part of the *Systeme Interna-tionale* and will be used throughout this book. The Ångstrom unit (Å) (named after the Swedish physicist) is 10^{-4} μm and was originally used for expressing wavelengths of light and X-rays. It has been widely used in electron mircoscope studies of cells since it conveniently covers the size range involved. However, there is now a tendency to use the nanometre when describing objects of this order of magnitude and so this is the unit which will be used in this book.

The size of plant cells varies considerably within quite broad limits. The smallest, such as the cells of some blue-green algae, are about 0·5 μm in diameter, whereas the largest, such as the green algae *Valonia* or *Nitella* may be several centimetres in diameter or length respectively. These large cells are atypical having many nuclei embed-ded in the peripheral cytoplasm (coenocytic). There may also be considerable variation in cell size within a particular tissue. This is illustrated in Fig. 1.1 which shows the dimensions of cells along the axis of the root tip of *Zea mays*. The dividing cells in the root meristem are the smallest and there is a considerable increase in all of the parameters measured within a few millimetres of this region.

However, when considered on the scale of dimensions of all living matter, cell size falls within a relatively narrow band. The majority of higher plant cells fall within the range 20–300 μm in length or

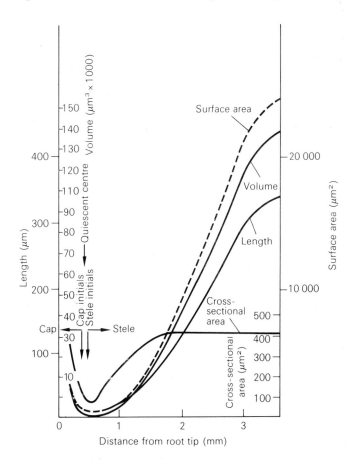

Fig. 1.1   The dimensions of cells along the axis of a maize root tip. Reproduced from Clowes and Juniper (1968).

3

diameter, and it is interesting to consider the factors which may operate to control cell size. The most important seem to be concerned with the ratio of surface area to volume for various cell parts which, in turn, is directly related to the exchange of nutrients and metabolites between the cell and its environment and between the various intracellular organelles. The most critical ratios are probably those between the surface area of the nucleus and the cytoplasmic volume, and between the cytoplasmic volume and the surface area of the cell. If we consider the cell as a sphere containing a spherical nucleus then, as the cell enlarges, the surface area of the cell and of the nucleus will increase with the square of their radii (area of sphere $= 4\pi r^2$) whereas the cytoplasmic volume will expand with the cube of the radius (volume of sphere $= \frac{4}{3}\pi r^3$). The nucleus is the major control centre of the cell, and this control depends upon the diffusion of specific molecules, produced in the nucleus, into the cytoplasm. As the cell volume increases the control exercised by the nucleus becomes limited by its surface area, and the cell tends to divide to restore the optimum nucleocytoplasmic ratio. Similarly, for active metabolism, there must be an efficient exchange of substances between the environment and the various regions of the cell: oxygen and nutrients must diffuse to all parts of the cell and carbon dioxide must not be allowed to accumulate. These considerations will clearly be governed by the volume of the cell and its surface area. In general, it is found that the smaller the cell the more active is its metabolism. Plant cells can overcome these limitations to some extent. For example, the shape of the nucleus can show considerable irregularity resulting in an increase in surface area, while cytoplasmic streaming can help in the movement of substances throughout the cell. The path of diffusion to the cell surface can be kept to a minimum (by changes in cell shape) as the cell volume increases. Cells can become flattened or elongated, and the presence of a large vacuole pushing the cytoplams to a narrow region against the cell surface must aid the efficient exchange of metabolites with the exterior and reduce cytoplasmic volume relative to cell size. Otherwise, when cells reach a certain size they must divide to restore the optimum surface to volume ratios.

Shape

As we have indicated above, the shape of plant cells can have an extremely important bearing on cell metabolism. The shape of some cells, such as Amoeba, is variable but most plant cell types have a typical, more or less rigid, shape which is closely related to their function and to the stresses operating on them during development while they are still plastic. Many unicellular plants and isolated cells grown in culture have a roughly spherical shape, but in tissues they frequently become polyhedral as a result of contact with other cells. Similar shapes can be seen in groups of soap bubbles pressed together. Calculations and observations on soap films have shown that the maximum utilization of space when spheres are compressed is produced by a fourteen-sided tetrakaidecahedron (Fig. 1.2). Measure-

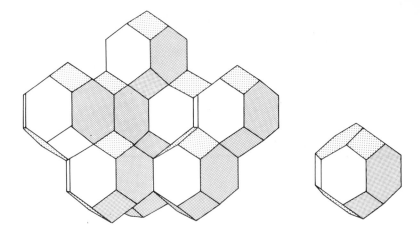

Fig. 1.2   The utilization of space by orthic tetrakaidecahedra. Reproduced from Clowes and Juniper (1968).

ments on roughly isodiametric cells in a number of different plant tissues, such as root apices and stem pith, have shown that although there is considerable variation in the number of faces per cell, fourteen-sided cells are the most prominent and the average number of faces per cell is close to fourteen. However it is clear from an examination of plant tissues that considerable deviation exists in cell shape from the ideal predicted by the compression of spheres. The reasons for this are numerous. Cell division is not synchronous and may vary in rate between cells throughout a tissue, producing considerable differences in cell size and lack of uniformity in the pressure exerted on neighbouring cells; this will lead to divergences from the fourteen-sided figure. In addition, there may be variation in the plasticity of the surface of a single cell. Growth tends to show polarity in the longitudinal direction and a further departure from the predicted polyhedral form.

Cell types

Plant tissues contain a variety of cell types as illustrated by the longitudinal and transverse sections of a maize root tip shown in Fig. 1.3. A brief description of the major types is desirable since they are frequently mentioned in relation to organelle function in the later chapters. All the cells shown in Fig. 1.3 arise from the small group of actively dividing cells which constitute the *meristem*. The meristematic cells are generally considered to be the least specialized of cells although they do have the very essential property of active cell division. Meristematic cells are usually small and thin-walled with a relatively large nucleus and no prominent central vacuole. The exceptions to this rule are the meristematic cells of the cambium that produce the lateral thickening of shoots and roots; these cells often tend to be elongated with a large central vacuole. The growth and differentiation of meristmatic cells produce all the other cell types found in plant tissues (Fig. 1.4). The gross changes that occur are largely due to increase in size, development of the vacuole, and to differences in the structure and composition of the cell wall, although,

Fig. 1.3   Transverse section cut
11 mm behind the tip (×150) and
longitudinal section (×84) of maize
root tip. E, epidermis; C, cortex; EN,
endodermis; X, differentiating xylem;
S, stele; M, meristematic region; RC,
root cap.

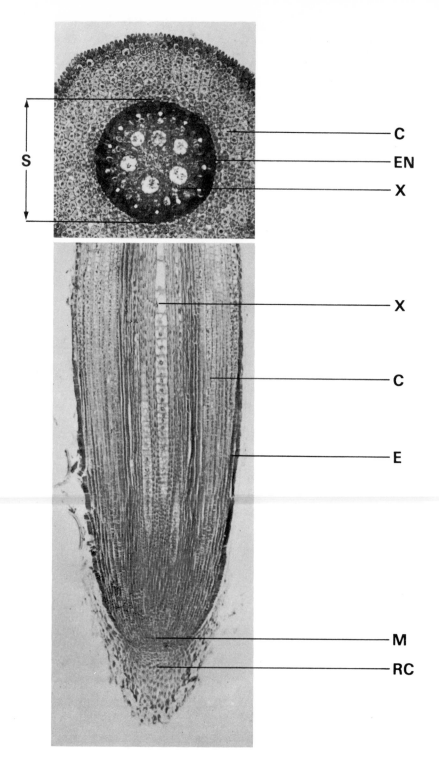

Fig. 1.4 Diagrammatic representation of some plant cell types. Not drawn to scale.

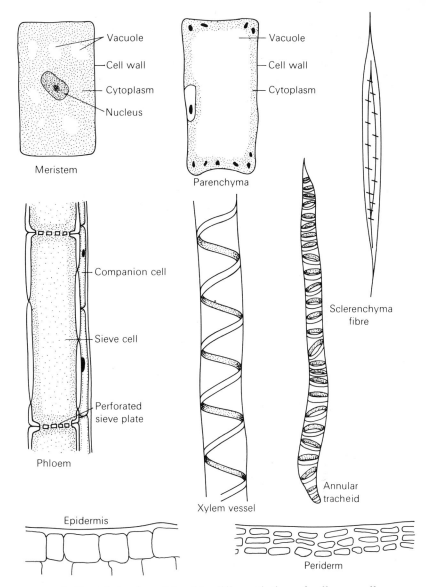

as we shall see later, considerable differentiation of cell organelles may also take place. This observation is particularly important when considering the results of cell fractionation experiments which are frequently carried out using whole plant organs, such as roots or shoots. It must be remembered that any isolated cell fraction will have been derived from many cell types, each with their distinct metabolism. The picture that results will be an amalgamation of characteristics, and the features of a minor cell type may be masked by those of the most abundant cells.

Other cell types found in plant tissues can, for simplicity, be divided into a few major groups. *Parenchymatous* cells are the most common cell type and may form up to 80% of the total cell complement in many higher plant organs. They are usually large when mature, relatively undifferentiated and consist of a thin wall, peripheral cytoplasm and central vacuole (Fig. 1.4). They may be adapted for food storage and contain large numbers of starch grains or lipid droplets, or for photosynthesis and contain chloroplasts, as in the palisade cells of the leaf. Parenchymatous cells are the dominant type in most roots, shoots, leaves and fruits.

A second major group can be described as conducting and structural cells. Of these the most important constitute the tissues of the *phloem* and *xylem*. The phloem is the major pathway for the transport of sugars and other photosynthates from the leaves to the other parts of the plant. It consists of parenchyma, sclerenchyma and sieve elements. *Sclerenchyma* consists of elongated cells with walls heavily impregnated with lignin and with no cytoplasm (Fig. 1.4). The cells are often known as fibres and form part of the support system of the plant. The *sieve tubes* function in translocation and consist of rows of elongated cells with perforated end walls known as sieve plates (Fig. 1.4). When mature they contain no nuclei and are associated with smaller *companion cells* which contain nuclei. The xylem forms the water-conducting tissue of the plant and consists of parenchyma, sclerenchyma, vessels and tracheids. *Vessels* are formed when rows of elongated cells become characteristically thickened with lignin to form ringed, spiral or reticulate patterns (Fig. 1.4). At maturity, the end walls and cytoplasm break down so that long tubes of dead cells are produced. *Tracheids* develop in a similar manner to vessels except that the end walls do not break down, although they remain permeable.

The other major cell types make up the protective and reproductive tissues. Protective cells are usually characterized by the impregnation of the cell walls with water-impermeable substances such as cutin and suberin. These include the cells of the *epidermis*, which forms the other layer of primary tissues, and of the *periderm*, or cork, produced in secondary tissues. The reproductive cells form only a small part of the total plant body and are produced only at certain times in the life cycle. They show many adaptations to their particular rôle but these will not be discussed more fully here.

Ultrastructure

Under the light microscope the largest intracellular structures of plant cells, such as the nucleus, large central vacuole and chloroplasts, can easily be recognized. The remainder of the cytoplasm has a somewhat granular appearance suggesting that it is not simply an homogeneous matrix. Electron micrographs, such as that in Fig. 1.5, have shown that a complex system of membranes and membrane-bound units, or organelles, permeates the whole matrix of the cytoplasm. In addition, the groundplasm surrounding the membranes contains a series of granular structures ranging from the ribosomes, which are about 20 nm in diameter, down to particles which are below the resolution of the electron microscope. As we have seen, plant cells vary considerably in their gross structure, and this is often reflected in differences in their ultrastructure as seen by the electron microscope, although there are many features which are common to almost all plant cells.

Fig. 1.5 Electron micrograph of a cell from the root of *Suaeda maritima* fixed in glutaraldehyde and osmium. N, nucleus; Nuo, nucleolus; NM, nuclear membrane; M, mitochondria; GB, Golgi body; ER, endoplasmic reticulum; CW, cell wall; Pd, plasmodesmata; V, vacuole; ×11 500.

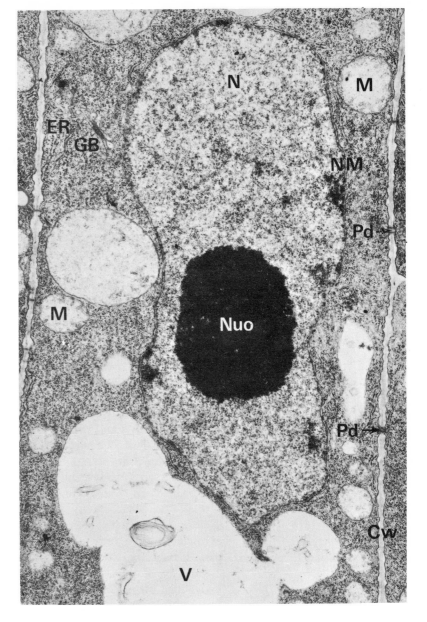

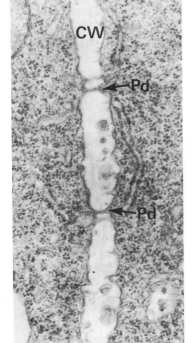

Fig. 1.6 Electron micrograph of plasmodesmata in the cell wall of a cell from the root tip of *Suaeda maritima*. Fixed in glutaraldehyde and osmium. CW, cell wall, Pd, plasmodesmata; ×32 000.

Every cell is delimited from its environment by the outer cell membrane, the *plasma membrane* or *plasmalemma*. This is a semi-permeable barrier; water can pass through freely whereas some selectivity is shown to the passage of solutes. The plasmalemma is the most important control point in the movement of substances between the cell and its surroundings. Outside the plasmalemma plant cells have an additional layer which is known as the *cell wall*. This gives the plant cell rigidity and functions primarily in support and also as a protection against attack by pathogens. An interesting feature of plant cells is the presence of *plasmodesmata* (see Fig. 1.6). These are threads of cytoplasm which are lined by plasmalemma and extend through the cell wall connecting adjacent cells. These pores are usually 50–80 nm in diameter and, in some cells, contain a central core although there is considerable debate as to the nature of this core. Likewise, the function of the plasmodesmata is not fully understood although they are assumed to provide a major pathway for the movement of materials and for communication between cells.

The degree of organization of the cytoplasm inside the plasmalemma is used as a basis for dividing living cells into two major groups, the *prokaryotes* and the *eukaryotes*. This division was originally based on differences in the organization of the nucleic acid material carrying the genetic information. Prokaryotic cells, which include the bacteria and blue-green algae, have no clearly defined nucleus although the DNA is usually found in a certain zone. Eukaryotic cells, which include the cells of all higher plants and animals, are generally larger and more complex with their DNA organized into a distinct membrane-bound nucleus. The electron microscope has shown that there are further differences between these cell types in the complexity of the organization of the cytoplasm. In eukaryotes, various metabolic processes such as photosynthesis, respiration and protein synthesis are compartmentalized into discrete structural units or *organelles* which in most cases are separated from the rest of the cell by semipermeable membranes. Thus we have a division of labour, or specialization of parts, at the intracellular level, just as there is division of labour between the cells of a multicellular organism. Clearly, differing processes, such as respiration and photosynthesis, require very different environments and the separation of cells into distinct regions by membranes allows the particular internal environment to be controlled and maintained at the optimum for the particular process. Again this compartmentation allows incompatible processes such as the synthesis and breakdown of lipids to occur at the same time in the same cell. Prokaryotic cells contain relatively few membranes and processes such as respiration and photosynthesis are not localized in specific organelles although they probably do occur within multienzyme complexes associated with membranes. We are chiefly concerned with eukaryotes in this book although many of the advances in our knowledge of cell biology have been achieved by studying bacterial systems. The short life cycle, ease of growth and manipulation, and the ready production of uniform populations make bacteria such as *Escherichia coli* excellent experimental tools.

The majority of cellular organelles are bounded by membranes, either single or double, and include the nucleus, mitochondria, chloroplasts, microbodies, Golgi bodies and lysosomes. These membranes do not function simply as selective barriers controlling the movement of substances between the organelle and the surrounding cytoplasm since, as we shall see in later chapters, they are also the sites of many complex biochemical processes. Functions such as solute transport, energy conversion in respiration and photosynthesis, and conductivity in nerve cells have been assigned to membranes. They provide sites for the ordered arrangement of many enzymes, greatly increasing the overall efficiency of sequential biochemical reactions. The complex network of membranes in the cytoplasm, known as the *endoplasmic reticulum* (ER), probably provides a foundation for the ordered attachment of many enzymes in the living cell, including those '*soluble*' enzymes which are not associated with specific organelles after cell fractionation, such as those involved in the synthesis of lipids (see Ch. 6). However some cell organelles are not bounded by membranes as seen in the case of the *ribosomes*, the sites of protein synthesis in both prokaryotic and eukaryotic cells. There are many other cell components which have characteristic structures and specific functions and yet are not membrane-bound. These include the *nucleoli* and *chromosomes* found in the nucleus, the *microtubules* in the cytoplasm and the *flagella* found at the exterior of many unicellular plants. These observations raise the question of the definition of a cellular organelle. For example, multienzyme complexes have been isolated from living cells and consist of a number of different enzymes, perhaps as many as two dozen, bound tightly together in a cluster. These complexes catalyse a series of ordered reactions in a metabolic pathway; the separated enzymes are normally inactive. Perhaps such a unit should be considered to be an organelle. As mentioned before, the cytoplasm has a very granular appearance under the electron microscope and this could be due to a variety of functional molecular aggregates. Clearly there is no hard and fast definition of an organelle; it must be an arbitrary decision applied to the particular discussion in progress.

What then is the relationship between the variety of metabolic processes which occur in plant cells and the major structural units seen with the electron microscope? This is, of course, the main theme of this book. The most prominent cell organelle is the *nucleus* which contains the nucleolus and chromosomes and most of the DNA of the cell. It is the major site of genetic information and is concerned largely with the conservation and duplication of this information and its transmission to the cytoplasm. The *mitochondria* and *chloroplasts* are organelles delimited by more than one membrane. The former are the site of respiratory enzymes concerned with the final stages of carbohydrate oxidation while the chloroplasts contain chlorophyll and other pigments, and are concerned with the process of photosynthesis in green plant cells. The cytoplasm also contains a number of organelles which are bounded by a single membrane. Differentiated plant cells generally possess a single large *vacuole* which may occupy over 90% of the total cell volume and is bounded by a membrane known as the *tonoplast*. Vacuoles are important storage compartments and may also possess digestive or *lysosome-like* activity. Other organelles include the *Golgi bodies* or *dictyosomes* which function primarily in the secretion of materials, particularly carbohydrates, to the exterior of the cell; the *lysosomes* which are digestive centres in the cell; and the *microbodies* which have oxidative properties and are associated with processes such as photorespiration and fat metabolism. In addition, there are the non-membrane bound organelles of which the most prominent are the *ribosomes*, the centres of protein synthesis, which are frequently associated with the membranes of the endoplasmic reticulum although the significance of this association is uncertain.

These discoveries have led to the construction of models for the ideal or generalized cell. One such model, proposed by Robertson (1959, 1967), is shown in Fig. 1.7(*a*), while a diagram of the fine structure of a plant cell is shown in Fig. 1.7(*b*). The Robertson model is based on the hypothesis that all cellular membranes are of a similar structure. The cell is seen as a three-phase system consisting of:

1. the membrane phase which is continuous from the plasmalemma to the nuclear membrane and includes the membranes of the endoplasmic reticulum;
2. the nucleo-cytoplasmic matrix continuous through the nuclear pores;
3. the system of anastamosing membrane-bound channels and vacuoles which permeate the cytoplasm but are essentially outside the cell.

Fig. 1.7(*a*) Model of a generalized animal cell reproduced from Robertson (1959). (b) Diagrammatic representation of a vacuolated plant cell based on observations made with the electron microscope. Not to scale. Abbreviations: ch, chromatin; chl, chloroplast; cw, cell wall; gb, Golgi body; gv, Golgi vesicles; is, intercellular space; 1, lipid droplet; mb, microbody; m, mitochondrion; mt, microtubule; n, nucleus; ne, nuclear envelope; np, nuclear pore; nu, nucleolus; pd, plasmodesmata; pm, plasmalemma; rer, rough endoplasmic reticulum; rib, ribosomes; ser, smooth endoplasmic reticulum; t, tonoplast; v, vacuole; vi, vacuolar inclusion.

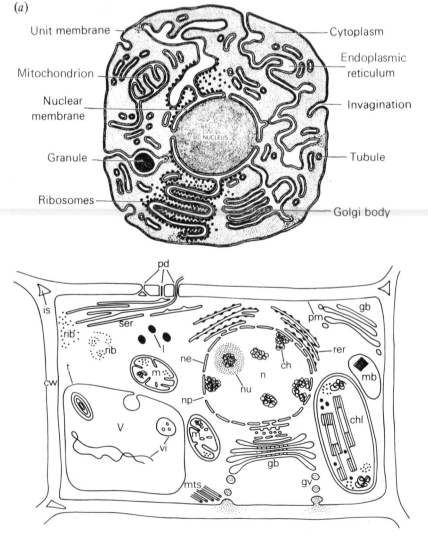

This is an interesting proposal since it would allow for the penetration of extracellular materials to the space surrounding the nucleus and would greatly increase the surface area for the absorption of these materials. However, such a model lacks sound experimental support. Although a connection between the nuclear membrane and the endoplasmic reticulum has been observed in many cells, we are still not certain as to the extent of discontinuities in the endoplasmic reticulum while connections between the endoplasmic reticulum and plasmalemma are observed only rarely. Again, such a model depends on a common structure for all biological membranes yet, as we shall see in Chapter 3, there is considerable evidence of differences in both chemical composition and structure between the various membranes found in the cell.

Methods of studying cells

We have already mentioned the importance of instrumentation and technique in the study of cells and this has been aptly summed up by Albert Claude who, in 1950, wrote:

'In the history of cytology, it is repeatedly found that further advance had to await the accident of technical progress.'

In this section we will consider the most important methods used in modern cell biology and, in particular, those that are mentioned in the study of the various cellular components described in the later chapters of this book.

Microscopy

Almost all cells are too small to be examined directly with the human eye and so our knowledge of cells has depended very much on microscopic techniques for magnifying them. The history of cell biology is an excellent example of the effect of one scientific discipline on another. The improvements in microscopy produced by developments in physics have been closely correlated with the expansion of cell biology. It is interesting to compare the appearance of the microscope used by Hooke in the seventeenth century with that of a modern light and electron microscope, shown in Fig. 1.8. The increasing complexity and power of the microscope is reflected in the diagrammatic representations of cell structure as described first by Hooke, then by Guilliermond in 1941 when light microscopy had been perfected to give maximum possible resolution (Fig. 1.9), and finally by the models shown in Fig. 1.7 which are based on electron microscope studies. Before we can consider the many facets of microscopy, it is important to understand a fundamental property of microscopes: resolution or resolving power.

Resolving power

The problem of the magnification of an object is best considered in relation to the property known as *resolution*. The ability of a lens

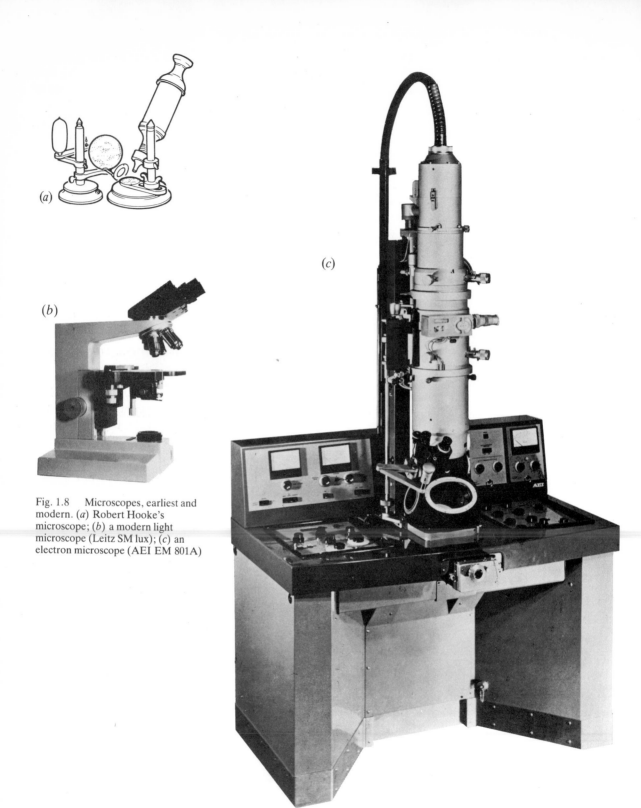

Fig. 1.8   Microscopes, earliest and modern. (*a*) Robert Hooke's microscope; (*b*) a modern light microscope (Leitz SM lux); (*c*) an electron microscope (AEI EM 801A)

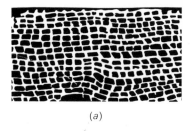

(a)

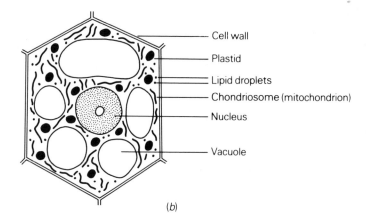

- Cell wall
- Plastid
- Lipid droplets
- Chondriosome (mitochondrion)
- Nucleus
- Vacuole

(b)

Fig. 1.9   (a) Drawings of cells in cork made by Hooke (1665); (b) diagrammatic representation of plant cell structure redrawn from Guilliermond (1941).

system to show fine detail is termed its resolving power and this may be defined as the ability of an optical system to distinguish between adjacent objects as separate entities. The human eye has a fundamental limitation in that it cannot discern points closer together than about 0·1 mm; it has a resolving power of about 0·1 mm. Points closer together than this will be seen as a single image. The resolving power of the eye is, in fact, less than that which is theoretically possible, due to the diffraction or scattering of light, and this is illustrated in Fig. 1.10. Consider two points and the image that they produce at the retina of the eye. Ideally, in the absence of diffraction, these would produce two sharp peaks of light intensity as shown in Fig. 1.10(a). However, because diffraction occurs, a more diffuse area of light will be produced and the two images will overlap (Fig. 1.10(b)). The diffuse area of light is known as an Airy disc and if the centres of the two discs are closer together than the radius of each, the eye will not resolve the two points and will record only a single region of light. The radius of an

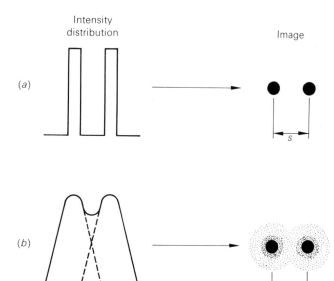

Intensity distribution

Image

(a)

(b)

Fig. 1.10   Effect of diffraction on the image of two point objects. (a) Ideal intensity distribution in the absence of diffraction; (b) actual distribution which produces a blurred image caused by diffraction.

15

Airy disc can be calculated approximately from the equation

$$R \simeq (\lambda/2) \sin \alpha,$$

where λ is the wavelength of illumination used and α is an inverse function of the aperture of the eye. The light microscope is able to increase the resolution of the eye because, essentially, it increases the aperture of the eye. This reduces the radius of the Airy dics and so allows the eye to distinguish points closer together.

By the end of the last century great improvements had been made in light microscopy and it was possible to obtain a resolving power of about 0.2 μm, which is considerably better than that of the unaided eye. However, if we refer to the above equation, it is clear that the resolution of any microscope is fundamentally limited by the wavelength (λ) of the illumination employed. The average wavelength of visible light is about 550 nm, and so to improve on the resolution of a normal light microscope shorter wavelength illumination must be used. In some microscopes, ultra-violet light with a wavelength of about 250 nm has been utilized to give a resolving power of about 0.1 μm but this is still well above the resolving power required to see many cell structures. This problem has now been overcome with the development, in the last forty years or so, of the electron microscope which makes use of the wave-like properties of a beam of electrons. The electron beam in an electron microscope normally has a wavelength of about 4 nm, some sixty times less than that of ultra-violet light, and this has enabled resolving powers of about 0.2 nm to be obtained, about 1 000 times better than that of the light microscope. This, in fact, is still much below the theoretically possible resolving power of the electron-microscope and this and other limitations will be discussed more fully in the following sections.

Light and electron microscopes

The basic principles of microscopy, whether it be light, ultra-violet or electron microscopy, are the same and are illustrated in Fig. 1.11. The specimen under examination is illuminated by a source of radiation which is directed onto it by a condenser lens, the function of which is to illuminate the specimen uniformly so that it can be examined by the objective lens. The objective lens is the most important since this produces a magnified image of the specimen. The resolving power of the objective determines the detail of structure that will be produced in the final image. The image produced by the objective is further magnified by the eyepiece or projector lens and finally examined by the eye or projected onto a photographic plate.

There are, of course, a number of very important differences between light and electron microscopes. The illumination source of a light microscope is a light bulb, but in the electron microscope, the source is a tungsten filament which is at high voltage (usually about 80 kV) and heated to about 3 500 K, at which temperature electrons are given off from the filament and accelerated towards a structure known as the anode plate which is at a positive potential relative to the

Fig. 1.11   Comparison between the light microscope and electron microscope.

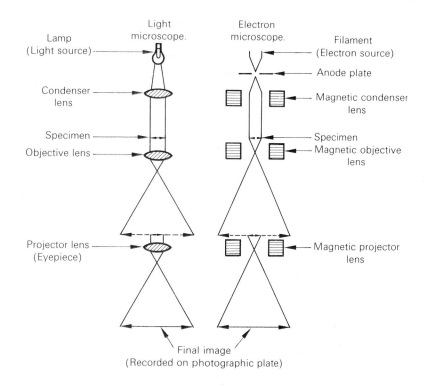

Light microscope.

Electron microscope.

Lamp (Light source)

Filament (Electron source)

Anode plate

Condenser lens

Magnetic condenser lens

Specimen

Specimen

Objective lens

Magnetic objective lens

Projector lens (Eyepiece)

Magnetic projector lens

Final image
(Recorded on photographic plate)

filament. The electron beam passes through a small hole in the anode plate and then down the column of the microscope. The filament and the anode plate together are known as the electron gun. The two types of microscope also differ in the modes of control of their illumination source. The light microscope uses glass lenses whereas electromagnetic lenses serve in the electron microscope, focusing being achieved by variation in the current passing through the lens. The human eye is not sensitive to electrons and so the image produced by the electron microscope cannot be examined directly. Instead the image is projected either onto a photographic plate or onto a screen which fluoresces in proportion to the number of electrons hitting it and so produces a visible image. Another major difference is that the column of the electron microscope must be under a high vacuum since electrons have a low penetration power and, because of collisions with air molecules, will travel only a few millimetres in air. This means that the specimens to be examined must be completely dehydrated as water vapour would destroy the vacuum in the column. Thus living specimens cannot be examined with the electron microscope.

Phase contrast microscopy

When living cells are examined by normal light microscopy it is frequently difficult to see cell contents. This is because the human eye normally detects contrast by differences in colour or in intensity of illumination, and living cells are normally colourless and more or less transparent. One common method of overcoming this difficulty is to

17

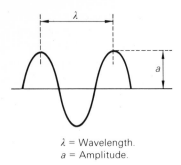

λ = Wavelength.
a = Amplitude.

Fig. 1.12   Characteristics of a light wave.

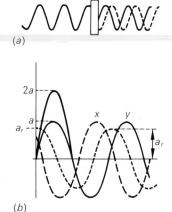

(a)

(b)

Fig. 1.13   Retardation of light waves. (a) Diagram showing the phase change (retardation) caused by the passage of a light wave through a transparent material. (b) Diagram showing interference between two light waves. The wave (x) is retarded in relation to wave (y) resulting in a reduced amplitude (a_r). A much larger amplitude ($2a$) would be produced if the waves were in phase.

stain cells with various dyes but, as this usually results in death or disruption, it cannot be used to examine living cells. One technique which is now widely used to solve this problem is known as phase contrast microscopy.

Consider the simple characteristics of a light wave (Fig. 1.12). The wavelength (λ) is a measure of colour whilst amplitude (a) is a measure of brightness. When a specimen is examined with a light microscope, the light passing through the specimen will be affected in two ways in relation to the light passing through the surroundings, and will produce the contrast in the image. Firstly, the light will be diffracted or scattered, and may be lost to the image; the greater the diffraction, the darker the image. Secondly, the diffracted rays that do pass through the microscope will be retarded in relation to the light that did not pass through the specimen, the degree of retardation depending on the thickness and density of the specimen (Fig. 1.13). Thus two sets of rays will arrive at the eye, the diffracted and undiffracted. The slight phase change produced in the diffracted rays will interfere with the other rays resulting in a net reduction in amplitude, and so a lowering of brightness. The image seen by the eye is, in fact, a complex interference pattern of diffracted and undiffracted rays. Normally the change in phase, and so amplitude, is slight, and produces little contrast. The phase contrast microscope functions by further retarding the diffracted rays producing a greater change in amplitude on interference, and so a greater reduction in brightness. The microscope is illustrated in Fig. 1.14. The illumination is in the form of an annulus and, if undiffracted by the specimen, passes through the annular groove in the glass phase plate. The diffracted rays will pass through the thicker part of the phase plate and so suffer a greater retardation, enhancing the phase difference produced by passage through the specimen. In addition, further light will be lost by the greater scatter produced by the thick part of the plate, and so contrast will be enhanced by both greater amplitude changes and loss of light.

Phase contrast microscopy is now widely used to study living cells and tissues. Considerable cytological detail of cell movement, changes in nuclei and other organelles, and cytoplasmic flow have been observed although resolution is still limited by the wavelength of visible light. Phase microscopy has been particularly useful when used in conjunction with time-lapse motion photography of living cells.

Fixation

The phase contrast microscope is a useful method for enhancing the contrast of living cells so that greater detail may be observed and has proved to be extremely useful for some purposes. However, the examination of living cells is extremely limited in many ways. Living cells and cell organelles move and are quite thick, and this tends to obscure cellular detail. Since resolution is dependent on specimen thickness, sections of living cells and tissues are generally employed in both light and electron microscopy. Futhermore, it is often desirable to stain or process cells in various ways, although this may cause

18

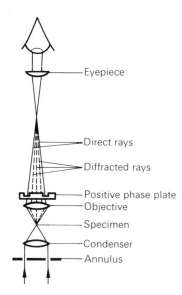

Eyepiece

Direct rays

Diffracted rays

Positive phase plate
Objective

Specimen

Condenser

Annulus

Fig. 1.14   The phase contrast microscope showing the paths taken by the directly transmitted light and the light diffracted by a specimen. The design is essentially the same as that of an ordinary light microscope except that it has a special diaphragm (the annulus) and a phase plate which may be positive (as in the diagram) or negative (⌐⌐⌐).

morphological damage since living cells are fragile structures. Thus for the majority of light and electron microscope studies of cell structure, the first step is a process called *fixation*. The purpose of fixation is to kill and stabilize cells so that the structure is preserved in the subsequent treatments. Normally a chemical, known as a fixative, is employed. Ideally the fixative should penetrate and kill the cell instantly, preserving the structure of the cell as it was in the living state. However, the ideal fixative has yet to be found, nor will it be, since if the cell is to die its molecular structure must clearly be altered in some way. However, it is important that structural alterations from the living state, which are known as artefacts, should be kept to a minimum. Major damage produced by poor fixation, such as membrane swelling, plasmolysis, and the extraction of certain compounds or structures, is usually readily seen. Finer changes may be more difficult to recognize and establish as artefacts and may be the subject of considerable debate.

A large number of chemicals have now been tested as fixatives for both light and electron microscopy. They include ethyl alcohol, acetone, mercuric chloride, acetic acid, chromic acid, potassium permanganate, osmium tetroxide and glutaraldehyde. The last three are those most widely used at present in electron microscope studies. There is clearly considerable variation in the chemical nature of these fixatives and the chemistry of fixation is far from being fully understood. It seems probable, however, that the majority of fixatives function by forming stable cross-links between various protein and lipid molecules. For example, osmium tetroxide (OsO_4) probably reacts with double bonds in unsaturated fats or proteins to form cross-links as follows:

$$R_1-CH \atop R_2-CH \quad + \quad {O \atop O}{>}OsO_2 \quad \longrightarrow \quad {R_1-CH-O \atop R_2-CH-O}{>}OsO_2$$

$$+$$

$$R_1-CH-O \atop R_2-CH-O {>}Os{<} {O-CH-R_3 \atop O-CH-R_4} \quad \longleftarrow \quad {CH-R_3 \atop CH-R_4}$$

Similarly, a dialdehyde fixative such as glutaraldehyde may form cross-links by reacting with amino groups on different protein molecules:

$$R-NH_2 + OHC-CH_2-CH_2-CH_2-CHO \rightarrow R-N=CH-CH_2-CH_2-CH_2-CHO$$
Glutaraldehyde

$$+$$

$$R-NH=CH-CH_2-CH_2-CH_2 \quad \leftarrow \quad R-NH_2$$

The effect of different fixatives on the appearance of cells under the electron microscope is shown in Fig. 1.15. The variation in appearance produced by different methods of fixation suggests that considerable caution must be used in the interpretation of fixed preparations. There seems to be no universally satisfactory fixative and it is now usual to use specific fixatives for the examination of specific structures. For

Fig. 1.15   The effect of different fixation procedures on the appearance of cells under the electron microscope. Electron micrographs of cells from similar regions of pea root tips (a) fixed in glutaraldehyde and osmium tetroxide (×9000) and (b) fixed in potassium permanganate (×9000). N, nucleus; P, plastid; M, mitochondrion; GB, Golgi body; CW, cell wall. In particular, note the disappearance of nuclear contents and cytoplasmic ribosomes and the enhanced appearance of membranes in (b).

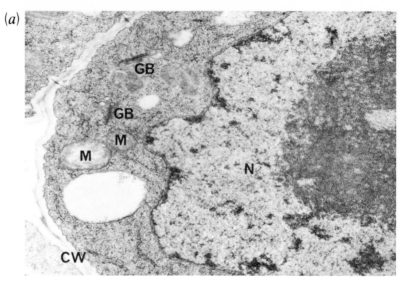

(a)

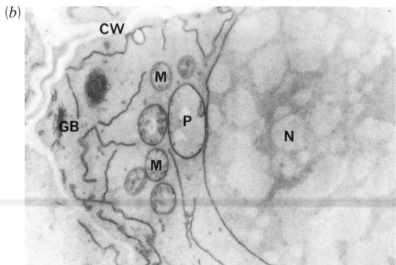

(b)

example, potassium permanganate is a good fixative for the examination of membranes although it is destructive to nucleic acids and so does not preserve ribosomes.

Although chemical fixation is widely used in both light and electron microscopy, the possibility that artefacts are produced by the chemical alteration of the cell constituents has led to attempts to fix tissues by freezing without the introduction of chemicals. The aim is to stop all cellular activity by subjecting the tissue to rapid, intense freezing when it is hoped that the structure of the living state will be retained. Rapid freezing is usually accomplished by immersing pieces of tissue in liquid nitrogen (about −160°C) or in certain solvents cooled by liquid nitrogen. This method should result in homogeneous fixation with no loss of soluble substances and little chemical change. Water must then be removed before the cells can be used for examination by electron microscopy. This is done by freeze-drying, in which the ice is removed by sublimation in a vacuum at about −40°C, or by freeze-substitution, in which the ice is dissolved in a solvent such as ethanol or acetone at a low temperature. Although the concept of fixation by freezing is a simple one, the methods are less straightforward in practice. They are less convenient to use than chemical fixation, freezing damage can occur, and the tissues become extremely fragile. However one particular development of the use of freezing, known as freeze-etching, has become widely used in electron microscopy of late and will be described in the next section.

Preparation of cells for electron microscopy

Since the discoveries of electron microscopy have been crucial to our knowledge of cell biology it is worth examining the usual methods employed to prepare cells for observation under the electron microscope.

As already mentioned, an electron beam has a low penetrating power and so the column of the microscope is kept under high vacuum. This means that water must be removed from the specimens or the vacuum will be destroyed. Therefore specimens are dehydrated after fixation, usually by immersion in a series of ethanol–water mixtures of increasing alcohol concentration up to 100% alcohol.

Associated with the low penetration property of the electron beam is the question of specimen thickness. The formation of the image in an electron microscope depends upon the electrons passing through the specimen. If the specimen is too thick all of the electrons will be scattered to some extent, resulting in an image with poor contrast and considerable loss of detail. Therefore it is necessary to cut thin slices or sections of the material to be examined, and these must be of the order of 50 nm thick. To cut these ultra-thin sections the tissue must first be embedded in a hard support medium since biological material has little mechanical strength. The embedding media normally used are plastic resins such as Epikote or Araldite. Since these are not easily miscible with ethanol, the tissue is first transferred to acetone. The unpolymerized resin is mixed with the acetone and its concentration slowly increased, perhaps over a period of days, so that it fully infiltrates the specimen. The resin is then polymerized to a hard plastic by heating in an oven at about 60°C for 24 h or so. The specimen, fully embedded and supported in the plastic, can be cut into ultra-thin sections by glass or diamond knives using a special apparatus called a microtome. The knife is held still and the specimen advanced on an oscillating arm towards the knife at a constant controlled rate, cutting sections at each stroke.

When sections of biological materials are examined under the electron microscope, little detail of structure is usually seen since there is little basic contrast, although potassium permanganate especially, and osmium tetroxide, do stain to some extent during fixation. Therefore it is frequently necessary to stain sections to enhance contrast. This is analogous to staining sections for light microscopy, except that the stains are not dyes but solutions of heavy metals which are differentially adsorbed by different cell structures and so increase the scattering of electrons from these parts. Two commonly used stains are salts of lead and uranium. The former enhance the contrast of membranes and nucleic acids, the latter of proteins and nucleic acids. This is known as positive staining and may be contrasted with the *negative staining* technique which is frequently used in the examination of fragments of isolated organelles and membranes and of small particles such as viruses and ribosomes. Negative staining produces an electron-dense background against which the less dense specimens are observed. Negative stains consist of the salts of heavy metals and include potassium phosphotungstate, uranyl acetate and ammonium molybdate. The electron-dense stain appears to dry more rapidly on the support film than on the specimen and forms a sharp boundary, the specimen appearing white on a dark background (see Figs. 7.3, 11.4). In some cases, the stain may penetrate any surface relief of the specimen to show considerable detail of the fine structure.

Finally, some further mention should be made of the technique known as *freeze-etching*. This has been developed for preparing material for electron microscopy as an alternative method to the standard procedure of chemical fixation, dehydration, embedding and sectioning. Freeze-etching is based on fixation by rapid freezing as discussed earlier and the procedure is illustrated in Fig. 1.16. Firstly the specimen is frozen rapidly and transferred to a special vacuum chamber.

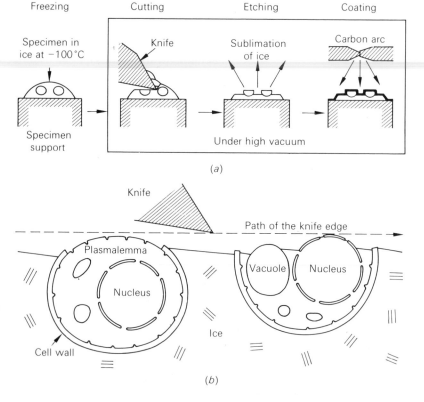

Fig. 1.16 Freeze-etching.
(a) Diagrammatic representation of the stages in the preparation of cells by freeze-etching (b) schematic example of the way the cutting process in the freeze-etch procedure can expose various cellular membranes.
Reproduced from Moor (1965).

The specimen is then fractured to expose the inner structures. This is followed by a short period of freeze-drying or etching in a vacuum which frees the exposed surface of the specimen from the ice in which it is embedded. A replica of the exposed surface is produced by coating the surface with a film of heavy metal and carbon vapour. The specimen is thawed and the replica floated off on water to be collected and examined under the electron microscope. An example of how this process reveals surface views of cell organelles is shown in Fig. 1.16 and an electron micrograph of a freeze-etched plant cell is seen in Fig. 1.17. This method is thus a complete alternative to the method of chemical fixation and embedding, the two procedures being excellent complements to each other for the examination of cell structures and the interpretation of cell artefacts. The freeze-etch technique has a number of limitations. The picture produced is a purely morphological one and, as yet, certain techniques such as enzyme cytochemistry which can be used with sections, have not been developed for freeze-etching. In addition, the resolution is limited by the thickness and structure of the heavy metal coat and is at present about 2.5 nm, considerably less than that obtainable with thin sections. Finally, as with any microscopic technique, the results may be open to several interpretations and the possibility of artefacts must always be considered.

Limitations and problems of electron microscopy

Electron microscopy has told us in the last decade or so a great deal about the structure and function of cells, although it is important to realize that this technique is subject to a number of important limitations which are often forgotten when the significance of the results is being considered. There are limitations both to the electron microscope itself and in its application to the examination of biological specimens.

From a simple consideration of wavelength, the resolution theoretically obtainable by the electron microscope is about 0·002 nm for an instrument operated at 100 kV. In practice, however, it has been found that the best resolution obtainable is about 0·2 nm. This is largely due to limitations in the electromagnetic lenses which suffer from various aberrations. In addition, resolution is limited by both mechanical and electrical stability. For example, when working at a resolution of 0·3 nm which is the distance apart of the atoms in many metals, slight vibrations or the effect of external magnetic fields can cause blurring of the specimen image. Clearly there is room for improvement in the microscope itself. This will probably come with the development of better electromagnetic lenses and also with the development of high voltage electron microscopes. In the latter, which use electron voltages of up to 1 MV, the penetrating power of the electrons is greatly increased, resulting in reduced lens aberration and, consequently, increased resolution.

Many of the problems associated with the examination of biological material with the electron microscope have already been briefly

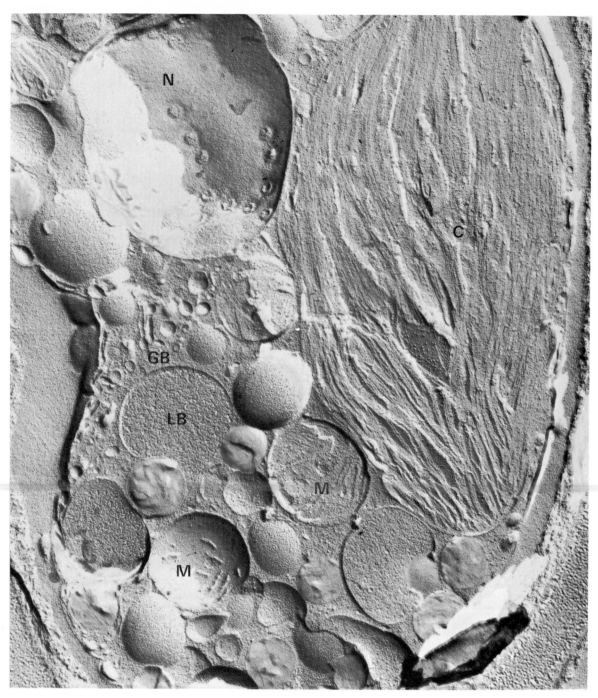

Fig. 1.17　Freeze-etch view of the
unicellular green alga *Oocystis
marssonii* showing nucleus with pores
(N), mitochondria (M), Golgi body
(GB), lipid body (LB) and chloroplast
with exposed membrane lamellae (C).
×56 800. Provided by J. C. Pendland
and H. C. Aldrich, Department of
Botany, University of Florida.

mentioned. Because of the low penetration power of the electron
beam, the specimens must be fixed, dehydrated, embedded in plastic,
cut into ultra-thin sections and stained with heavy metals; severe
distortion of structure may occur at any stage in this process. With light
microscopy, some estimate of the effect of the preparative procedure
can be made by comparison with living cells of the same material

examined by phase contrast microscopy. This, of course, is not possible with electron microscopy and so the best check that can be made is to employ variations in the procedures, such as the use of different fixatives or dehydration solvents, and to notice significant differences in structure produced by these changes. In general, it has been found that these variations in procedure give rise to essentially similar results although important details of structure may vary. The most fundamental changes are produced by variation in the fixatives used and this is perhaps not surprising since we have seen that they may have very different chemical reactions resulting in the preservation of different cell structures. It is interesting, however, that even a fundamentally different technique such as freeze-etching yields a similar overall picture of cell structure to that produced by chemical fixation. This gives confidence that the image seen with the electron microscope is broadly representative of that in the living state.

Microscopic cytochemistry

The term microscopic cytochemistry refers to a wide range of techniques which all attempt to localize, visually, specific molecules within cells and tissues. The simplest of these methods are those which involve the staining of certain molecules with colouring reagents for examination by light microscopy. Many of these stains have been in common use for many years and include the iodine reaction for starch and the methyl green and Feulgen procedures for DNA. The coloured reaction product of these methods is rarely electron dense however and so they cannot be applied to electron microscopy. The heavy metal stains used to enhance contrast in electron microscopy show some selective deposition and so are further examples of these simple staining procedures. Two further examples of microscopic cytochemical procedures which are now widely used in plant cell biology are autoradiography and enzyme cytochemistry. These will be frequently referred to in the later chapters of this book and so will now be described in some detail.

Autoradiography

This technique, which is now widely used in plant cell biology, involves the visual localization of radioactive substances in cells and tissues and is illustrated in Fig. 1.18. Tissues are incubated in solutions containing radioactive isotopes which become incorporated into the cells. The tissue is sectioned and the sites of incorporation are traced by placing a photographic emulsion over the section. The radioactive particles emitted from the isotope sensitize silver halide crystals in the emulsion and these are converted to metallic silver by normal photographic development. The unsensitized crystals are washed away during the subsequent fixation and the site of radioactive incorporation is then seen as a deposit of silver grains on the sections. The most widely used isotopes in biological research are tritium (^{3}H), carbon (^{14}C), phosphorus (^{32}P), calcium (^{45}Ca) and sulphur (^{35}S). These can be used to 'label'

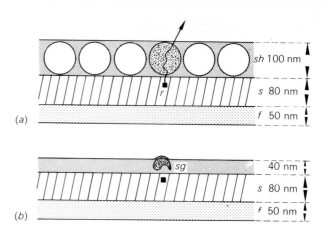

Fig. 1.18   Autoradiography for electron microscopy. (*a*) An ultra-thin section (*s*) containing radioactive atoms (*r*) is mounted on a support film (*f*) and coated with a carbon film. A monolayer of silver halide grains (*sh*) is deposited on top. This preparation is kept in the darkness, usually for several weeks, while the radioactive atoms emit β-particles that sensitize the silver halide grains above. (*b*) The process of photographic development reduces the sensitized grains to silver grains (*sg*) while the unsensitized grains are washed away during fixation. The whole preparation is examined in the electron microscope where the sites of incorporation are shown by the electron-dense silver grains.
Reproduced from Juniper *et al.* (1970).

a wide range of molecules which, when incubated with plant tissues, are incorporated into various cellular constituents. Autoradiography can then be applied at various levels. For example, on the large scale, isotopes can be fed to leaves and the subsequent distribution to various plant organs followed by covering the whole plant with a photographic film. On the microscopic level, the tissues can be sectioned, covered with emulsion and the isotope localized in cells or cell organelles.

The resolution obtainable by autoradiography depends upon a number of factors. The isotopes mentioned above are all emitters of β-particles although there is considerable variation in the particle energies, as shown below:

^{3}H	0·013 MeV
^{14}C	0·155 MeV
^{35}S	0·167 MeV
^{45}Ca	0·255 MeV
^{32}P	1·718 MeV

The greater the particle energy, the greater the distance the particles travel in the emulsion, and the larger the number of silver grains produced. This will clearly effect the resolution, as illustrated in Fig. 1.19. The higher the energy source, the larger the halo of silver grains produced, and so the lower the resolution. Resolution will also be greatly influenced by the thickness and quality of the emulsion and the thickness of the section. For example, the quality of the emulsion will determine the size, concentration and uniformity of the silver halide crystals. The more concentrated the crystals, the greater is the possibility that the β-particles will sensitize a crystal close to the point of emission and so produce a precise localization of the radioactive source. The thickness of the emulsion, which will affect the number of silver grains formed, and the thickness of the section, which will affect the scatter of β-particles, will also have a marked influence on resolution. Finally, the closeness of contact between specimen and emulsion is an important factor. The closer the contact, the higher the resolution since the β-particles are more likely to sensitize a silver

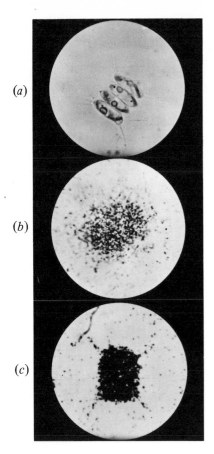

(a)

(b)

(c)

Fig. 1.19 The effect of particle energy on the resolution obtainable by autoradiography in biological material. (*a*) *Scenedesmus*, unlabelled; (*b*) the alga labelled with ^{14}C; (*c*) the alga labelled with ^{3}H. Provided by Dr C. Chapman-Andresen and reproduced from *Botanical Histochemistry* by William A. Jensen. W. H. Freeman and Company. Copyright © 1962.

halide crystal close to the source of radioactivity. At the cellular level, the best resolution is obtained using tritium, the emitter of β-particles with the lowest energy, and this is the only isotope which has been used routinely with the electron microscope. Even then the resolution is severely limited since the developed silver grain is larger than many of the underlying cellular structures.

Enzyme cytochemistry

A number of techniques have now been developed for the detection of the sites of activity of certain enymes in cells and tissues using both light and electron microscopy. The basic requirement of an enzyme localisation procedure is that the enzyme is presented with a substrate molecule and that the product of the reaction between enzyme and substrate immediately forms an insoluble precipitate by reaction with another substance present. The site of enzyme activity is then localized by the deposit of insoluble precipitate. The method is best illustrated by consideration of a commonly used cytochemical procedure, that for the detection of acid phosphatase activity. These enzymes catalyse the hydrolysis of mono-esters of phosphoric acid with the release of inorganic phosphate:

$$ROPO_3H_2 + H_2O \rightleftharpoons ROH + H_3PO_4$$

and work most effectively when the pH is on the acid side of neutrality. For the localization of the sites of this activity, sections of plant tissues are incubated in a medium which contains a phosphate ester as substrate, usually sodium glycerophosphate, lead nitrate, and a buffer to maintain an acid pH, usually at about pH 5.5. The enzymes hydrolyse the substrate with the release of inorganic phosphate (P_i) which is immediately precipitated by the lead ions as insoluble lead phosphate. This is white precipitate not easily visible by light micros-copy and so the sections are dipped into a dilute solution of ammonium sulphide which converts the lead phosphate to insoluble lead sulphide, which is dark brown in colour and easily visible by light microscopy (see Figs. 12.6 and 12.7).

$$Na\ Glycerophosphate \xrightarrow{\text{enzyme}} Na\ glycerate + \overset{\overset{Pb^{++}}{+}}{P_i}$$

$$PbS \xleftarrow{(NH_4)_2S} PbP_i$$

Another widely used reaction employs 3,3′-diaminobenzidine (DAB) to localize various oxidase activities. For example, when tissue sections are incubated in a medium containing DAB and hydrogen peroxide at a suitable pH, DAB may be oxidized by the enzymes peroxidase and catalase to produce a dark brown, insoluble, osmio-philic polymer at the sites of enzymic activity.

These techniques, and many others, depend upon the production of a coloured, insoluble precipitate to visualize the sites of enzyme activity with the light microscope. Only a limited number of these procedures, however, can be used for electron microscopy since the precipitate must be electron dense. The lead phosphate produced by the phosphatase procedure is electron dense and is readily seen as dark deposits in the electron microscope (see Figs. 12.3 and 12.8). Again, the osmiophilic nature of the DAB polymer produced by oxidase activity means that subsequent exposure of the incubated sections to OsO_4 produces an electron dense complex, known as osmium black, which is readily identifiable in electron micrographs (see Figs. 9.3 and 9.4).

As in any enzymic or morphological study, it is important in enzymic cytochemical procedures to maintain both enzymic activity and cellular structure. These aims are often contradictory: to preserve structure, fixation is necessary but this destroys or severely inhibits many enzymes. Fixation and sectioning of plant tissues may cause the loss of enzymes by diffusion into the incubation medium and give a false picture of enzyme localization. Thus the investigator must be very aware of possible artefacts and great care must be taken in the interpretation of the results of these methods. These techniques are most useful when used in conjunction with cell fractionation studies (described in the next section) in which enzyme activity is examined in isolated cell parts which have been separated by centrifugation. These methods complement each other since, in general, cell fractionation ignores cell and tissue differences. For example, an enzyme associated with a particular cell part, such as the nucleus, can be studied in preparations of isolated nuclei, say from roots, and its properties fully characterized. This method gives no indication of changes occurring in this enzyme between the various cells and tissues of the root. An indication of this variation can be obtained if the enzyme can be localized by a microscopic cytochemical stain. Unfortunately, at present only a limited number of enzymes can be studied in the latter way, and only a few of these stains can be made quantitative.

Scanning electron microscopy

In the type of electron microscope described earlier, the beam of electrons passes through the section and is scattered by the denser regions of the specimen; the instrument is usually referred to as a transmission electron microscope (TEM). However, in the last ten years or so, another type of microscope, the scanning electron microscope (SEM), has become widely used in biological research. The main advantage of the SEM is that it has a great depth of focus and so provides a three-dimensional image of the specimen. Both its depth of focus and resolution are far superior to the light microscope, although the resolution is not as good as that of the TEM. The SEM has so far not been widely used in cell biology although its range of use is expanding rapidly (e.g. the study of the topography of the cell surface) and so the principles of the microscope will be outlined here.

In the SEM, a fine beam of electrons scans to-and-fro across the specimen and this produces a number of interactions. The SEM makes use of the secondary electrons that are dislodged from atoms at the specimen surface, their intensity depending on the surface topography of the specimen. They are collected, converted to photons via a scintillator, and then passed through a photomultiplier to produce an image on a cathode ray tube. Variation in the intensity of emission of secondary electrons is reflected in the brightness of the corresponding spot on the screen and so an image is built up as the beam scans the specimen.

The methods used to prepare specimens for the SEM vary widely. As with the TEM, specimens have to be dehydrated since the column of the microscope is under vacuum. Little preparation is needed for tough, dry specimens (e.g. pollen grains), while others are fixed, dehydrated and dried in various ways depending on the nature of the material. Finally specimens must be coated with a thin, conducting film (e.g. carbon or gold) to prevent the build up of electrical charge or heat which can impair the image and even damage the specimen. An example of the application of the SEM is shown in Fig. 1.20.

Cell fractionation

The various microscopic techniques described in the last section have given us a detailed picture of the structure of cells and, in some cases,

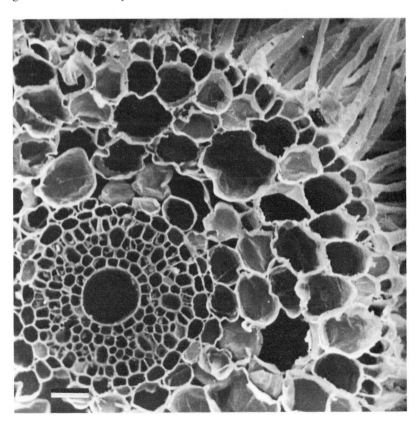

Fig. 1.20   Transverse fracture of part of a mature region of barley (*Hordeum vulgare*) root. The tissue was fixed, dehydrated and critical point dried using CO_2. The stele and cortical cells are clearly seen and may be compared with the micrograph produced by light microscopy in Fig. 1.3 ×250.
Bar = 50 μm. Provided by Drs A. Wilson and A. W. Robards, University of York.

considerable insight into the function of the various cell components. However, these techniques are limited since, in general, they give a static and non-quantitative picture of the cell with little information concerning its chemical activity. The principal method that has been developed to study the chemistry and functions of the various cell parts and organelles is known as *cell fractionation*, and this has proved a particularly fruitful technique when used in conjunction with microscopic techniques as we shall see in a number of chapters later in this book. All cell fractionation techniques consist of three basic steps: disruption of the tissue to produce a suspension of cell fragments, a process often known as *homogenization*; *fractionation* or separation of the various cell fragments, usually by centrifugation; and *analysis* of the chemical activity of the separated fragments.

Homogenization

A number of techniques are now available to produce a suspension or homogenate of cell fragments, ranging from a simple pestle and mortar and the glass homogenizer illustrated in Fig. 1.21 to sophisticated mechanical blenders. The major aim is to carry out this procedure with as little loss or distortion of cellular activity as possible. The presence of a cell wall in plant cells may mean that the tissue requires more severe treatment than would be needed for animal cells and this in turn can result in greater damage to cell organelles. The method of disruption may have important consequences for the final recovery of a particular cell fragment. This is illustrated by an experiment which was

Fig. 1.21 Schematic representation of the main steps in cell fractionation. Not all cell structures are depicted in this diagram.

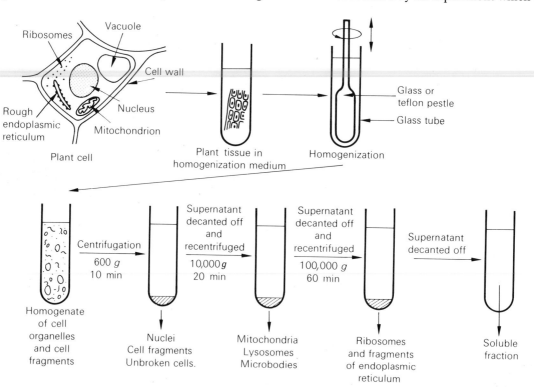

carried out on tobacco leaves using a pestle and mortar and a blender as alternative methods of tissue disintegration (Jagendorf and Wildman, 1954). With the former, 90% of the total DNA of the cells was sedimentable in a centrifuge at a speed giving 100 g for 12 min, indicating that the nuclei were relatively undamaged since these large organelles are sedimented at that speed in that time. However, after blending, only 80% was sedimentable using 600 g. In contrast, 90% of the total chlorophyll was recovered using 600 g after blending, suggesting good preservation of the chloroplasts, whereas only 50% was sedimented when the pestle and mortar was used. This clearly shows not only that the method of disruption has an important effect on the subsequent recovery of a particular cell organelle, but that organelles show a differential sensitivity to the method of homogenization; this must be adapted to suit the particular organelle under study.

A number of other precautions are normally taken in the routine disruption of plant tissues in an attempt to reduce damage and loss of activity which might occur when substances which are normally rigidly separated in the living tissued are mixed by homogenization. Firstly, the homogenization is carried out in the cold, usually at temperatures of 1–3°C, either by working in a special cold room or by using precooled apparatus and solutions and keeping all materials at the temperature of ice whenever possible. It is hoped that these cold conditions will minimize chemical changes which might occur between compounds normally separated in the intact cell. In addition, the tissue is homogenized in a medium which usually contains the following ingredients. A buffer is included to reduce changes in the pH which might result when organic acids, normally stored in the plant cell vacuole, are released. Large changes in pH can result in considerable loss in enzymic activity. Sucrose, or some other osmoticum, is added to the medium to minimize structural damage to membrane-bound organelles caused by sudden changes in osmotic pressure when cells are disrupted. In addition to these general precautions, certain protective compounds are frequently added to the medium in an attempt to reduce specific deleterious chemical reactions. For example, most plant tissues contain a wide variety of phenolic compounds, sometimes at very high concentrations. When the tissues are disrupted, the phenols are usually rapidly oxidized to quinones by a series of enzymes known as phenoloxidases which, in the intact cells, are normally spatially separated from the phenols. The oxidation products may severely inhibit certain enzymic and metabolic processes since they are able to form complexes with proteins by hydrogen and covalent bonding. Attempts are therefore made, by the addition of certain compounds to the extraction medium, to remove phenols and to prevent oxidation occurring. These additions include polymers such as polyvinylpyrrolidone which remove phenolic substrates by adsorption; inhibitors of phenoloxidase activity, such as diethyldithiocarbamate, which prevent the formation of oxidation products; and reducing agents, such as thiols or ascorbate, which reduce quinones as they form and so prevent deleterious accumulations. Thiols, such as β-mercaptoethanol, have an additional protective function. Many enzymes have sulphydryl-containing amino acids as an essential part of their catalytic site and the presence of a suphydryl-containing thiol ensures that the sulphydryl group of the enzyme remains in the reduced state.

It should be emphasized that, at present, the approach to homogenization is still largely empirical. Each plant material presents its own particular problems, and the choice of homogenizer and extraction medium is arrived at largely by trial and error. There is no one general procedure that can be employed for the isolation of cell fractions. The procedure must be modified to suit the particular tissue and particular cell fraction under study.

Fractionation

Once the homogenate has been prepared it is then necessary to fractionate or separate the various intracellular components from each other. The most widely used method of doing this is by *centrifugation*. When a centrifugal force is applied to a suspension of cellular particles the rate of sedimentation of any given particle will depend on its size, density and shape. For example, for spherical particles the rate of sedimentation is determined by Stokes's law,

$$V = \frac{2r^2(D - d)g}{9\eta},$$

where
V = rate of sedimentation in cm/s
r = radius of particles (cm)
D = density of particles
d = density of suspension medium
η = viscosity of suspension medium (poise)
g = gravitational force (981·2 cm/s)

Shape is not taken into account in this equation but it will clearly affect the resistance the particle encounters in moving through the medium, and thus the sedimentation rate. Thus the various cellular components can be separated on the basis of differences in their size, density and shape; this is known as differential centrifugation. A typical fractionation procedure is illustrated in Fig. 1.21. A suspension of cell fragments produced by homogenization is first centrifuged at 600 g for 10 min sedimenting the largest fragments including nuclei, cell walls and any remaining intact cells. The supernatant is decanted off and re-centrifuged at 100 000 g for 20 min sedimenting mitochondria and other similar sized organelles. The remaining supernatant is finally centrifuged at 100 000 g for 60 min to sediment the ribosomes and pieces of endoplasmic reticulum (known as the microsomal fraction). The remaining supernatant is usually referred to as the 'soluble' fraction and contains a variety of smaller particles, enzymes and other molecules. The fractions obtained by single centrifugations as described above tend to be mixtures of various particles although one particular component will predominate in any given fraction. By repeated resuspension and recentrifugation at particular speeds, relative pure preparations of particular cell organelles or parts can be obtained. The nature of the fractions can be established by microscopy, and their chemical activity studied by various analytical methods.

There have been a number of interesting developments of this basic concept of separation by centrifugation. In *density gradient centrifugation*, the homogeneous medium in the centrifuge tube is replaced by a medium with a gradient of densities, usually produced by layering sucrose solutions of differing concentrations, one on top of the other, with the heaviest at the bottom. If a suspension of cell particles is placed on top of the tube and centrifuged, particles of different densities, shapes and sizes (different sedimentation coefficients) will move down the tube at different rates and will be concentrated in distinct zones as illustrated in Fig. 1.22. When centrifugation is stopped there will be distinct zones of similar particles along the tube which can be carefully removed and analyzed. A further modification of this technique is known as *isopycnic* or *equilibrium density centrifugation*, in which the gradient is designed to include the density of the particles under investigation. If centrifugation is allowed to proceed to equilibrium, particles will come to lie in a band centred at their own buoyant density. Separation will depend on density only and so this technique has a greater potential ability to separate particles of similar size and shape.

The centrifuges used in the techniques described above are known as preparative centrifuges. A further refinement is the analytical ultra-centrifuge which can be used to separate and characterize very small particles and macromolecules, such as ribosomes and proteins. The specimen is mixed with a sucrose or caesium chloride medium in a quartz centrifuge cell and centrifuged at very high speeds to generate forces of up to 500 000 *g*. Such extreme forces cause small particles and macromolecules to sediment from the solution leaving pure

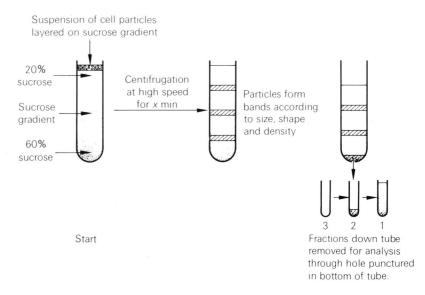

Fig. 1.22   Density gradient centrifugation.

solvent at the top of the cell and forming a sharp boundary between the pure solvent and solvent containing the specimen. The rate of movement of this boundary and the position of the concentration peak of the specimen in the centrifuge cell during the experimental run is recorded by optical methods such as that known as the Schlieren method which measures the refractive index gradient along the cell (Fig. 1.23). The sedimentation rate is expressed as the sedimentation coefficient s, where:

$$s = \frac{\mathrm{d}x/\mathrm{d}t}{\omega^2 x},$$

where
x = distance from centre of rotation (cm)
ω = angular velocity of centrifuge head in radians per second
t = time in seconds

Fig. 1.23   Analytical ultracentrifugation showing how the boundary formed between the pure solvent and the solute is recorded by optical methods while the sample is being centrifuged. Redrawn by permission of Beckman Instruments, Inc., California.

Most proteins have sedimentation coefficients in the range of 1–100 $\times$ 10^{13} s. For convenience, the basic unit is taken as 1×10^{13} s and is called the *Svedberg unit* S. Thus most proteins have sedimentation coefficients of between 1 and 100 S, while ribosomes and viruses lie in the range of 50–200 S. The sedimentation coefficient increases

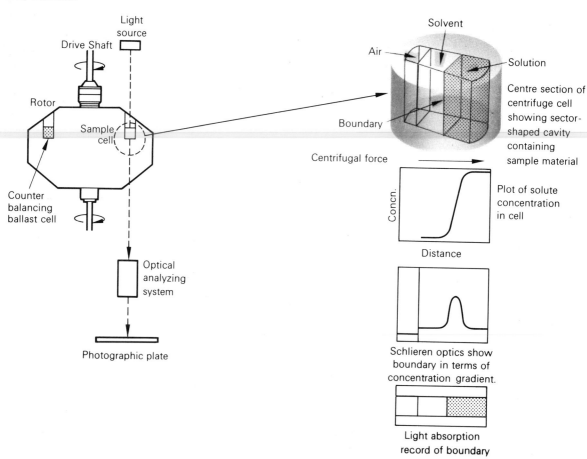

with molecular weight but not proportionally, since the rate of sedimentation is also affected by the shape of the particle.

These cell fractionation techniques have been the major instruments in demonstrating that specific chemical reactions are located in specific cell components, and, in fact, have led to the discovery of at least one major cell organelle which is not easily visible by microscopy (see Ch. 12). However, there are a number of potential pitfalls in these techniques, and artefacts can arise just as in microscopy. The first possibility is that a particular subcellular fraction is contaminated with other components. This is very possible since many cell organelles, e.g. mitochondria, lysosomes and microbodies, are of similar size, and there is considerable variation in the size and shape of some organelles. The degree of contamination can be checked by microscopic examination, and also by chemical tests for a component that is known to be restricted to one organelle. For example, the presence of chlorophyll may be used as a test for the presence of chloroplast material, while the enzyme succinic dehydrogenase is thought to be restricted to mitochondria and so can be used as a 'marker' for these organelles.

A second difficulty is that the morphological structure and chemical activity found in an isolated cell fraction may differ from that in the living cell. For example, when cells are homogenized there is a tendency for broken cell membranes to round off and form vesicular structures which sediment with other cell fractions during centrifugation. Again, there may be considerable distortion of enzyme activity during homogenization and fractionation. Enzymes show considerable variation in the ease that they can be removed from cell organelles. Enzymes may be lost from or adsorbed onto certain cell fractions during separation, and so their absence or presence in the isolated cell fraction may not be characteristic of the situation in the intact cell. Plant cell walls are particularly susceptible in this respect since they are highly charged and readily adsorb soluble cytoplasmic enzymes during homogenization.

A further problem arises due to the presence of bacteria in many isolated subcellular fractions. The major source of contamination comes from the surface of plant mateial itself which may contain a wide range of bacterial types. Bacteria and bacterial fragments may adhere to and sediment with various subcellular organelles and be a considerable source of error. The level of bacterial infection can be considerably reduced by the use of sterile media, of antibiotics to control bacterial growth, and of sterile tissue. Thus although the techniques of homogenization and fractionation are subject to a number of difficulties, they have been a major instrument in the elucidation of cell function in recent years, and many examples will be described in the following chapters.

Analysis

Once cell fractions have been isolated they can then be examined by standard biochemical procedures, such as spectrophotometry or radioisotopic methods, in an attempt to characterize their biochemical activities. The range of these techniques is wide and it is beyond the scope of this book to discuss them all fully. However, three of these techniques, spectrophotometry, the use of radioisotopes, and chromatography are so widely used and form the basis of so many of the investigations described in the following chapters, that they will be described in some detail here.

Spectrophotometry. Many biologically important molecules, including the substrates and products of many enzymic reactions, absorb light and have a characteristic absorption spectrum (see Figs. 2.23, 2.44). This can be an important means of identifying a compound in any biological sample. If the absorbance of a standard solution of the particular molecule at a given wavelength is known, then the concentration of that compound can be determined from an absorbance measurement—providing no contaminating molecules are present. In relation to enzymic reactions, the substrate and product of any given reaction usually have different absorption spectra. For example, the enzyme fumarate hydratase catalyses the conversion of fumarate to malate in the TCA cycle (see p. 289). Fumarate absorbs strongly at 300 nm whereas malate does not absorb at that wavelength, and so the rate of the reaction can be measured by following the utilization of fumarate at 300 nm. Again, the interconversion of the coenzyme NAD and its reduced form NADH is involved in a wide range of enzymic reactions (see p. 000). NADH has an absorption maximum at 340 nm whereas the oxidized form NAD does not absorb light at this wavelength. This means that the rate of many oxidation-reduction reactions can be determined by measuring the production or loss of NADH at 340 nm.

The instrument normally used for these absorbance measurements is known as a spectrophotometer and is based on the fundamental laws of Lambert and Beer. Lambert's law states that light absorbed is directly proportional to the thickness of the solution under analysis, and Beer's law that light absorbed is directly proportional to the concentration of solute in solution. These two laws can be combined to give the following expression:

$$A = Ecl,$$

where

A = absorbance or optical density
E = molar extinction coefficient (absorbance of a 1 M solution having a 1 cm light path)
c = the concentration of the solution moles per litre
l = the cell or sample length in cm

A typical spectrophotometer is shown diagrammatically in Fig. 1.24. It contains a light source and a monochromator (a means of selecting particular wavelengths of light) using either a grating or a prism. The selected wavelength is directed through the sample contained in a glass

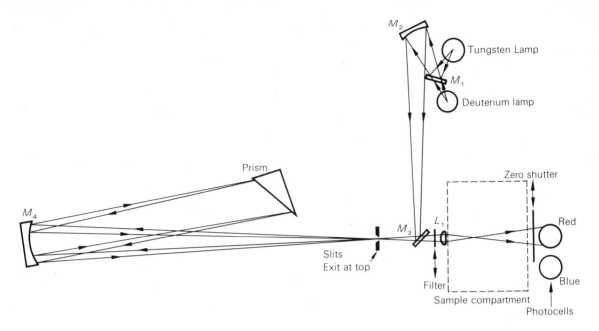

Fig. 1.24 The optical system of a Unicam SP500 spectrophotometer. Light from a tungsten lamp (source of visible light absorbed by coloured solutions such as chlorophyll) or from a deuterium lamp (source of ultra-violet light absorbed by colourless solutions such as nucleic acids or phenolic compounds) is directed by the mirrors (M_1, M_2, M_3) onto the entrance slit of the monochromator. The beam passing through the slit is collimated by the mirror M_4 and then dispersed by the 30° silica prism. After further reflection the spectrum is brought to a focus at the exit slit. The position of this spectrum and so the wavelength transmitted by the slit, is adjusted by rotating the prism assembly. This selected wavelength of light is focused by the lens L_1 into the sample compartment containing the sample cuvette. The transmitted light is detected by the appropriate photocell and converted into a measurable electrical signal.

or silica cell or cuvette, and the emergent light intensity measured by a photocell, enabling the absorbance or optical density of the sample to be calculated. It is possible with modern spectrophotometers to make a continuous recording of optical density at a given wavelength as a function of time.

Uses of radioisotopes. One of the most important techniques in biochemistry is the use of radioactive isotopes to 'label' a particular compound so that its subsequent enzymic or metabolic fate can be traced. These methods can be applied to intact organisms, to isolated tissues or cell fractions, and to purified enzyme preparations. Some of the most useful radioisotopes for biological research have been listed on p. 26. Most of these are β-emitters which means that when the nuclei of these atoms disintegrate, electrons (or β-rays) are ejected with an energy characteristic of that nucleus. The various methods used for the detection and measurement of radioactivity depend upon the radiation ionizing or exciting molecules in the detector. With the Geiger–Müller tube, a radioactive particle ionizes gas molecules in the tube causing the release of further electrons. These electrons ionize other gas molecules producing even more electrons. The tube contains two electrodes with a potential difference applied between them, causing the electrons to flow towards the anode. When a shower of electrons impinges on the anode, a surge of current results which is easily detected and recorded as one count. The scintillation counter is a more sensitive detection system. The radioisotope is dissolved in an organic solvent containing a phosphor compound which will emit a light flash or scintillation when struck by a radioactive particle and which can be detected by a sensitive photomultiplier. This system is particularly useful in detecting low energy radiation such as that from ^3H and ^{14}C, isotopes that are much used in biological research.

The uses of radioisotopes in biochemical research are now extremely numerous and varied and there are now available a wide range of compounds labelled in specific positions with various isotopes. The chemical activity of an element is, of course, determined by its atomic number, so that compounds differing only with respect to certain isotopes would be expected to have similar metabolic properties, although some slight changes in reaction rate may be found due to the change in atomic weight. The elucidation of the pathway of carbon dioxide fixation in photosynthesis described in Chapter 8 (p. 359) is a good illustration of the use of this technique. Another approach is in the determination of intracellular sites of a particular biochemical mechanism, and an excellent example is furnished by the discovery of the site of protein synthesis which is described in Chapter 5. Labelled amino acids were injected into rats and later the liver was removed, homogenized and various cell fractions isolated by centrifugation. When the radioactivity which had been incorporated into protein was determined, it was found that incorporation occurred largely in the cell fraction containing the ribosomes. Other approaches have since confirmed that the ribosomes are the sites of protein synthesis in the cell.

Chromatography. The general term chromatography is used to describe a wide range of techniques which are used for the separation and identification of substances in a mixture which are frequently present only at very low concentrations. Chromatography means colour diagram or scheme, and the term was first used by Tswett in 1903 to describe the separation of leaf pigments into various coloured zones that is effected by passing a petroleum ether leaf extract through a column of adsorbent calcium carbonate. The separation of the various pigments depends on their different affinities for the adsorbent material. We now recognize a number of different forms of chromatography which depend upon differences in adsorption, partition, ion-exchange or molecular size, and the separation achieved by some procedures involves a combination of these factors.

1. Adsorption chromatography. The separation of leaf pigments described above is an example of adsorption chromatography. The various components of a mixture are separated by their differential adsorptions onto certain materials (e.g. calcium carbonate, alumina gel, charcoal) which are packed into columns in glass tubes. The different zones, which are not necessarily coloured, can be separated by cutting up the column or by eluting separately with different solvents.

2. Partition chromatography. This method relies on the differential separation of substances between two immiscible phases, i.e. on differences in their *partition coefficients*. If a solution of a substance is shaken with an immiscible solvent, the solute will distribute itself between the two phases. At equilibrium, the ratio of the concentration of the solute in the two phases is known as the partition coefficient. *Paper chromatography* is probably the most well known form of partition chromatography and is widely used for the separation of amino acids, sugars and pigments. Paper contains a high proportion of cellulose which becomes hydrated and forms a stationary aqueous phase. The mixture to be separated is applied to one end of a paper strip. A solvent, usually consisting of water and an organic solvent, ascends (or descends) the strip by capillary action and so produces a stationary aqueous phase and a mobile organic phase. Depending on their partition coefficients, substances which are more soluble in the organic phase move to the front of the paper while those more soluble in the aqueous phase remain near the origin. Greater separation of the components of a mixture is frequently obtained by two-dimensional paper chromatography in which the chromatogram is first developed in one direction as described above, turned through 90°, and developed a second time with a different solvent system. The paper usually has to be sprayed with a reagent to give specific colours to the separated substances. An excellent example of this technique is described in Chapter 8 where it has been used to separate and identify the various products of photosynthesis.

There are a number of other forms of partition chromatography that are widely used to separate biologically important molecules. *Thin-layer chromatography*, which has proved to be extremely useful in the separation of lipids, employs a glass plate which is covered with a thin layer of dry, adsorbent material such as silica gel or cellulose. Otherwise, the procedures are similar to those employed for paper chromatography. In both these techniques, adsorption phenomena may play an important role in the separation process. *Gas–liquid chromatography* also depends on partition effects and is used for the separation of a wide range of organic compounds including fatty acids. An inert gas such as nitrogen is used to carry a vaporized mixture of compounds along a heated tube which is packed with a powdered support coated with the liquid phase. The differences in the partition coefficients of the gaseous components between the gaseous and liquid phases, and to some extent their adsorption to the support, form the basis for separation.

3. Ion-exchange chromatography. This depends on the acid–base properties of molecules such as amino acids and organic acids and is usually carried out in columns which are filled with ion-exchange resins. For the separation of amino acids, cation-exchangers are used, such as sulphonated polystyrene charged with sodium by treatment with sodium hydroxide solution. The amino acids are loaded onto the column in an acid solution (about pH 3·0) which makes them positively charged (see p. 82). As the amino acids move down the column they will displace some of the sodium ions:

$$\text{Resin}-\text{SO}_3^-\text{Na}^+ + \text{NH}_3^+\text{R} \rightleftharpoons \text{Resin}-\text{SO}_3^-\text{NH}_3^+\text{R} + \text{Na}^+$$

The more basic the amino acid, the more tightly it will be bound to the resin. The amino acids can then be differently eluted and collected from the column by washing with buffers of increasing pH and ionic strength. This method also depends on adsorption to some extent and can be carried out by thin-layer techniques.

4. Molecular sieve chromatography. This method which is also known as *gel filtration* and *molecular-exclusion chromatography* separates molecules according to their molecular size and has been particularly useful in the purification of proteins. The mixture of proteins is passed through a column which has been packed with a gel of a particular porosity range. As the solution percolates through the column, the larger molecules which cannot penetrate into the pores of the gel particles move down more rapidly than the small molecules which penetrate the pores to varying degrees according to their size. The different proteins therefore appear in the eluate in a sequence of decreasing molecular size. The gels themselves are usually comprised either of highly cross-linked polymeric carbohydrates or polyacryl-amide beads which are swollen in solvent before packing in the column. The more highly cross-linked the gel, the lower is the molecular size of the solutes it excludes. As a further refinement, ionic groups can be introduced onto the polymeric matrix so that separation depends upon a combination of charge and molecular size.

5. Affinity chromatography. A fifth type of chromatography which has had particular recent value in protein purification is affinity chroma-tography. This depends upon stereospecific adsorption in order to pick out a specific macromolecule from a complex mixture. In general, a ligand which binds with high affinity to the macromolecule is selected and attached covalently to an inert polysaccharide or polyacrylamide support. Frequently a spacer molecule is used to separate the ligand from the adsorbent matrix (Fig. 1.25). The ligand itself may be a substrate or competitive inhibitor which binds relatively tightly to the active site of one particular enzyme or enzyme class but not to most others. The protein mixture is passed through the column bearing the attached ligand and the enzyme(s) of interest retained. In order to

Fig. 1.25   A diagrammatic representation of affinity chromatography. A ligand which binds to a specific protein is attached to an inert, adsorbent matrix through an intermediate spacer molecule. The latter extends the ligand away from the surface of the support, thus allowing easier access to the binding site of the protein. The tightness of binding of the protein can be manipulated by variations in the ligand, by the length and nature of the supporting spacer arm, the quality of the adsorbent, and the conditions for loading and eluting the column (e.g. pH and salt concentration).

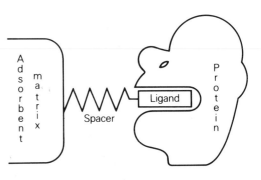

elute the protein from the columns a variety of approaches may be attempted. Thus, excess substate or a competitive inhibitor, dissolved in buffer, can often be used to elute the column. Alternatively, high salt concentration or lowered pH are successful ploys since they often relax the hold of the enzyme to the immobile ligand. Indeed, the art of affinity chromatography is to select a ligand which binds the macromolecule of interest sufficiently tightly to retard it, but which does not retain it with such avidity that only denaturing conditions can be employed for its elution. Examples of affinity chromatography of enzymes include purification of nucleases on immobilized nucleic acids and retention of dehydrogenases on nicotinamide derivatives or on various dyes. Probably the earliest use of affinity chromatography in plant biochemistry was the purification of amylases by adsorbing them to insoluble starch granules. In this case, the enzymes were released by allowing them to digest the grains.

Molecules other than enzymes can also be purified in this way. A good example is provided by the plant lectins which can often be selectively retained on Sepharose columns substituted with sugars. Concanavalin A, a lectin from Jack beans (see p. 141), can be purified directly in one step merely by passing the seed extract through a column of Sephadex G-100, a polymer of glucose. The protein is eluted with either D-glucose or α-methyl D-mannoside.

One final chromatographic method which is now widely used is known as *high pressure liquid chromatography* or HPLC. In this, the process of column chromatography, whether it be of the partition, ion-exchange or molecular sieve type, is speeded up enormously by conducting the elution by liquid solvents under high pressure. The key to these instruments is the use of high precision pumps and suitable column supports which can withstand extremely high pressures. HPLC has revolutionized the rapid separation of low molecular weight metabolites and is now beginning to have an impact on the field of protein and nucleic acid chemistry as well.

Further reading

Clowes, F. A. L. and Juniper, B. E. (1968) *Plant Cells*. Blackwell Scientific Publications, Oxford.

Cutter, E. G. (1978) *Plant Anatomy*. 2nd edn. Edward Arnold, London.

Grimstone, A. V. (1976) *The Electron Microscope in Biology*. 2nd edn. Edward Arnold, London.

Gunning, B. E. S. and Steer, M. W. (1975) *Ultrastructure and the Biology of Plant Cells*. Edward Arnold, London.

Hall, J. L. (1978) *Electron Microscopy and Cytochemistry of Plant Cells*. Elsevier/North-Holland, Amsterdam.

Hall J. L. (1982) Cells and their organization: current concepts. In *Plant Physiology*, Vol. 7. Ed. R. G. S. Bidwell and F. C. Steward. Academic Press, New York.

Hall, J. L., Al-Azzawi, M. J. and Fielding, J. L. (1977) Microscopic cytochemistry in enzyme localization and development. In *Regulation of Enzyme Synthesis and Activity in Higher Plants*. Ed. H. Smith. Academic Press.

Leech, R. M. (1977) Subcellular fractionation techniques in enzyme distribution studies. In *Regulation of Enzyme Synthesis and Activity in Highter Plants*. Ed. H. Smith. Academic Press, London.

Meek, G. A. (1976) *Practical Electron Microscopy for Biologists*. 2nd edn. John Wiley and Sons, London.

Quail, P. H. (1979) Plant Cell Fractionation. *Ann. Rev. Plant Physiol.,* **30,** 425.

Rhodes, M. J. C. (1977) The extraction and purification of enzymes from plant tissues. In *Regulation of Enzyme Synthesis and Activity in Higher Plants.* Ed. H. Smith, Academic Press, London.

Reid, E. (1979) *Plant Organelles.* Methodological Surveys, Vol. 9. Ellis Horwood Ltd, Chichester.

Troughton, J. H. and Sampson, F. B. (1973). *Plants. A Scanning Electron Microscope Survey.* John Wiley and Sons, Sydney.

White, G. M. (1966) *Introduction to Microscopy.* Butterworths, London.

Literature cited

Guilliermond, A. (1941) *The Cytoplasm of the Plant Cell.* Chronica Botanica Co., Waltham, Mass.

Jagendorf, A. T. and Wildman, S. G. (1954) The proteins of green leaves. VI Centrifugal fractionation of tobacco leaf homogenates and some properties of isolated chloroplasts, *Plant Physiol.,* **29,** 270.

Lwoff, A. (1965) *Biological Order.* The M.I.T. Press, Cambridge, Mass.

Moor, H. (1965) Freeze-etching, *Balzers High Vacuum Report* 2.

Robertson, J. D. (1959) The ultrastructure of cell membranes and their derivatives, *Biochem. Soc. Symp.,* **16,** 3.

Robertson, J. D. (1967) The organization of cellular membranes. In *Molecular Organization and Biological Function.* Ed. J. M. Allen. Harper and Row, New York.

2 The molecules of cells

We have already mentioned in Chapter 1 that living cells show considerable unity in relation to their biochemical characteristics. As the techniques of microscopy, and so our knowledge of cell structure, were developing, organic chemists were steadily isolating and characterizing the chemical components of living cells. It was shown that the great diversity of living things could be considered in terms of the various combinations of a small number of elements and that isolated biological molecules were inanimate entities which showed considerable complexity in structure and properties but lacked, in themselves, the characteristics of life. It is the aim of biochemists and cell biologists to understand how these molecules interact and are organized to produce, maintain and perpetuate the living state. All living cells are composed of similar macromolecules which are themselves made up of the same small building-block molecules. The great diversity in the macromolecules found in living systems arises from the varied combination of a limited number of an even smaller number of elements. For example, bacterial cells contain about 3 000 different types of protein molecules. These are made up from various combinations of only twenty different amino acids, the building blocks, while the amino acids are largely composed of four elements. These twenty amino acids which combine to form proteins are the same in all cells whether they be plant, animal or microbial. The existence of a vast range of form and function in living cells and organisms is a result of small differences among the basically similar kinds of macromolecules.

The main constituent of living cells is water, which makes up 80–95% of the total fresh weight of most organisms. Of the dry matter, about 90% is usually composed of the four major classes of macromolecules, the rest consisting of free building-block molecules, other metabolic intermediates and inorganic ions. The four major classes of macromolecules are the proteins, nucleic acids, polysaccharides and lipids and they are all based on the element carbon. They all contain hydrogen and oxygen and many contain nitrogen. The four elements make up over 90% of the atoms of most organisms (Table 2.1). A number of other elements such as phosphorus, sulphur, potassium, magnesium and chloride are known to be essential for life, although only twenty-two of the 100 different elements found on the earth are known to be essential when all living organisms are considered. The importance of some elements is restricted and their function is not always understood. For example, cobalt is required by some organisms but this requirement has not been generally established for higher

Table 2.1 Inorganic composition of corn and man (%). Data from Epstein (1972)

Element	Corn plant, Zea mays	Man, Homo sapiens
O	44·43	14·62
C	43·57	55·99
H	6·24	7·46
N	1·46	9·33
Si	1·17	0·005
K	0·92	1·09
Ca	0·23	4·67
P	0·20	3·11
Mg	0·18	0·16
S	0·17	0·78
Cl	0·14	0·47
Al	0·11	–
Fe	0·08	0·012
Mn	0·04	–
Na	–	0·47
Zn	–	0·010
Rb	–	0·005

plants, although it appears to be essential for the nitrogen-fixing bacteria that form nodules on the roots of the leguminous plants.

What then is the reason for the importance of the four major elements: carbon, hydrogen, oxygen and nitrogen? This is thought to be because they are the smallest elements in the periodic table that can form strong, stable covalent bonds by the sharing of a pair of electrons: this bond formation is the first step in the formation of macromolecules. Since the stability of both single and multiple bonds increases with a decrease in atomic size, carbon, oxygen and nitrogen are able to interact to form stable double bonds while carbon can also form triple bonds with nitrogen and with other carbon atoms. Taken together, the four major elements of living cells possess a considerable range of possibilities as regards bond formation.

Of particular importance are the bonding properties of the element carbon which forms the basis of all biological molecules. Carbon is a small atom which is situated in the middle of the second horizontal row of the periodic table and requires four electrons to achieve a stable electronic configuration. It is able to react with electronegative elements like oxygen, nitrogen and sulphur, and with the electropositive hydrogen atom. It forms single, double or triple bonds with other carbon atoms and so gives rise to straight or branched chains, rings, or combinations of these structures. These various carbon bonds are stable and so form the basis of a vast number of chemical compounds which show considerable variation in size, structure and chemical properties. Silicon, which is situated in the same vertical row of the periodic table, shows a similar tendency to form covalent bonds and to combine with other silicon atoms. However, although silicon is many times more abundant than carbon in the earth's crust, it is apparently not as suitable as carbon in relation to the properties of living matter. The probable reason is that silicon–silicon bonds are much less stable than carbon–carbon bonds, particularly in the presence of oxygen or water when they form silicates and large, insoluble polymers of silicon dioxide.

Thus the major classes of biological macromolecules are composed of smaller molecules (e.g. amino acids, simple sugars) with molecular weights generally in the range of 100–400, while those of the macromolecules range from 1 000 or so to well over one million in the case of some proteins and nucleic acids. Whether large molecular size in biological systems is an advantage is a difficult question to answer but, presumably, these molecules have specific properties which are not possessed by the smaller molecules of which they are composed. These macromolecules, in turn, form the building blocks of cellular fine structure and so the properties of cells must be related to the properties of their constituent macromolecules. Just as all cells are composed of the same types of macromolecules, then these same molecules have similar functions in all living cells. Proteins are the major organic components of cells and perform a variety of functions, both as important structural molecules and as specific catalysts, the enzymes, which are responsible for the expression of genetic information and for the regulation of cellular metabolism. Proteins are synthesized under the direction of the nucleic acids which are the macromolecules concerned with the storage and transmission of the genetic information. Proteins and nucleic acids may be grouped together as informational molecules; their different building-blocks are arranged in specific sequences which are used to convey information. In this respect they differ from the polysaccharides and lipids which do not carry information in this way since they are composed of only one or two repeating units. Polysaccharides have important rôles, both as structural components (e.g. cellulose and pectin in plant cell walls, chitin in insect cuticles), and as major food reserves (e.g. starch, glycogen). They also exhibit some biological specificity. For example, in animal cells, cell surface antigens, blood group specificity, pathogen receptor sites and cell adhesion, are all functions of carbohydrates. The lipids play somewhat similar rôles to the polysaccharides. They are involved in cell structure, being a major component of cell membranes, and also act as a major energy reserve for cellular metabolism.

In the following sections we will describe the major features of the various molecules found in plant cells and will emphasize those concepts that are important to the understanding of the remaining chapters in this book.

Carbohydrates

A large portion (usually from 60 to 90%) of the dry matter of plant material is composed of carbohydrates. Of these, the water soluble, sweet-tasting sugars have been recognized since antiquity. Later it was found that certain other substances such as starch and cellulose could be converted into sugars when heated with dilute mineral acids. Sugars were consistently shown to contain carbon, hydrogen and oxygen in the ratio of $1:2:1$ and when heated slowly gave off water and left a black residue of carbon. For this reason, the compounds were called carbohydrates, meaning 'watered carbon' and given the empirical formula $C_x(H_2O)_x$. We now know that they are not truly hydrates of carbon at all and that there are several exceptions to the generalized empirical formula. Deoxyribose, for example, a sugar of common biological occurrence, has the formula $C_5H_{10}O_4$, while lactic acid ($C_3H_6O_3$) and *myo*inositol ($C_6H_{12}O_6$) are not carbohydrates at all. Further, there are a whole group of compounds such as the amino sugars and the uronic acids which are clearly related to the commoner sugars yet would be excluded from the group by a rigorous definition based on C, H, O composition.

A more satisfactory definition is that carbohydrates are *polyhydroxyaldehydes or polyhydroxyketones, or compounds which can be hydrolysed to these substances using dilute mineral acid*. Almost all of the important reactions that carbohydrates undergo involve either their reducing groups or their hydroxyl (or secondary alcohol) groups.

Optical isomerism. Carbohydrates, like the amino acids discussed later, are optically active in solution. That is, they are able to rotate the plane of polarized light because they have asymmetric carbon atoms (i.e. ones with four different substituents). The reasons why such asymmetry causes such an effect is incompletely understood and a detailed discussion of optical isomerism and specific optical rotation [α] will not be presented here. Nevertheless, when we consider glyceraldehyde, which provides us with our initial example of a carbohydrate, we can see that it has a single asymmetric carbon atom, and that it can, therefore, exist in the form of two optical isomers

D-Glyceraldehyde   L-Glyceraldehyde

$$
\begin{array}{cc}
\text{CHO} & \text{CHO} \\
| & | \\
\text{C} & \text{C} \\
\text{HOH}_2\text{C} \quad \text{OH} & \text{HO} \quad \text{CH}_2\text{OH} \\
\text{H} & \text{H}
\end{array}
$$

which are mirror images of each other and are known as *en-antiomorphs*. Emil Fischer decided arbitrarily that the formula with the —OH to the right should be designated D- (or dextrorotary) and the one with the —OH to the left L- (or laevorotary). It was fortunate that this empirical designation did in fact denote the actual direction of rotation. As we shall see, however, some sugars whose structures are based on D-glyceraldehyde and which are therefore in the D-structural series rotate the plane of polarized light to the left and it is necessary to have the prefix D (+) or D (−) present if the true direction of rotation (whether dextro- or laevo-) is to be designated.

Glyceraldehyde is known as a triose because it consists of only three carbon atoms. Carbon atom 1 is the aldehydic carbon, and, as we have pointed out, there is only one centre of asymmetry and therefore only two possible isomers. Because it bears an aldehyde group, it is termed an *aldose* sugar. Its sister compound, dihydroxyacetone, has a ketone group, and bears its carbonyl function on C-2 and therefore a *ketose* sugar. Note that it has no asymmetric carbon atom so that only one

$$
\begin{array}{l}
\text{CH}_2\text{OH} \\
| \\
\text{C=O} \\
| \\
\text{CH}_2\text{OH}
\end{array}
$$

isomer exists. Other simple sugars exist with more than three carbon atoms and are named according to the length of their carbon chains (Table 2.2). Clearly, the larger the number of carbon atoms, the greater the number of optical isomers that are possible, although only a few members of each series occur naturally.

Structural relationships among sugars. It is possible to increase the chain length of the aldose sugars one carbon atom at a time by the so-called Kiliani synthesis, a series of reactions which is summarized in Fig. 2.1. Each C-1 addition produces one further asymmetric carbon atom and hence two isomeric forms of the new compound are possible. Application of these reactions to D-glyceraldehyde can give a whole series of related sugars (Fig. 2.2). Note that the penultimate carbon

Table 2.2 The principal monosaccharides found in plants

Group	Name	Empirical formulae
Trioses	D-Glyceraldehyde	$C_3H_6O_3$
	Dihydroxyacetone	
Tetroses	D-Erythrose	$C_4H_8O_4$
	D-Erythulose	
Pentoses	L-Arabinose	
	D-Xylose	
	D-Xylulose	$C_5H_{10}O_5$
	D-Ribose	
	D-Ribulose	
	D-Deoxyribose	
Hexoses	D-Glucose	
	D-Mannose	
	D-Galactose	$C_6H_{12}O_6$
	L-Sorbose	
	D-Fructose	
	L-Rhamnose	$C_6H_{12}O_5$
	L-Fucose	
Heptoses	D-Sedoheptulose	$C_7H_{14}O_7$

Fig. 2.1 The Kiliani-Fischer synthesis for increasing the chain length of aldose sugars.

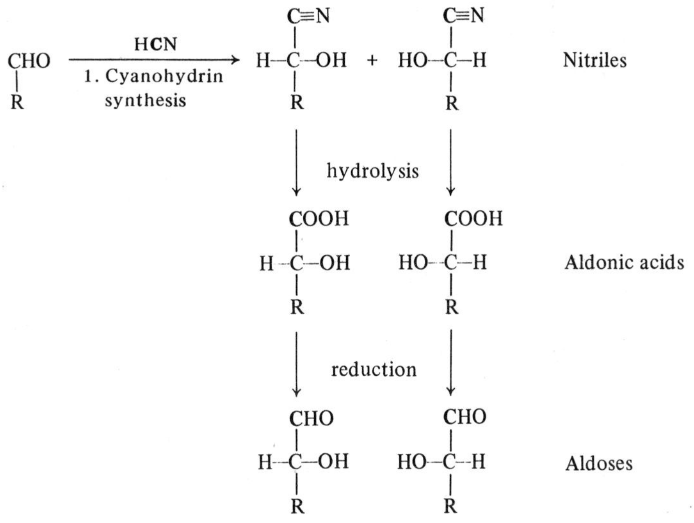

atom (C-3 in a tetrose, C-4 in a pentose, etc.) always has its hydroxyl to the right and is of course identical with the penultimate, or C-2 carbon, of D-glyceraldehyde. The series is known, therefore, as the D-series, while that based on L-glyceraldehyde is the L-series.

Clearly, when there are two asymmetric carbon atoms as in the tetroses, four isomers are possible:

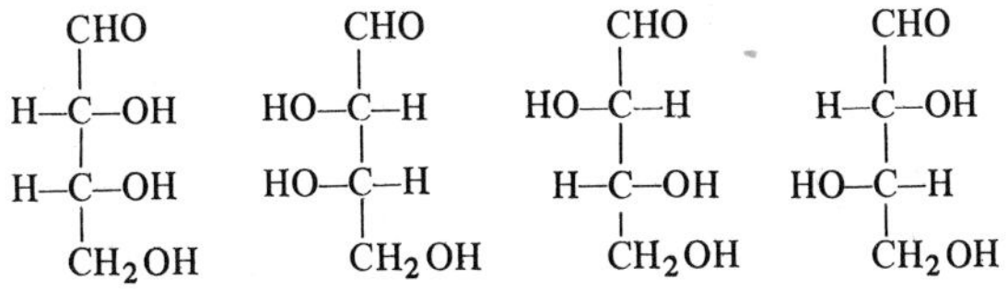

D-Erythrose   L-Erythrose   D-Threose   L-Threose

Among the aldopentoses, eight isomers can be synthesized. As a

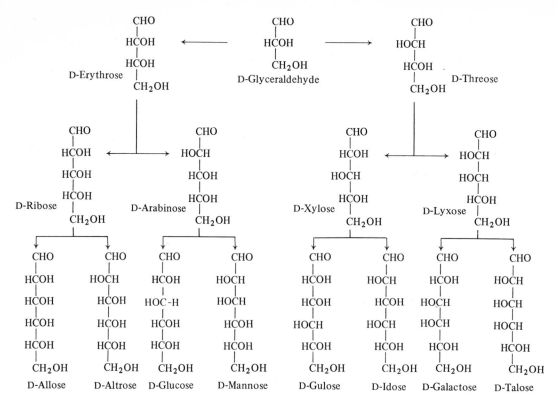

Fig. 2.2   Relationship of the D-aldoses.

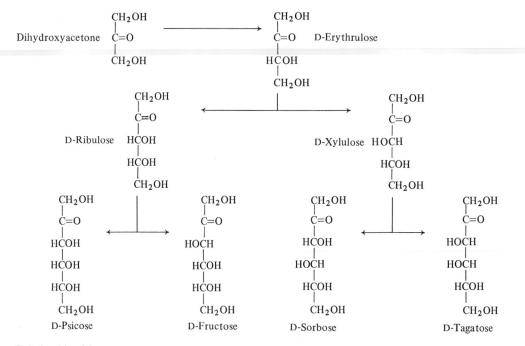

Fig. 2.3   Relationship of the D-ketoses.

general rule (the so-called van't Hoff rule), the maximum possible number of isomers is 2^n, where n is the number of asymmetric carbon atoms. Again it must be recalled that the D- or L- designation refers to the structural series and not to the direction of optical rotation.

D- and L-erythrose are mirror images of each other, and like D- and L-glyceraldehyde are known as *enantiomorphs*. They have identical physical properties (except for the direction of rotation). They are also similar chemically, although they can be distinguished in biological systems by enzymes. Note that threose and erythrose are not mirror images although they are optical isomers. Pairs of compounds such as these that are optical isomers but not enantiomorphs are called *diastereoisomers* and they usually differ in such properties as melting point, solubility and optical rotation. By a prudent choice of solvents, they can usually be distinguished by paper chromatography and other separation techniques.

Ketose sugars such as D-fructose cannot be found in the family of sugars originating from glyceraldehyde by Kiliani synthesis. However, we continue to retain the convention that D-sugars have their penultimate carbon on the right, even though in the case of fructose, the sugar is strongly laevorotary with a specific rotation of $-92°$. The D-series of ketose sugars (up to the hexoses) is shown in Fig. 2.3.

Monosaccharides

The sugars with which we have dealt so far are called monosaccharides. They cannot be converted to simpler sugars by acid hydrolysis and are, therefore, the basic building blocks of all the more complex carbohydrates. Of the many monosaccharides that are known the 5-carbon pentose sugars and the 6-carbon hexose sugars are the most abundant. The principal monosaccharides found in plants are listed in Table 2.2.

Pentoses. The commonest of these, D-xylose and L-arabinose, are particularly important components of the complex polymers of cell walls, and they are easily prepared by hydrolysis of appropriate plant material. Other pentoses of importance to biochemists are D-ribose and D-2-deoxyribose which are components of ribonucleic acid and deoxyribonucleic acid, respectively.

Hexoses. Eight aldohexoses can be prepared from D-glyceraldehyde. Of these, the one that occurs most commonly and the only one that can usually be detected in uncombined form is *D-glucose* (or dextrose). This sugar is particularly common in fruits but is probably ubiquitous in plant cells. It is also the principal product of starch or cellulose hydrolysis. Other important aldohexoses are *D-galactose* and *D-mannose*. These sugars are both constituents of cell wall polysaccharides and often occur together as components of the complex galactomannans of seeds which are important storage forms of carbohydrate in such families as the Leguminosae. Galactose is also found as a component sugar in the oligosaccharides, raffinose and stachyose, and of the milk sugar lactose.

D-*Fructose* (or laevulose because it is highly laevorotary). This is the only example of a ketohexose with which we shall deal. Like glucose, it is found in the free state, and as one of the component sugars of sucrose, it is also present in a number of polymeric forms, the most common of which is the storage polysaccharide inulin.

The ring structure of monosaccharides

In this section, we shall deal with D-glucose, but the same argument applies to all the pentoses and hexoses. Until now, we have considered monosaccharides to have an open-chain or acyclic structure and this is consistent with a number of their reactions, including

1.   the reduction of their carbonyl function to an alcohol

$$\begin{array}{c} CHO \\ | \\ (CHOH)_4 \\ | \\ CH_2OH \end{array} \quad \xrightarrow[\text{borohydride}]{\text{Sodium}} \quad \begin{array}{c} CH_2OH \\ | \\ (CHOH)_4 \\ | \\ CH_2OH \end{array}$$

Glucose   Glucitol (or sorbitol)

2.   their oxidation to aldonic acids

$$\begin{array}{c} CHO \\ | \\ (CHOH)_4 \\ | \\ CH_2OH \end{array} \quad \xrightarrow[\text{water}]{\text{Bromine}} \quad \begin{array}{c} COOH \\ | \\ (CHOH)_4 \\ | \\ CH_2OH \end{array}$$

3.   the reaction of their carbonyl functions with phenylhydrazine and hydrogen cyanide to give phenylhydrazones and cyanhydrins, respectively.

Nevertheless, although glucose shows some reducing properties it does not colour Schiff's reagent (which reacts with aldehydes) suggesting that the aldehyde group is masked in some way. Solutions of D-glucose also show changes in specific optical rotation $[\alpha]_D$ upon standing which suggest that more than one species of compound is present in solution. This phenomenon is known as *mutarotation*. Indeed, two forms of anhydrous D-glycose can be recognized which differ in their initial $[\alpha]_D$ when freshly dissolved in water. The more common form which is easy to crystallize and is known as α-D-glucose, has an initial $[\alpha]_D$ of $+113°$ which falls off to a constant value of $+52.5$, while the other form, β-D-glucose has an $[\alpha]_D$ of $+19°$, which rises to the same constant value of $+52.5°$. Clearly, an equilibrium mixture is formed in solution which contains the two forms.

These observations can now be explained if we take into account the interaction between functional groups on the same molecule.

Aldehydes are known to undergo reactions with alcohols. First, a hemiacetal is formed (1) which is usually unstable and normally reacts with a second alcohol group to give a full acetal (2) with the elimination of water.

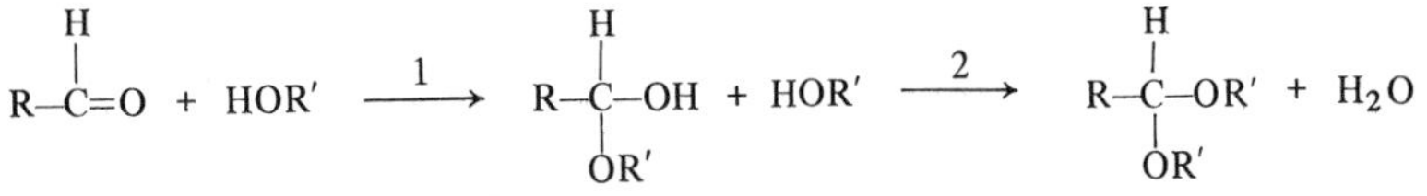

$$\begin{array}{ccc}
\overset{\displaystyle H}{\underset{\displaystyle}{R-C=O}} + HOR' & \xrightarrow{\ 1\ } & \overset{\displaystyle H}{\underset{\displaystyle OR'}{R-C-OH}} + HOR' & \xrightarrow{\ 2\ } & \overset{\displaystyle H}{\underset{\displaystyle OR'}{R-C-OR'}} + H_2O
\end{array}$$

In the case of glucose, the chain of the molecule twists so that the hydroxyl group on carbon atom five can react with the exposed aldehyde group to form a stable hemiacetal and hence an oxygen bridge with C-1 (Fig. 2.4). As a result, a 6-membered ring is produced which has five carbon atoms and one oxygen atom. Further, carbon 1 now becomes asymmetric so that two additional forms of D-glucose, called *anomers*, become possible. The anomeric form with the —OH on C-1 to the right (or pointing downward in the planar formula) is

α-Form   β-Form

Fig. 2.4   The mutarotation of D-glucose.

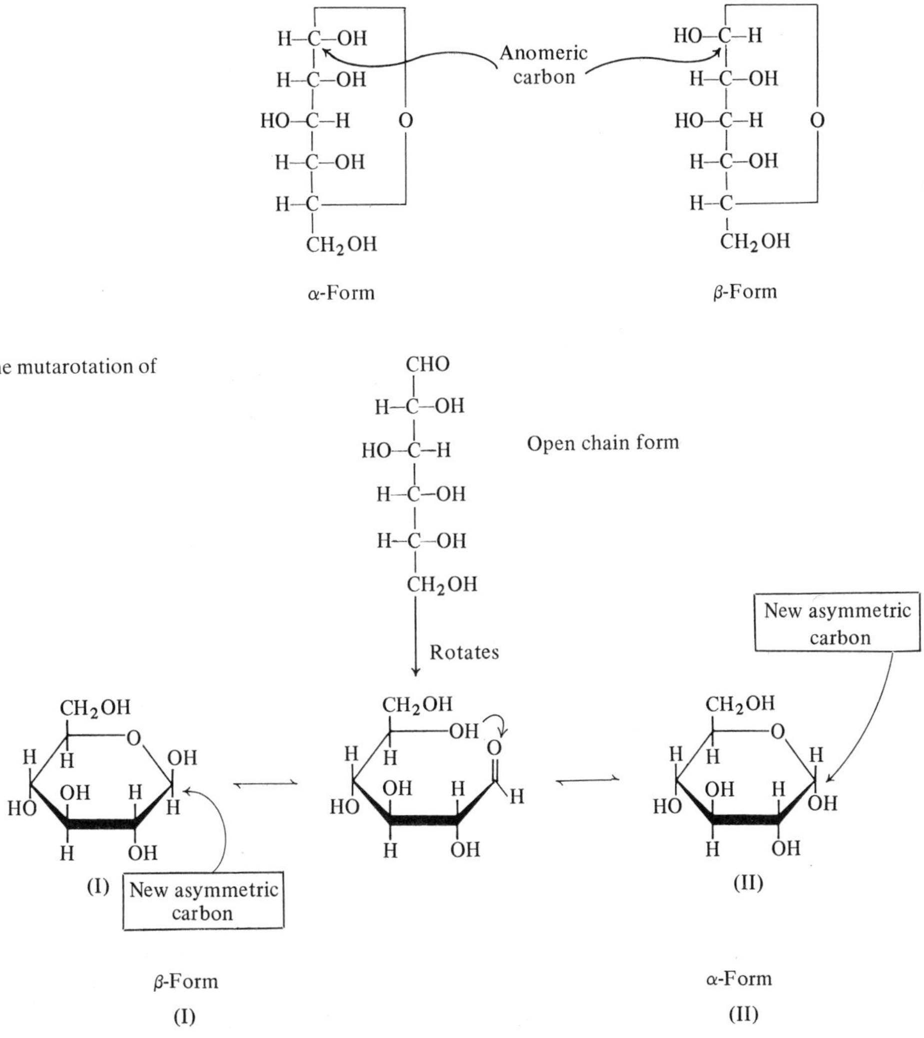

Open chain form

Rotates

New asymmetric carbon

(I)   New asymmetric carbon

β-Form
(I)

α-Form
(II)

51

known as α-D-glucose, the other with the —OH to the left, β-D-glucose. The former has a specific optical rotation of +113°, the latter +19°. In solution, the two forms come into equilibrium with each other and hence produce the observed mutarotation. The intermediate open chain aldehydic form is present in only trace amounts (0·024%), and has never been isolated. The two ring forms, however, can be crystallized separately by varying the temperature appropriately. Nevertheless, glucose is a potential aldehyde and for this reason will undergo many of the reactions of aldehydes.

The planar ring

Haworth pointed out that a more realistic portrayal of the ring structure of glucose is as a hexagon. He suggested that it should be drawn as a hexagonal plane perpendicular to the plane of the paper. The front portion of the ring often is represented by thicker lines, indicating it is pointing towards the reader (Fig. 2.5). This can be best visualized by cutting out a hexagon from stiff cardboard, and sticking small pegs into it to represent the various substituent groups. The —H or —OH groups are placed appropriately above or below the plane of the ring. Those substituents placed to the right in the open-chain formula are placed below the plane of the ring, those to the left above it. However, in the case of glucose, the D- or L-determining —OH group is the one which reacts with the aldehyde function. If this is to the right as in the D-form, it will react so that the —CH₂OH is above the right as shown in Figs. 2.4 and 2.5. However, if it is in the L-form, the —CH₂OH will be forced below the plane of the ring. Thus, the D- or L-form in 6-membered rings of this kind are recognized by whether

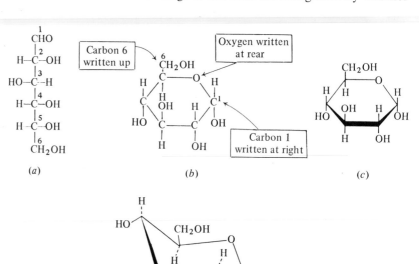

(a)   (b)   (c)

(d)

Fig 2.5   Representative structures of D-glucose. (a) Open chain linear structure (after Fischer); (b) ring form with all carbons shown (after Haworth); (c) ring shown in abbreviated perspective form; (d) conformational ring structure of α-D-glucose in chair form. (Axial groups are shown with broken line, equatorial groups with unbroken line.)

the hydroxymethyl group is above or below the plane of the ring. Haworth formulae of some of the commoner aldohexoses and pentoses are shown in Fig. 2.11.

The 6-membered rings encountered most commonly with the aldohexoses are known as *pyranoses* because they resemble pyran (Fig. 2.6). Five-membered rings (or *furanose* forms) are known for certain sugars, which are similar in structure to *furan* (Fig. 2.6). In the case of a ketohexose such as D-fructose, this type of ring exists because the exposed carbonyl function is on C-2, and the oxygen bridge is formed most easily with C-5 (Fig. 2.7) (though the pyranose form is known in which there is interaction between C-2 and C-6). In the case of L-arabinose and D-galactose, C-4 as well as C-5 is in a position to react easily with the aldehyde function, so that furanose and pyranose ring forms are known (each with an α- or β-species). Because of this, their mutarotation in water is particularly complicated.

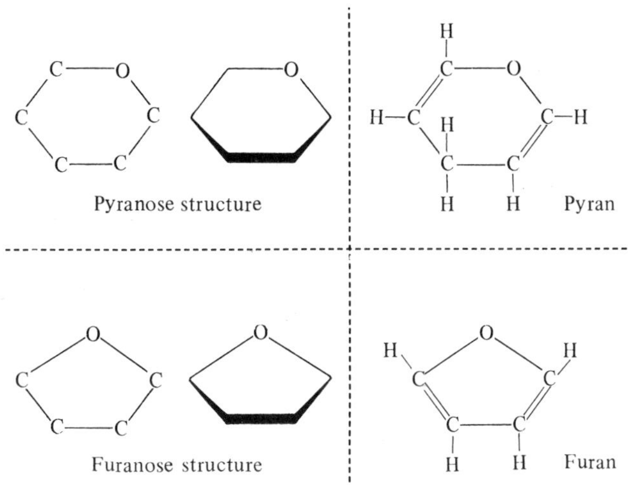

Fig. 2.6 The resemblance of the pyranose and furanose ring structures of sugars to pyran and furan.

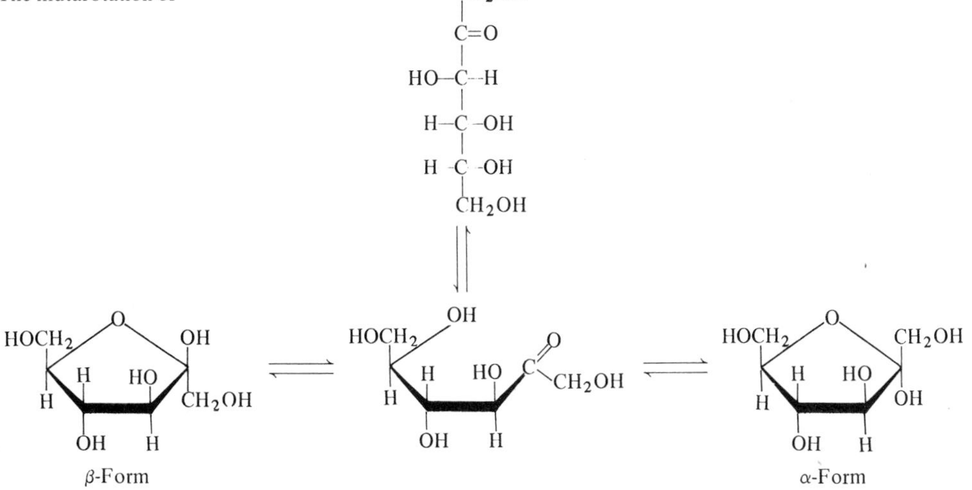

Pyranose structure   Pyran

Furanose structure   Furan

Fig. 2.7 The mutarotation of D-fructose.

β-Form   α-Form

The puckered ring

The 6-membered ring as found in glucopyranose is not in reality a flat hexagon, but exists in the form of a flattened, puckered ring, resembling a chair (Fig. 2.5). This is the most stable of a number of strainless forms and can be best appreciated by referring to atomic models which have the bond angles of each carbon atom accurately depicted. Substituent groups are either approximately axial to this ring (meaning sticking out from it) or equatorial (meaning in the approximate plane of the ring) and not simply above or below as in the simple Haworth representation. The most stable form is always the one that keeps to a minimum non-bonded interactions with neighbouring molecules. In sugars, this is best achieved if the hydroxyl groups and the bulkier hydroxymethyl groups are placed approximately equatorial to the ring. The chair form of β-D-glucopyranose has all of its substituent groups placed in such a manner and it is in this form that it can be crystallized. Other sugars have both equatorial and axial substituents.

Glycosides

Glycoside is the name given to the product obtained from a sugar by reaction with an alcohol or a phenol. Glucose, for example, will react with methanol in the presence of HCl to give two products, the α- and β-methyl glucosides (Fig. 2.8). In each case, the hydroxyl function at position 1 has been replaced by a methoxyl group to establish the full acetal or *glycosidic* linkage. Once the full acetal has been formed, the ring cannot reopen and so both glycosides are stable and not interconvertible. Hydrolysis of either compound with hot, dilute mineral acid will yield glucose and methanol.

If one replaces the methanol in the equation in Fig. 2.8 with another molecule of glucose (which is also an alcohol), a compound would be formed in which two sugars are linked glycosidically. This linkage would be from either the α- or β-position on the first sugar and to any one of a number of hydroxyls on the second. Such reactions between

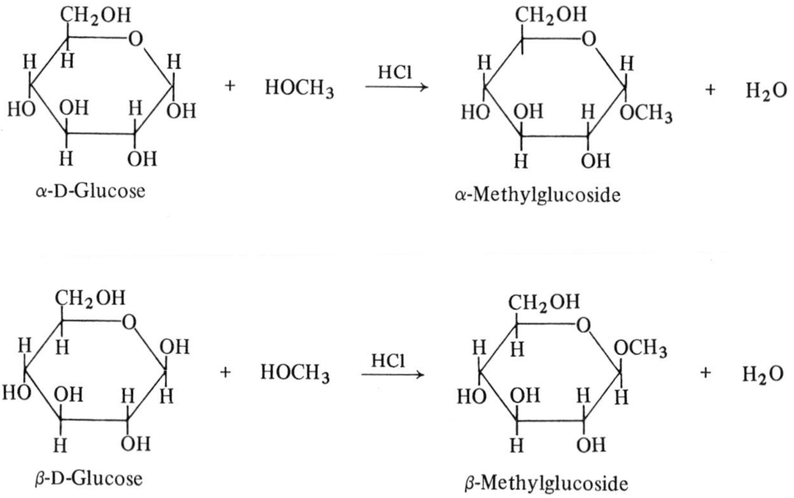

Fig. 2.8   Formation of methyl glucosides from methanol and glucose.

sugars do not occur spontaneously, however, but they can be established biologically from activated derivatives when catalysed by enzymes (see Ch. 10). Complex sugars which are made up of a limited number of monosaccharides linked together by glycosidic bonds are known as *oligosaccharides*. When more than about ten such units are linked together, the compound is usually referred to as a *polysaccharide*. The latter are usually less sweet, less soluble and less easy to crystallize than the shorter chain sugars. Moreover, they are often variable in molecular weight and highly complex both in terms of their constituent monosaccharides and the types of linkages involved.

Oligosaccharides

Disaccharides

We shall refer here to only two plant disaccharides, sucrose and maltose. Their structures are shown in Fig. 2.9. They also illustrate the nomenclature of complex carbohydrates although we shall not discuss any of the governing rules.

Sucrose is known commercially as cane-sugar (because it is often extracted from sugar cane) and is found abundantly in plants both as a storage and transport form of carbohydrate. It is made up of one unit of glucose and one of fructose, but is non-reducing, indicating that the two sugars are linked through their reducing centres. Sucrose has been shown to be α-D-glucopyranosyl β-D-fructofuranoside (or alternatively

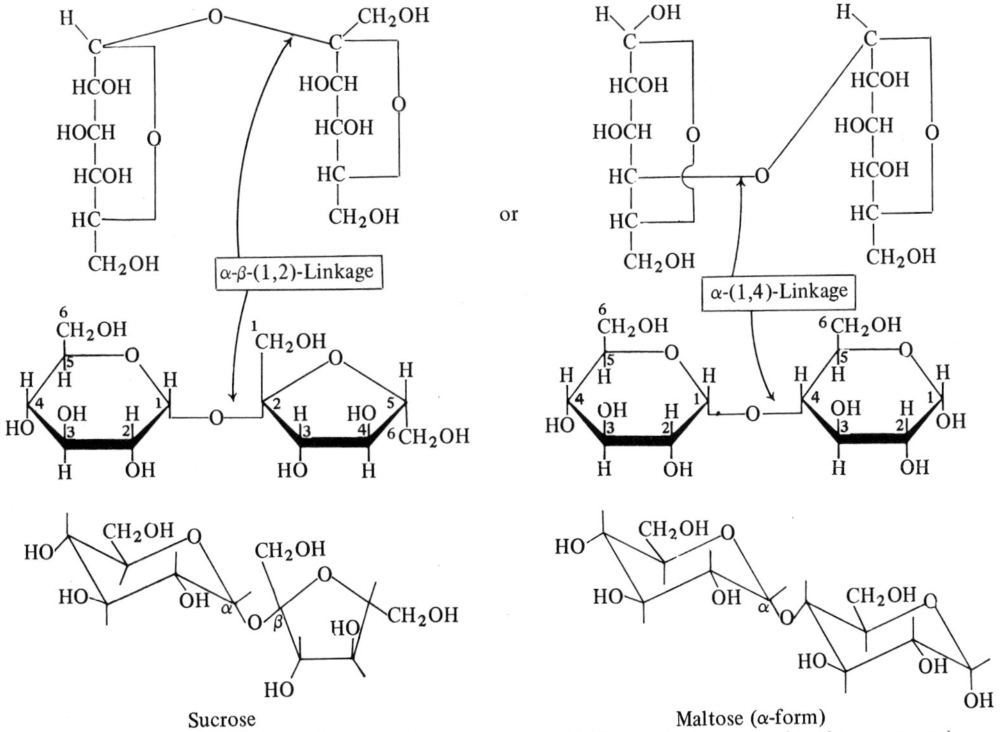

Fig. 2.9   Representative structures of (a) sucrose and (b) maltose. Note that in sucrose neither hemiacetal hydroxyl is free so that it is non-reducing. In the case of maltose, the hemiacetal hydroxyl on the second glucose unit may be in either the α-(down) or β-(up) position.

Sucrose
(α-D-Glucopyranosyl-β-D-fructofuranoside)

(a)

Maltose (α-form)
(4-O-α-D-Glucopyranosyl-D-glucopyranose)

(b)

β-D-fructofuranosyl α-D-glucopyranoside). Each name designates the monosaccharides involved in their ring form, and the hydroxyl group to which the glycosidic bond is joined.

Maltose is relatively uncommon, but is produced when starch is hydrolysed by amylases. It is comprised of two units of glucose, but in this case, the hemiacetal hydroxyl group of the second glucose (drawn on the right) is free and it is, therefore, a reducing sugar. The linkage is $\alpha(1 \rightarrow 4)$ and its structure, therefore, 4-O-(α-D-glucopyranosyl)-D-glucopyranose.

Higher oligosaccharides

Raffinose is the most common trisaccharide found in plants. On hydrolysis, it yields glucose, fructose and galactose in equimolar amounts. Its structure is shown in Fig. 2.10. The tetrasaccharide, stachyose, has one additional galactosyl unit and is often found with raffinose as a storage sugar in seeds. In addition, stachyose often seems to replace sucrose as the principal transport carbohydrate of a number of herbaceous and woody plants.

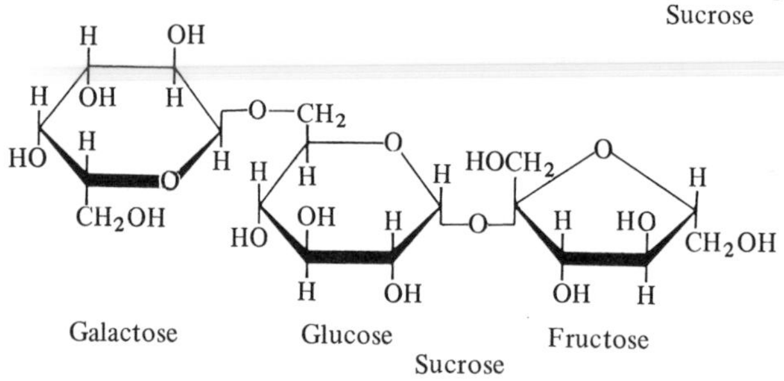

Fig. 2.10 Structures of stachyose and raffinose.

Stachyose

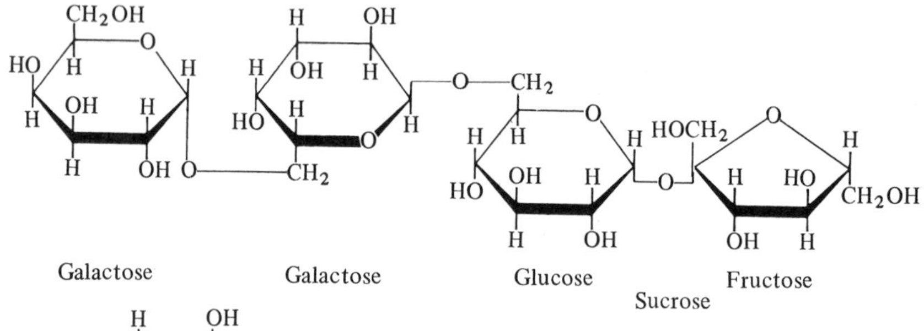

Galactose — Galactose — Glucose — Fructose (Sucrose)

Raffinose

Galactose — Glucose — Fructose (Sucrose)

Polysaccharides

Polysaccharides are widely distributed in plants both as structural components of the cell wall and as storage products in seeds, fruits and leaves. They are of high molecular weight being made up of many monosaccharides linked together by glycosidic bonds. Some, such as cellulose (Ch. 10), are linear, made up of only a single species of monosaccharide (glucose) and have only one type of linkage. Others are branched with a more variable linkage, while the more complex forms may also contain many different types of monosaccharide building block. Some of the commoner monosaccharides isolated from plant cell wall polysaccharides are illustrated in Fig. 2.11.

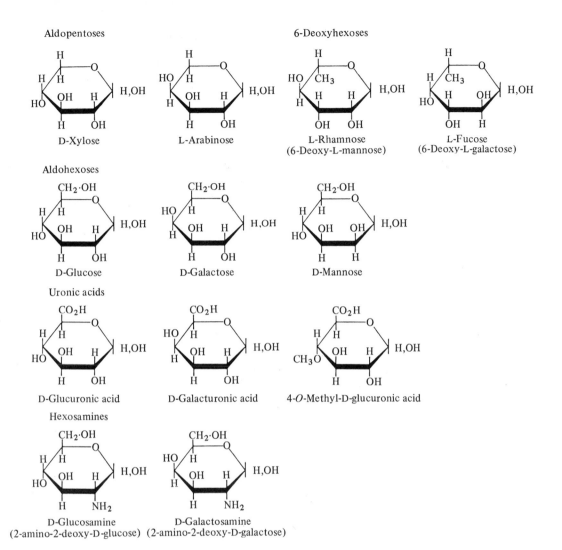

Aldopentoses

D-Xylose

L-Arabinose

6-Deoxyhexoses

L-Rhamnose
(6-Deoxy-L-mannose)

L-Fucose
(6-Deoxy-L-galactose)

Aldohexoses

D-Glucose

D-Galactose

D-Mannose

Uronic acids

D-Glucuronic acid

D-Galacturonic acid

4-O-Methyl-D-glucuronic acid

Hexosamines

D-Glucosamine
(2-amino-2-deoxy-D-glucose)

D-Galactosamine
(2-amino-2-deoxy-D-galactose)

Fig. 2.11   Some of the sugars encountered in the polysaccharides and glycoproteins of plant cells.

Because of their complexity, very few polysaccharides have been characterized in any detail. In this chapter we shall only consider the two common storage polysaccharides, starch and inulin, although the structures of a number of cell wall polysaccharides are discussed in Chapter 10. Starch is extremely widespread; it is commonly stored in growing roots, in perennating organs, and in chloroplasts during active photosynthesis. It is made up entirely of glucose units and occurs in two forms, amylose and amylopectin. The former is a long, straight chain molecule with several hundred identical $\alpha(1 \rightarrow 4)$ bonds (Fig. 2.12). Because the α-bond is axial with respect to the pyranose ring, it produces a helical molecule with about six D-glucose units in each turn. It is stained blue by iodine because the iodine atoms occupy the space within the axis of the helix. Amylopectin is a branched molecule with $\alpha(1 \rightarrow 6)$ branch chains (Ch. 6). The outer tiers of the molecule are less highly branched than the core.

Inulin is also a reserve polysaccharide found in high amounts in the roots and storage tissues of composites. It is made up of fructofuranose

Fig. 2.12   Structures of amylose and inulin.

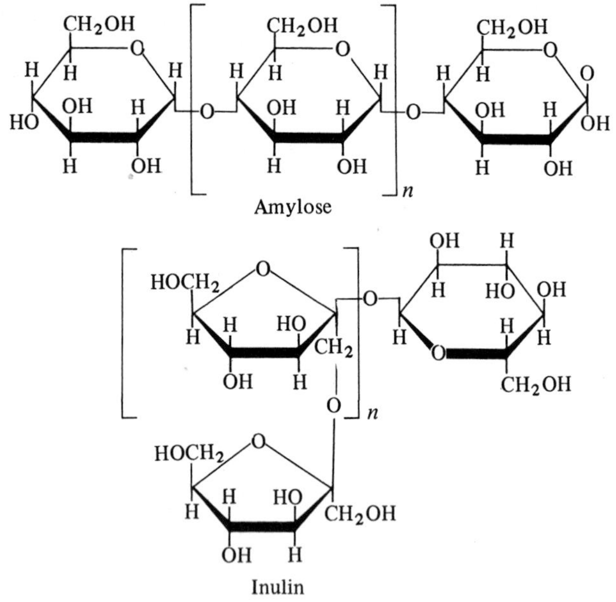

Amylose

Inulin

units linked from C-2 (the glycosidic hydroxyl) to C-1. A single glucose molecule occurs at one end of each polysaccharide chain suggesting that inulin is built up by repetitive transfer of fructose to a starting molecule of sucrose.

Glycoproteins are large molecules in which carbohydrate is linked covalently to a protein and are widespread in plants. Some of these are enzymes, while others have an unknown function but have received great attention because of their specific interactions with mammalian cells. Thus, certain allergens of pollen and some of the storage proteins of plant seeds are glycoproteins. Cell walls have also been shown to contain a hydroxyproline-rich glycoprotein.

Derivatives of monosaccharides

Suffixes used in the nomenclature of some of the commoner sugar derivatives are shown in Table 2.3. When the carbonyl group of a monosaccharide is reduced either chemically or in reactions catalysed

Table 2.3   Suffixes used in nomenclature of sugar derivatives

Suffix	*Structural characteristics*	*Derivative*
1. -itol	$-C^1H_2OH$	Alditol
2. -onic acid	$-C^1OOH$	Aldonic acid
3. -uronic acid	$HOOC^n-C^1HO$	Uronic acid
4. -aric acid	$HOOC^n-C^1OOH$	Dicarboxylic acid
5. -amine	$-C^2H-C^1HO$ $\quad\quad\mid$ $\quad\quad NH_2$	Amino sugar

by enzymes, a *sugar alcohol* or alditol is produced. Two common sugar alcohols found in nature are mannitol and glycerol.

Three main types of sugar acids are found in living cells, the *aldonic acids,* the *uronic acids* and the *aldaric acids*. The former are produced by oxidation of the aldehyde groups of aldose sugars, either enzymically or chemically. D-Gluconic acid, for example, can be formed from D-glucose by treatment with bromine water or by using D-glucose dehydrogenase in presence of NAD^+. In the case of the uronic acids, it is the primary hydroxyl group (that on C-6 in the case of hexoses) which has been oxidized. The two uronic acids occurring most commonly in plants are D-glucuronic acid and D-galacturonic acid (Fig. 2.11) which are both components of the cell wall.

The aldaric acids are dicarboxylic acids having both their aldehydic carbon and the carbon atom bearing the primary hydroxyl group oxidized. This can be accomplished chemically by oxidation of the monosaccharide using nitric acid or other strong oxidizing agents.

Just as the reducing group of a sugar can react internally with one of its hydroxyl groups to form a hemiacetal, so a sugar acid can form an internal ester or *lactone* with the loss of one molecule of water. When more than one hydroxyl group is in a position to react with the carboxyl group, more than one lactone can be formed.

Amino sugars are also found in plants, usually as components of glycoproteins. These compounds have one of their hydroxyl groups replaced by an amino group. The two commonest are D-glucosamine and D-galactosamine (Fig. 2.11).

Lipids

Of all the groups of compounds dealt with in this chapter, the lipids are the most difficult to classify. This is because they are so heterogeneous for, unlike the carbohydrates, proteins or nucleic acids, they cannot be defined on the basis of chemical structure alone, and are best recognized by their solubility characteristics. Lipids are compounds which can be extracted with one of several organic solvents such as ether, acetone or chloroform, but which are insoluble in water. Because they lack a common structural feature, it has become convenient to subdivide them into a number of main groups, although even here the subdivisions are not very precise.

The triglycerides and fatty acids

The commonest lipids found in nature are the triglycerides which are esters of the trihydric alcohol glycerol and long chain fatty acids (Fig. 2.13). The triglycerides of plants are primarily storage materials, found typically as intracellular oil droplets in storage organs, in seeds and fruits and in chloroplasts. Although the three fatty acid residues combined with a single molecule of glycerol may be identical, usually more than one type is present and the triglycerides are said to be mixed. A complex population of such mixed triglycerides is always

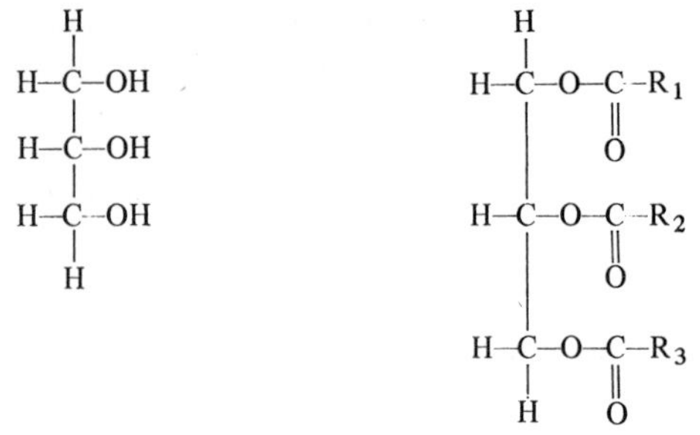

Fig. 2.13 The structures of glycerol and a triglyceride. The R groups represent the hydrocarbon chains of fatty acids.

Glycerol   Triaclyglycerol

found in natural fats and oils. The long chain fatty acids, which are themselves lipids on the basis of their solubility characteristics, are also an exceedingly diverse group of compounds (Table 2.4). With reference to Table 2.4, the carbon chains of fatty acids are numbered from the carbonyl group, and the position of the double bond is indicated by a single number which is that of the lowest numbered carbon atom participating in the bond.

As well as containing the two common saturated acids, palmitic and stearic acids, plant triglycerides usually contain large amounts of

Table 2.4 Some of the principal fatty acids of higher plants

A. *Saturated*

Formula	Trivial name	Systematic name	Structure
$C_2H_4O_2$	Acetic acid	Ethanoic acid	CH_3COOH
$C_4H_8O_2$	Butyric acid	Tetranoic acid	$CH_3(CH_2)_2COOH$
$C_6H_{12}O_2$	Caproic acid	Hexanoic acid	$CH_3(CH_2)_4COOH$
$C_8H_{16}O_2$	Caprylic acid	Octanoic acid	$CH_3(CH_2)_6COOH$
$C_{10}H_{20}O_2$	Capric acid	Decanoic acid	$CH_3(CH_2)_8COOH$
$C_{12}H_{24}O_2$	Lauric acid	Undecanoic acid	$CH_3(CH_2)_{10}COOH$
$C_{14}H_{28}O_2$	Myristic acid	Tetradecanoic	$CH_3(CH_2)_{12}COOH$
$C_{16}H_{32}O_2$	Palmitic acid	Hexadecanoic	$CH_3(CH_2)_{14}COOH$
$C_{18}H_{36}O_2$	Stearic acid	Octadecanoic	$CH_3(CH_2)_{16}COOH$
$C_{20}H_{40}O_2$	Arachidic acid	Eicosanoic	$CH_3(CH_2)_{18}COOH$

B. *Unsaturated*

Unsaturated acids containing one double bond

Formula	Trivial name	Systematic name	Structure
$C_{14}H_{26}O_2$	Myristoleic	9-Tetradecenoic	$CH_3(CH_2)_3CH:CH(CH_2)_7COOH$
$C_{16}H_{30}O_2$	Palmitoleic	9-Hexadecenoic	$CH_3(CH_2)_5CH:CH(CH_2)_7COOH$
$C_{18}H_{34}O_2$	Oleic	9-Octadecenoic	$CH_3(CH_2)_7CH:CH(CH_2)_7COOH$
$C_{18}H_{34}O_2$	Petroselinic	6-Octadecenoic	$CH_3(CH_2)_{10}CH:CH(CH_2)_4COOH$
$C_{18}H_{34}O_2$	Ricinoleic	12-Hydroxy-9-octadecenoic	$CH_3(CH_2)_5CHOHCH_2CH:CH(CH_2)_7COOH$
$C_{22}H_{42}O_2$	Erucic	13-Docosenoic	$CH_3(CH_2)_7CH:CH(CH_2)_{11}COOH$

Unsaturated acids containing one triple bond

$C_{18}H_{32}O_2$	Tariric	6-Octadecinoic	$CH_3(CH_2)_{10}C:C(CH_2)_4COOH$

Unsaturated acids containing two double bonds

$C_{18}H_{32}O_2$	Linoleic	*cis,cis*-9,12-Octadecadienoic	$CH_3(CH_2)_4CH:CHCH_2CH:CH(CH_2)_7COOH$

Unsaturated acids containing three double bonds

$C_{18}H_{30}O_2$	Linolenic	*cis*-9,12,15-Octadecatrienoic	$CH_3(CH_2)_3CH:CHCH_2CH:CHCH_2CH:CH(CH_2)_7COOH$
$C_{18}H_{30}O_2$	Elaeostearic	*cis*-9,11,13-Octadecatrienoic	$CH_3(CH_2)_3CH:CHCH:CHCH:CH(CH_2)_7COOH$

Table 2.5 Fatty acid components of triglycerides from seeds

Species	Trivial Name	Saturated acids		Unsaturated acids	
		Acid	%	Acid	%
Juglans regia	Walnut	Palmitic	5·1	Oleic	28·9
		Stearic	2·5	Linoleic	47·6
				Linolenic	15·9
Linum usitatissimum	Flax, linseed	Palmitic	5·4	Oleic	9·9
		Stearic	3·5	Linoleic	42·6
				Linolenic	38·1
Olea europaea	Olive	Palmitic	6·0	Oleic	83·0
		Stearic	4·0	Linoleic	7·0
Sesamum indicum	Sesame	Palmitic	9·1	Oleic	45·4
		Stearic	4·3	Linoleic	40·4
Helianthus annuus	Sunflower	Palmitic	3·5	Oleic	34·1
		Stearic	2·9	Linoleic	58·5
Anacardium occidentale	Cashew	Palmitic	6·4	Oleic	74·1
		Stearic	11·3	Linoleic	7·7
Gossypium hirsutum	Upland cotton	Palmitic	21·9	Oleic	30·7
		Stearic	1·9	Linoleic	44·9
Bertholletia excelsa	Brazil nut	Palmitic	14·3	Oleic	58·3
		Stearic	2·7	Linoleic	22·8
Zea mays	Maize	Palmitic	7·8	Oleic	46·3
		Stearic	3·5	Linoleic	41·8
Arachis hypogaea	Peanut	Palmitic	6·3	Oleic	61·1
		Stearic	4·3	Linoleic	21·8
		Arachidic, etc.	5·9		
Glycine max	Soybean	Palmitic	6·8	Oleic	33·7
		Stearic	4·4	Linoleic	52·0
		Arachidic	0·7		

Source: Hilditch, T. P. (1947).

unsaturated acids, in particular oleic, linoleic and linolenic acids (Table 2.5). Other fatty acids which are often peculiar to the particular species or genus may also be detected, sometimes in very large amounts (Table 2.6). The oily rather than lardy nature of plant triglycerides is due to their higher content of double bonds and hence low melting points. The animal fats, which are comparatively rich in stearic or palmitic acids, melt at higher temperatures. It is the plant kingdom which is primarily responsible for introducing the second

Table 2.6 Unusual fatty acid components of certain plant seeds

Species	Trivial name	Unusual fatty acid	% Total fatty acid
Aleurites sp.	Tung	Elaeostearic	79·7
Ricinus communis	Castor bean	Ricinoleic	87·8
Apium graveolens	Celery	Petroselinic	51·0
Brassica campestris	Rape	Erucic	50·0

Source: Hilditch, T. P. (1947).

double bond into oleic to form linolenic acid. Animals cannot accomplish this and have to depend upon ingested fats, ultimately derived from plants, to satisfy their dietary requirements.

Waxes are chemically quite similar to the triglycerides, but do not contain glycerol. They are esters of higher fatty acids with long chain monohydric alcohols and, as their name suggests, they are pliable materials which become relatively hard when cold. Waxes are found as an extracellular layer on leaves and stems, forming a protective waterproof coating to the plant.

Phosphoglycerides

Phosphoglycerides along with the sphingolipids are often classified as part of a larger group of lipids, called *phospholipids*, all of which give inorganic phosphate as one of their products of hydrolysis. Like the other phospholipids, phosphoglycerides are also found predominantly in cellular membranes and are not usually deposited in storage form. The parent compound of this series is L-glycerol-3-phosphate (Fig. 2.14). Two fatty acid residues are esterified to the hydroxyl groups at carbon atoms 1 and 2. An unsaturated fatty acid (usually

Fig. 2.14 Examples of the commoner phosphoglycerides.

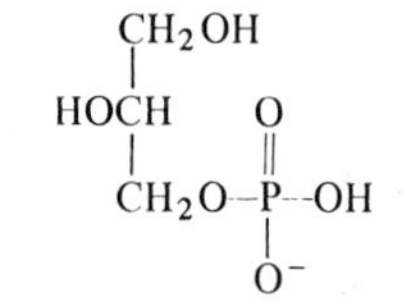

Glycerol-3-phosphate

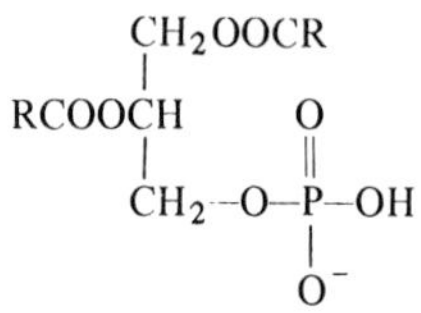

Phosphatidic acid

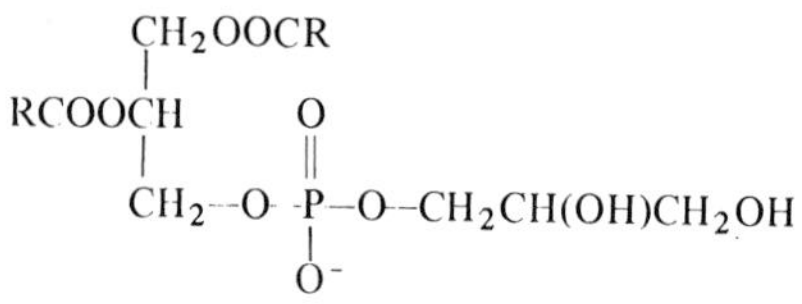

Phosphatidyl glycerol

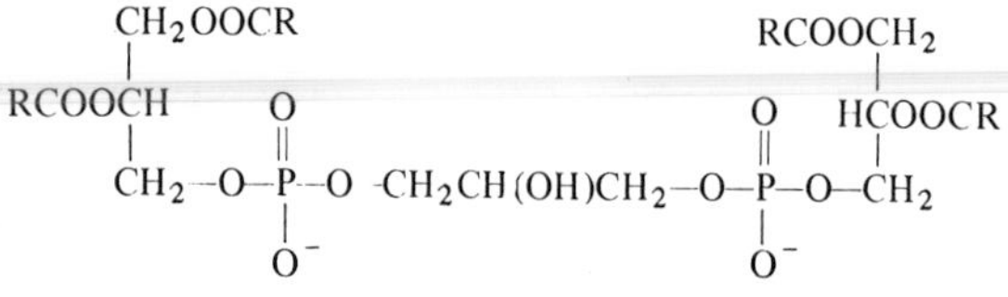

Cardiolipin (diphosphatidyl glycerol)

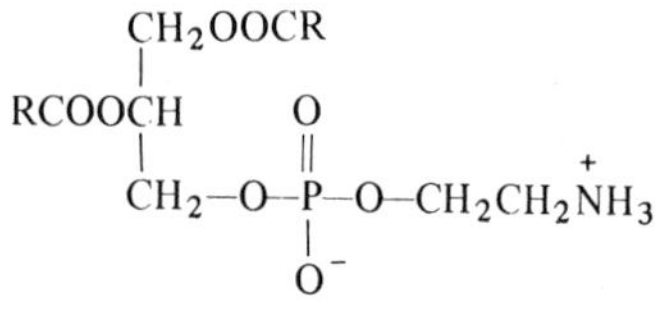

Phosphatidyl ethanolamine

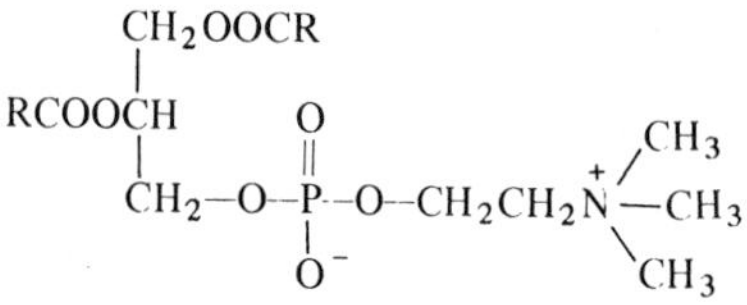

Phosphatidyl choline

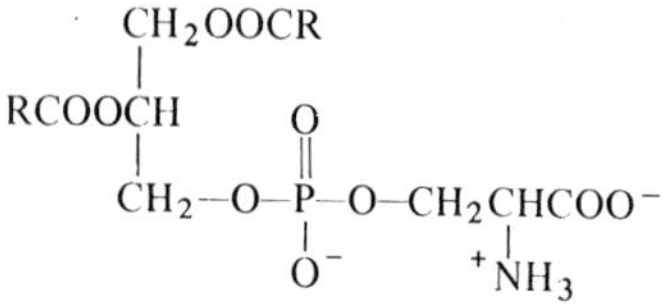

Phosphatidyl serine

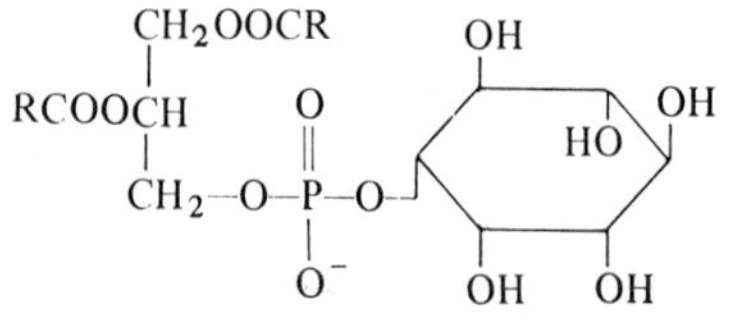

Phosphatidyl inositol

oleic, linoleic or linolenic acid) normally occupies the central or 2' position while a saturated acid tends to be associated with the α or 1' position, the one furthest away from the polar end of the molecule. Thus the simplest phosphoglyceride is phosphatidic acid. This is a relatively rare compound, although it is an intermediate in the biosynthesis of other phosphoglycerides which contain an additional alcohol component esterified to the phosphoric acid (Fig. 2.14).

The two commonest phosphoglycerides are *phosphatidyl ethanolamine* and *phosphatidyl choline* (known previously as lecithin) which contain the amino alcohols ethanolamine and choline, respectively. Phosphatidyl choline is used commercially as an emulsifying agent and is particularly abundant in soybeans. A third common membrane phosphoglyceride which contains nitrogen is *phosphatidyl serine*. This compound is also the precursor of phosphatidyl ethanolamine *in vivo*.

Phosphatidyl glycerol is formed when the phosphatidic acid is esterified with glycerol. Closely related to this compound is *cardiolipin*, a complex lipid consisting of a molecule of phosphatidyl glycerol in which the 3'-hydroxyl group of the second glycerol moiety is esterified to the phosphate group of a second molecule of phosphatidic acid.

In other phosphoglycerides the phosphate group is esterified to *myo*-inositol, a hexahydroxycyclohexane. This compound is known as phosphatidyl inositol and has been purified from peanuts. Phosphoinositides can be highly charged as the cyclitol ring may be further substituted in one or more positions with additional phosphate ester groups.

Sphingolipids

When sphingolipids are hydrolysed they yield one molecule of phosphoric acid, one molecule of a fatty acid and one molecule of sphingosine (or more commonly in plants its saturated, hydroxylated analogue phytosphingosine) (Fig. 2.15). An alcohol (usually a sugar, *myo*-inositol, or choline) is also released and this represents the X-group esterified to sphingosine phosphate (Fig. 2.15). The fatty acid (R') of these sphingolipids is attached in an amide linkage to the primary amino group of sphingosine.

The structure of a glucocerebroside which is a common type of sphingolipid of chloroplast membrane is shown in Fig. 2.15. Cerebrosides are those sphingolipids in which the X-group is a sugar, whereas the sphingomyelins, by definition, contain choline.

Some properties of the phospholipids

Phospholipids *in situ* are relatively insoluble in most organic solvents and are usually extracted from tissues using mixtures of hot chloroform and methanol. Once isolated they can be taken up in water as though they were soluble. However, like the salts of fatty acids, but unlike triglycerides, they are *amphipathic* molecules possessing both marked

Fig. 2.15   Sphingosine phosphate, phytosphingosine and an example of a cerebroside found in plant membranes.

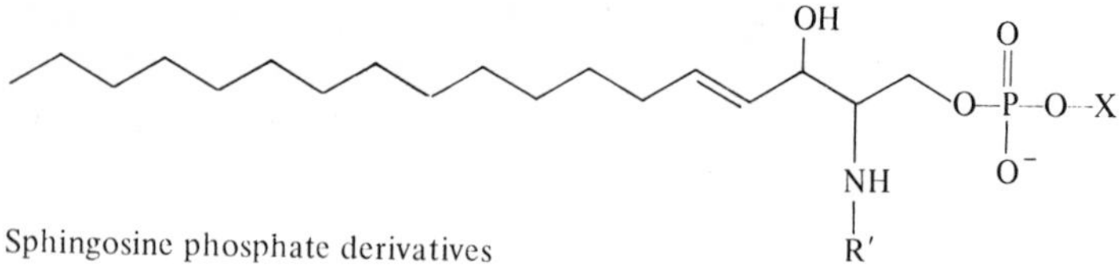

Sphingosine phosphate derivatives

R' is a fatty acid
X is an esterfied alcohol

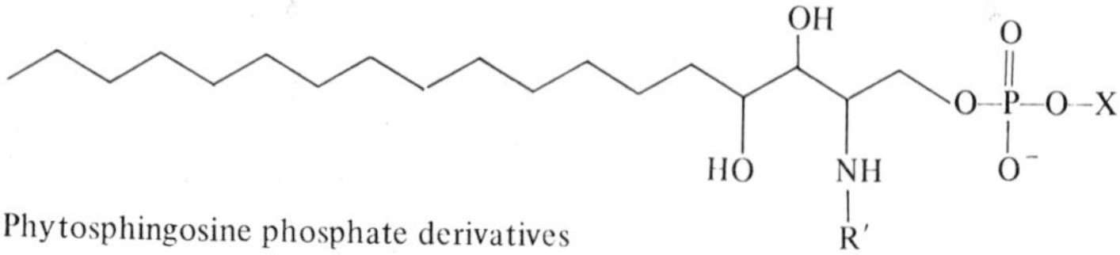

Phytosphingosine phosphate derivatives

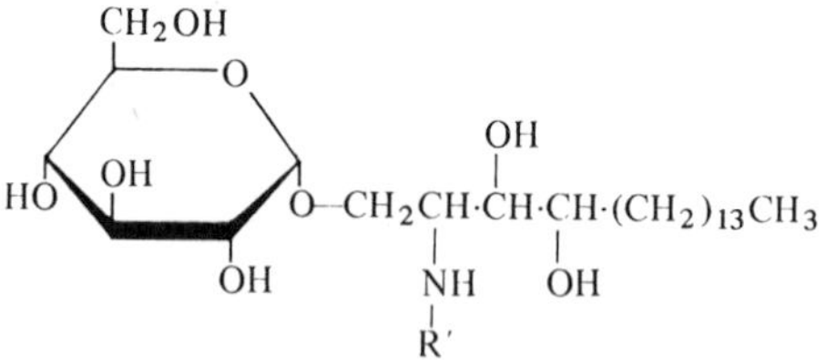

Glucocerebroside
N-Acyl-α-D-glucosyl-(1-1')-
phytosphingosine

hydrophobic and hydrophilic properties. The long chain of methylene residues found on their fatty acids or on sphingosine are hydrophobic, while the substitutions on the 3' position of glycerol or on the head group of the sphingosine are highly polar (Fig. 2.16). While the hydrocarbon chain prefers to exist out of the aqueous phase, the polar groups prefer to interact with water. One way of satisfying these demands is to form spherical *micelles* which are structures containing many lipid molecules arranged with their hydrophobic chains on the inside of the sphere, hidden within the micellar structure. For this reason, phospholipids and salts of long chain fatty acids disperse readily in aqueous solution even though they do not form a truly molecular solution. Combinations of different phospholipids will form 'mixed' micelles.

Phospholipids also act as natural detergents and emulsifying agents. The non-polar end of the molecule becomes preferentially soluble in a layer of grease or other hydrophobic substance while the polar head continues to interact with the aqueous phase. Assisted by mechanical agitation, the grease eventually becomes 'solubilized' in the form of small droplets at the centre of a micelle (Fig. 2.17). Similar associations between surfactant lipids of this kind and the hydrophobic portions of proteins are thought to be involved in stabilizing biological membranes.

A second way in which phospholipids can interact with water is to form a surface monolayer where the hydrocarbon chains stick up into the air while the polar heads point downward into the water.

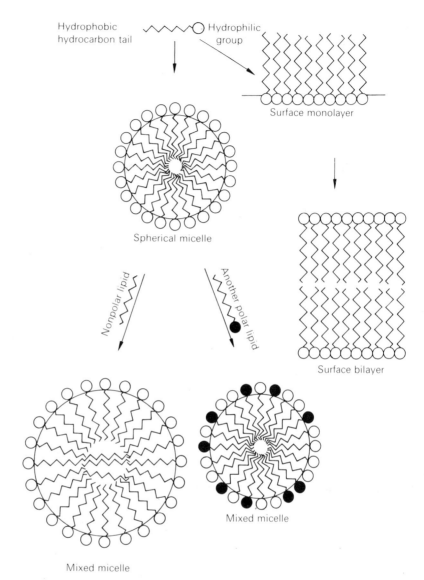

Fig. 2.16   The formation of micelles of polar lipid molecules. The spherical micelles are shown in cross-section and for simplicity the lipids are drawn with only a single hydrophobic 'tail'.

Hydrophobic hydrocarbon tail

Hydrophilic group

Surface monolayer

Spherical micelle

Nonpolar lipid

Another polar lipid

Surface bilayer

Mixed micelle

Mixed micelle

Fig. 2.17   An emulsion droplet with polar lipids serving as emulsifying agents.

Oil

Bimolecular micelles will also form readily if such a lipid monolayer is compressed (Fig. 2.16). It was on such a model that Danielli and Davson based their original theory concerning the structure of cell membranes (see Ch. 3). The variations in size, shape, polarity and charge on the phospholipids are expected to play an important rôle in the structure of such natural membrane systems.

Glycolipids

Glycolipids contain no phosphate, but like many of the phospholipids have polar carbohydrate groups and are therefore amphipathic substances. The simplest are glycosyldiglycerides such as the mono- and digalactosylidiglycerides shown in Fig. 2.18. However, many other glycolipids containing a variety of glycosyl groups have been reported as components of plant membrane systems.

Fig. 2.18   Two glycolipids isolated
from chloroplast membranes.

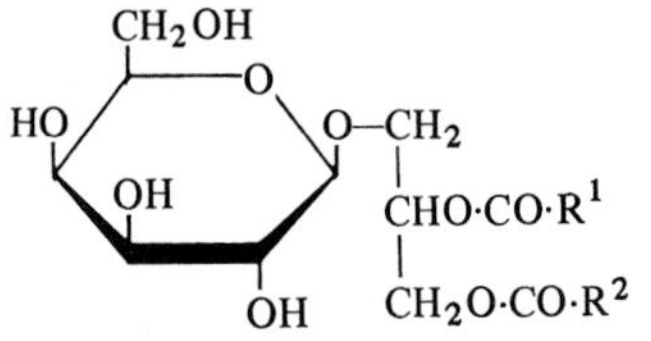

Monogalactosyldiglyceride
[β-D-galactosyl-(1 → 1')-2', 3'-
diacyl-D-glycerol]

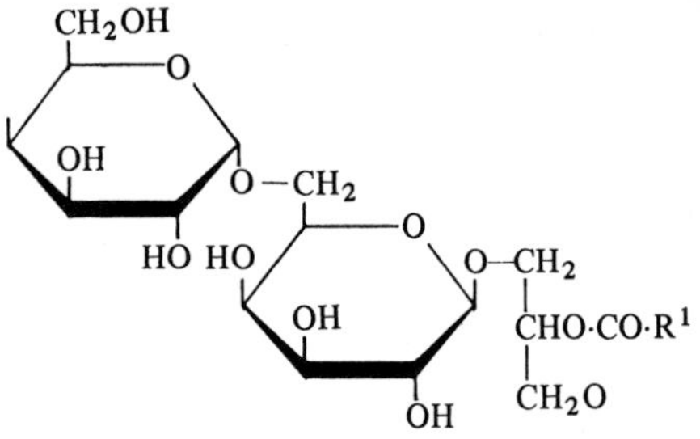

Digalactosyldiglyceride [α-D-galactosyl-
(1 → 6)-β-D-galactosyl-(1-1')-2',3'-diacyl-
D-glycerol]

Lipids derived from Δ^3-isopentenyl pyrophosphate

A large number of lipids are derived from a common 5-carbon building block Δ^3-isopentenyl pyrophosphate:

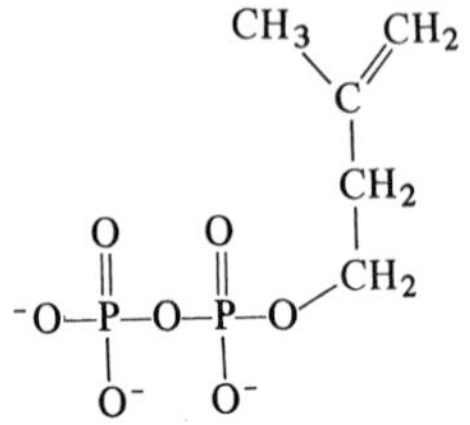

This is the precursor of isoprene 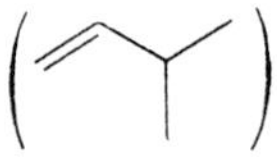 units in biological molecules. The lipids are of two major types, the *terpenes and steroids*. The various classes of terpene are designated mono- (having two isoprene units and ten carbon atoms), sesqui- (with three isoprene units), di (with four isoprene units and triterpenes (with six isoprene units), etc. (Fig. 2.19). They may be linear or cyclic molecules or combinations of both. The successive isoprene units are usually linked in a head to tail arrangement, and the double bonds are usually in the more stable *trans* configurations.

Large numbers of mono- and sesquiterpenes have been identified in plants and a few are shown in Fig. 2.19. These compounds are often responsible for the characteristic flavours and odours of plants, and the diterpene, phytol, is a component of chlorophyll. Gibberellic acid and its analogues are also diterpenes. Higher terpenes include the carotenoids, and the lipids believed to be involved as intermediates in polysaccharide biosynthesis in plants, animals and bacteria. Natural rubber is a polyisoprenoid consisting of many hundreds of isoprene units in long, linear chains.

The steroids are all derivatives of a ring compound called perhydrocyclo-pentanophenanthrene which contains three fused cyclohexane rings and a single cyclopentane ring:

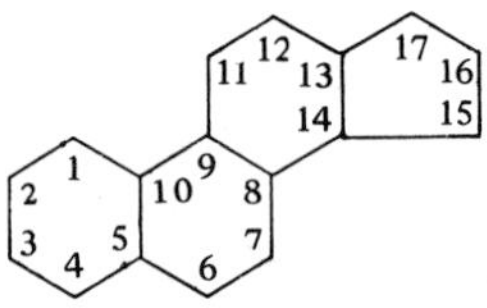

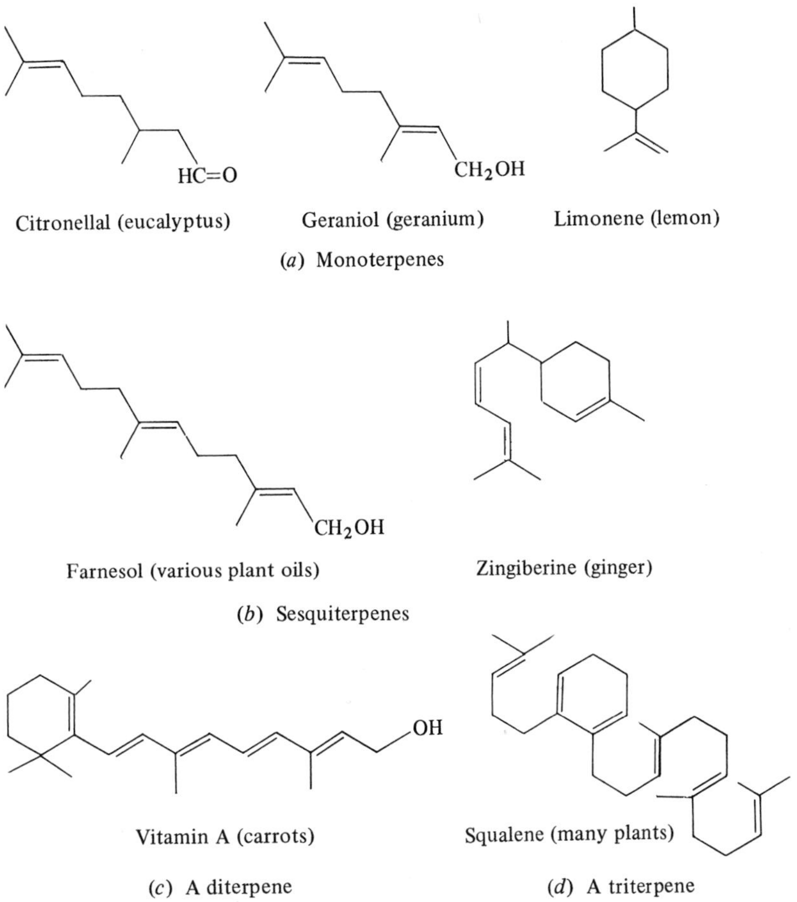

Fig. 2.19   Examples of different
classes of plant terpene.

Citronellal (eucalyptus)   Geraniol (geranium)   Limonene (lemon)

(a) Monoterpenes

Farnesol (various plant oils)   Zingiberine (ginger)

(b) Sesquiterpenes

Vitamin A (carrots)   Squalene (many plants)

(c) A diterpene   (d) A triterpene

The linear triterpene squalene is the precursor of all steroids and like most aliphatic terpenes readily cyclizes to form ring structures. The steroids are important biological molecules and in animals include a number of sex hormones, the corticosteroids, bile acids and the sterols. The latter, as typified by cholesterol, contain a hydroxyl group at C-3 and a long, aliphatic chain of eight or more carbon atoms at C-17. Those that occur in plants are known as the phytosterols and, although they are widespread, their functions are not understood.

Nucleic acids

Nucleic acids are found in all living cells and form the genetic material, thus being responsible for the storage and replication of genetic information. In addition, nucleic acids are closely associated with protein synthesis and hence with the expression of this genetic information. Nucleic acids are high molecular weight polymers composed of structures called *nucleotides* as repeating units. Nucleotides will be discussed later in this chapter in relation to enzyme cofactors, since a number of important coenzymes are composed wholly or in part of nucleotides. Nucleotides are composed of a nitrogenous base, five-carbon sugar, and phosphoric acid which are present in equimolar

67

proportions. A unit composed of a sugar and base linked by a β-glycosidic bond is known as a *nucleoside*. A nucleotide is a sugar-phosphate ester of a nucleoside. Nucleic acids are *polynucleotides* and are divided into two groups depending on the five-carbon sugar contained in the nucleotides. Those containing D-*ribose* are *ribonucleic acids* (RNA) and those containing 2-*deoxy*-D-*ribose* are *deoxyribonucleic acids* (DNA):

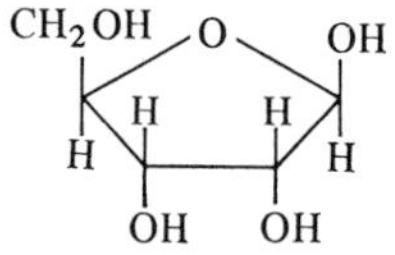

D-Ribose 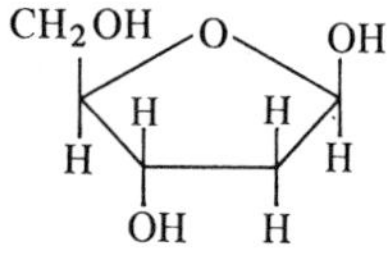   D-Deoxyribose

Nucleic acid bases

Two types of nitrogenous bases are found in nucleic acids, substituted *pyrimidines* and substituted *purines*. A purine is a pyrimidine fused with an imidazole ring. The most important pyrimidines are *thymine* (T), *cytosine* (C) and *uracil* (U) and their structures with that of the parent compound are shown below:

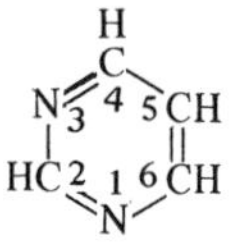

Pyrimidine 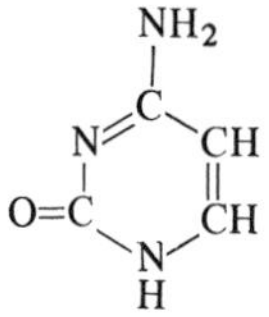   Cytosine 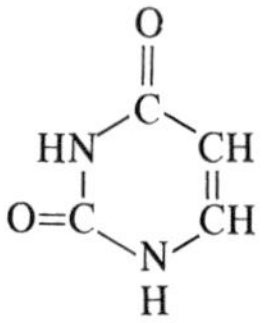   Uracil 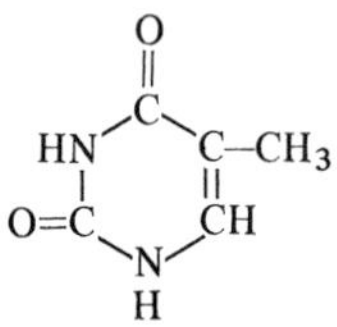   Thymine

In relation to higher plants and animals, thymine is found almost exclusively in DNA while uracil is confined almost entirely to RNA. Some higher plants, e.g. wheat germ, contain significant quantities of 5-methylcytosine in their DNA.

The purines found in most DNA and RNA molecules are *adenine* (A) and *guanine* (G). In some RNA molecules, particularly the low molecular weight transfer RNA, a number of unusual bases are found including various methylated purines and hypoxanthine, which is similar to adenine but with the amino group replaced by an hydroxyl group. The structures of the common purines are as follows:

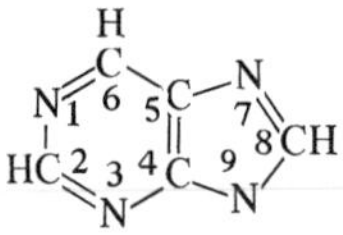

Purine 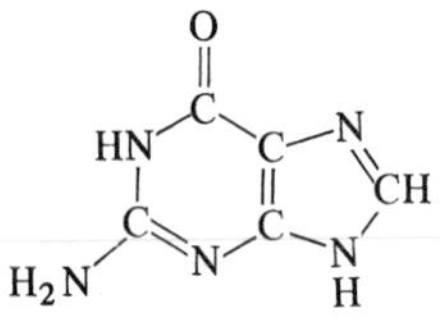   Guanine 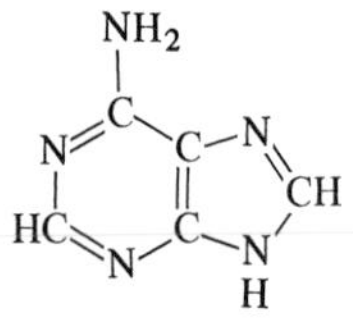   Adenine

Nucleosides and nucleotides

Nucleosides consist of a purine or pyrimidine attached through an N-glycosidic bond to the carbon-one position of the five-carbon sugar, ribose or deoxyribose. The nitrogen atom involved in this bond is

Table 2.7 Some common nucleosides and nucleotides

Base	Nucleoside	Nucleotide	Abbreviation
Adenine	Adenosine	Adenosine-5'-phosphoric acid (adenylic acid)	5'AMP (AMP)
		Adenosine-5'-triphosphate	5'-ATP (AMP)
	Deoxyadenosine	Deoxyadenosine-5'-phosphoric acid (deoxyadenylic acid)	5'-dAMP (dAMP)
Guanine	Guanosine	Guanosine-5'-phosphoric acid (guanylic acid)	5'-GMP (GMP)
		Guanosine-5'-triphosphate	5'-GTP (GTP)
Cytosine	Cytidine	Cytidine-5'-phosphoric acid (cytidylic acid)	5'-CMP (CMP)
	Deoxycytidine	Deoxycytidine-5'-phosphoric acid (deoxycytidylic acid)	5'-dCMP (dCMP)
Uracil	Uridine	Uridine-5'-phosphoric acid (uridylic acid)	5'-UMP (UMP)
Thymine	Deoxythymidine	Deoxythymidine-5'-phosphoric acid (deoxythymidylic acid)	5'dTMP (dTMP)

N-1 in the case of the pyrimidines and N-9 with the purines. The trivial names of nucleosides are somewhat confusing but generally end in -idine with pyrimidines and -osine with purines (Table 2.7).

Nucleotides are the sugar-phosphate esters of the nucleosides. The ribose ring has three possible positions that may be esterified, at carbons 2, 3 and 5, whereas the deoxyribose group has only two potential sites, at carbons 3 and 5. However, the majority of nucleotides found in living cells are esterified in the 5'-position. The structure and nomenclature of nucleotides is illustrated by the following two examples:

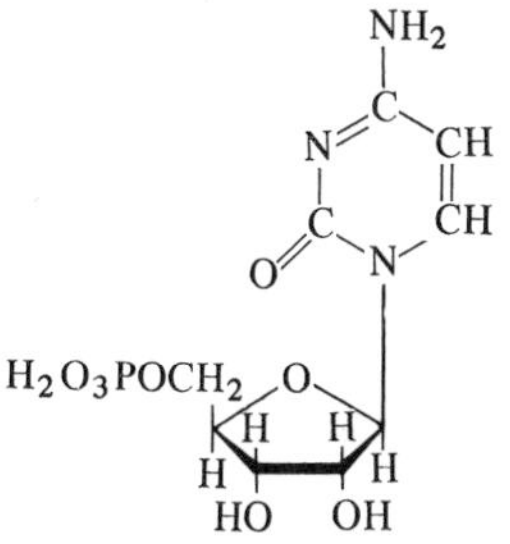

Cytidine 5'-phosphoric acid
(5'-cytidylic acid; 5'-CMP)

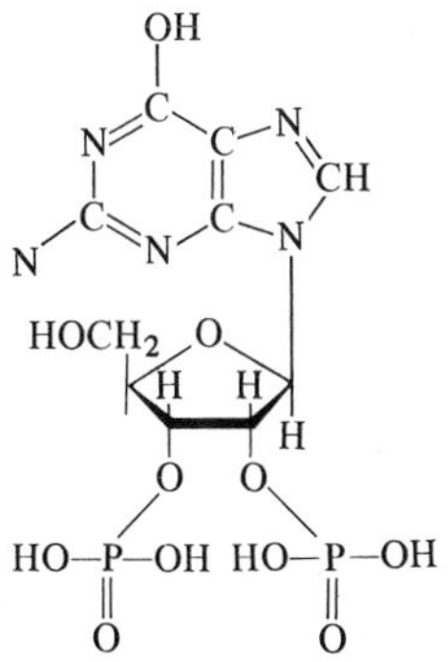

Guanosine-2'-3'-diphosphoric acid

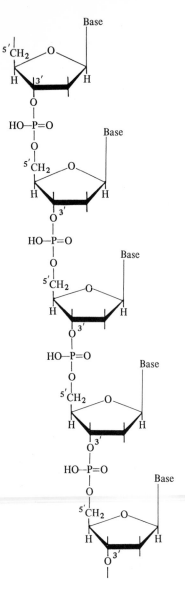

Fig. 2.20   Polynucleotide structure:
3'–5' phosphodiester linkages.

Commonly the terms CMP, AMP, etc., are used and in this case it is assumed that the monophosphate is linked to the 5'-position. All of the common 5'-mono-phosphates occur in cells as the 5'-diphosphates and 5' triphosphates (see structure of ATP and UDP-glucose, pp. 114, 116). A list of the most important nucleotides is given in Table 2.7. Apart from their importance as the building blocks of nucleic acids, various nucleotides serve extremely important rôles in metabolism as coenzymes, where they may constitute the whole coenzyme (e.g. ATP) or part of the coenzyme molecule (e.g. NAD). The function of these coenzymes will be discussed later in this chapter. Finally, mention should be made of cyclic AMP which functions as a versatile regulatory agent in animals and microorganisms. However, its presence in higher plant cells has not been fully established and so its function, if any, remains obscure (see Amrhein, 1977; Johnson *et al.*, 1981).

Polynucleotides

The nucleic acids are high molecular weight unbranched polynucleotides in which the successive nucleotide units are covalently linked by phosphodiester bonds between the 3'-position of one pentose and the 5'-position of the next (Fig. 2.20).

   The molecule may be envisaged as consisting of a sugar phosphate backbone with the nitrogenous bases arranged as side groups along the chain. The number of bases in some DNA molecules may be as high as 10^8 although it can be as low as eighty in transfer RNA.

DNA

DNA is a large polymer composed of deoxyribonucleotides with a molecular weight that may be as high as 10^{12}. In all eukaryotic cells it is found largely in the nucleus although some DNA is localized in the chloroplasts and mitochondria. The evidence pointing to DNA as the genetic material is summarized in Chapter 4.

Base composition

The base composition of DNA from a wide variety of sources has now been determined and provides valuable evidence both of the structure of the DNA molecule and of its rôle as the genetic material. The four major bases found in DNA are adenine, guanine, cytosine and thymine, although 5-methylcytosine is also found in some species; the highest levels of this base have been recorded in certain higher plant tissues such as wheat germ and peanut. The base compositions of DNA from a number of plant and animal species are shown in Table 2.8 and they show a number of important features:

1. the base composition of DNA is characteristic of a given species and is the same in different tissues of the same species;
2. the sum of the purines (A + G) equals the sum of the pyrimidines (C + T), and the sum of amino bases (A + C) equals the sum of the keto bases (G + T);

3. considering the individual bases, then adenine and thymine are found in equimolar amounts, as are guanine and cytosine.

Table 2.8   Base ratios of DNA from various sources (as molar percentages)

Source	Adenine	Guanine	Cytosine	5-Methylcytosine	Thymine
Allium cepa (onion seed)	31·8	18·4	12·8	5·4	31·3
Phaseolus vulgaris (bean seed)	29·7	20·6	14·9	5·2	29·6
Daucus carota (carrot leaf)	26·7	23·2	17·3	6·0	26·8
Gossypium hirsutum (cotton embryos)	32·8	16·9	12·7	4·6	32·9
Cucurbita pepo (pumpkin seed)	30·2	21·0	16·1	3·7	29·0
Sheep thymus	29·3	21·4	21·0	—	28·3
Arbacia punctulata (sea urchin sperm)	28·4	19·5	19·3	—	32·8
Clostridium perfringens (bacterium)	36·9	14·0	12·8	—	36·3

With the exception of some viral RNA molecules, simple relations between purines and pyrimidines and between amino and keto bases do not exist for RNA.

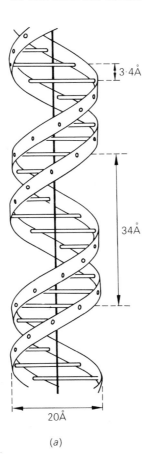

3·4Å

34Å

20Å

(a)

(b)

Fig. 2.21   The structure of DNA.
(a) Diagrammatic representation of the DNA molecule as proposed by Watson and Crick (1953). The phosphate sugar chains are shown as ribbons and the pairs of bases holding the chains together as horizontal rods.
(b) Drawing of a model of the DNA helix. The dotted lines represent the hydrogen bonds between the bases. Reproduced from Davidson (1972).

Structure of DNA

As in the case of protein structure, the technique of X-ray diffraction analysis has played a major part in the understanding of the structure of DNA. Such studies, initiated by Astbury in the 1930s and later extended by Wilkins and others, showed that DNA fibres from different sources gave a very similar and regular X-ray diffraction pattern, with reflections corresponding to regular spacings of 3·4 Å and 34 Å along the fibre axis. From this and other evidence Watson and Crick in 1953 proposed that the DNA molecule was a double helix with two right-handed polynucleotide chains coiled around a common axis (Fig. 2.21). The deoxyribose phosphate chain is on the outside with the bases directed inwards. The two chains run in opposite directions and are held together by hydrogen bonding between the bases (Fig. 2.22). Adenine pairs with thymine, and guanine with

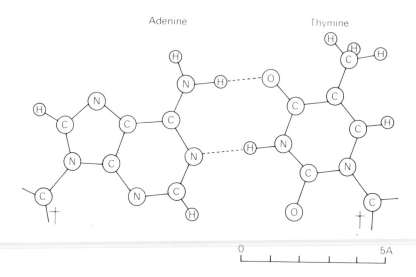

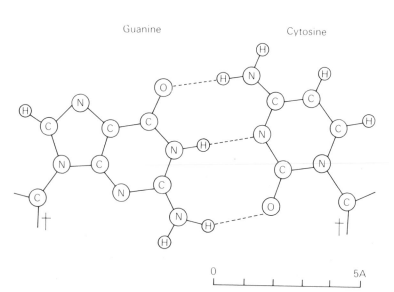

Fig. 2.22 The pairing of adenine and thymine and guanine and cytosine. The dotted lines indicate the hydrogen bonds. The carbon atoms marked † belong to the sugar rings. Reproduced from Davidson (1972).

cytosine which, of course, corresponds with the molar proportions of the bases obtained by chemical analysis of DNA. The two polynucleotide chains are therefore complementary so that the base sequence in one determines that in the other. The 3·4 Å spacing shown by X-ray analysis corresponds to the vertical distance between the base pairs while the 34 Å repeat distance corresponds to a complete turn of the double helix since this contains ten bases.

Properties of DNA

The characteristic structure of DNA confers upon it a number of important properties which are relevant to the discussions in the later chapters of this book.

Molecular weight. The molecular weight of DNA is often quoted as about $10^6 - 10^7$, but DNA in the intact cell is probably very much larger than this. It is very difficult to obtain reliable estimates of the size and length of DNA molecules because they are very easily broken by the forces required to isolate DNA from the intact cell. However DNA from the bacterium *Escherichia coli* appears to have a molecular weight of about $2·8 \times 10^9$ while DNA from higher cell chromosomes may have a MW of around 10^{12}.

Acid-base properties. Because of the highly polar phosphate groups located on the outside of the double helix, DNA is a polyanion and readily binds various cations to maintain charge neutrality. These phosphate groups bind monovalent and divalent cations, various organic molecules such as amines and, in higher cell chromosomes, the basic proteins known as histones (see Ch. 4). All of these interactions probably affect the conformation of the DNA molecule to a greater or lesser extent.

Buoyant density. The difference in density between different types of DNA in concentrated salt solutions is widely used as a basis for DNA fractionation. The method usually involves centrifugation at high speed for long periods in concentrated (8 M) caesium chloride (see p. 33). This produces a linear concentration gradient of the salt and if DNA is also present, it will migrate to a position corresponding to its own buoyant density. The buoyant density of DNA is closely related to the base composition since G–C pairs are more compact than A–T pairs, buoyant density is a linear function of the ratio of G–C to A–T pairs, unless 5-methyl-cytosine is present which results in a decrease in density. Most nuclear DNAs from higher plants have buoyant densities in the range $1·69-1·71$g cm^{-3}.

Determination of the buoyant density of nuclear DNA has shown that, although the bulk of DNA sediments as a homogenous band, some of the DNA is sufficiently different in base composition to form a shoulder of the main peak or even a separate peak. This is known as *satellite DNA* and, in plants, is usually denser than the main band. Satellite DNAs usually consist of repeated base sequences and it is thought that it may have a structural rather than informational rôle (see p. 187).

Absorption and hypochromism. DNA (and RNA) has a characteristic absorption spectrum in the ultra-violet region with a peak around 260 nm. This is similar to the absorption properties of the constituent bases which all absorb in the 260–290 nm region (Fig. 2.23). An interesting feature of this absorbance is an effect known as *hypochromism* which refers to the observation that, quantitatively, the double helix of DNA absorbs less ultra-violet light than would be expected from the summation of the light absorption of the individual bases. For example, if the two component strands of the double helix are separated by heating to 60°C, the absorbance at 260 nm increases by up to 40%. This effect is reversed if the double coil is reformed by slow cooling.

Fig. 2.23   Absorption curves for purine and pyrimidine bases at pH 7·0.

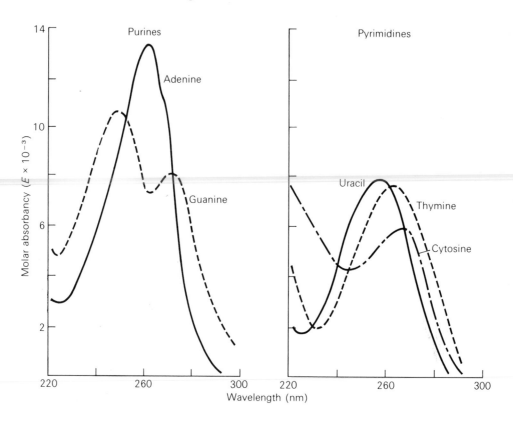

Denaturation. Denaturation involves a change in the conformation of DNA produced by a rupture of the complementary hydrogen bonds between bases. This rupture may be reversible or irreversible. Denaturation may be produced by heat, exposure to extremes of pH (2·5

or 11), reduction in ionic strength (probably due to increased repulsion between the backbone phosphate groups), and addition of reagents which from hydrogen bonds with the bases and so destroy inter-base bonding (e.g. urea). These treatments result in the uncoiling of the double helix, wholly or in part, to produce randomly coiled DNA strands. However, the complementary strands which have been denatured by heat or alkali will reassociate if the pH is restored to neutrality or the temperature is lowered; this can be followed, for example, by measuring the decrease in absorbance at 260 nm (see p. 74).

The tendency of DNA to unwind or dissociate under certain conditions and then to reassociate if favourable conditions are restored is utilized in a technique known as *molecular hybridization*. This enables an estimate to be made of the degree of complementation between the base sequence of two samples of DNA or DNA and RNA. The reassociation reaction is primarily dependent on the DNA concentration and the sequence content. The greater the tendency of a given pair of DNA strands to reassociate, the greater the complementarity of their sequences. For example, two specimens of DNA, one of which is labelled with radioactive phosphorus, are mixed, heated to a temperature high enough for denaturation to occur (known as the melting point) and then slowly cooled causing complementary strands to reassociate (Fig. 2.24). Single and double strands may be

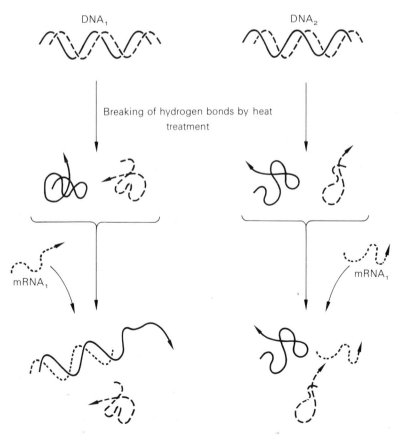

Fig. 2.24 Hybridization of nucleic acids. The use of DNA–RNA hybrids to show the complementarity in nucleotide sequences between an RNA molecule and one of the two strands of its DNA template. The left side of the diagram shows the formation of a hybrid molecule between an RNA molecule and one of the two strands of the template. The specificity of this method is shown on the right side. Here the same RNA molecule is mixed with unrelated DNA. No hybrid molecules are formed. Reproduced from James D. Watson, *Molecular Biology of the Gene*, Second Edition. Copyright © 1970 by J. D. Watson, W. A. Benjamin, Inc., Menlo Park, California.

75

separated by centrifugation or filtration and the degree of hybridization determined by a measure of the radioactivity in the double stranded fraction. Hybridization may be used to study the degree of relatedness between the DNA from various species; the tendency of the DNA from two given species to hybridize is related to how close they stand taxonomically. Similarly, the genetic origin of RNA can be studied by DNA–RNA hybridization. The uses of this technique are discussed further in Ch. 4.

RNA

RNA is a single strand polymer of ribonucleotide units and is found universally in living cells. It is found in almost all cell fractions including the nucleus, mitochondria, chloroplasts, cell wall, soluble phase and cytoplasmic ribosomes, the latter containing the largest part of the total cellular RNA. There are three major kinds of RNA and all are intimately associated with the processes of protein synthesis as described in Chapter 5. As we shall see below, the molecular weights of the various RNA molecules vary from about 25 000 to over one million, and as can be seen from Table 2.9, the base composition does not follow the same rules as outlined for DNA although, in general, the proportion of A + C roughly equals G + U. Some RNA molecules contain significant proportions of some methylated bases and an unusual nucleoside known as pseudouridine (ψU), in which the glycosidic bond is associated with position 5 of uracil rather than position 1. The three major kinds of RNA found in plant and animal cells are known as ribosomal RNA, messenger RNA and transfer RNA.

Table 2.9 Base ratios of RNA from various sources (as molar percentages)

Source	Adenine	Guanine	Cytosine	Uracil
Allium cepa (onion seed)	24·9	29·8	24·7	20·6
Phaseolus vulgaris (bean seed)	24·9	31·4	24·1	19·6
Cucurbito pepo (pumpkin seed)	25·2	30·6	24·8	19·4
Ox liver	17·1	27·3	33·9	21·7
Yeast	25·4	24·6	22·6	27·4
E. coli	25·3	28·8	24·7	21·2

Ribosomal RNA

Ribosomal RNA (rRNA) constitutes the largest part (up to 80%) of the total cellular RNA. It is found primarily in the ribosomes although, since it is synthesized in the nucleus it is also detected in that fraction. The base composition of rRNA has no particularly distinctive features. It contains the four major RNA bases with a slight degree of methylation, and shows differences in the relative proportions of the bases between species.

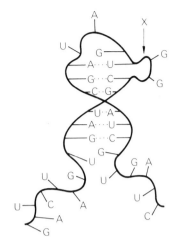

Fig. 2.25 The secondary structure of RNA showing a helical region with complementary base pairing and a looped out region of the helix at X. Reproduced from Davidson (1972).

As we shall see later (Ch. 5), ribsosomes may be dissociated into a large and a small sub-unit, each with its own characteristic RNA. The size of the major RNA molecules from plant cytoplasmic ribosomes is typical of eukaryotes. The RNA from the larger sub-unit has a molecular weight of about $1 \cdot 3 \times 10^6$, while that from the smaller sub-unit is about $0 \cdot 7 \times 10^6$. One major problem is that RNA is quite readily degraded by hydrolysis and so significant changes in size may occur during the isolation procedures.

The secondary structure of rRNA in solution is very dependent on the conditions, particularly the temperature and ionic strength. The molecules appear to be single polynucleotide strands which are unbranched and highly flexible. At low ionic strength, rRNA behaves as a random coil, but with increasing ionic strength the molecule shows helical regions produced by base pairing between adenine and uracil and guanine and cytosine. Since the sequences involved in pairing may not be entirely complementary, non-bonded nucleotides within the helix can be accommodated in loops outside of the helix (Fig. 2.25). Because of this base pairing rRNA exhibits hypochromism but not as markedly as does DNA. The rôle of rRNA in relation to the structure of the ribosome is discussed in Chapter 5.

Messenger RNA

As we shall see in Chapters 4 and 5, messenger RNA (mRNA) carries the information needed for the synthesis of proteins from the nucleus to the sites of protein synthesis in the cytoplasm, the ribosomes. The information contained in the chromosomal DNA is transcribed into mRNA, and so the base composition of mRNA, made up of just A, C, G and U, is similar to that of DNA.

Messenger RNAs appear to be large molecules with molecular weights of around one million. Since they made up less than 5% of the total cellular RNA and are usually quite labile, mRNAs are difficult to isolate. Presumably, however, cells must contain many thousands of mRNAs carrying information for the synthesis of a wide range of proteins. Nevertheless mRNA (i.e. RNA capable of directing protein synthesis *in vitro*) has been isolated from a number of plants and, in a few cases, specific mRNAs have been identified by virtue of the proteins coded for e.g. leghaemoglobin, seed globulins, α-amylase and phenylalanine ammonia lyase (see p. 177). The stability of mRNA appears to be quite variable. It has been calculated that in some bacterial cells mRNA has a life of only two minutes before being degraded, and that during this time it may function up to twenty times in protein synthesis. In higher cells, mRNA normally has a much longer life and may function for many hours or even several days before being broken down.

Transfer RNA

Transfer RNA (tRNA) or soluble RNA, which makes up about 15% of the total cellular RNA, is the smallest of the RNA molecules with a molecular weight of around 25 000. Transfer RNAs function as amino acid-carrying molecules during protein synthesis and so there is at least one specific tRNA for each of the twenty protein amino acids. They consist of a single polynucleotide chain containing up to ninety nucleotide units and are characterized by the presence of a high proportion of guanine, cytosine, pseudouridine and some uncommon methylated bases (Fig. 2.26). The composition of tRNA is therefore quite distinct from the larger RNA molecules from the same species.

Fig. 2.26   Structure of pseudouridine and some uncommon methylated bases found in tRNA.

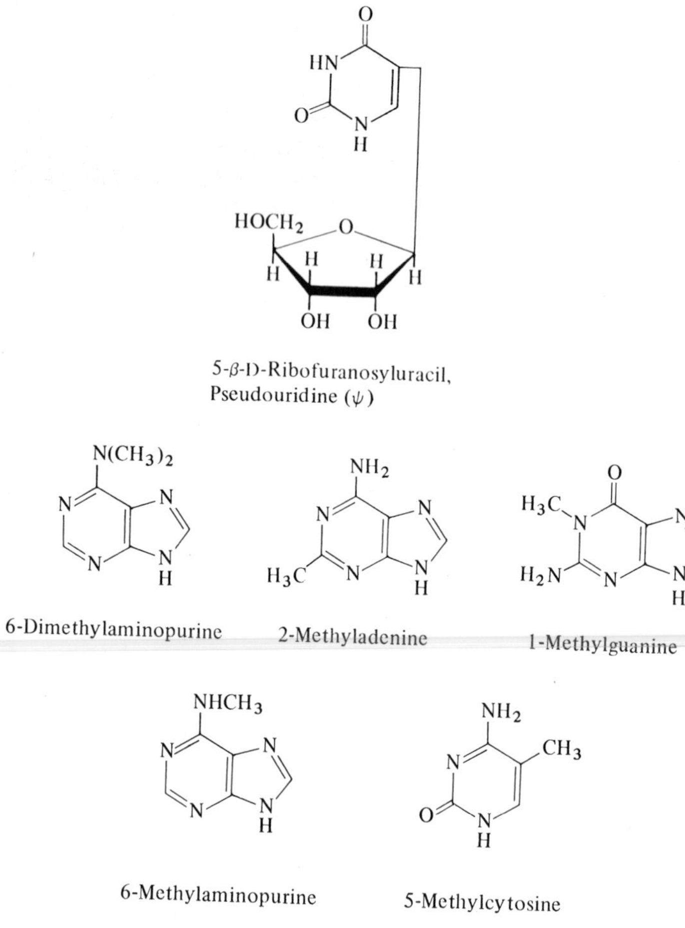

5-β-D-Ribofuranosyluracil,
Pseudouridine (ψ)

6-Dimethylaminopurine   2-Methyladenine   1-Methylguanine

6-Methylaminopurine   5-Methylcytosine

The nucleotide sequence of tRNA isolated from some microorganisms has now been determined. It has been shown that tRNA exhibits hypochromism when subjected to heating and cooling indicating that many of the bases are paired. If the nucleotide chains of tRNAs of known base sequence are arranged to give maximum base pairing then a characteristic 'clover leaf' structure is produced. This structure and the nucleotide sequences of tRNA are discussed more fully in Chapter 5 in relation to the various binding sites involved in the specific rôle of these molecules in protein synthesis.

Proteins are ubiquitous and indispensable components of all living cells since they are intimately concerned with all metabolic processes. For example, they play an essential rôle in the structure of cellular membranes, viral coats, and connective tissues; they function as hormones (e.g. insulin) and perhaps most important of all, form the enzymes which catalyse the reactions of intermediary metabolism. Proteins are large molecules with molecular weights ranging from about 5 000 to over one million, built up from simpler organic molecules, known as *amino acids*, linked to form long unbranched chains. The lower size limit of about 5 000 is arbitrarily set and molecules below this are usually called polypeptides although they are composed of similar units linked together in the same way. When proteins are hydrolysed, say in strong acid, they break down into about twenty different amino acids. Since the average molecular weight of an amino acid is about 140, proteins clearly consist of a large number of amino acids which are normally arranged in linear chains. Since any position in the chain can be occupied by any one of the twenty or so amino acids it is clear that there is potentially a vast number of structurally different proteins, and this will allow for the wide range of highly specific biological functions with which they are associated.

Amino acids

Most of the amino acids found in proteins are termed α-amino acids since they have a primary amino group and a carboxyl group linked to the same carbon atom and are of the general structure:

$$R-\underset{\underset{H}{|}}{\overset{\overset{NH_2}{|}}{C}}-COOH$$

This general group includes two amides, asparagine and glutamine, derived from the α-amino acids aspartic and glutamic and which are widely found in proteins. The exceptions to this rule are the amino acids proline and hydroxyproline which have a substituted α-amino group and so are α-imino acids. These protein amino acids are normally classified in relation to the R-group and Lehninger (1970) has proposed that the most meaningful classification is that based on the polarity of R. We can then distinguish four major classes (Fig. 2.27): (1) non-polar or hydrophobic, (2) polar but uncharged, (3) positively charged and (4) negatively charged. The last three are based on the charge at pH 6-7 since as we shall see later the net charge carried by an amino acid is related to the pH at which it is determined. Within these broad divisions there is considerable variation in the size and structure of R-groups, which include amino acids with aromatic and sulphur-containing side chains.

In addition to the twenty or so common protein amino acids, a few unusual forms have been found in proteins, while over 100 non-protein amino acids have so far been isolated from plant sources, although their function is largely unknown. They have been described more

Amino acids with
nonpolar R-groups

Alanine
(Ala)

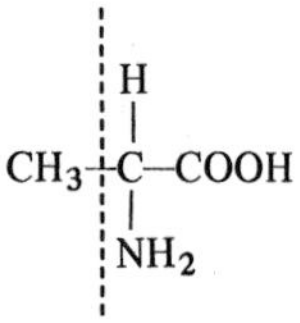

$$CH_3-\overset{\overset{\displaystyle H}{|}}{\underset{\underset{\displaystyle NH_2}{|}}{C}}-COOH$$

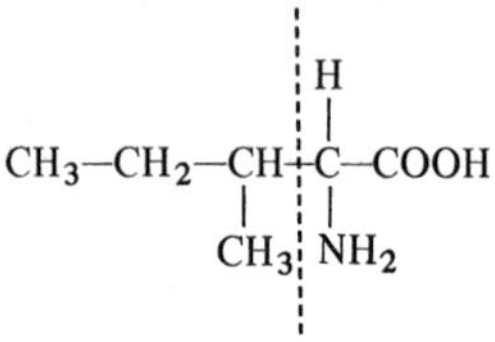

Proline
(Pro)

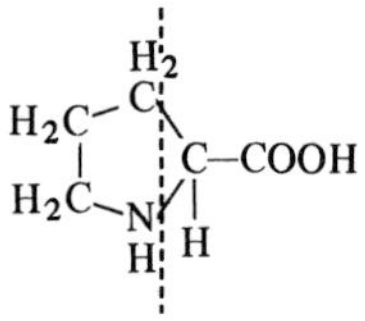

Valine
(Val)

$$\overset{\displaystyle CH_3}{\underset{\displaystyle CH_3}{\diagdown}}CH-\overset{\overset{\displaystyle H}{|}}{\underset{\underset{\displaystyle NH_2}{|}}{C}}-COOH$$

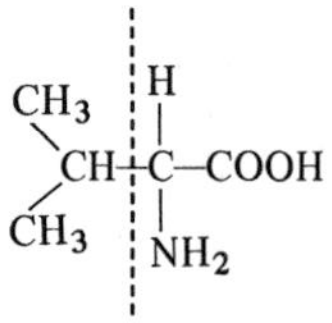

Phenylalanine
(Phe)

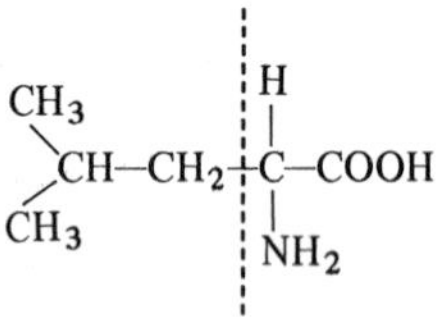

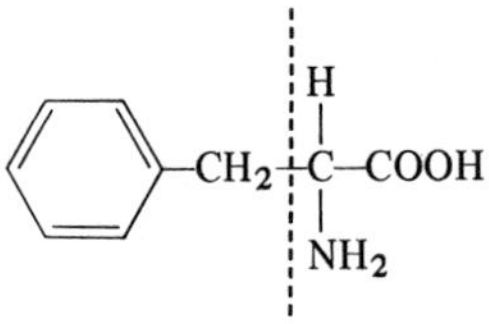

Leucine
(Leu)

$$\overset{\displaystyle CH_3}{\underset{\displaystyle CH_3}{\diagdown}}CH-CH_2-\overset{\overset{\displaystyle H}{|}}{\underset{\underset{\displaystyle NH_2}{|}}{C}}-COOH$$

Tryptophan
(Trp)

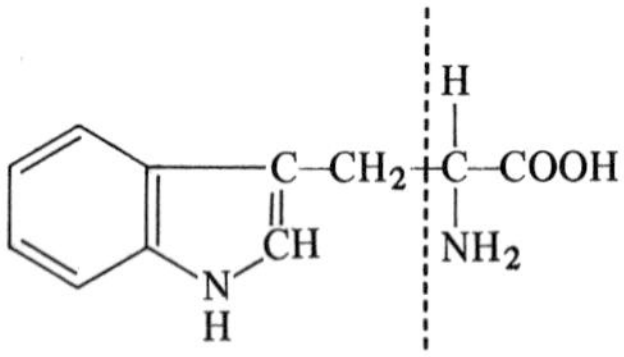

Isoleucine
(Ile)

$$CH_3-CH_2-\underset{\underset{\displaystyle CH_3}{|}}{CH}-\overset{\overset{\displaystyle H}{|}}{\underset{\underset{\displaystyle NH_2}{|}}{C}}-COOH$$

Methionine
(Met)

$$CH_3-S-CH_2-CH_2-\overset{\overset{\displaystyle H}{|}}{\underset{\underset{\displaystyle NH_2}{|}}{C}}-COOH$$

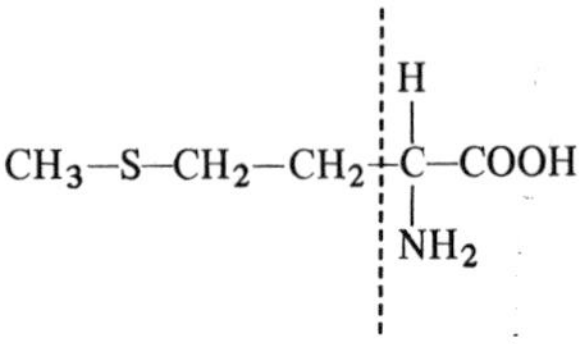

Amino acids with
uncharged polar R groups

Glycine
(Gly)

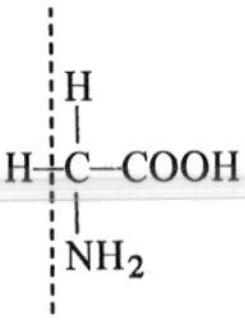

$$H-\overset{\overset{\displaystyle H}{|}}{\underset{\underset{\displaystyle NH_2}{|}}{C}}-COOH$$

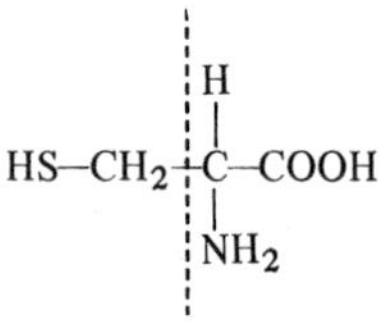

Tyrosine
(Tyr)

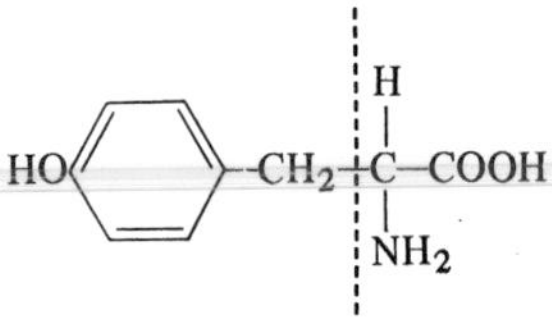

$$HO-\!\!\!\bigcirc\!\!\!-CH_2-\overset{\overset{\displaystyle H}{|}}{\underset{\underset{\displaystyle NH_2}{|}}{C}}-COOH$$

Serine
(Ser)

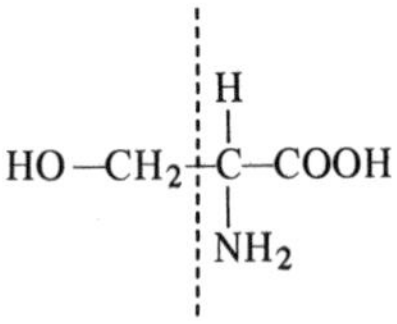

$$HO-CH_2-\overset{\overset{\displaystyle H}{|}}{\underset{\underset{\displaystyle NH_2}{|}}{C}}-COOH$$

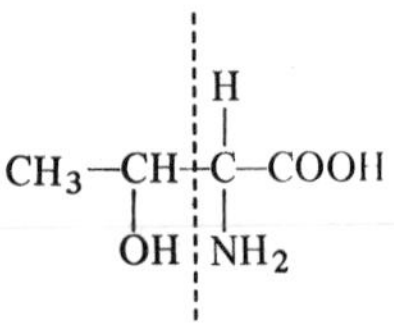

Asparagine
(Asn)

Threonine
(Thr)

$$CH_3-\underset{\underset{\displaystyle OH}{|}}{CH}-\overset{\overset{\displaystyle H}{|}}{\underset{\underset{\displaystyle NH_2}{|}}{C}}-COOH$$

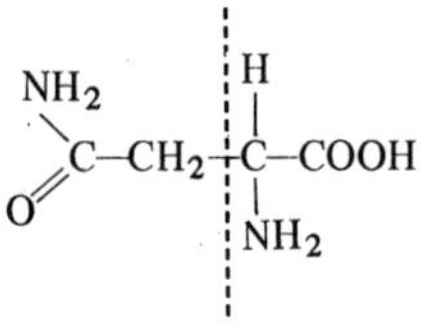

Glutamine
(Gln)

Cysteine
(Cys)

$$HS-CH_2-\overset{\overset{\displaystyle H}{|}}{\underset{\underset{\displaystyle NH_2}{|}}{C}}-COOH$$

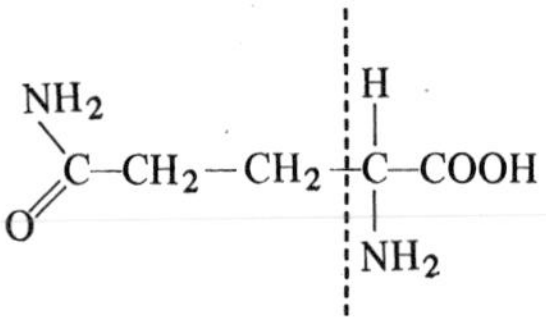

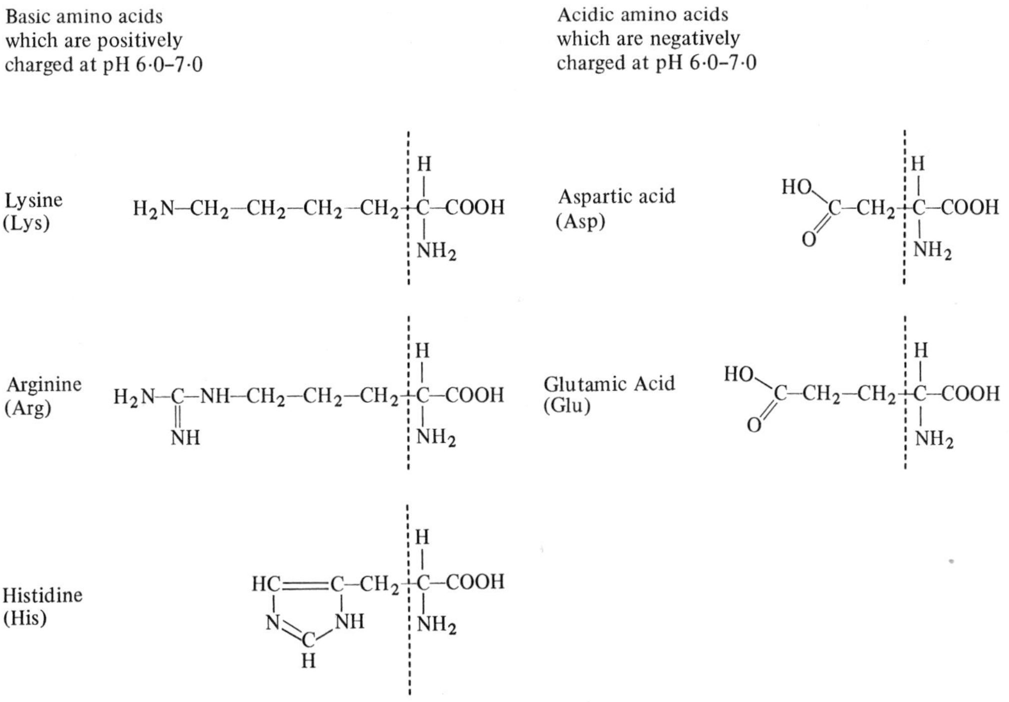

Basic amino acids
which are positively
charged at pH 6·0–7·0

Acidic amino acids
which are negatively
charged at pH 6·0–7·0

Lysine
(Lys)

$$H_2N-CH_2-CH_2-CH_2-CH_2 \overset{|}{\underset{|}{C}}-COOH$$

(R group: $H_2N-CH_2-CH_2-CH_2-CH_2-$; H and NH_2 on the central carbon)

Aspartic acid
(Asp)

$$\overset{HO}{\underset{O}{}}C-CH_2-\overset{H}{\underset{NH_2}{C}}-COOH$$

Arginine
(Arg)

$$H_2N-\overset{|}{\underset{NH}{C}}-NH-CH_2-CH_2-CH_2-\overset{H}{\underset{NH_2}{C}}-COOH$$

Glutamic Acid
(Glu)

$$\overset{HO}{\underset{O}{}}C-CH_2-CH_2-\overset{H}{\underset{NH_2}{C}}-COOH$$

Histidine
(His)

$$HC=C-CH_2-\overset{H}{\underset{NH_2}{C}}-COOH$$

$$\underset{\underset{H}{C}}{N}\diagdown\diagup NH$$

Fig. 2.27   Classification and structure of the protein amino acids. The R groups are shown to the left of the vertical dotted line. However, this division is somewhat arbitrary. For example, glycine is a borderline case and is sometimes classified as a nonpolar amino acid. It should be noted that cysteine often occurs in proteins in its oxidized form cystine in which the thiol groups of two molecules of cysteine have been oxidized to give a disulphide cross-link between them (and perhaps between two polypeptide chains).

fully by Bell (1981) and some of these structures are shown in Table 2.10. Some, such as homoserine and pipecolic acid, are of widespread distribution in plants and act as intermediates in the metabolism of protein amino acids. Many of the others, however, such as azetidine-2-carboxylic acid which is restricted to the Liliaceae, have a much narrower distribution and it is difficult to believe that they play a rôle in basic intermediary metabolism. Others such as homoarginine and canavanine occur in high concentrations in some seeds although their levels fall markedly some days after germination, perhaps suggesting that they act as nitrogen-storage molecules in the seed. Some of the unusual amino acids appear to be toxic to other organisms and so may play a protective rôle, but there is no clear evidence to show that these amino acids in general confer any protection against plant pathogens and predators.

Properties of amino acids

Since amino acids are the basic units of proteins, an understanding of the properties of amino acids will tell us something of the properties of proteins. A number of their properties are not consistent with the general structural formula shown before which shows the amino and carboxyl groups as uncharged. For example, they have high melting points and are generally soluble in water but insoluble in non-polar organic solvents, which suggests the presence of charged, polar groups.

The presence of such charged groups is further demonstrated by a consideration of the acid-base properties of amino acids. The presence

Table 2.10 Structure and occurrence of some non-protein amino acids isolated from higher plants

Non-protein amino acid	Corresponding protein amino acid	Structure	Distribution
Homoserine	Serine, threonine	$HO \cdot CH_2 \cdot CH_2 \cdot CH(NH_2) \cdot COOH$	Many plants
Homoarginine	Arginine	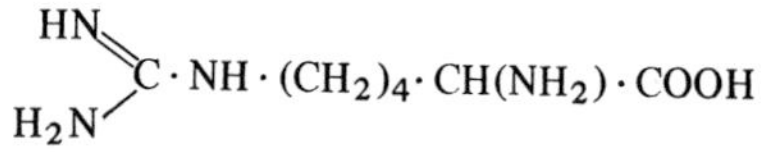 $\begin{array}{c} HN \\ H_2N \end{array} \!\!\!\! C \cdot NH \cdot (CH_2)_4 \cdot CH(NH_2) \cdot COOH$	Some *Lathyrus* species
Canavanine		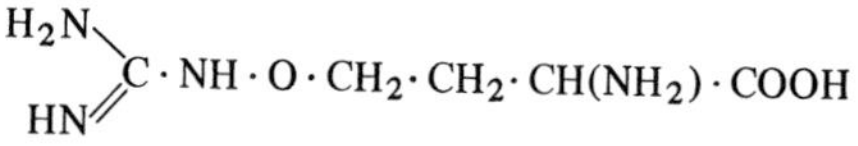 $\begin{array}{c} H_2N \\ HN \end{array} \!\!\!\! C \cdot NH \cdot O \cdot CH_2 \cdot CH_2 \cdot CH(NH_2) \cdot COOH$	
Azetidine-2-carboxylic acid	Proline	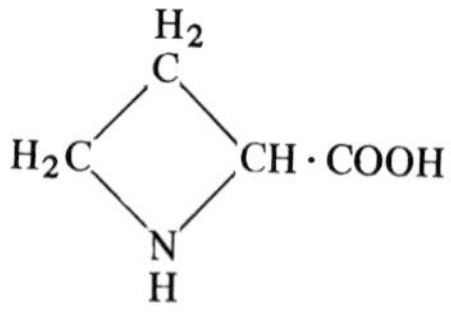 (azetidine ring structure) $CH \cdot COOH$	Members of the family Liliaceae
Pipecolic acid	Proline	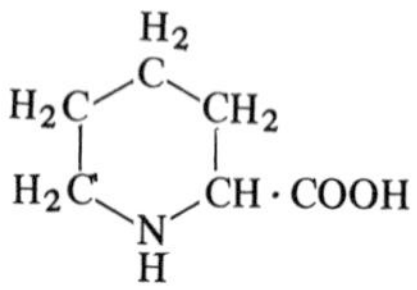 (piperidine ring structure) $CH \cdot COOH$	Many plants, especially legume seeds
m-Carboxy-phenylalanine	Phenylalanine	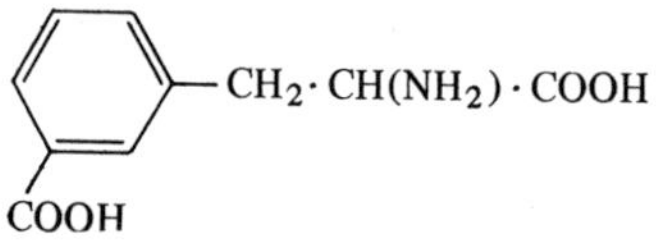 (benzene ring with COOH) $- CH_2 \cdot CH(NH_2) \cdot COOH$	*Iris*, *Reseda*

of both amino and carboxyl groups means that amino acids are *amphoteric substances*, able to react with both acids and bases. The carboxyl group will have a pK value (pH at which the group is half associated and half dissociated) of 2–3, while that of the amino group will be about 10. Consider an amino acid with two ionizable groups, the amino and carboxyl. This means that under acid conditions the amino acid will become positively charged, at alkaline pH values it will be negatively charged, but at pH values around neutrality it is thought that it will behave as a dipolar molecule with no net charge, commonly called a *zwitterion*:

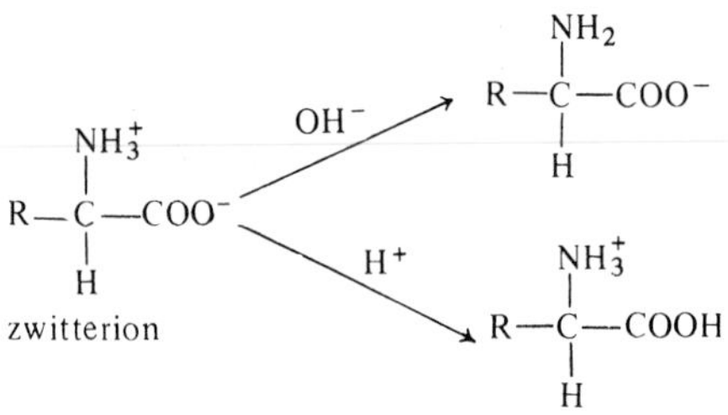

This behaviour is illustrated by the titration curve for the amino acid glycine (Fig. 2.28). When HCl is added to a solution of glycine the

Fig. 2.28   Titration curve of the amino acid glycine.

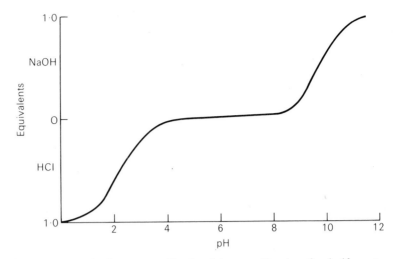

carboxyl group is first neutralized, giving a pK value for half neutra-lization of pH 2·3. When NaOH is added, the amino group is neutralized to give a pK value of 9·6. At a pH value around neutrality the molecule will have no net charge and this is known as the *isoelectric point*, which will be approximately midway between the two pK values. Some amino acids have more than one amino or carboxyl group, while others have additional ionizable groups, such as the sulphydryl group of cysteine, all of which will affect the acid-base titration curve. These properties are extremely important in that the charge differences allow for the separation of amino acids by elec-trophoresis or ion-exchange chromatography, an essential step in the determination of the composition and structure of proteins.

Another important property concerns the *stereochemistry* of amino acids. Since all protein amino acids, with the exception of glycine, contain as asymmetric carbon atom, they are optically active. This means that, like sugars, they are able to rotate the plane of plane-polarized light, a property shown by all compounds which can exist as two structures and are nonsuperimposable mirror images of each other. This rotation may be dextro- or laevorotatory and is again denoted by the signs $(+)$ or $(-)$ respectively. However, the absolute stereochemistry of any optically active compound is related to an arbitrarily chosen standard, the sugar glyceraldehyde (see p. 46). Without exception, all amino acids isolated from proteins by methods which do not allow racemization or conversions to occur have the L configuration:

COOH   COOH
| |
NH₂—C—H   NH₂—C—H
| |
CH₂OH   CH₃

L-Serine   L-Alanine

This designation is not related to the optical rotatory activity of the amino acids. Some D-amino acids are found in peptides isolated from the cell walls of certain bacteria.

The peptide bond and peptides

All amino acids have a number of chemical properties in common, which are related to the presence of the α-carboxyl and α-amino group, and, in addition, may have other characteristics due to the presence of a reactive side chain. These reactions are thoroughly discussed in biochemistry texts and will not be detailed here. The most important reaction is that between the α-carboxyl group of one amino acid and the α-amino group of another to form a secondary amide linkage known as *a peptide bond*. This bond is the basic linkage in the structure of proteins and leads to the formation of chains of amino acids known as *peptides*. This reaction and the nature of the peptide bond is shown below:

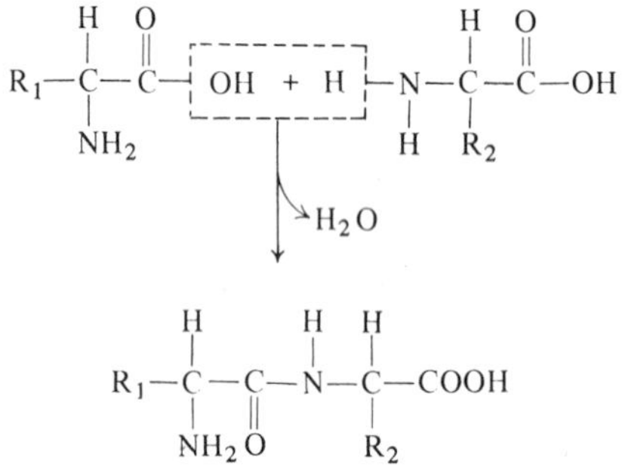

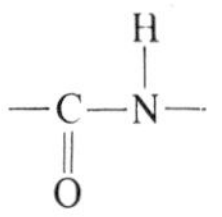

The peptide bond

The constituent amino acids of peptides are called *residues,* and the peptide may be a dipeptide, tripeptide or tetrapeptide depending on the number of residues present. Those containing large numbers of amino acids are known as polypeptides. Peptides are usually written with their terminal amino group on the left and the terminal carboxyl group on the right. The distinction between polypeptides and proteins is not clear cut but peptide chains with molecular weights above about 5 000 are usually called proteins.

The chemical properties of peptides reflect those of the constituent amino acids to a considerabe degree. They have high melting points, show optical activity and their acid-base behaviour is governed by both the free terminal amino and carboxyl groups and ionizing side chain (R) groups. The titration curves of short peptides are similar to those of the free amino acids and mixtures of peptides may be separated by electrophoresis or ion-exchange chromatography making use of differences in the acid-base properties of the different peptides.

It appears that most living tissues contain a number of low molecular weight peptides. One of universal occurrence is the tripeptide known as glutathione (λ-L-glutamyl-L cysteinylglycine) and although its biological rôle is not fully understood, it appears to function as a coenzyme in a few enzymic reactions. The synthesis of some small peptides involves a mechanism very different from that associated with protein synthesis. No RNA is involved, the formation of the peptide linkages being entirely under the control of enzymes, as shown for the synthesis of glutathione below:

L-Glutamic acid + L-cysteine + ATP $\rightleftharpoons$ γ-L-glutamyl-L-cysteine + ADP + P_i

γ-L-Glutamyl-L-cysteine + glycine + ATP $\rightleftharpoons$ glutathione + ADP + P_i

A number of larger peptides are known which show either physiological or antibacterial activity. These include the animal hormone vasopressin (containing nine residues), the adrenocorticotropic hormone (containing thirty-nine residues), and the antibiotics gramicidin A and bacitracin which contain ten and twelve residues respectively.

Classification and diversity of proteins

There is no really satisfactory scheme for the classification of proteins. The earliest attempt, which dates back to 1907, is based on their conformation or three-dimensional strucure and on their solubility properties. This divides proteins into two major groups, the *fibrous* and the *globular*. Fibrous proteins usually have polypeptide chains in parallel rows and joined by various cross-linkages to form sheets or fibres which are tough and insoluble in water and dilute salt solutions. Many are structural proteins such as α-keratin found in skin and feathers, and collagen from various connective tissues. In contrast, globular proteins contain polypeptide chains which are considerably folded to produce roughly spherical shapes. They are usually soluble in aqueous solutions and have a dynamic function in living systems. They include the enzymes, some animal hormones, transport proteins such as haemoglobin, and also some seed storage proteins.

Another method of classification divides proteins into *simple* and *conjugated* types. The former are proteins which yield only α-amino acids or their derivatives on hydrolysis. Conjugated proteins consist of a protein molecule which is bound to a non-protein molecule known as a prosthetic group. Simple proteins include both fibrous and globular proteins. Conjugated proteins are sub-divided according to the nature of the prosthetic group and include *nucleoproteins* (bound to nucleic acids), *glyco-* or *mucoproteins* (which have carbohydrate prosthetic groups), *lipoproteins* (bound to lipids, such as cholesterol and phosphatide), *chromoproteins* (which contain various metallo-pigments) and *phosphoproteins* (which contain phosphate groups other than those found in phospholipids of nucleic acids).

As we have seen above, there is a wide variation in the types of protein molecule found in living systems and this is reflected in the great diversity in biological functions attributed to proteins. Perhaps the most important class on a functional basis are the *enzymes* which catalyse a wide range of chemical reactions within cells; over a thousand have so far been described. The other major functional groups are: *storage* proteins, e.g. albumin from eggs and glutelins from wheat seeds; *transport* proteins, e.g. haemoglobin which transports oxygen in vertebrate blood; *structural* proteins, e.g. membrane proteins, glycoproteins found in cell walls of bacteria, collagen from connective tissues; certain *hormones*, e.g. insulin; proteins which function in *contractile* systems, e.g. dynein of flagella, myosin from muscle.

The structure of proteins

Protein molecules show four levels of structural organization which are given specific names. The *primary structure* is governed by the covalent bonding between the amino acids in the polypeptide chain, and refers to the number and sequence of amino acids in that chain. The term *secondary structure* relates to conformational changes in the polypeptide chain which produce extended or coiled structures by the formation of hydrogen bonds between the carbonyl oxygen and amide nitrogen atoms in the polypeptide chain. *Tertiary structure* refers to the folding and bending of the polypeptide chain produced by interactions between the side chains of the amino acids and results in the compact structure of globular proteins. Finally, the term *quaternary structure* is used to describe aggregates of individual polypeptide chains or subunits produced by intermolecular interactions. Most larger proteins contain two or more polypeptide chains.

Primary structure

The primary structure of proteins concerns the sequence of amino acids in the polypeptide chain and this has now been determined for a number of proteins after the pioneering work of Sanger in 1955. He determined the primary structure of insulin. Such determinations involve a number of standard procedures. Firstly, a protein may contain a number of polypeptides which are separated and the NH_2 and COOH terminal residues identified by various chemical tagging techniques. For example, Sanger used 2,4-dinitrofluorobenzene which reacts with α-amino groups to give yellow 2,4-dinitrophenyl derivatives which can be identified later by chromatography, although more sensitive techniques are now available. The intact polypeptide is fragmented into smaller peptides by the action of proteases such as trypsin, the peptides are separated and their terminal residues and amino acid content determined by successive degradation. Another sample of the polypeptide is then degraded by other proteolytic enzymes and the different group of peptides that result are analysed as before. From analysis of the 'overlap' produced by the two methods of cleavage, the sequence of the chain can be elucidated. The primary structure of the enzyme ribonuclease which has been determined by these techniques is shown in Fig. 2.29. Today, with the use of automated amino acid analysers, the complete determination of the amino acid composition of protein hydrolysates can be achieved within several hours, a process which earlier took many months or even years to complete. In addition, amino acid sequencers are also available which allow the automatic determination of the amino acid sequence

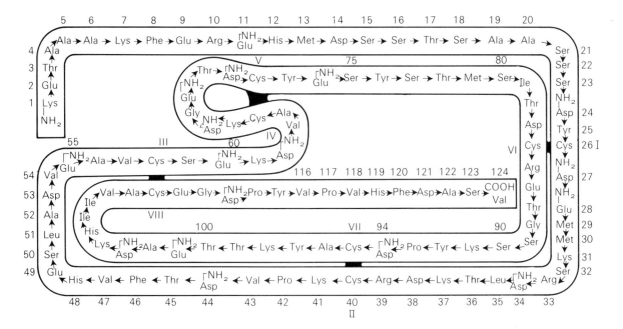

Fig. 2.29   The amino acid sequence and covalent structure of ribonuclease. Reproduced from Smyth, Stein and Moore (1963).

of peptides of some twenty to thirty residues. Such technology has made possible the determination of the primary structure of a large number of proteins.

From these studies a number of generalizations may be made concerning the primary structure of proteins. Each protein has a characteristic sequence of amino acids, which are all of the L-configuration. Not all proteins contain the twenty amino acids normally found in proteins, and there is little obvious regularity in the amino acid sequences. Functionally homologous proteins, such as the cytochromes, from different species contain the same amino acids at certain positions in the chain, whereas the other residues vary from species to species. However, polypeptides are unique structures and cannot be placed in general classes which differ only in random substitutions along the chain. Repeating sequences of amino acids in the chain occur only rarely. Some proteins such as histones contain high proportions of positively charged R-groups and are basic, while others which show a predominance of the negatively charged groups of glutamic and aspartic acids are acidic.

Secondary structure

The term secondary structure relates to the manner in which the peptide chain is folded and to the nature of the bonds which stabilize this structure. It has largely been determined by X-ray diffraction analysis initiated by Astbury in the 1930s and followed up later by Pauling and Corey. These studies showed that various fibrous proteins possessed a major repeating unit at periods of about 0·55 nm along their axes, together with a minor periodicity of about 0·15 nm. X-ray studies also showed that the C–N link of the peptide bond was not able to rotate freely although free rotation was possible around the bonds

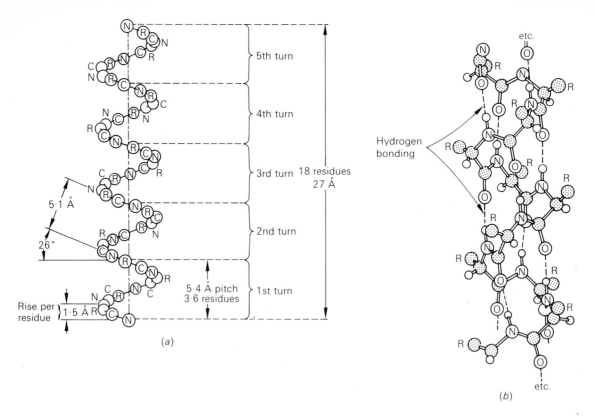

Fig. 2.30 The α-helix.
(a) Representation of a polypeptide chain in an α-helical configuration.
(b) Stabilization of an α-helical configuration by hydrogen bonding. The shaded spheres represent carbon-atoms or residues (R) of amino acids. From Pauling and Corey as redrawn by Anfinsen (1959).

Labels in figure (a): 5th turn, 4th turn, 3rd turn, 2nd turn, 1st turn, 18 residues 27 Å, 5·4 Å pitch 3·6 residues, 5·1 Å, 26°, Rise per residue 1·5 Å

Labels in figure (b): etc., Hydrogen bonding, etc.

of the α-carbon atoms. Pauling and Corey suggested that these observations could most easily be accounted for by a structure called the *α-helix* (Fig. 2.30). This helical arrangement contains approximately 3·6 amino acids per turn, with a spacing of 0·15 nm between each residue and 0·55 nm per turn. The structure is stabilized by the formation of hydrogen bonds between the hydrogen atom attached to the amide nitrogen and the carbonyl oxygen atom three residues back in the polypeptide chain. Both fibrous and globular proteins contain α-helix spontaneously.

However, not all proteins are able to assume an α-helix. Owing to the nature of their R-group, some proteins exist as a *random coil* which describes a flexible, changing structure. Other proteins, such as silk fibroin, show a different X-ray pattern with a repeating unit at every 0·7 nm. These are thought to be characteristic of extended polypeptide chains occurring in a parallel series and stabilized by intermolecular hydrogen bonds oriented at right angles to the long axis of the polypeptide chains (Fig. 2.31). This is called the *pleated sheet structure*, and may be termed parallel or antiparallel depending on whether the peptides all run in the same direction or have adjacent chains running in opposite directions.

Tertiary structure

The term tertiary structure describes the folding and bending of the polypeptide chain to give the tight, compact stuctures characteristic of

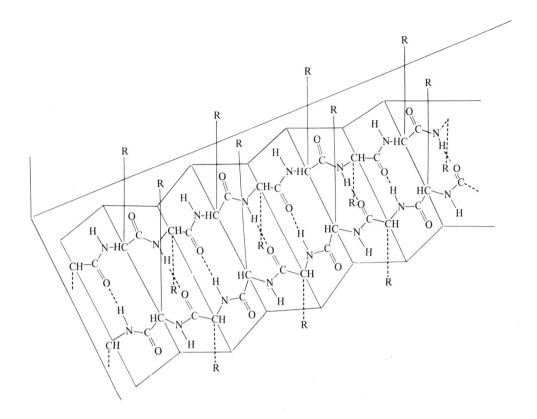

Fig. 2.31 The pleated-sheet
structure of proteins. Redrawn from
Karlson (1968).

most proteins. Again X-ray analysis, pioneered by Kendrew and
Perutz at Cambridge, has played a major part in the elucidation of
these structures. Interpretation of X-ray data is much more difficult in
the case of tertiary structures since the helical coils are orientated in
different planes, and it is necessary to label the proteins with highly
diffracting heavy metals to provide reference points for later interpretation. However, beginning with the studies on myoglobin by
Kendrew and Perutz, the tertiary structure of a number of globular
proteins has now been elucidated. Myoglobin is an oxygen-carrying
protein, found in muscle, which has a molecular weight of about
17 000 made up of a single polypeptide chain of 153 residues and an
associated iron-containing haem group. The tertiary structure of this
protein (Fig. 2.32) contains a number of features which are probably
common to many globular proteins. The molecule is compact with
room in its interior for four water molecules at most. All the polar
groups are at the surface of the molecule and are hydrated, whereas
the non-polar groups are shielded in the interior. Proline residues
which cause bends to occur in α-helical coils are present only at these
bends.

The characteristic tertiary structure of proteins is probably a
reflection of a number of interactions between residues to produce the
most stable thermodynamic conformation. These forces include hydrogen bonding between peptide linkages and between various side
chains, ionic bonding, apolar bonding between hydrocarbons, and

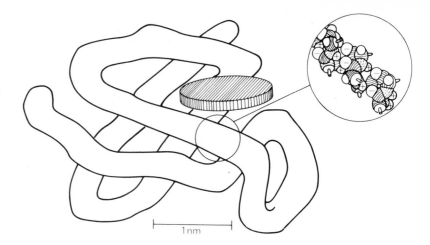

Fig. 2.32 The tertiary structure of myoglobin. Most of the long peptide chain is folded in an α-helical configuration as shown in the inset. The disc represents the haem group which is conjugated to the protein. Reproduced from Haggis et al. (1964).

1 nm

covalent bonding between cysteine residues to form disulphide linkages (Fig. 2.33). Hydrophobic interactions between non-polar R groups (apolar bonds) are by far the most important. This implies that the three-dimensional structure of proteins is solely determined by the primary structure, i.e. by the nature and sequence of the amino acids in the polypeptide chain. Some convincing evidence for this suggestion comes from the phenomenon known as the reversible denaturation of proteins. For example, the enzyme ribonuclease may be caused to unfold by treatment with mercaptoethanol and urea, which reduces the four disulphide bridges of the sulphydryl groups. However reoxidation of the molecule under controlled conditions can result in 100% recovery of the original conformation (Fig. 2.34). These results do not necessarily mean that this structure is formed spontaneously in the living cell but do indicate that the natural conformation is probably the most stable one under biological conditions.

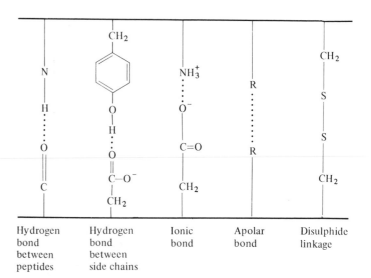

Fig. 2.33 Bonds responsible for the stabilization of the secondary and tertiary structure of proteins. Not all these bonds are present in all proteins.

Hydrogen bond between peptides	Hydrogen bond between side chains	Ionic bond	Apolar bond	Disulphide linkage

Fig. 2.34   Denaturation and renaturation of a globular protein. A native protein, with intramolecular disulphide bonds, is reduced and unfolded by treatment with β-mercaptoethanol and urea. After removal of these reagents, the protein will undergo spontaneous refolding and reoxidation. Reproduced from Anfinsen (1967).

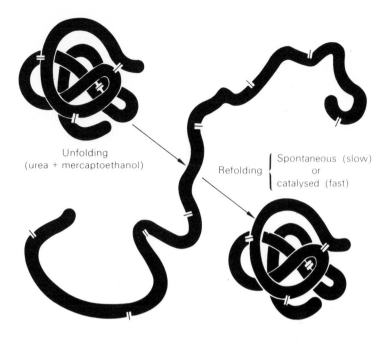

Unfolding
(urea + mercaptoethanol)

Refolding {
Spontaneous (slow)
or
catalysed (fast)
}

Quaternary structure

Most proteins with molecular weights of about 50 000 or above contain several polypeptide chains, each with their own primary, secondary and tertiary structures. The individual polypeptides are known as *promoters* or *sub-units,* the protein is called an *oligomeric* protein, and the arrangement of protomers to give the native conformation of the oligomer is termed its quaternary structure. The sub-units of an oligomeric protein are generally independent in so far as they are not normally linked by covalent bonds; they are probably held together by electrostatic attraction between charges at the surface of the sub-units and by the attraction between a few hydrocarbon residues not shielded in the interior of the sub-unit. These properties endow many oligomeric proteins with considerable stability and with self-assembling abilities when returned to favourable conditions after dissociation by exposure to extreme pH values or high salt concentrations. Thus the primary structure of the protein determines both the conformation of the peptide chain itself and the arrangement of sub-units within the quaternary structure.

Some proteins show a more complex organization at the level of quaternary structure. For example, the enzyme glutamic dehydrogenase has a molecular weight of over one million. It may be dissociated into sub-units, each with a molecular weight of about 300 000 made up from a number of polypeptide chains. A further level of organization still is shown by the fatty acid synthetase system isolated from yeast. This contains seven different enzymes in a single complex, each of these enzymes containing three sub-unit polypeptide chains. Thus the amino acid sequence of the polypeptide chains in this complex specifies an additional level of information above that so far considered; that is, the organization of the seven enzymes into a multienzyme complex.

Enzymes

Enzymes are protein molecules that catalyse the many thousands of chemical reactions that occur in living cells and so they may be considered as the primary instruments involved in the expression of the genetic information contained in nucleic acids. Less than sixty years ago the protein nature of enzymes was still in doubt. The first enzyme to be prepared in crystalline form was urease and this was achieved by Sumner in 1926. He concluded that the crystals were protein, but it was not until the work of Northrop and his associates in the early 1930s, who isolated and crystallized a number of proteolytic enzymes, that the protein nature of these compounds was firmly established. Today over one thousand enzymes have been described and many have been purified and crystallized. All have been shown to be proteins, and in recent years the amino acid sequence and three-dimensional structure of many enzymes have been elucidated. We will not attempt in this section to give a full and detailed account of the many aspects of enzymology, but merely attempt to summarize the most important features that relate to the rest of this book. For more detailed information of the reader is referred to the classic text of Dixon and Webb (1979) and recent texts on enzymes by Ferdinand (1976) and Fersht (1977).

Characterization of enzymes

Dixon and Webb (1964) have proposed that perhaps the most satisfactory definition of an enzyme is as follows: 'A protein with catalytic properties due to its power of specific activation'. It is first perhaps worth describing these and other terms that are widely used in relation to enzyme activity.

Firstly, what is meant by catalysis? In any population of molecules the kinetic energies will not be uniform. In order for a chemical reaction to occur, the molecules must attain a certain minimum energy value for chemical bonds to be altered to give the product of the reaction. This minimum energy value is known as the *activation energy* and the molecules are said to reach a *transition state*. This concept is illustrated in Fig. 2.35. One way of increasing the number of molecules that obtain the activation energy is by increasing the temperature and this, of course, results in an increase in the rate of the reaction. Catalysts are able to increase the rate of a chemical reaction without an increase in temperature. Since this rate increase cannot be due to a rise in the kinetic energy of the molecules, catalysts must function by lowering the level of the activation energy. They achieve this by combining with the reactants to produce a transition state complex which has a lower energy of activation than the uncatalysed reactants (Fig. 2.35). When the products are formed, the catalyst is regenerated

Fig. 2.35   Energy diagrams of catalysed and uncatalysed chemical reactions.

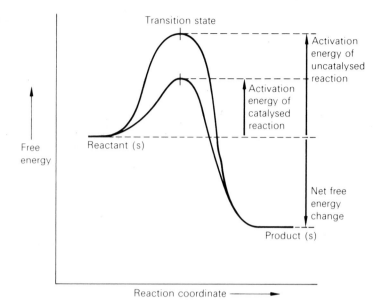

unchanged and can be used again. Chemical reactions may, of course, be catalysed by inorganic compounds as well as by enzymes, but, in general, enzymes are the more effective in reducing the activation energy. Most enzymic reactions, in fact, occur at 10^8–10^{11} times the rate of the corresponding non-enzymic reactions.

A number of other terms must be clearly defined at this point although they will be discussed more fully further on. The term *substrate* is used to describe the substance on which the enzyme acts and which is thereby activated. A number of enzymic reactions require, in addition to the enzyme and substrate, a non-protein structure or *cofactor* for the efficient catalysis of the reaction. These cofactors are frequently divided into two groups: *prosthetic groups* are tightly bound to the enzyme structure, while those cofactors that are readily separable from the enzyme are called *coenzymes*. Coenzymes usually undergo some structural alteration during the enzymic reaction but are regenerated in later reactions. The term *inhibitor* is used to describe a wide range of compounds that reduce the rate of an enzymic reaction. It is usually reserved for substances that act at low concentrations and may be quite specific for certain reactions. The term would not normally be employed to describe strong acids or bases which simply destroy the structure of the enzyme.

For the precise characterization of an enzyme it is clearly important to standardize the expression of the activity of an enzyme and the International Commission on Enzymes have recommended the following definitions:

'One *unit* (U) of any enzyme is defined as that amount which will catalyse the transformation of 1 micromole of substrate per minute, or, where more than one bond of each substrate molecule is attacked, 1 microequivalent of the group concerned per minute under defined conditions. The temperature should be stated, and it is suggested that where practicable it should be 30°C. The other conditions, including pH and substrate concentration should, where practicable, be optimal. Where inconvenient numbers would otherwise be involved, terms such as milli-unit (mU), kilo-unit (kU), etc., may be used.'

'*Specific activity* is expressed as units of enzyme per milligram of protein.'

'*Molecular activity* is defined as units per micromole of enzyme at optimal substrate concentration, that is, as the number of molecules of substrate transformed per minute per molecule of enzyme.'

As we can see from the above definitions it is important to state the conditions under which enzymic activity is measured. Enzymic activity is affected by a variety of factors of which the most important are the concentrations of substrate, cofactors and enzyme, and the pH and temperature at which the reaction is proceeding. The effect of these factors has been extensively studied and has provided considerable information about the nature and mechanism of enzyme-catalysed reactions.

Effect of substrate concentration

Whereas a non-catalysed chemical reaction shows a linear relationship with substrate concentration, the rate of an enzyme-catalysed reaction shows a saturation effect with increasing substrate concentration (Fig. 2.36). At low substrate concentrations the reaction rate is directly proportional to substrate concentration, but as the substrate

Fig. 2.36   Effect of increasing substrate concentration on the rate of an enzyme-catalysed reaction.

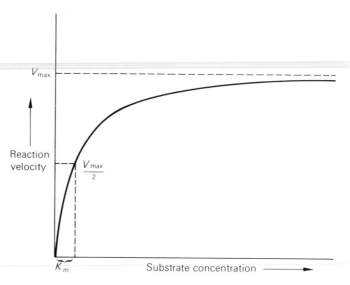

concentration is increased, the reaction rate declines giving a characteristic hyperbolic relationship. All enzyme reactions exhibit this saturation effect, although there is considerable variation in the substrate concentration needed to produce saturation. This observation is often consistent with a reaction that is controlled by a simple dissociation, i.e. $AB \rightleftharpoons A + B$, and led Michaelis and Menten in 1913 to propose the following formulation:

$$E \;+\; S \;\underset{k_2}{\overset{k_1}{\rightleftharpoons}}\; ES \;\overset{k_3}{\longrightarrow}\; E \;+\; P$$

enzyme   substrate   enzyme-substrate   enzyme   product
complex (1)

Since the initial velocity of the reaction is considered then the rate of formation of ES from $E + P$ is negligible and so is ignored. k_1, k_2 and k_3 are the velocity constants for the three reactions involved, and e, s and p are the concentrations of the *total* enzyme, free substrate and complex respectively. In the original derivation of the Michaelis–Menten equation, it was assumed that the rate of formation and dissociation of ES was always much greater than the breakdown of ES to $E + P$ and so any effect of k_3 on the equilibrium could be ignored. At equilibrium:

$$k_2 p = k_1(e - p)s \tag{2}$$

$$\frac{k_2}{k_1} = K_s(\text{dissociation constant of } ES) = \frac{(e - p)s}{p} \tag{3}$$

Rearranging,

$$p = \frac{es}{K_s + s} \tag{4}$$

The rate of the reaction, v, will be determined by the rate of breakdown of the complex and will be given by

$$v = k_3 p \tag{5}$$

Substituting for p in (4),

$$v = \frac{k_3 es}{K_s + s} \tag{6}$$

The maximum velocity, V_{max}, for the reaction will be reached when the enzyme is saturated with substrate, i.e. when $e = p$.

$$V_{max} = k_3 e \tag{7}$$

$$v = \frac{V_{max}\, s}{K_s + s} \tag{8}$$

This is known as the *Michaelis–Menten equation* and describes the relationship between the enzyme reaction rate and the substrate concentration. When s is equal to K_s, V is equal to $\frac{1}{2}V_{max}$. The substrate concentration that is found to give half the maximum velocity is known as the *Michaelis constant* (K_m), and under these conditions $K_m = K_s$. However, in other enzymic reactions the relationship between the dissociation constant and the velocity constants may be more complex and the K_m may not equal the dissociation constant of the ES complex. K_m is defined as the substrate concentration which gives half maximum velocity and is normally measured in moles/litre.

The above derivation assumes that ES always remains in equilibrium with E and S during the reaction and that the reaction velocity k_3

may be ignored. However this may not always be the case, particularly with enzymes having high catalytic values. Briggs and Haldane (1925) discussed the case in which the velocities k_1, k_2 and k_3 are similar, and considered the steady state phase in which the rates of formation and breakdown of ES are equal, so that p remains constant. The rate of formation of ES is given by

$$\frac{dp}{dt} = k_1(e - p)s \qquad (9)$$

and the rate of breakdown of ES by

$$\frac{-dp}{dt} = k_2p + k_3p \qquad (10)$$

At the steady state position

$$k_1(e - p)s = k_2p + k_3p \qquad (11)$$

$$\frac{k_2 + k_3}{k_1} = \frac{(e - p)s}{p} \qquad (12)$$

Rearranging,

$$p = \frac{es}{(k_2 + k_3)/k_1 + s} \qquad (13)$$

which is similar to equation (4) for the simple equilibrium state. The term $(k_2 + k_3)/k_1$ is called the Michaelis constant K_m and development as before gives:

$$v = \frac{V_{max}s}{K_m + s} \qquad (14)$$

In many reactions involving one substrate and one intermediate complex, k_3 can be neglected so that the Michaelis–Menten and Briggs–Haldane treatments give identical results, since K_m will then equal k_2/k_1, the dissociation constant K_s. However there are many more complicated systems in which the steady state concept gives complicated expressions and these have been dealt with in detail by Dixon and Webb (1979).

Determination of K_m. K_m is a characteristic of the enzyme which is independent of enzyme concentration but may be effected by pH and temperature. If an enzyme catalyses the transformation of more than one substrate, it shows a characteristic K_m for each substrate. The K_m value can be extrapolated from a simple plot of reaction velocity against substrate concentration (Fig. 2.36) although more accurate determinations can be made by transformations of the Michaelis–Menten equation (14) into a linear form. One such transformation is made by taking the reciprocal of both sides.

$$\frac{1}{v} = \frac{K_m + s}{V_{max}s}$$

and rearranging to give

$$\frac{1}{v} = \frac{K_m}{V_{max}} \cdot \frac{1}{s} + \frac{1}{V_{max}}$$

This is the *Lineweaver–Burk equation* and if $1/v$ is plotted against $1/s$, a straight line is obtained with slope of K_m/V_{max} which cuts the abscissa at a point corresponding to $-1/K_m$ (Fig. 2.37). This is probably the most widely used method for determining K_m values, but a number of other linear plots are described by Dixon and Webb (1979). The determination of K_m and V_{max} for enzymic reactions gives a measure of the formation of the *ES* complex and of the rate of its breakdown respectively. Although such determinations are essential to the study of enzyme kinetics, many enzymic reactions have been shown to be far more complex than the simple systems described above. For example, not one but a number of complexes may be involved, including those between enzyme and substrate, enzyme and product, and a transition state complex. In addition, many enzymic reactions involve more than one substrate and produce several products, and complexes may be formed in which more than one substrate or product is found. Clearly, considerable caution must be exercised in the interpretation of K_m and V_{max} determinations.

Fig. 2.37   Lineweaver–Burk reciprocal plot for the determination of K_m values.

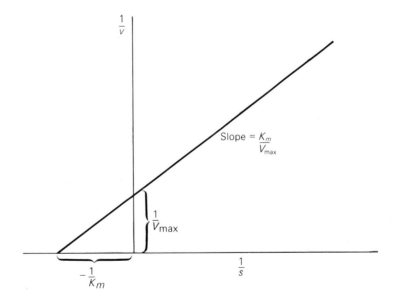

Effects of pH

Almost all enzymes are sensitive to changes in pH. They have a *pH optimum* at which activity is maximal and are only active over a limited pH range on either side of this optimum. The pH curves of a number of enzymes from a higher plant species are shown in Fig. 2.38; both the position and sharpness of the optimum vary considerably. pH in-

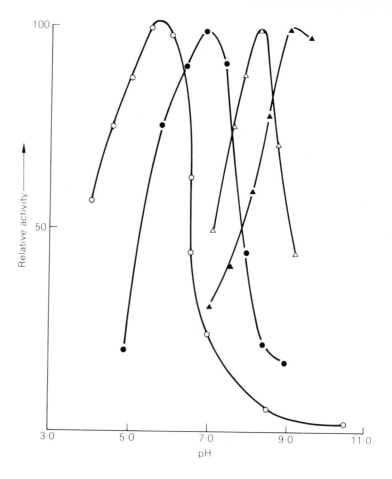

Fig. 2.38   pH-Activity curves for a number of enzymes assayed in pea root homogenates. ○—○, peroxidase; ●—●, ATP-ase; △—△, triose phosphate dehydrogenase; ▲–▲, malic dehydrogenase (malate as substrate). Data provided by Dr. R Sexton, University of Stirling.

fluences enzyme activity in a number of ways:

1. at extreme values, it brings about changes in protein structure that can irreversibly alter the stability of the enzyme;
2. it changes the ionization of the substrate, which may affect the binding of the substrate to the enzyme;
3. it changes the ionization of various groups in the enzyme molecule, which may affect the affinity of the enzyme for the substrate;
4. it changes the ionzation of the enzyme-substrate complex.

Factors (2) and (3) will affect the K_m value while factor (4) will influence the V_{max} of the reaction but have no effect on K_m.

Effect of temperature

In general, a rise in temperature results in an increase in the rate of enzymic reactions until a temperature is reached when the rate falls sharply. This effect is a result of the interaction of two factors. The first is a direct kinetic effect on the reaction rates and results in an increase in rate with an increase in temperature. The second is related to the stability of the enzyme and involves an inactivation of the enzyme with increasing temperature due to denaturation of the enzyme protein. The optimum temperature of an enzymic reaction is therefore a result of the balance between these two processes. Most enzymes are denatured rapidly at temperatures well below 70°C, although a few enzymes have been found to tolerate temperatures of 100°C.

Enzyme inhibition

An inhibitor may be defined as any substance that decreases the rate of a biochemical reaction, and the use of such substances has provided an important tool in the investigation of various biochemical processes. A study of the nature and action characteristics of inhibitors has given valuable information regarding the nature of the enzyme-substrate interactions and the mechanism of catalysis of enzymic reactions. The use of inhibitors to block specific steps in metabolic sequences, thus producing changes in levels of biochemical intermediates, has provided crucial evidence leading to the discovery of various metabolic pathways such as glycolysis and the TCA cycle.

Enzyme inhibitors may be roughly divided into two classes: *reversible* and *irreversible*. The term reversible inhibition usually means that enzyme activity may be restored by removing the inhibitor by methods such as dialysis. This means that there is an equilibrium relationship between inhibitor and free enzyme which can be analysed by Michaelis–Menten kinetics. Irreversible inhibitors cannot be removed by dialysis and this usually means that the functional groups of the enzyme molecule have been altered or destroyed. Examples of irreversible inhibitors are alkylating reagents such as iodoacetamide which react irreversibly with —SH groups of a number of enzymes, and the nerve gases which are organophosphorus compounds that combine irreversibly with the serine residues essential for the action of the enzyme acetylcholine esterase.

The most widely studied inhibitors are those which cause reversible inhibition, and these are classified into two main types: *competitive* and *non-competitive*. Competitive inhibitors compete with the substrate for the enzyme and so their effect can be reduced by increasing the substrate concentration. Non-competitive inhibitors combine with the enzyme in some way to reduce its activity, but are not affected by an increase in the substrate concentration. In addition, *uncompetitive* and *mixed* inhibitions occur.

Competitive inhibition. In competitive inhibition, the inhibitor binds with the enzyme producing an *EI* complex instead of the usual *ES*. The extent of inhibition depends on the relative concentrations of substrate and inhibitor, and may be negligible if the substrate concentration is high enough. Because of the affinity of enzyme for inhibitor this usually means that the inhibitor resembles the substrate in chemical structure. Perhaps the classic example of competitive inhibition is the inhibition of the enzyme succinic dehydrogenase by malonate. This enzyme catalyses the oxidation of succinate to fumarate and the similarity in the structure of succinate and malonate is shown below:

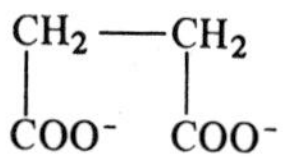

Succinate 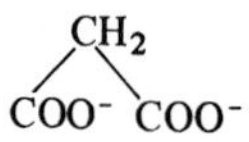   Malonate

The two molecules each have two ionizable carboxyl groups and compete for the same site on the enzyme molecule. Some other dibasic acids such as oxaloacetate may also act as competitive inhibitors of this enzyme, implying that the enzyme has two approximately spaced positively charged groups which attract the two carboxyl groups of the substrate. Other examples of competitive inhibitors are provided by fluorocitrate, which inhibits citrate utilization by the enzyme aconitase and the sulphonamide drugs which inhibit p-aminobenzoate utilization of folic acid formation. The similarity in structure of these compounds is illustrated below:

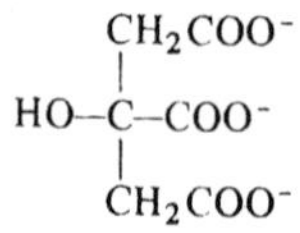

Citrate 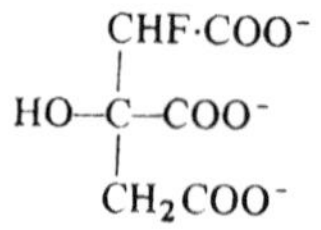   Fluorocitrate

p-Aminobenzoic acid 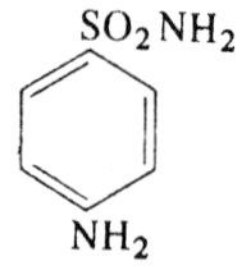   Sulphanilamide

Competitive inhibition can be studied by the Michaelis–Menten procedure by assuming that the inhibitor reacts with the enzyme to form a reversible *complex:*

$$E + I \underset{k_2}{\overset{k_1}{\rightleftharpoons}} EI$$

This cannot break down to form reaction products so that we can define an inhibitor constant K_i as $k_i = k_2/k_1$.

The full treatment of this relationship by both Michaelis–Menten and Briggs–Haldane kinetics can be found in Dixon and Webb (1979). Competitive inhibition may be demonstrated by using Lineweaver–Burk plots in the presence or absence of varying concentrations of inhibitor (Fig. 2.39). With fully competitive inhibition, the $V_{\max}$ will not be altered since the substrate concentration can be increased to completely overcome the effect of the inhibitor. However, there is an apparent increase in the K_m value as shown by the intercept on the $1/s$ axis, and this will increase without limit as the inhibitor concentration increases.

Non-competitive inhibition. Non-competitive inhibitors do not affect the formation of the enzyme-substrate complex but produce a lowering

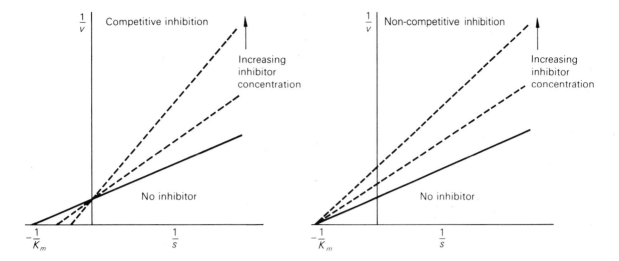

Fig. 2.39 The demonstration of competitive and non-competitive inhibition by Lineweaver–Burk plots.

in the V_{max}. The inhibitor presumably binds to the enzyme but not at the site of substrate attachment. It may bind to the free enzyme or to the enzyme-substrate complex, and may act by preventing the breakdown of the complex completely or reducing its rate of breakdown. The following reactions are therefore possible:

$$E + S \rightleftharpoons ES$$
$$E + I \rightleftharpoons EI$$
$$EI + S \rightleftharpoons EIS$$
$$ES + I \rightleftharpoons ESI$$

As with competitive inhibition, non-competitive inhibition may be demonstrated by Lineweaver–Burk plots in the presence of increasing concentrations of inhibitor (Fig. 2.39). Since the inhibitor does not affect the affinity of enzyme for substrate, there is no change in the K_m value. However the rate of breakdown of the ES complex is reduced, giving a reduction in the value for V_{max}.

Many non-competitive inhibitors act by combining reversibly with the —SH groups on enzymes that play an essential rôle in their catalytic activity. Many heavy metal ions such as silver and mercury act in this manner. This inhibition is usually reversible by the addition of metal-complexing agents such as ethylenediaminetetraacetic acid (EDTA). Other enzymes are inhibited by reagents that form inactive complexes with the metal of the prosthetic group which is essential for the catalytic activity. For example, cyanide and hydrogen sulphide are potent inhibitors of iron and copper-containing enzymes such as cytochrome oxidase and polyphenol oxidase, due to the inactivation or removal of the essential metal by the formation of a complex.

Mixed and uncompetitive inhibition. Mixed inhibition arises from the combination of partially competitive inhibition with a non-competitive type. The dissociation constant of S from EIS is different from that from ES and both the V_{max} and K_m are altered. Uncompetitive inhibition is rare with single substrate reactions but is more common in more complex reactions. The inhibitor combines only with the enzyme–substrate complex to give an inactive complex which is not reversed by increasing the substrate concentration. Uncompetitive inhibition is recognized in double reciprocal plots by a constant slope but decreasing V_{max} with increasing inhibitor concentrations.

Classification of enzymes

The discovery of an ever increasing number of enzymes makes it essential to standardize the classification and nomenclature of enzymes, and a standard system has now been recommended by the International Commission on Enzymes. This scheme divides enzymes into six major classes based on the chemical reaction that they catalyse, this being the specific property that distinguishes one enzyme from another. Each enzyme has a name made up from the name of the substrate and a word ending in -ase which describes the kind of reaction catalysed by the enzyme, e.g. orthophosphoric monoester phosphohydrolase. In addition to these systematic names which tend to be too long, many enzymes have shorter trivial names for everyday use. The Commission also recommends a numbering scheme which is closely related to the classification. The six major classes are given numbers from 1 to 6, followed by three other numbers denoting the various sub-classes. Thus the enzyme orthophosphoric monoester phosphohydrolase is given the number 3.1.3.2 and has the trivial name of acid phosphatase. The first number refers to the major class (the hydrolases), the second number to the sub-class of the hydrolases that acts on ester bonds, the third number to the sub-sub-class that hydrolyses phosphoric monoesters; the fourth figure is the serial number of the enzyme in its sub-sub-class. The complete classification of enzymes is detailed in Dixon and Webb, and we will be content here with a brief description of the six major classes which are as follows:

1. *Oxidoreductases*. These are enzymes that catalyse oxidation–reduction reactions and so are closely related to respiratory processes in the cell. They include the dehydrogenases, which bring about oxidation in conjuction with coenzymes such as NAD and NADP, which act as hydrogen acceptors; the peroxidases, which use H_2O_2 as oxidant; and the oxygenases, which use molecular oxygen directly to oxidize their substrates.
2. *Transferases*. These catalyse the transfer of various groups such as one-carbon groups (e.g. CH_3, CHO), glycosyl groups, or phosphate groups from substrate to acceptor molecule.
3. *Hydrolases*. These enzymes catalyse a wide range of hydrolytic reactions which involve the introduction of a molecule of water. They include enzymes that hydrolyse ester linkages, glycosidic bonds and peptide bonds.
4. *Lyases*. These enzymes catalyse the addition of groups to double bonds or the reverse. They include decarboxylases that release CO_2 from various substrates, and the aldolases that catalyse aldol condensations or the reverse and have an important rôle in

carbohydrate metabolism, including the breakdown of hexoses into 3-carbon units.

5. *Isomerases*. These enzymes bring about isomerizations and include the racemases and epimerases which catalyse the intercoversion of the sterioisomers of amino acids and sugars respectively.

6. *Ligases*. These enzymes catalyse the linking of two molecules coupled with the breakdown of ATP or other triphosphates. The energy liberated by the triphosphate supplies the energy for the condensation. These enzymes are involved in such important reactions as the linkage of amino acids to tRNA, the first stage in the synthesis of proteins.

Enzyme specificity

One of the most striking characteristics of enzymes is their marked substrate specificity, that is, their action is limited to one substrate or a few closely related substances. This property of enzymes is clearly essential for the construction and maintenance of an ordered metabolism from the wide range of chemical reactions occurring in living cells. The degree of specificity exhibited by enzymes shows wide variation. In many reactions, the enzyme shows nearly *absolute specificity*, acting on only one substrate. For example, most dehydrogenases have only one substrate and only one hydrogen acceptor. Other enzymes show a far broader *group specificity*, acting on a number of substrates which have a particular structural feature in common. This property is shown by the phosphatases which catalyse the hydrolysis of a wide range of phosphate esters. In addition to chemical specificity, enzymes also show a marked stereospecificity, attacking only one of the optically active isomers. Thus D-amino acid oxidase oxidizes only D-amino acids and glutamic dehydrogenase reacts only with L-glutamate.

What is perhaps of special interest is the ability of some enzymes to differentiate between chemically identical groups in symmetrical molecules; that is, the molecule behaves asymmetrically when acting as a substrate for the enzymic reaction. For example, consider the action of the enzyme aconitase which metabolizes citric acid in the TCA cycle (see Ch. 7). If citric acid is prepared from radioactively labelled oxaloacetate and unlabelled acetate, the α-ketoglutarate which is formed enzymically is found to be labelled only in the carboxyl group next to the carbonyl group (Fig. 2.40). Since citrate is a completely symmetrical molecule, its two terminal carboxyl groups would be expected to be indistinguishable chemically, resulting in the formation of α-ketolglutarate labelled equally in both carboxyl groups. However Ogston has proposed that this asymmetrical treatment of the substrate by the enzyme can be explained in terms of a three-point interaction between substrate and enzyme, meaning that binding occurs in only one orientation which places each of the identical groups in a unique position (Fig. 2.40).

A study of enzyme specificity is an important tool in the study of the nature of the enzyme–substrate complex; the higher the degree of

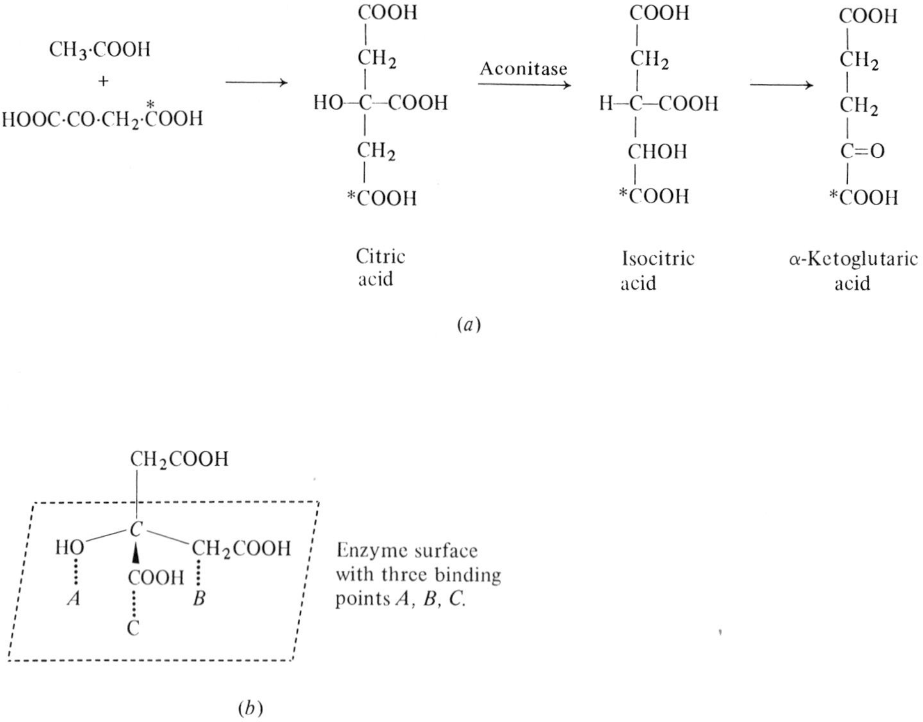

$$CH_3 \cdot COOH + HOOC \cdot CO \cdot CH_2 \cdot {}^*COOH \longrightarrow$$

Citric acid	Aconitase $\rightarrow$	Isocitric acid	$\rightarrow$	α-Ketoglutaric acid
COOH CH$_2$ HO—C—COOH CH$_2$ *COOH		COOH CH$_2$ H—C—COOH CHOH *COOH		COOH CH$_2$ CH$_2$ C=O *COOH

(a)

CH$_2$COOH
C
HO — CH$_2$COOH
COOH
A — B
C

Enzyme surface with three binding points A, B, C.

(b)

Fig. 2.40 Asymmetric behaviour of the enzyme aconitase. (a) Labelling pattern of α-ketoglutarate derived enzymically from citrate which in turn was prepared from unlabelled acetate and labelled oxaloacetate. Only the carboxyl group next to the carbonyl group is labelled. (b) Diagrammatic representation of the three-point attachment of citrate to the active site of aconitase allowing it to react asymmetrically.

specificity, the greater the chance of deducing the binding sites between the two structures. The nature of the functional groups can be investigated by replacing specific groups on the substrate molecule by a series of other groups. These sorts of investigation have shown that the enzyme specificity is determined by two structural features of the substrate molecule. The substrate possesses a distinct chemical bond that is attacked by the enzyme and also other functional groups that attach and orientate the substrate molecule at the catalytic sites of the enzyme.

Enzyme structure and mechanism of action

Enzymes are thus protein molecules which sometimes have an additional prosthetic group attached and that act as extremely efficient catalysts for specific chemical reactions. Their molecular weights range from about 10 000 to over one million and so they may contain many thousands of amino acids. Therefore it is clear that the substrate molecules are very small compared to the enzymes which presumably means that only a small part of the polypeptide chain is in contact with the substrate in the enzyme–substrate complex. This observation has given rise to the concept of the *active centre* or *site* of the enzyme, which may be described as the part of the enzyme molecule at which the process of activation and reaction of the substrate occurs and which is designed to accommodate the specific substrate molecule. Anfinsen (1967) considers that enzymes may contain three major aspects of structure: the components of the functionally active centres, the residues that support and stabilize these centres, and the portions of the amino acid sequence that provide a structural framework on which the functional portions are arranged. Many enzymes, however, appear to be unnecessarily large since large portions may be removed or modified with little alteration in their activity or specificity.

Since relatively few amino acids are associated with the active centre, it is essential that these be identified if we are to understand the mechanism of enzyme reaction. The functional amino acids may be identified by a number of methods.

1. It is sometimes possible to label the active centre, giving a product that is sufficiently stable to survive subsequent degradation of the enzyme. A good example is provided by the use of the organophosphorous compound diisopropylfluorophosphate (DFP) to label the active centres of various proteases and esterases. This compound phosphorylates serine resides and so inhibits a number of enzymes which have a serine residue at the active centres. The covalent derivative so formed is able to withstand the subsequent digestion of the enzyme molecule and is recovered as a small peptide which enables the amino acids around the serine residue to be identified. Table 2.11 shows the amino acid sequence surrounding the reactive serine residues of a number of enzymes studied by these methods.

Table 2.11 Amino acid sequences around the reactive, DFP-sensitive serine residues of some enzymes

Enzyme	Sequence
Chymotrypsin (ox pancreas)	Met-Gly-Asp-*Ser*-Gly-Gly-Pro
Trypsin (ox pancreas)	Glu-Gly-Gly-Asp-*Ser*-Gly-Pro-Val-Cys
Cholinesterase (horse serum)	Phe-Gly-Glu-*Ser*-Ala-Gly
Protease (*Aspergillus*)	Thr-*Ser*-Met-Ala
Alkaline phosphatase (*E. coli*)	Tyr-Val-Thr-Asp-*Ser*-Ala-Ala-Ser

There are a range of other procedures which aim to inactivate an enzyme by the modification of certain residues. For example, alkylation or *p*-hydroxymercuribenzoate are used to modify certain sulphydryl groups. These treatments presumably so change the character of the active centre that substrate binding does not occur. Interpretation of such experiments is however not always clear, since inactivation of the enzyme could result either from blocking of the active centre or from modification of the structural backbone which leads to a critical conformational change.

2. The lack of specificity mentioned above may be overcome in certain cases by a technique known as *affinity labelling*. The

enzyme is treated with a chemically reactive molecule which has been synthetically prepared to resemble the substrate. The label binds specifically to the active centre and forms covalent bonds with proteins residues on or near this site. For example, tosyl-L-phenylalanine chloromethyl ketone (TPCK) resembles the normal substrates for the enzyme chymotrypsin but contains a chloromethyl ketone group which is a potent alkylating agent. TPCK binds to the active centre but causes irreversible aklylation of an essential histidine residue and so inactivation of the enzyme. This residue may be isolated after complete hydrolysis of the peptide.

3. Kinetic studies can also yield important information regarding the nature of the active centre. These include studies of substrate specificity, the nature of competitive inhibitors, and of the effects of pH on V_{max} and K_m values. The latter enables the pK value of ionizable groups to be calculated and their identity deduced from known pK values of amino acids.

4. Finally, mention must be made of X-ray diffraction methods which have been crucial not only in the study of protein (enzyme) structure but also in the elucidation of the mechanism of enzyme action. Although it is clearly very difficult to study the structure of enzyme–substrate complexes due to the nature of the enzymic reaction, good approximations can be achieved by using substrate analogues or inhibitors. The substrate or its equivalent is 'seen' fitting into a single region on the protein surface, thus supporting the concept of specific binding sites.

Studies of the type described above have yielded a great deal of information about the nature of the functionally active residues in enzyme molecules. The side groups of serine, histidine, lysine and cysteine residues are frequently involved in the catalytic activity. These groups may be far apart in the linear sequence of the polypeptide chain but are brought close together at the active centre by coiling and folding of the chain. This is illustrated by the structure of the enzyme chymotrysin shown in Fig. 2.41.

Two generalized theories have been proposed to explain the relationship between enzyme and substrate at the active site. The *lock-key hypothesis*, proposed by Fischer in 1894, considers that the enzyme has a fairly rigid conformation into which only very specifically structured molecules may fit. A conformational change in the enzyme could then impose a strain on certain bonds in the substrate molecule which encourages formation of the transition state (see p. 92). However, it is difficult to see how a rigid structure could accommodate both the substrate and product of a reversible reaction. One modification of this concept proposes that it is the transition state complex that exactly complements the enzyme structure; certainly, for some reactions, synthetic and stable transition state analogues have been shown to bind more tightly to their enzymes than either substrate or product.

An alternative theory proposed by Koshland (1959) is known as the *induced-fit hypothesis*. This proposes that enzymes undergo a change in their three-dimensional structure as the substrate is bound to the

Fig. 2.41   A schematic drawing representing the three-dimensional structure of the enzyme α-chymotrypsin as determined by X-ray analysis. The enzyme consists of three polypeptides (A, B and C) linked by disulphide bridges. The residues linked by these bridges are numbered. Residues His (57) and Ser (195) function in the catalytic cycle. Redrawn after Matthews *et al.* (1967)

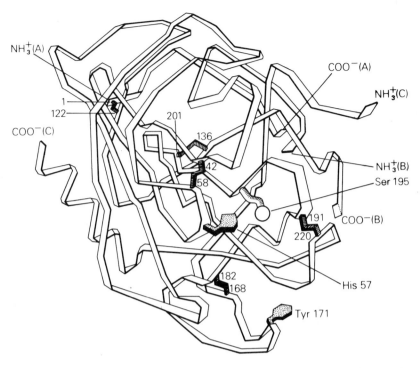

active centre. This conformational change produced by the substrate is thought to cause a precise alignment of the catalytic groups on the enzyme molecule to provide the optimal orientation for catalysis. X-ray diffraction studies of several enzymes (e.g. lysozyme, hexokinase, carboxypeptidase) support this idea in that their three-dimensional structure is significantly altered when a substrate analogue is bound at the active centre. Some other enzymes however show no evidence of induced fit. The relative contribution of factors such as strain (mentioned above) and induced fit may well vary between different enzymic reactions.

Chemical aspects of catalysis such as acid-base, nucleophilic and metal ion catalysis must also be considered in a detailed discussion of enzyme reaction mechanisms, although this will not be attempted here. The approximate details of the reation mechanisms for a number of enzymes are now known and the reader is referred to a number of useful texts on enzymes listed at the end of this chapter.

Enzyme systems and the control of enzymic reactions

As mentioned previously, the high specificity of enzymic reactions is a necessary requirement for an ordered metabolic process in which the product of one reaction becomes the substrate of the next. If any enzyme in the chain is lost or inhibited the whole process is disrupted. Each enzyme in the chain is therefore linked to another enzyme through the existence of a common substrate, since most enzymes catalyse both directions of a particular reaction. Thus, even in a system of mixed soluble enzymes there will be a degree of organization imposed on the system by the chemical specificity of the constituent enzymes. The product of one enzymic reaction will fall within the specificity limits of another enzyme, which will therefore be capable of carrying on the transformation of the initial metabolite. The relatively rapid rates of diffusion of small substrate molecules in an aqueous medium will enable the metabolites to move from one enzyme to another very quickly.

A higher level of organization is shown by some *multienzyme complexes* in which the enzymes involved in a particular metabolic process are physically bound together into a larger unit. Perhaps the most thoroughly investigated complex of this kind is the fatty acid synthetase system isolated from yeast, in which seven enzymes responsible for the synthesis of fatty acids are associated together in a large molecular weight globular unit. Any attempt to dissociate the complex results in a complete loss of synthetase activity. Such a system presumably has an advantage in that the substrate molecules have shorter distances to diffuse during the sequence of reactions.

The highest level of organization is exhibited by the localization of enzyme systems in discrete organelles, and in particular, when these enzymes show further structural organization at specific sites on the organelle membranes. The mitochondria, for example, show a high degree of structural organization. These organelles contain the enzymes of the TCA cycle responsible for the oxidation of acetyl groups derived from the catabolism of carbohydrates and lipids, and the enzymes of the respiratory chain responsible for transferring electrons from the oxidized substrates to oxygen. Thus these closely related enzyme systems are kept close to one another in a discrete unit. In addition, the enzymes of the respiratory chain show a high degree of organization within the inner membrane of the mitochondria, presumably for an efficient transfer of electrons from one carrier molecule to another (see Ch. 7).

The control of enzymic reactions, particularly of multienzyme systems, is clearly a very complex area of discussion involving the interaction of a large number of factors. It can be easily discussed by considering several distinct areas of control, although the overall rate of a particular series of enzymic reactions will clearly be a function of these various regulatory mechanisms.

The simplest consideration of the regulation of enzymic reactions involves the kinetic characteristics of the individual enzymes. As we have seen, each enzyme has its own distinct characteristics in relation to specificity and affinity for the substrate, pH, and cofactor and inhibitor effects. The overall rate of a metabolic process involving a number of enzymic reactions will therefore be a function of such factors as the concentration of the individual enzymes, temperature, pH and the concentration of substrates, cofactors and inhibitors. Almost all of these factors will in turn be influenced by a variety of other intracellular processes which are taking place at the same time. For example, the concentration of enzymes will be influenced by their rate of synthesis and breakdown, and will be discussed more fully later on. The intracellular pH will be a function of a wide variety of processes which free or absorb hydrogen ions. The levels of substrates will be influenced by any other process in the cell which produces or utilizes the various intermediates of the metabolic sequence in question. Again, the levels of cofactors will be strongly influenced by the requirements of other reactions using common cofactors. An example of this is seen in Chapter 6 where the rate of utilization of NADPH by various synthetic reactions probably has a marked effect on the rate of glucose breakdown by the pentose phosphate pathway which produces NADPH. Similarly there are a number of examples where ATP utilization has a marked effect on ATP synthesizing reactions (see Ch. 7). The overall energy status of a cell (the *energy charge*, see p. 254) may regulate pathways that utilize or produce high energy bonds, either by a mass action effect or, more subtly, by allosteric control (see below). In some cases it has been found that specific intermediates of one metabolic process act as inhibitors of another pathway and so may exert a major controlling influence in this manner. Thus the overall rate of a metabolic pathway will be a function of the prevailing conditions in the intracellular environment acting on a series of enzymic reactions.

A second major control mechanism operates through the function of special *regulatory or allosteric enzymes*. These enzymes are usually the first enzymes in a metabolic pathway and are specifically and effectively inhibited by some product, usually the end product, of the pathway. These end products are usually structurally unrelated to the substrate of the regulatory enzyme but are able to bind to the enzyme and reduce its catalytic activity. This is called *end-product* or *feedback inhibition* and the metabolite that acts as the inhibitor is known as an *effector* or *determinant*. Regulatory enzymes are thought to possess an active site which binds the substrate, together with a separate distinct *allosteric site* which binds the effector. The latter is thought to produce a change in the conformation of the enzyme resulting in a less catalytically active enzyme molecule. There are a number of types of allosteric effects which show that there is considerable deviation from the interaction described above. The effector may be a molecule other than the substrate and this is known as an *heterotropic* effect; this may be inhibitory or may produce an activation of the enzyme. In other cases, the substrate acts as an effector and results in an increased rate of reaction. This is known as an *homotropic* effect. In some instances, allosteric enzymes exhibit both these effects, being stimulated by the substrate acting as an effector and inhibited by a second, structurally different, effector.

Allosteric enzymes do not usually obey the classical Michaelis–Menten relationship between substrate concentration and velocity (see p. 94). Many, particularly homotropic enzymes, show a sigmoid curve for substrate saturation rather than the normal hyperbolic one (Fig. 2.42.) This is interpreted as positive cooperativity where the binding of one substrate molecule facilitates the binding of subsequent substrate molecules. Negative cooperativity can also occur and results in a flattened curve for initial velocity against substrate concentration. (Fig. 2.42). All allosteric enzymes are assumed to consist of two or

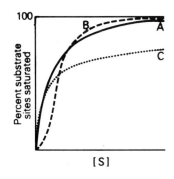

Fig. 2.42   Comparison of idealized plots for substrate saturation for a non-regulatory enzyme obeying Michaelis–Menten kinetics (A), a regulatory enzyme showing positive co-operativity (B), and a regulatory enzyme showing negative cooperativity (C).

more identical sub-units which can exist in at least two different conformational states (e.g. high- and low affinity forms). Two major models have been proposed to explain the cooperative interactions which occur. In the *symmetry* model proposed by Monod, binding of a substrate molecule to one sub-unit causes all of the sub-units to simultaneously change their conformational state. In contrast, the *sequential* model of Koshland postulates that sub-units undergo individual changes so that intermediate positions between all-on or all-off states may occur. Other possibilities have also been proposed and perhaps no single model will adequately explain the range of allosteric interactions.

The effector molecules may clearly exert a rapid and powerful controlling influence at certain important points in metabolism. A number of examples of allosteric control will be mentioned in the following chapters and a few are listed in Table 2.12. There are a number of cases in which ATP acts as an allosteric inhibitor of enzyme systems involved in the production of ATP (see 'Control of Glycolysis', p. 253). The pathway is controlled not only by the availability of substrate but, in addition, by the concentration of the product.

Table 2.12   Characteristics of some allosteric enzyme systems. Data taken from Monod, Wyman and Changeux (1965)

Enzyme	Substrate	Inhibitor	Activator
Phosphofructokinase (guinea pig heart)	Fructose 6-P ATP	ATP	5'-AMP
NAD-Isocitric dehydrogenase (*Neurospora*)	L-Isocitrate NAD$^+$	–	Citrate
Glycogen synthetase (yeast, lamb muscle)	UDP-glucose	–	Glucose 6-P
Glutamate dehydrogenase (beef liver)	Glutamate	ATP, GTP, NADH Estrogens Thyroxine	ADP, Leucine Methionine
Phosphorylase b (rabbit muscle)	Glucose 1-P Glycogen P_i	ATP	5'-AMP
Biosynthetic L-threonine deaminase (yeast)	L-Threonine	L-Isoleucine	L-Valine
Aspartate transcarbamylase (*E. coli*)	Aspartate Carbamyl phosphate	CTP	ATP
Deoxythymidine kinase (*E. coli*)	Deoxythymidine ATP or GTP	dTTP	dCDP

The third major area of enzyme regulation concerns the control of the amount of an enzyme within a cell. This may be achieved in various ways: by the degradation or inactivation of an enzyme, or by its activation or synthesis. As with other macromolecules, enzymes undergo metabolic turnover and estimates for the half-life of various enzymes range from several hours to several days. These estimates are usually made by measuring the fall which occurs in enzyme activity in the presence of inhibitors of protein synthesis; this approach is open to criticism since many related processes may be affected. In some cases the presence of the inhibitor *prevents* a fall in enzyme activity, and it is assumed that this is due to an inhibition of the synthesis of proteolytic enzymes which specifically remove other enzymes (see Wallace, 1977). Examples from plants are limited although there is evidence of a specific protease from maize roots which inactivates nitrate reductase.

Enzymes may be activated or inactivated by chemical modifications that are catalysed by other enzymes, and it is sometimes difficult to distinguish these changes from synthesis or degradation. The best known examples of modification involve phosphorylation or adenylation, i.e. the transfer of phosphate or adenylyl groups from ATP to the enzyme. Over twenty enzymes are now known to undergo phosphorylation and dephosphorylation (Krebs and Beavo, 1979). For example, mammalian glycogen phosphlorylase may be dephosphorylated (inactivated) or rephosphorylated (activated) by the enzymes phosphlorylase phosphatase and phosphorylase kinase respectively and, for this process to have a regulatory rôle, these enzymes in turn are subject to hormonal or allosteric control. Evidence for similar regulatory mechanisms in plants is sparse although the light-activation of enzymes of photosynthetic carbon metabolism could well involve chemical modification by reduction or phosphorylation (see Ch. 8).

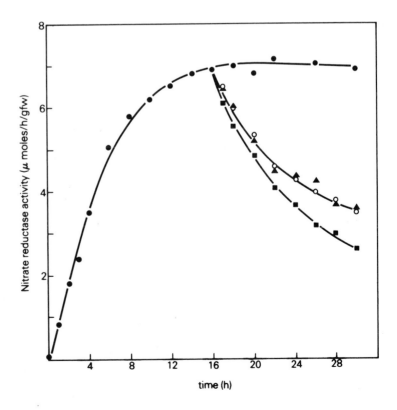

Fig. 2.43 Time course of nitrate reductase induction and decay in *Lemna minor*. Plants, which had been grown on 1 mM asparagine as a nitrogen source, were transferred to 5 mM NO₃ at zero time. After 16 hours, some plants were transferred back to the asparagine medium (▲), while others were treated with 6-methyl purine which blocks RNA synthesis (O) or with cyclohexamide which inhibits protein synthesis (■). Reproduced from Stewart and Rhodes (1977)

111

The control of enzyme synthesis must presumably be exerted through factors in the external environment and so is a much slower mechanism. Enzymes that show a marked increase in their rate of synthesis in response to the addition of substrate are known as *induced enzymes*. In contrast, enzymes that show little response to environmental changes, being invariably present at nearly constant levels, are known as *constitutive enzymes*. Numerous examples of inducible (and repressible) enzymes have been found in bacteria, fungi, algae, and certain mammalian systems, but there are far fewer examples so far from higher plants (see Marcus, 1971; Stewart and Rhodes, 1977). The only well established enzymes to be substrate induced in higher plants are nitrate and nitrite reductase, which are primary enzymes of nitrogen assimilation. These enzymes are induced by the addition of nitrate, and this induction may be repressed by certain amino acids, amides and ammonia (Fig. 2.43). There is also evidence that certain enzymes in some higher plant systems may be induced by hormones. Thus, in certain cereal seeds, gibberellic acid appears to induce a series of hydrolytic enzymes in the aleurone cells (see Ch. 4), while ethylene may induce cellulase synthesis during leaf abscission in some species (see Ch. 10). This subject of enzyme induction is clearly an important area of future research for plant biochemists.

Enzyme cofactors

Many enzymic reactions require, in addition to the enzyme and substrate, a non-protein substance for the catalytic reaction to proceed. These substances are known as *cofactors* and are usually divided into two major groups:

1. *Activators* are simple inorganic substances such as magnesium and potassium ions, which probably affect the conformation of the enzyme producing a more catalytically active state or interact with the substrate in substrate binding; Table 2.13 shows the ion activation requirements of some plant enzymes.

2. *Coenzymes* are generally more complex organic compounds which show considerable specificity and play a more active rôle in the reaction itself, usually as carriers of specific chemical groups. They may be changed structurally during the course of the reaction but are invariably regenerated in subsequent reactions. Frequently a further distinction is made between coenzymes, those which are tightly bound to the enzyme molecule being called *prosthetic groups*. These include haem groups which are covalently bound to the polypeptide chains of a number of enzyme molecules and function as electron carriers undergoing alternate oxidation and reduction. Many coenzymes, however, are more loosely bound and there is no sharp division in these cases between prosthetic groups and coenzymes. The only clear distinction that can be made is between a true prosthetic group

which remains attached to one enzyme throughout its whole catalytic activity and those coenzymes, like NAD and ATP, which move from one enzyme to another to fulfil their catalytic functions. In some reactions, however, NAD may also be firmly bound to the enzyme, e.g. glyceraldehyde-3-phosphate dehydrogenase (see p. 246).

Table 2.13 Ion requirements of some enzymes. The asterisk indicates that the enzyme contains the metal. Otherwise the ion is added as an activator

Ion	Enzyme
Zn^{++}	Alcohol dehydrogenase* Lactate dehydrogenase* Glyceraldehydephosphate dehydrogenase*
Cu^{++}	Catechol oxidase* Cytochrome oxidase* Ascorbate oxidase*
Mg^{++}	Many phosphotransferases (e.g. hexokinase) Many phosphohydrolases (e.g. phytase) Many ligases (e.g. succinyl-CoA synthetase)
Mn^{++}	Isocitrate dehydrogenase Some phosphotransferases
Ca^{++}	Apyrase Phospholipase Lipase
Fe^{++} or Fe^{+++}	Cytochromes* Peroxidase* Catalase* Ferredoxin*
K^+	Pyruvate Kinase Acetyl-CoA synthetase Aldolase γ-Glutamyl-cysteine synthetase

More detailed discussions of enzyme cofactors can be found in the texts on enzymes and biochemistry mentioned at the end of this chapter. We will confine ourselves here to a brief description of the properties of the coenzymes most frequently mentioned in the rest of this book.

Adenosine triphosphate

The energy available from biological oxidations is largely transformed into chemical energy by conversion of adenosine diphosphate (ADP) into the energy-rich adenosine triphosphate (ATP). The energy available from the hydrolysis of ATP may then be utilized in the various energy-requiring biosynthetic processes which take place in living cells. ATP is a nucleotide (see p. 69) which is composed of a purine base (adenine), a five-carbon sugar (ribose) and three phosphoric acid groups attached to position 5 of the ribose:

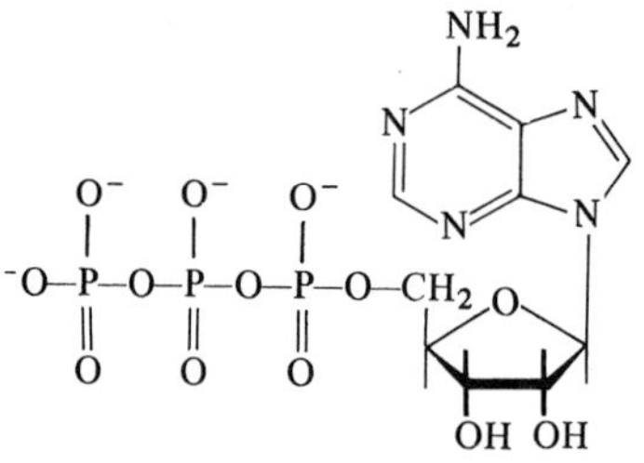

ATP is the most well known example of an *energy-rich* compound. The term energy-rich bond refers to the large negative free energy that is released on hydrolysis and makes these compounds highly reactive. In practice, organic compounds exhibit a wide range of free energy changes on hydrolysis and there is no clear quantitative distinction between energy-rich and energy-poor bonds. Generally, however, those molecules which show a standard free energy change (see p. 126) on hydrolysis of more than 29 kJ/mole are considered to be energy-rich (Table 2.14). On these terms, both of the terminal pyrophosphate linkages of ATP are energy-rich, so that this compound constitutes a versatile carrier of energy which may be transferred from one process to another by the transfer of phosphate groups catalysed by a group of enzymes known as the kinases. Although ATP is generally considered to be the most important phosphate carrier, the other nucleoside triphosphates may also function in this manner.

Table 2.14 Standard free energy of hydrolysis of some important biological compounds

Compounds	Standard free energy charge ($\Delta G°$) at pH 7.0	
	kcal mol^{-1}	kJ mol^{-1}
Phosphoenolpyruvate	-12.8	-53.6
Creatine phosphate	-10.3	-43.1
Acetyl phosphate	-10.1	-42.3
Amino acid esters	-8.4	-35.2
Acetyl-CoA	-8.2	-34.3
Uridine diphosphoglucose	-7.6	-31.8
Adenosine triphosphate (ATP)	-7.3	-30.6
Glucose-1-phosphate	-5.0	-20.9
Glutamine	-3.4	-14.2
Glucose 6-phosphate	-3.3	-13.8
Glycerol 1-phosphate	-2.2	-9.2
Peptides	-0.5	-2.1

ATP and the other adenosine derivatives are now known to participate in a wide range of enzymic reactions (many of which will be described in the later chapters) and to be involved in respiration, photosynthesis, and the synthesis of all major macromolecules found in plant cells. These ATP-linked reactions can be broadly divided into two classes. There are those reactions in which ATP is hydrolysed to produce the driving force for energy-requiring reactions, usually releasing ADP, which can then be coupled to energy-conserving processes and ATP regenerated:

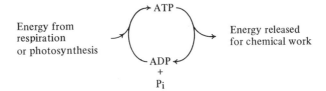

The enzymes responsible for the utilization of ATP as an energy source include the ligases and are illustrated by the reactions of a peptide synthetase and a carboxylase shown below:

ATP + L-glutamate + L-cysteine = ADP + P_i + γ-L-glutamyl-L-cysteine

ATP + pyruvate + CO_2 + H_2O = ADP + P_i + oxaloacetate

Alternatively, ATP may be involved in the transfer of some part of its molecule to an acceptor molecule, catalysed by the transferase enzymes known as the kinases. This group may be further sub-divided depending on the size of the group donated by the ATP. There are four types of transfer involving ATP.

1. Transfer of the orthophosphate group with the release of ADP, e.g.

   ATP + D-glucose = ADP + D-glucose 6-phosphate

2. Transfer of the pyrophosphate group with the release of AMP; e.g.

   ATP + thiamine = AMP + thiamine pyrophosphate

3. Transfer of adenosyl monophosphate group with the release of pyrophosphate, e.g.

   ATP + RNA = pyrophosphate + RNA_{n+1}

   ATP + FMN = pyrophosphate + FAD

4. Transfer of adenosyl group with release of both orthophosphate and pyrophosphate, e.g.

   ATP + L-methionine + H_2O = P_i + pyrophosphate + S-adenosyl-methionine

Uridine nucleotides

Uridine nucleotide coenzymes have a central rôle in carbohydrate metabolism in plant cells (see Chs. 6 and 10); the third phosphate group of UTP can be exchanged for a sugar molecule giving a variety of uridine diphosphate sugars containing an energy rich bond as follows:

UTP + sugar 1-phosphate = UDP-sugar + pyrophosphate

The structure of UDP-glucose is shown below:

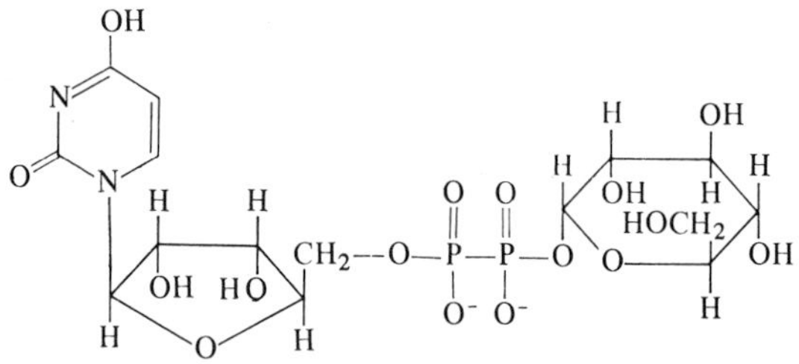

These UDP-sugar coenzymes have a number of important rôles in carbohydrate metabolism. They are involved in the transformation of the sugar molecules themselves through the action of the epimerases, e.g.

UDP–Glucose $\rightleftharpoons$ UDP-galactose

They also function as glycosyl group donors in the synthesis of a variety of carbohydrates, e.g.

UDP-Glucose + fructose = sucrose + UDP

UDP-Glucose + $(glucose)_n$ = $(glucose)_{n+1}$ + UDP

Pyridine (nicotinamide) nucleotides

There are two pyridine nucleotide coenzymes which act as hydrogen carriers in a number of oxidation-reduction reactions catalysed by the dehydrogenase enzymes. One is called *nicotinamide adenine dinucleotide* (NAD) and is composed of one molecule of nicotinamide, two molecules of ribose, one molecule of adenine and two phosphate molecules arranged in the following structure:

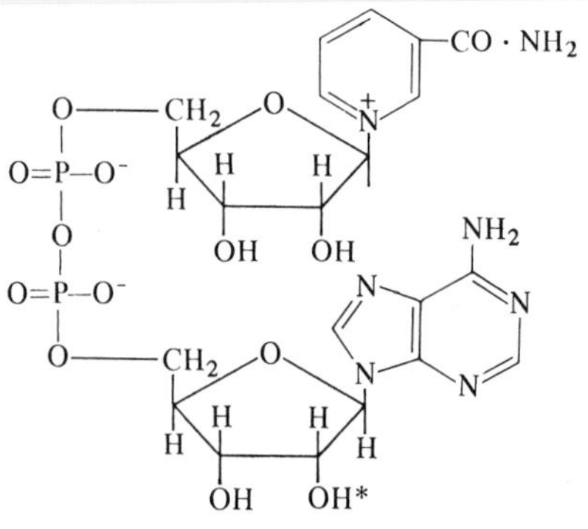

Sometimes this coenzyme is referred to by the older name of diphosphopyridine nucleotide (DPN) or coenzyme I (CoI). The pyridine nucleotides were, in fact, the first coenzymes to be discovered.

The second of these coenzymes is *nicotinamide adenine dinucleotide phosphate* (NADP). This carries an additional phosphate group at the C-2 position of the ribose portion of the adenosine group (marked*) and is also known as triphosphopyridine nucleotide (TPN) or coenzyme II (CoII).

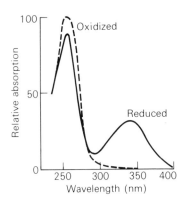

Fig. 2.44   Absorption spectra of NAD⁺ and NADH.

These coenzymes have been found in all types of cell and participate in oxidation-reduction reactions of a wide range of substrates. A typical reaction, that catalysed by malate dehydrogenase, is shown below:

$$\begin{array}{c} CH_2 \cdot COOH \\ | \\ CHOH \cdot COOH \end{array} + NAD^+ = \begin{array}{c} CH_2 \cdot COOH \\ | \\ CO \cdot COOH \end{array} + NADH + H^+$$

Malate   Oxaloacetate

Normally the enzymes are specific for either NAD or NADP but there are a few exceptions, such as glutamate dehydrogenase, that can be coupled to either coenzyme. Generally, NAD-linked dehydrogenases are associated with respiratory reactions whereas NADP-linked reactions are frequently involved in reductive steps in biosynthetic processes, such as those found in photosynthesis and the biosynthesis of lipids.

The oxidized forms of these coenzymes have a characteristic absorption spectrum in the ultra-violet region with a maximum at a wavelength of about 260 nm (Fig. 2.44); when reduced an additional peak appears at 340 nm. This property is extremely important for two reasons. Firstly it provides a means of assaying dehydrogenase activity by measuring changes in absorbance at 340 nm in a spectrophotometer (see p. 36). Secondly, it is important evidence regarding the part of the coenzyme involved in the enzymic reaction. Absorption at 260 nm is due largely to the adenine moiety, whereas the peak at 340 nm is characteristic of a dihydronicotinamide, indicating that this part of the coenzyme is involved in the oxidation-reduction reaction. The transfer of hydrogen is now known to involve the following changes in the pyridine ring of the nicotinamide:

NAD⁺ + 2H ⇌ NADH + H⁺

In addition to specificity for either NAD⁺ or NADP⁺, enzymes also show stereospecificity towards these coenzymes, as demonstrated by the use of tritium- or deuterium-labelled substrates. The oxidation of a deuterium-labelled substrate has shown that dehydrogenases transfer hydrogen to either the A or B side of the ring, but not both:

A-form   B-form

Only one enzyme, dihydrolipoyl dehydrogenase, is known that shows no stereospecificity and so produces both types of reduced coenzyme.

Flavin coenzymes

Like the pyridine nucleotides, these coenzymes are also involved in oxidation–reduction reactions, but are usually much more tightly bound to the enzymes and are regarded as prosthetic groups. They normally remain attached to the enzyme during the catalytic reaction and during purification of the enzyme. There are two flavin coenzymes both containing riboflavin (vitamin B_2) as part of the molecule: these are *flavin mononucleotide* (FMN) and *flavin adenine dinucleotide* (FAD). The former is incorrectly named since it contains no nucleotide as shown by the structures below, although the name is still commonly used.

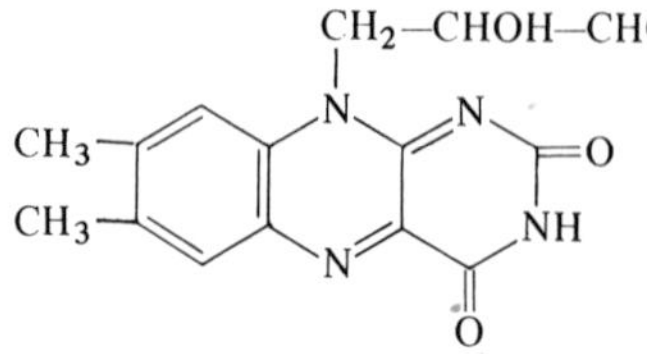

Flavin mononucleotide (FMN)

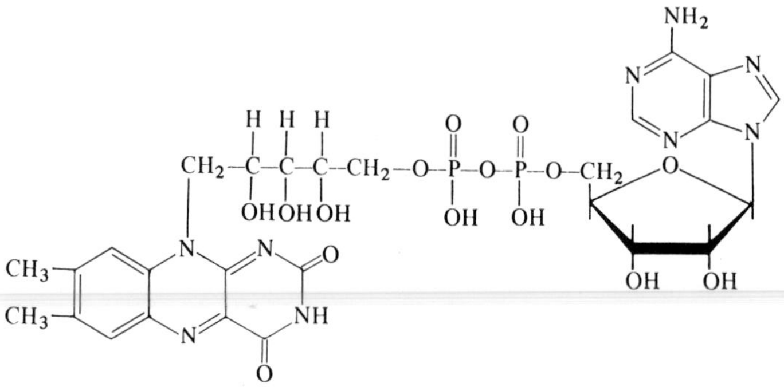

Flavin adenine dinucleotide (FAD)

About forty flavoprotein enzymes have so far been described involving a wide range of substrates; about one-quarter of these utilize FMN and the rest FAD. A list of some important enzymes utilizing flavin coenzymes is shown in Table 2.15. The oxidation–reduction reaction is known to involve the isoalloxazine ring system of the riboflavin part

Table 2.15 Some example of flavin-requiring enzymes

FAD containing	FMN containing
Pyruvate dehydrogenase (plants and animals)	
Succinate dehydrogenase (plants and animals)	Glycollate oxidase (plants)
NADPH diaphorase (plants)	L-Amino acid oxidase (animals)
Glutathione reductase (plants and animals)	Quinone reductase (bacteria)
NADH nitrate reductase (plants and animals)	
Nitrite reductase (bacteria)	

of the coenzyme as shown below, although the reaction probably proceeds by two consecutive one-electron transfer reactions:

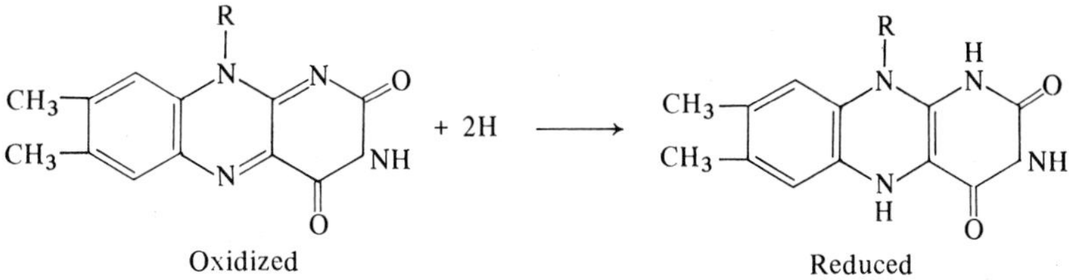

Oxidized   + 2H ⟶   Reduced

Haem coenzymes

A number of enzymes and biologically important substances contain haem groups as an essential functional part of the molecule; these include the cytochromes (discussed fully in Chs. 7 and 8), the enzymes catalase and peroxidase, the chlorophyll pigments (see Ch. 8) and haemoglobin. Haem groups consist of a porphyrin ring complexed with various metal ions. The porphyrin system is derived from four pyrrole rings linked by methine ($=CH—$) groups:

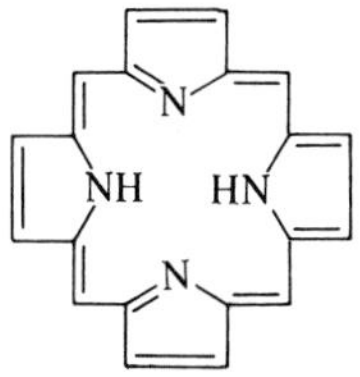

A series of porphyrins are derived from this parent tetrapyrrole compound by the substitution of various side chains and by the chelation of a variety of metal ions such as iron, magnesium and copper. The most abundant derivatives are the protoporphyrins, and the structure of protoporphyrin IX which is found in haemoglobin and many cytochromes is shown below:

Protoporphyrin IX

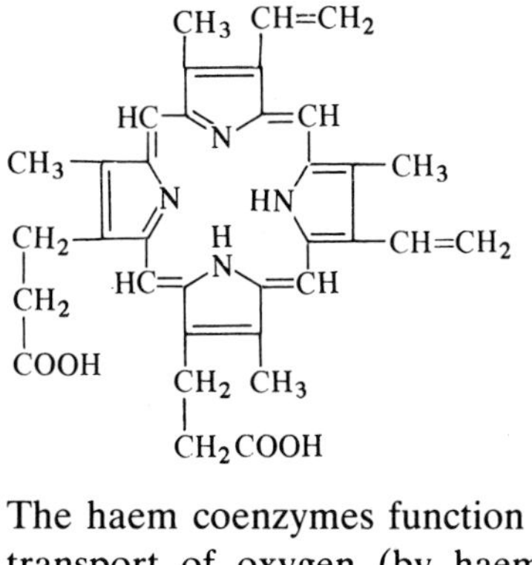

The haem coenzymes function in a variety of reactions including the transport of oxygen (by haemoglobin), oxidation of substrates by hydrogen peroxide (by catalase and peroxidase), and the transport of electrons to molecular oxygen in biological oxidation reactions (by the cytochromes). All of these particular processes involve the iron protoporphyrin IX, again demonstrating that the specificity both for substrate and for the type of reaction is dependent upon the protein and not upon the prosthetic group. These different functions require different involvements of the metal ion. In the case of haemoglobin, the iron shows no valency change in the course of the reaction, remaining in the ferrous state throughout. With the cytochromes the iron undergoes reversible changes between the ferrous and ferric states which is clearly associated with the transport of electrons. In the case of the peroxidases, the catalytic reaction may involve a transient oxidation of the iron to an even higher (quadrivalent Fe^{++++} or ferryl $(FeO)^{++}$) valency state.

Coenzyme A

Coenzyme A or 3'-phospho-ADP-pantoyl-β-alanyl-cysteamine has been found in all living cells and participates in a wide range of reactions involving the transfer of acyl groups. It has the rather complex structure shown below:

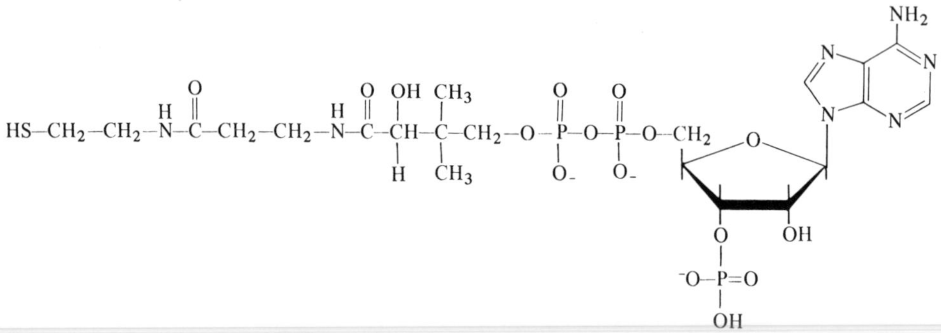

The molecule bears certain resemblances to the pyridine nucleotides and FAD, and like these other coenzymes the functional group, which is the —SH group for this compound, is the moiety farthest from the adenine ring. The coenzyme functions as an acyl carrier according to the general scheme.

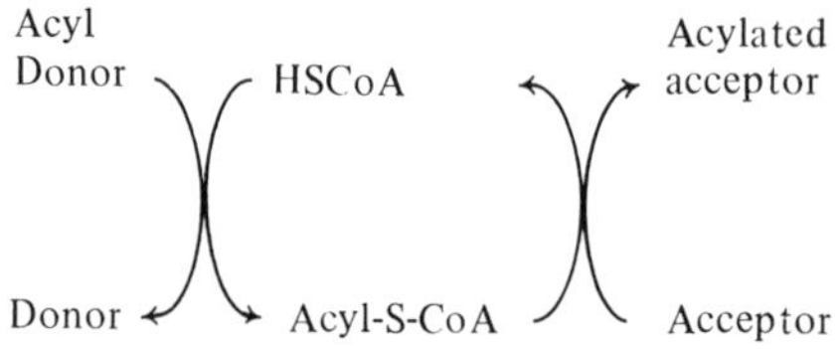

It is involved in a number of important biochemical processes including the oxidation of pyruvate by the TCA cycle (see Ch. 7), fatty acid oxidation (see Ch. 9) and synthesis of fatty acids (see Ch. 6).

Biotin

Biotin is found in both plant and animal tissues and is also known as vitamin H. It is known to function as a coenzyme in a number of carboxylation reactins involving the transfer or incorporation of carbon dioxide.

The coenzyme is usually firmly bound by covalent linkage to a lysine residue at the active centre of the enzyme protein; this biotin-lysine peptide is known as biocytin.

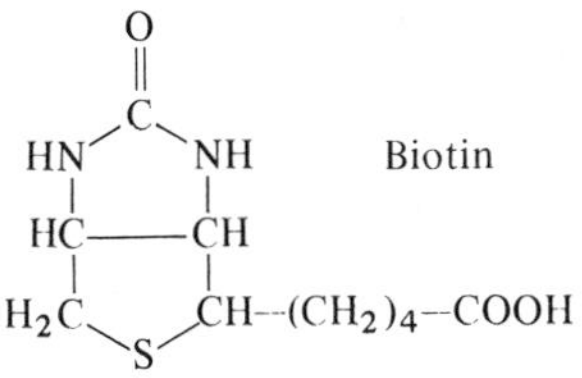

Biotin

Biotin functions as a coenzyme in both ATP-dependent carboxylation reactions, such as that catalysed by the mammalian enzyme pyruvate carboxylase:

ATP + pyruvate + CO_2 + H_2O = ADP + P_i + oxaloacetate

and in carboxyl group transferase reactions, illustrated by the reaction catalysed by methyl malonyl-CoA carboxytransferase:

Methyl malonyl-CoA + pyruvate = propionyl-CoA + oxaloacetate

In the ATP-dependent reactions the carboxylation is thought to involve the formation of a labile carboxy–biotin compound by combination of CO_2 with a ring nitrogen of biotin:

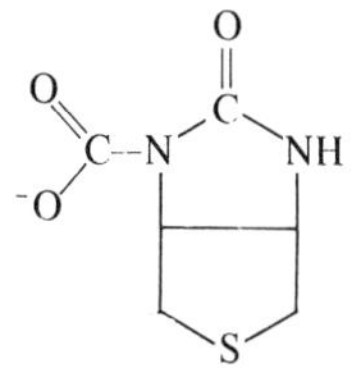

This carboxy–biotinyl–enzyme complex probably functions as an intermediate carrier donating CO_2 to the acceptor molecule in an overall two-step reaction.

Thiamine pyrophosphate

This is another universally-occurring vitamin-associated coenzyme. Thiamine or vitamin B, is an important growth factor for many microorganisms and animals and functions as a coenzyme in the form of its pyrophosphate ester, thiamine pyrophosphate (TPP):

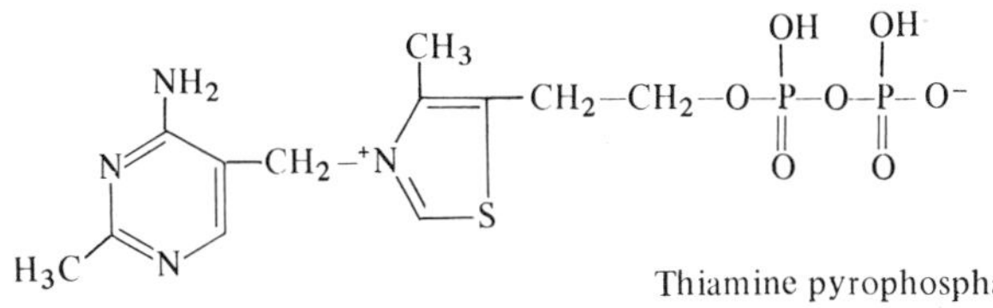

Thiamine pyrophosphate

This coenzyme is associated with a variety of reactions involving the synthesis or breakdown of carbon–carbon bonds adjacent to a keto group and include:

1. the oxidative and non-oxidative decarboxylation of α-keto acids, e.g.

$$CH_3CO \cdot COOH + \tfrac{1}{2}O_2 \;\rightarrow\; CH_3 \cdot COOH + CO_2$$

Pyruvate   Acetate

$$CH_3CO \cdot COOH \;\rightarrow\; CH_3 \cdot CHO + CO_2$$

Pyruvate   Acetaldehyde

2. transketolase activity which plays an important rôle in the pentose phosphate pathway (see Ch. 6) and the photosynthetic fixation of CO_2 (see Ch. 8), e.g.

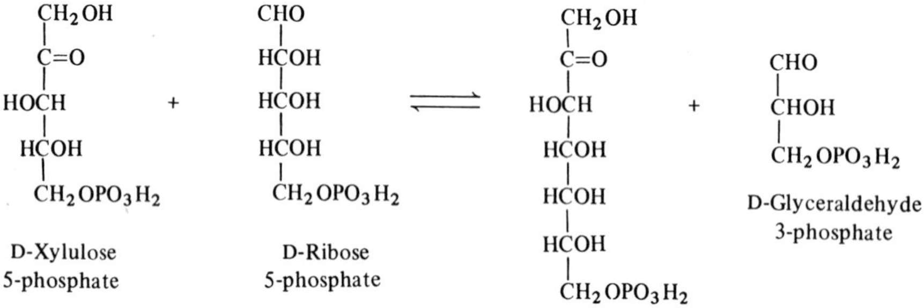

D-Xylulose 5-phosphate + D-Ribose 5-phosphate ⇌ D-Sedoheptulose 7-phosphate + D-Glyceraldehyde 3-phosphate

The site of reaction in all these cases is the carbon atom between the S and N atoms of the thiazole ring: For example, in the decarboxylation of pyruvate:

Thiazole ring of TPP

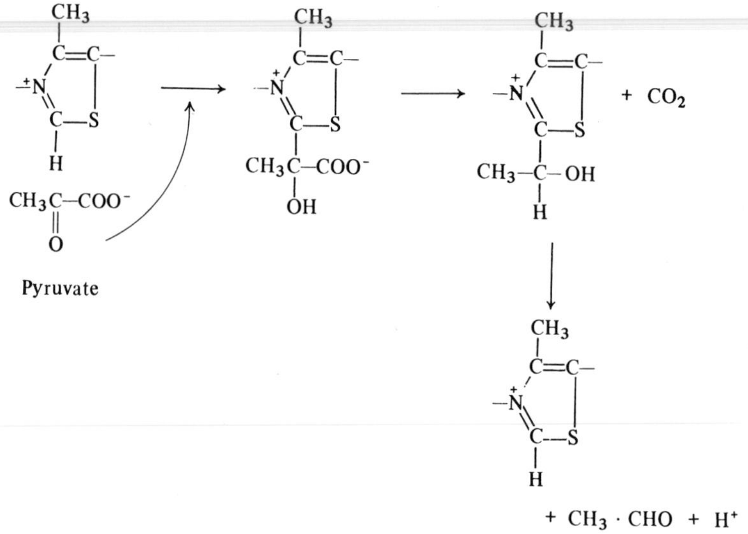

+ CH$_3$ · CHO + H$^+$

Acetaldehyde

The hydrogen atom attached to the reactive carbon in the thiazole ring readily dissociates to form a carbanion with an unshared pair of electrons; this pair reacts with the carbonyl carbon atom of pyruvate. The derivative so formed undergoes decarboxylation to leave an α-hydroxyethyl–TPP complex which may be regarded as active acetaldehyde. Finally the carbanion reacts with a proton to yield free acetaldehyde and regenerate the co-enzyme.

Tetrahydrofolic acid

Tetrahydrofolic acid (FH_4) is an important coenzyme that participates in reactions involving the transfer of one-carbon fragments such as methyl, hydroxymethyl, formyl and formimino groups and is particularly important in the various interconversions found in amino acid and purine metabolism. FH_4 consists of a reduced pteridine, p-aminobenzoic acid, and L-glutamic acid and its structure is shown below:

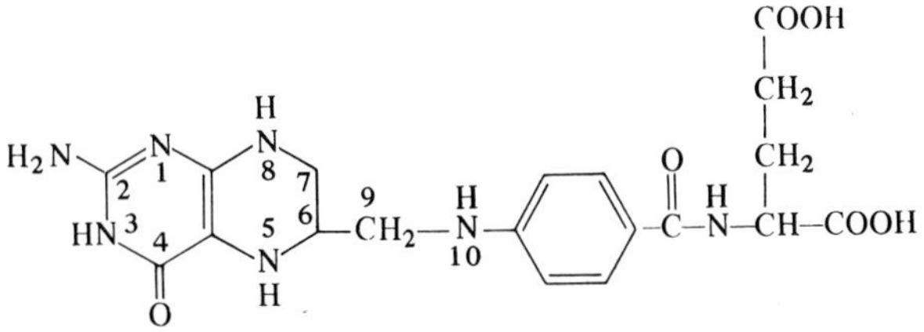

Tetrahydrofolic acid

The transfer of one-carbon groups involves the nitrogen atoms in the 5 and/or 10 positions, and some examples of the reactions between these groups and the carrier FH_4 are shown in Table 2.16. An example

Table 2.16   Reactions between one-carbon groups and FH_4

Group transferred		Derivative formed	
Formyl	H–C– ‖ O	10-Formyl- 5,6,7, 8-tetrahydrofolate	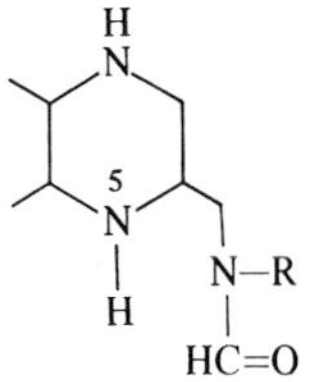
Formimino	H–C– ‖ NH	5-Formimino-5,6,7, 8-tetrahydrofolate	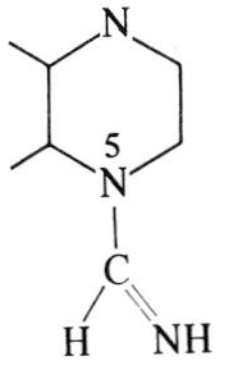
Methyl	H_3C–	5-Methyl-5,6,7, 8-tetrahydrofolate	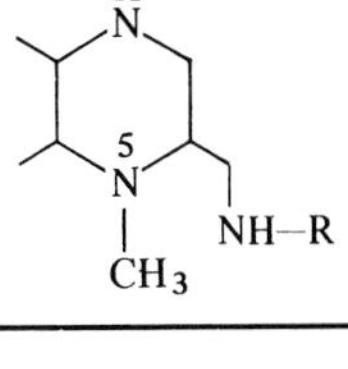

123

of the rôle of FH_4 in amino acid metabolism is seen in the conversion of serine to glycine (or the reverse, see p. 420):

$$\text{Serine} + FH_4 \longrightarrow \text{glycine} + N^5,N^{10}\text{-methylene } FH_4$$

The enzyme that catalyses this reaction, hydroxymethyltransferase, contains pyridoxal phosphate as a prosthetic group and also requires manganese ions. The nature of the group transfer reaction in this conversion is shown below:

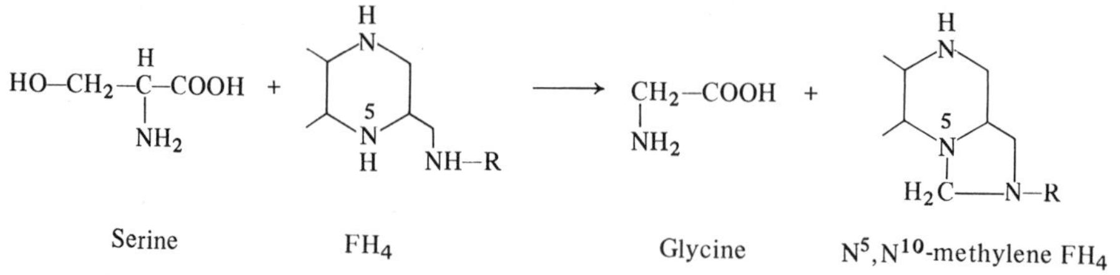

| Serine | FH₄ | Glycine | N⁵,N¹⁰-methylene FH₄ |

Bioenergetics

A typical plant cell may contain as many as 50×10^8 enzyme molecules which catalyse a wide variety of chemical reactions upon which life is dependent. Basically they are responsible for the replication and expression of genetic information and there would clearly be no life without enzymes. Enzymic reactions, however, do not take place in isolation within a disorganized mixture. As discussed earlier, the property of enzyme substrate specificity means that enzymic reactions are linked in sequences, the product of one reaction becoming the substrate of the next; this results in a series of metabolic pathways or cycles which make up intermediary metabolism. These pathways are not independent, but rather are highly integrated through the presence of common intermediates and through common co-enzymes such as ATP and NAD, which act as carriers of energy and reducing power between various metabolic sequences. Essentially, metabolism may be divided into *catabolism* and *anabolism*. Catabolism involves the enzymic breakdown of the large food macromolecules such as proteins, carbohydrates and lipids with the release of energy which is stored in molecules such as ATP. This energy is used to drive the various anabolic processes that are concerned with the synthesis of macromolecules. Life really depends on a number of unstable molecules, including the enzymes themselves, which are constantly being broken down and resynthesized, a process known as *turnover*. Energy is required for synthesis of molecules and cell growth, for osmotic work such as the accumulation of ions and, in some cases, for mechanical work. Catabolic and anabolic processes occur at the same time in the cell although, as we shall see in the later chapters of this book, they are usually spatially separated. The cell balances these two processes, maintaining a steady rate with the environment in which the energy and matter transferred into the cell is balanced by the energy and matter that is utilized and removed. For cells to live, a constant input of energy is required since otherwise they quickly exhaust their reserves and die. The ultimate source of this energy is the radiant

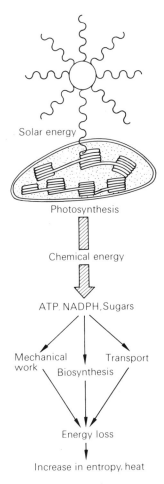

Solar energy

Photosynthesis

Chemical energy

ATP, NADPH, Sugars

Mechanical work   Transport

Biosynthesis

Energy loss

Increase in entropy, heat

Fig. 2.45   Energy flow in the biosphere.

energy from the sun, which is converted into chemical energy by the process of photosynthesis, occurring largely in green plant cells. The chemical energy stored in the various products of photosynthesis is then utilized by other cells in respiratory processes, to produce molecules such as ATP which can be used directly by energy-requiring reactions (Fig. 2.45). To understand the factors governing energy flow within cellular systems some knowledge of the laws of thermodynamics is required and these are briefly discussed below.

There are two fundamental principles of the thermodynamics relating to the physical universe which are known as the first and second laws. The first law is concerned with the conservation of energy and states that the total energy of any system and its environment remains constant. Energy is neither created nor destroyed although, as we have already suggested, it may be transformed from one form to another. The second is concerned with the direction in which a process may move and introduces the term *entropy*; this may be defined as the degree of disorganization or randomness of a system. This law states that any chemical or physical process will spontaneously move towards a state of maximum entropy, and so result in increased disorder. Biological processes appear to disobey this second law since much of cell metabolism is concerned with the synthesis of highly complex and ordered macromolecules to maintain a highly organized structure. Cell metabolism does not go in the direction of increased disorder. This is not however a transgression of the second law. Such a movement towards an increase in organization of a system is allowed if the *total free energy* of the system is increased. This term free energy is important to the understanding of energy changes in biological systems and may be defined as the useful energy, or the energy available in a system for doing work. In an isolated system at constant temperature and pressure, free energy is related to entropy by the following equation:

$$\Delta G = \Delta H - T\Delta S$$

where

ΔG = change in free energy
ΔH = change in heat content or enthalpy of the system
T = absolute temperature
ΔS = change in entropy

If no heat is exchanged between the system and its surroundings (i.e. $\Delta H = 0$), an increase in ΔG is accompanied by a decrease in $T\Delta S$. If heat is lost or gained by the system, then the changes in G and S will not be equal but will differ according to the amount of heat exchanged. If the supply of free energy to a living system is stopped it will proceed to a state of maximum entropy resulting in death. The high level of organization of living cells is maintained by a constant supply of free energy, which in turn must result in increased disorder elsewhere; there can be no overall decrease in entropy in the universe. The ultimate source of this free enegry is the sun. However, even the energy that is captured and utilized in biological processes is ultimately lost to the environment, leading to an increase in entropy.

How is the concept of free energy related to biochemical reactions? Consider the general equation

$$A + B \rightleftharpoons C + D$$

The free energy change for this reaction is given by the equation

$$\Delta G = \Delta G^0 + RT \ln \frac{[C][D]}{[A][B]} \qquad (15)$$

where

ΔG^0 = standard change in free energy (defined below)
R = gas constant ($1\cdot987$ cal deg^{-1} mol^{-1} or $8\cdot319$ joules deg^{-1} mol^{-1})
T = absolute temperature

When the reaction is at equilibrium there is no net conversion of $A + B$ to $C + D$ and so there will be no change in free energy ($\Delta G = 0$). The expression $[C][D]/[A][B]$ will be equal to the equilibrium constant, K_{eq}, and so we can write

$$\Delta G^0 = -RT \ln K_{eq} \qquad (16)$$

Thus for any given reaction the *standard change in free energy* may be calculated at any given temperature if the equilibrium constant is known, and will be a characteristic constant for that particular reaction. Having determined the standard free energy change, the actual free energy change may be calculated for any conditions by reference to equation (15) if the ratio of products to reactants is known. The free energy change of a reaction is basically the difference in free energy content of products and reactant

$$\Delta G = G_{products} - G_{reactants}$$

and, as mentioned earlier, is really a measure of the amount of useful energy that may be produced from the reaction. When the free energy change is positive for a proposed reaction, energy must be put into the system to drive the reaction. When the proposed free energy change is negative, the reaction can proceed spontaneously with the release of useful energy. It is worth noting however that a reaction with a standard free energy change which is positive may still proceed if the concentrations of reactants and products are arranged to give an actual free energy change that is negative. Reactions with a negative free energy change are called *exergonic* since they release energy to their surroundings. Those with a positive free energy change are called *endergonic* and need an energy input to go to completion. Living cells achieve a steady state by a balance of exergonic and endergonic reactions. As we shall see in the later chapters of this book, such reactions are frequently coupled, with an exergonic process supplying energy for an endergonic one. The standard free energy changes of a number of important biochemical reactions are shown in Table 2.17.

Table 2.17 Approximate values for standard free energy changes of some chemical reactions (pH 7·0, 25°C) In biochemical reactions, pH 7·0 is designated as the reference state and the standard free energy change at pH 7·0 is written $\Delta G^{\circ\prime}$

Reaction	$\Delta G^{\circ\prime}$	
	$kcal\ mol^{-1}$	$kJ\ mol^{-1}$
Oxidation		
Glucose + 6 $O_2 \rightarrow$ 6 CO_2 + 6 H_2O	−686	−2872
Pyruvate + $2\frac{1}{2}$ O_2 + $H^+ \rightarrow$ 3 CO_2 + 2 H_2O	−283	−1184
Palmitate + 23 $O_2 \rightarrow$ 16 CO_2 + 16 H_2O	−2338	−9789
Hydrolysis		
Sucrose + $H_2O \rightarrow$ glucose + fructose	−7	−29
Pyrophosphate + $H_2O \rightarrow$ 2 P_i	−8	−33
Acetic anhydride + $H_2O \rightarrow$ 2 acetate	−22	−92
Elimination		
Citrate $\rightarrow$ cis-aconitate + H_2O	+2·0	+8
Malate $\rightarrow$ fumarate + H_2O	+0·8	+3
Rearrangement		
Glucose 6-P $\rightarrow$ glucose1-P	+1·7	+7
Glucose 6-P $\rightarrow$ fructose 6-P	+0·5	+2

Particularly large changes are found in oxidation reactions, which as we shall see are extremely important in energy transformations in the cell.

Free energy changes and oxidation–reduction reactions

As we have seen, the maintenance of living systems requires a continual supply of free energy which is ultimately derived from various oxidation–reduction reactions. Except for photosynthetic and some bacterial chemosynthetic processes, which are themselves oxidation–reduction reactions, all other cells depend ultimately for their supply of free energy on oxidation reactions in respiratory processes. These involve the controlled breakdown of complex organic molecules by a series of metabolic pathways (to be described in detail in later chapters) in which the transfer of electrons and protons from these food molecules to molecular oxygen occurs by a series of oxidation–reduction reactions. Oxidation–reduction reactions involve the transmission of electrons from one molecule to another. The electron donor is known as a *reducing agent* and the acceptor molecule as an *oxidizing agent*. In some cases, biological oxidation involves the removal of hydrogen, a reaction catalysed by the dehydrogenases linked to specific coenzymes as has been discussed earlier. Clearly it is important to have a quantitative measure of the tendencies of various molecules to donate or accept electrons, i.e. their ability to act as oxidizing or reducing agents. Such a measure is given by the *standard oxidation–reduction potential* or *redox potential* (E_0) and a scale is drawn up with reference to a common standard which has an arbitrary potential value of zero. The standard chosen is the hydrogen electrode

$$H_2 + \ \rightleftharpoons\ 2H^+ + 2e^-$$

which is given an E_0 of 0 volts when measured at pH 0 ($1M$ H^+), 25°C, and with the pressure of hydrogen gas at 1 atm. Most enzyme systems, of course, are inactivated at such a low pH value and so the standard potential E_0' is usually determined at the more physiological pH value of 7. At this pH, the hydrogen electrode has a potential difference of -0.42 volts with respect to the same electrode at pH 0, as calculated from the equation below. Standard potentials, therefore, are calculated at pH 7.0, 25°C, and with the oxidized and reduced forms of the substance present at a concentration of $1M$, or, as is often done in practice, in equal concentrations. The actual potential observed is related to the standard potential by the following equation:

$$E = E_0' + \frac{RT}{nF} \ln \frac{a_{\text{oxid}}}{a_{\text{red}}} \tag{17}$$

where

E = observed potential
E_0' = standard potential
R = gas constant
T = absolute temperature
n = number of electrons transferred in the reaction
F = the Faraday (96 494 coulombs or joules $volt^{-1}$)
a_{oxid} = concentration of oxidized form
a_{red} = concentration of reduced form

When the concentration of reduced and oxidized forms are equal, then $E = E_0'$. In any other case, the observed potential will be related to the ratio of oxidized and reduced forms. Most biological oxidation–reduction reactions are treated as two-electron transfers so that equation (17) can be more simply written as

$$E = E_0' + 0.03 \log \frac{a_{\text{oxid}}}{a_{\text{red}}}$$

A list of E_0' values of some biologically important compounds is given in Table 2.18. These values permit the direction of flow of electrons

Table 2.18 Standard redox potentials (E_0') of some important biochemical systems

Oxidant	Reductant	$E_0'(V)$
$\frac{1}{2}O_2$	H_2O	+0.82
Cytochrome a (Fe^{3+})	Fe^{2+}	+0.29
Cytochrome c (Fe^{3+})	Fe^{2+}	+0.25
Dehydroascorbic acid	Ascorbic acid	+0.17
Cytochrome b (Fe^{3+})	Fe^{2+}	+0.04
Pyruvate	Lactate	−0.19
Riboflavin, oxidized	Reduced	−0.20
FAD^+	$FADH + H^+$	−0.22
NAD^+	$NADH + H^+$	−0.32
$NADP^+$	$NADPH + H^+$	−0.32
α-Ketoglutarate + CO_2	Isocitrate	−0.38
$2H^+$	H_2	−0.42

between two redox systems to be predicted. The more electronegative the compound, the greater the tendency to donate electrons.

We have already mentioned that the ultimate source of free energy for living systems is derived from oxidation–reduction reactions and so knowledge of the relationship between free energy change and redox potentials is clearly important in the understanding of energy transfer in biochemical reactions. From the relationships that we have already discussed it is possible to derive the following important expression:

$$\Delta G^{0'} = -nF \, \Delta E'_0 \qquad\qquad (19)$$

where

$\Delta G^{0'}$ = standard free energy change
n = number of electrons transferred
F = the Faraday
$\Delta E'_0$ = the difference in standard redox potential between the oxidizing and reducing agent

Thus when the redox potential change is positive the free energy change is negative and the reaction an exergonic one accompanied by the release of free energy.

Essentially biological oxidations involve the transfer of electrons and protons from organic molecules to molecular oxygen with the release of free energy. For example, the oxidation of glucose

$$C_6H_{12}O_6 + 6\,O_2 \rightarrow 6\,CO_2 + 6\,H_2O$$

is an exergonic reaction with a free energy change of 2 872 kJ/mole. In living cells, this combustion does not take place in one step with a large release of free energy but involves a number of stages with a controlled release of energy and its conservation in the chemical bonds of compounds such as ATP. Many of the oxidation steps in the degradation of carbohydrates and lipids in respiratory pathways involve the action of dehydrogenase enzymes and the participation of pyridine nucleotides as cofactors. If we consider the free energy change involved in the oxidation of reduced NAD which is a product of dehydrogenase activity we find the following:

$$NADH + H^+ + \tfrac{1}{2}O_2 \longrightarrow NAD^+ + H_2O$$

This reaction involves the transfer of two electrons and from Table 2.18 we see that the

$$\Delta E'_0 = 0\cdot816 - (-0\cdot32) = 1\cdot136V$$

and therefore

$$\begin{aligned}
\Delta G^{0'} &= -nF \, \Delta E'_0 \\
&= -2(9\,6494)(1\cdot136) \\
&= -219 \text{ kJ}
\end{aligned}$$

We have already seen (p. 114) that the free energy of hydrolysis of formation of the terminal pyrophosphate bond of ATP is about 29 kJ/mole, so there is clearly ample energy released by the oxidation of NADH to provide for the synthesis of a number of molecules of

ATP from ADP and inorganic phosphate. The mechanisms involved in such energy transformations are discussed fully in Chapters 7 and 8 in relation to ATP formation by oxidative metabolism in the mitochondrion and by photosynthetic reactions in the chloroplast.

Thus life is driven by a series of oxidation–reduction reactions which, as we shall see in Chapter 8, are ultimately dependent on the capture of radiant free energy from the sun by photosynthetic cells. This leads to the synthesis of large and complex organic compounds which are used to build up the highly organized structure of the cell. This movement apparently contradicts the second law of thermodynamics which states that systems in isolation move to a state of maximum entropy or disorder. Living cells, however, are open systems and are able to decrease their entropy producing an equivalent increase in entropy in their surroundings. This has been elegantly described by the physicist Schrödinger in his essay *What is Life?* (1951). He considers that living organisms survive by drawing negative entropy from their environment in the form of complex, highly ordered food molecules which will have a low entropy. These are metabolized, producing smaller molecules with higher entropy which are liberated into the environment. Living systems therefore are maintained by feeding on negative entropy.

Further reading

Barker, R. (1971) *Organic chemistry of Biological Compounds*. Prentice-Hall, New York.

Bonner, J. and Varner, J. E. (1976) *Plant Biochemistry*. 2nd edn. Academic Press, New York.

Davidson, J. N. (1972) *The Biochemistry of the Nucleic Acids*. Chapman and Hall, London.

Dixon, M. and Webb, E. C. (1979). *Enzymes*. 3rd edn. Longman Group Ltd., London.

Ferdinand, W. (1976) *The Enzyme Molecule*. John Wiley and Sons, London.

Fersht, A. (1977) *Enzyme Structure and Mechanism*. Freeman and Co., Reading and San Francisco.

Gurr, M. I. and James, A. T. (1980) *Lipid Biochemistry: An Introduction*. 3rd edn. Chapman and Hall, London.

Hitchcock, C. and Nichols, B. W. (1971) *Plant Lipid Biochemistry*. Academic Press, London.

Lehninger, A. L. (1975) *Biochemistry*. 2nd edn. Worth Publishers, New York.

Smith, H. (1977) *Regulation of Enzyme Synthesis and Activity in Higher Plants*. Academic Press, London.

Stryer, L. (1981) *Biochemistry*. 2nd edn. Freeman and Co., Reading and San Francisco.

Literature cited

Amrhein, N. (1977) The current status of cyclic AMP in higher plants. *Ann. Rev. Plant Physiol.* **28**, 123.

Anfinsen, C. B. (1959) *Molecular Basis of Evolution*. John Wiley, New York.

Anfinsen, C. B. (1967) Molecular structure and the function of proteins. In *Molecular Organization and Biological Function*. Ed. J. M. Allen. Harper and Row, New York.

Bell, E. A. (1981) The non-protein amino acids occurring in plants. In *Progress in Phytochemistry*. Vol. 7. Eds. L. Reinhold, J. B. Harborne and T. Swain. Pergamon Press, Oxford.

Briggs, G. E. and Haldane, J. B. S. (1925) *Biochem. J.*, **19**, 338.

Epstein, E. (1972) *Mineral Nutrition of Plants*. John Wiley, New York.

Haggis, G. H., Mickie, D., Muir, A. R., Roberts, K. B. and Walker, P. M. B. (1964) *Introduction to Molecular Biology*. Longmans, Green, London.

Hilditch, T. P. (1947) *The Chemical Constitution of Natural Fats*. Chapman and Hall, London.

Johnson, L. P., MacLeod, J. K., Parker, C. W., Letham, D. S. and Hunt, N. H. (1981) Identification and quantitation of adenosine-3':5'-cyclic monophosphate in plants using gas chromatography-mass spectrometry and high performance liquid chromatography. *Planta*, **152**, 195

Karlson, P. (1968) *Kurzes Lehrbuch der Biochemie*, Georg Thieme Verlag, Stuttgart.

Koshland, D. E. Jr. (1959) In *The Enzymes*. Vol. 1. Eds. P. D. Broyer, H. Lardy and K. Myrback, Academic Press, New York.

Krebs, E. G. and Beavo, J. A. (1979) Phosphorylation–dephosphorylation of enzymes. *Ann. Rev. Biochem.* **48**, 423.

Marcus, A. (1971) Enzyme induction in plants. *Ann. Rev. Plant Physiol.*, **22**, 313.

Matthews, B. W., Sigler, P. B., Henderson, R. and Blow, D. M. (1967) Three-dimensional Structure of Tosyl-α-chymotrypsin. *Nature, London* **214**, 652.

Monod, J., Wyman, J, and Changeux, J-P. (1965) On the nature of allosteric transitions: a plausible model. *J. molec. Biol.* **12**, 88.

Schrödinger, E. (1951) *What is Life?* Cambridge University Press, London.

Smyth, D. G., Stein, W. H. and Moore, S. (1963) The sequence of amino acid residues in bovine pancreatic ribonuclease: revisions and confirmations. *J. Biol. Chem.*, **238**, 227.

Stewart, G. R. and Rhodes, D. (1977) Control of enzyme levels in the regulation of nitrogen assimilation. In *Regulation of Enzyme Synthesis and Activity in Higher Plants*. Ed. H. Smith. Academic Press, London.

Wallace, W. (1977) Proteolytic inactivation of enzymes. In *Regulation of Enzyme Synthesis and Activity in Higher Plants*. Ed. H. Smith. Academic Press, London.

Watson, J. D. (1970) *Molecular Biology of the Gene*. W. A. Benjamin, New York.

Watson, J. D. and Crick, F. H. (1953) A structure for deoxyribose nucleic acid. *Nature, Lond.*, **171**, 737.

3   Cell membranes

As we have seen in Chapter 1, plant cells are delimited from their surroundings by a semipermeable membrane and intracellularly are sub-divided into a number of micro-environments, which are again delimited by membranes. The different functions and enzymic activities associated with these various membranes are described in the later chapters of this book. An understanding of the structure, chemistry and properties of membranes is essential therefore to an understanding of cell function and metabolism. Although the thickness of cellular membranes is well below the resolution of the light microscope, a great deal about their nature and characteristics was learnt long before the introduction of the electron microscope. This was largely based on studies of solute permeability and other properties of the surface of cells such as the algae *Nitella* and *Chara*, yeast and red blood cells, and so was really concerned with the nature of the outer cell membrane, the plasmalemma. However most of these findings are now known to apply to the other membranes of the cell.

Properties of membranes

Any postulated model of cell membrane structure must account for the following physiological and physical characteristics of membranes.

1. Cell membranes are highly permeable to water, the rate of penetration being very much greater than that for various solute molecules.
2. The permeability of the plasmalemma to non-electrolytes is, in general, found to be correlated with the lipid solubility of the solute; an increase in permeability is related to an increase in the partition coefficient between oil and water (Fig. 3.1)
3. The electrical resistance of cell membranes is high, normally with a resistivity in the range of $10^9 - 11^{11}$ Ω cm. This may be compared with the values for the resistivity of distilled water, 10^6 Ω cm, and for olive oil, 5×10^{12} Ω cm.
4. Studies on isolated cells and cell organelles have shown that the membrane surface usually has a net negative charge at physiological pH values, but that it also shows amphoteric properties, the net charge varying with changes in the external pH, which is a characteristic property of protein molecules (see p. 82).

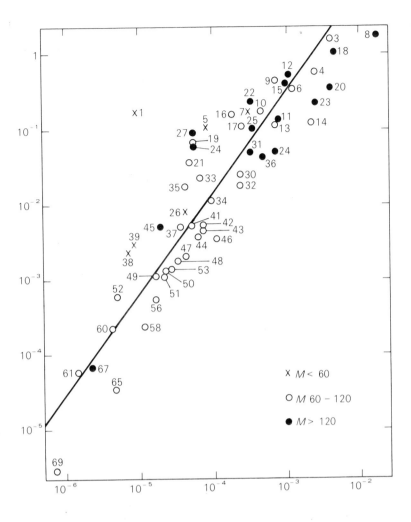

Fig. 3.1 Correlation between the permeation power of several non-electrolytes toward *Nitella mucronata* cells on the one hand and their relative oil solubility and molecular weight on the other. Ordinate: $PM^{1.5}$, where P = apparent permeability and M = molecular weight, abscissa: partition coefficient olive oil: water. The substances are: 1, HDO; 3, methyl acetate; 4, *sec*-butanol; 5, methanol; 6, *n*-propanol; 7, ethanol; 8, paraldehyde; 9, urethane; 10, isopropanol; 11, acetonylacetone; 12, diethylene glycol monobutyl ether; 13, dimethyl cyanamide; 14, *tert*-butanol; 15, glycerol diethyl ether; 16, ethoxyethanol; 17, methyl carbamate; 18, triethyl citrate; 19, methoxyethanol; 20, triacetin; 21, dimethylformamide; 22, triethylene glycol diacetate; 23, pyramidon; 24, diethylene glycol monoethyl ether; 25, caffeine; 26, cyanamide; 27 tetraethylene glycol dimethyl ether; 30, methylpentanediol; 31, antipyrene; 32, isovaleramide; 33, 1,6-hexanediol; 34, *n*-butyramide; 35, diethylene glycol monomethyl ether; 36, trimethyl citrate; 37, propionamide; 38, formamide, 39, acetamide, 41, succinimide; 42, glycerol monoethyl ether; 43, *N,N*-diethylurea; 44, 1,5-pentanediol; 45, dipropylene glycol; 46, glycerol monochlorohydrin; 47, 1,3-butanediol; 48, 2,3-butanediol; 49, 1,2-propanediol; 50, *N,N*-dimethylurea; 51, 1,4-butanediol; 52, ethylene glycol; 53, glycerol monomethyl ether; 56, ethylurea; 58, thiourea; 60, methylurea; 61, urea; 65, dicyanodiamide; 67 hexamethylenetetraamine; 69, glycerol. Reproduced from Collander (1954).

5. Measurements of the surface tension between cells and their aqueous surroundings give very low values, usually in the range 0.1–2 dynes cm^{-1}, which may be compared with a value of about 8 dynes cm^{-1} for oil drops in water. The values for cell membranes are not compatible with the presence of neutral lipids at the surface, but could be accounted for by a layer of protein at the exterior.

6. Cell membranes can frequently be lysed by treatment with lipid solvents, phospholipases or proteolytic enzymes.

Thus the various properties listed above clearly point to the presence of lipids and proteins as major components of cell membranes.

Chemical composition of cell membranes

The chemical composition of cell membranes is not easy to determine since membranes are difficult to isolate in reasonable quantities and in a pure state free from cytoplasmic contaminants. This is particularly true in the case of plant cells where membranes make up a smaller proportion of the total cell content than in animal cells. In addition, the cell wall presents a major problem, particularly in relation to analysis of the plasmalemma where the close interrelationship of the two make it difficult to isolate either in a pure form. Hence very much more information is available concerning the composition of animal cell membranes, such as those from red blood cells and nerve myelin, than from plant cells. Analyses show that most membranes consist almost entirely of protein and lipid, and so confirm the predictions of the various physiological and physical measurements described above. However, there is considerable variation in the proportions of protein and lipid, the ratio of protein to lipid varying from 4·0 in some bacteria to 0·25 in nerve myelin (Table 3.1). This variation might present a considerable objection to general models for membrane structure as they must account for this wide range of values between types and species.

Table 3.1 Protein and lipid content of membranes. From Korn (1969). The abbreviations are: Cer, cerebrosides; DPG, diphosphatidylglycerol; GalDG, galactosyldiglyceride; PA, phosphatidic acid; PC, phosphatidylcholine; PE, phosphatidylethanolamine; PGaa, amino acyl esters of phosphatidylglycerol; Plas, plasmalogen; SL, sulpholipid; Sph, sphingomyelin.

Membrane	Protein/lipid	Cholesterol/polar lipid	Major polar lipids
	wt/wt	mole/mole	
Myelin	0·25	0·7–1·2	Cer, PE, PC
Plasma membranes			
Liver cell	1·0–0·4	0·3–0·5	PC, PE, PS, Sph
Ehrlich ascites	2·2		
Intestinal villi	4·6	0·5–1·2	
Erythrocyte ghost	1·5–4·0	0·9–1·0	Sph, PE, PC, PS
Endoplasmic reticulum	0·7–1·2	0·03–0·08	PC, PE, Sph
Mitochondrion			DPG, PC, PE, Plas
Outer membrane	1·2	0·03–0·09	
Inner membrane	3·6	0·02–0·04	
Retinal rods	1·5	0·13	PC, PE, PS
Chloroplast lamellae	0·8	0	GalDG, SL, PS
Bacteria			
Gram-positive	2·0–4·0	0	DPG, PG, PE, PGaa
Gram-negative		0	PE, PG, DPG, PA
PPLO	2·3	0	
Halophilic	1·8	0	Ether analogue PGP

Lipids

In general cell membranes contain only very low levels of triglycerides, the lipids based on glycerol. Animal cell membranes contain high levels of phospholipids, such as lecithin, which are all polar molecules with one end hydrophobic and the other charged and hydrophilic, and of sterols, such as cholesterol. For example, phospholipid constitutes 60% and cholesterol 25% of the total lipid in the plasmalemma of mammalian red blood cells and phospholipid makes up 47% of the lipid in nerve myelin membranes, the ratio of these lipids being characteristic for different cell membranes (Table 3.1). Plant cell membranes also contain considerable levels of phospholipids, and the limited evidence available to date suggests that the plant plasmalemma contains a high sterol to phospholipid ratio. The other important lipids of cell membranes are the glycolipids, which have carbohydrate polar groups, usually the sugar galactose, and the sphingolipids, which contain phosphoric acid and an amino alcohol but no glycerol. An important group of membrane lipids, the cerebrosides, may be classified in either of these groups since they contain both a sugar and an amino alcohol. The structure and properties of these various lipids are described in Chapter 2. Glycolipids, including the cerebrosides, constitute 27% of nerve myelin lipid and 10% of that from red blood cell membranes. Thus although the lipids described are common to a wide variety of biological membranes, there is clearly considerable variation in the relative concentrations of the major groups of lipids between plant, animal and bacterial species, and this again must be carefully considered in relation to proposed models for a common structure of all biological membranes.

Proteins

With a few exceptions, the proteins of cell membranes are poorly characterized, particularly those of plant cells, and only recently has information been derived about the arrangement and function of particular proteins within membranes. The reasons for this paucity of knowledge relate to the considerable practical difficulties associated with the isolation and analysis of many membrane proteins. They have low solubilities in aqueous solutions and a high affinity for lipids, and, because of this, have been difficult to isolate in pure form by the conventional techniques of fractionation that have been applied successfully to 'soluble' proteins. Recently, however, it has been possible to render many such proteins soluble in detergent solutions, particularly of the non-ionic type, with retention of their native structure and biological activity. The detergents probably replace interacting lipid molecules. Molecular weights and polypeptide chain compositions have now been determined for several such proteins, including cytochrome b_5 which is a component of the endoplasmic reticulum.

Analyses of this kind have revealed that all integral proteins (i.e. those which penetrate into the lipid regions of the membrane) are *amphipathic* molecules (see Vanderkooi, 1974). That is, they have portions which are hydrophilic and exposed in the fluid phase surrounding the membrane and a hydrophobic region in contact with the lipid molecules themselves. Cytochrome b_5, an integral protein of the endoplasmic reticulum, is a good example of such a structure. Of its 152 amino acids residues, the aminoterminal 104 residues form a globular structure which protrudes out of the membrane, while the carboxy-terminus consists of a hydrophobic 'tail' which is embedded in the membrane (Strittmatter *et al.*, 1972) (Fig. 3.2). We shall discuss the structure of membrane proteins in greater detail when considering the fluid–mosaic model of membrane structure in the next section.

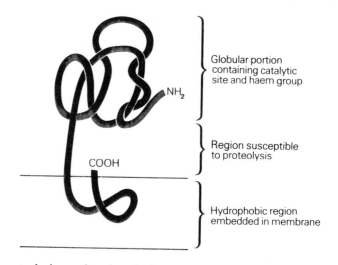

Fig. 3.2 Diagrammatic representation of the structural relationship between the amphipathic protein cytochrome b_5 and the membrane of the endoplasmic reticulum. The haem group is embedded in the exposed globular portion of the molecule. If the globular portion is cleaved from its hydrophobic tail by trypsin, the protein retains its full biological activity. The carboxyl terminal tail therefore serves primarily to anchor cytochrome b_5 to the membrane.

One technique that has had an enormous impact on studies of membrane proteins is high resolution, polyacrylamide gel electrophoresis carried out in the presence of the ionic detergent sodium dodecylsulphate (SDS) (see p. 40). Most membrane proteins are readily soluble in aqueous solutions of this detergent, particularly after they have been reduced using mercaptoethanol. They unfold, and the detergent binds to their surfaces displacing lipids and other molecules. Because the complexes are negatively charged, they migrate towards the anode when subjected to an electric field. Their migration rate is related to their molecular sizes and limited by the degree of cross-linking of the polyacrylamide matrix. This technique has allowed many different types of membrane to be analysed with regard to the numbers and proportions of different polypeptide components. A recent refinement in which the polypeptides are first subjected to isoelectric focusing in the presence of a denaturant (e.g. urea) and a non-ionic detergent (such as Nonidet P-40) and then, in the second dimension, to SDS-polyacrylamide gel electophoresis, greatly extends the resolution of this methodology, since the position of a protein in the gel will be determined by both its isoelectric point and its molecular weight. Figure 3.3 illustrates the two-dimensional electrophoretic analysis of the polypeptides present in the plasmalemma of Chinese hamster fibroblasts. Most have isoelectric points in the pH range 8 to 4 and molecular weights between 20 000 and 225 000.

Membrane structure

As all biological membranes have a number of basic properties in common, it has long been assumed that there may be a fundamental membrane structure which is common to all cells. Both indirect studies of membrane properties and direct chemical analysis of isolated membranes show that their major chemical components are proteins and lipids. The problem has been to decide how these molecules are arranged within the limits of the membrane.

It has been known since the classic work of Langmuir in 1917 that when certain polar lipids are spread on the surface of water, they

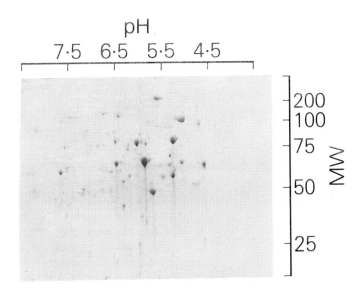

Fig. 3.3   Two dimensional
polyacrylamide gel electrophoresis of
the polypeptides associated with the
plasma membrane of Chinese hamster
fibroblasts. The membrane
polypeptides were solubilized in urea –
Nonidet P-40 (a non-ionic detergent)
and subjected to isoelectric focusing in
polyacrylamide gels cast in narrow bore
tubes. The gels contain a pH gradient
and a protein will migrate to a region
where the pH is identical with its
isoelectric point. It then has zero
mobility and will be concentrated or
'focused' to form a narrow zone. These
gels were extruded, equilibrated with
sodium dodecyl sulfate (SDS) and fixed
to the top of a polyacrylamide slab gel.
Electrophoresis was carried out in
presence of SDS in the second
dimension, thus separating the
polypeptides according to their size.
The horizontal scale indicates
approximate isoelectric points, the
vertical scale molecular weights
($\times\ 10^{-3}$). The gel was stained for
protein by Coomassie blue. (Courtesy
Drs M. N. Horst and R. M. Roberts,
University of Florida.)

become arranged in a monolayer with the charged hydrophilic groups
extending into the water and the hydrocarbon chains exposed to air at
right angles to the interphase. A little later, in 1925, Gorter and
Grendel analysed the lipid content of erythrocyte membranes and
calculated that there was sufficient to form a bimolecular film at the
cell surface with the polar groups facing to the outside of the bilayer.
However such a structure would not account for the low surface
tension and amphoteric properties of the surface of many cells,
although this could be explained if a film of protein were present at the
outside of the lipid. If the protein only covered one side of the bilayer
this would give an unstable structure, since the surface tension on one
side would be very much lower than on the other. This led Danielli and
Davson (1935) to propose that a stable membrane structure would be
produced by a symmetrical arrangement of a bimolecular layer of lipid
covered on either side by a layer of protein (Fig. 3.4). This model can

Fig. 3.4   Membrane model proposed
by Danielli and Davson (1935).

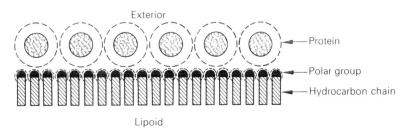

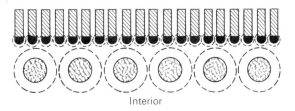

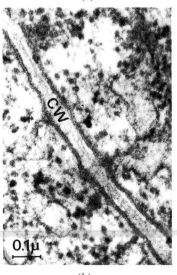

(a)

(b)

account for most of the properties of membranes listed previously although the existence of pores must be assumed to explain the rapid penetration of water and some other small molecules.

The use of the electron microscope and improvements in preparative techniques made possible the direct visualization of cell membranes. Figure 3.5 shows the typical appearance of a plasmalemma under the electron microscope after fixation and sectioning in the standard way. It appears as a three layered structure consisting of two electron-dense lines separated by a light space, the whole being about 7·5 nm wide; this structure is clearly seen after fixation with potassium permanganate and, although less clearly, with osmium tetroxide fixation. Examination of various membranes from both plant and animal cells suggests that this triple-layered appearance seen under the electron microscope is a characteristic one for all biological membranes and led Robertson (1959) to propose the '*unit membrane*' hypothesis. This considers that all biological membranes consist of a bimolecular layer of polar lipids arranged with their hydrophilic ends oriented outwards and covered with a layer of protein molecules on both sides, and having an overall width of about 7·5 nm; under the electron microscope this structure appears as two electron-dense bands, 2 nm thick, separated by a relatively unstained space 3·5 nm wide (Fig. 3.6). This is a generalized hypothesis based largely on the similar appearance of various cellular membranes under the electron microscope and is closely related to the Davson-Danielli model. This concept is essential to the generalized picture of the cell put forward by Robertson and described in an earlier chapter (p. 12) which assumes a continuity of cell membranes to produce a three-phase system of membranes, nucleocytoplasmic matrix and membrane-lined channels.

However the unit membrane hypothesis is open to a number of serious objections and is no longer generally accepted. Firstly, the functions of membranes vary considerably between different cell types and between different membranes in a given cell. This may be reflected in the considerable variation found in the chemical composition of membranes, both in the ratio of protein to lipid and in the proportions of the various types of lipids (Table 3.1), although as yet no direct

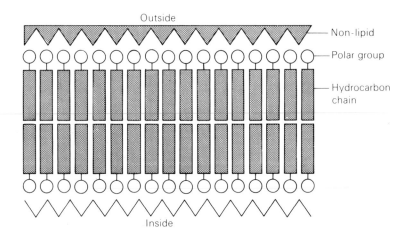

correlation has been established between a particular chemical composition and specific membrane function. Secondly, electron microscopy itself, which has been the basis of the unit membrane hypothesis, has raised a number of apparent contradictions to such a generalized concept. Studies on both plant and animal cells have shown that there may be large differences in membrane thickness. The plasmalemma is usually the widest membrane at about 9–10 nm, the Golgi cisternae are about 8 nm, and the inner mitochondrial membrane and tonoplast are usually the thinnest membranes at about 5–6 nm. In addition, some asymmetry in electron micrographs of membranes has been detected, particularly in relation to the plasmalemma. With potassium permanganate fixation the two dark lines are equally electron-dense but, after osmium tetroxide treatment, the outer line usually appears to be more weakly stained than the inner one, suggesting that some polarity in the structure may exist; this was not anticipated by the original Davson-Danielli model. This problem is made more complex by the uncertainty that exists concerning the chemistry of fixation and staining of membranes for electron microscopy. Very little is known of the reactions involved in potassium permanganate fixation, although there is more evidence concerning fixation with osmium tetroxide. Initially it was thought that osmium specifically reacted with the polar groups of phospholipids in the lipid bilayer and although this reaction may well occur, it does not seem to be important in producing the characteristic appearance of membranes seen in electron micrographs; the unit membrane appearance is found with osmium fixation even when all the lipids have been removed prior to fixation. These results strongly suggest that the dark lines of unit membrane represent the protein component but the degree of molecular disruption which might result from such fixation is not known; this information is obviously essential to the correct interpretation of the electron microscope image.

The result of these varied observations, which suggest that there may be considerable differentiation within the membrane matrix, has been the proposal of a series of models for membrane structure. Of these, the *fluid-mosaic* model, which was first proposed by Singer and Nicolson in 1972, and modified only slightly since then, is now the generally accepted working model for the molecular organization of the proteins and polar lipids of membranes. Its basis is a lipid bilayer, similar to that proposed by Davson and Danielli, and later by Robertson. This structure is proposed to be in equilibrium with an array of different types of protein molecule. The latter interact with the phospholipids of the membrane in a manner determined by their primary amino acid sequence and are accommodated by complex hydrophobic and hydrophilic interactions. A feature of the model is the ability of the proteins to 'float' freely within the fluid bilayer. Two main types of membrane proteins have been distinguished: *peripheral* (or extrinsic) and *integral* (or intrinsic). The former have a surface location and are more easily dissociated from the bilayer, often by using salt solutions or chelating agents such as EDTA to break up ionic interactions. The integral proteins, as their name implies, are embedded in the bilayer and can only be removed by means of detergents. A rather overused but apt analogy likens these proteins to icebergs in a sea of lipid. The fluid mosaic model can be portrayed diagrammatically in a number of ways. A three dimensional representation with a two dimensional 'section' is shown in Fig. 3.7.

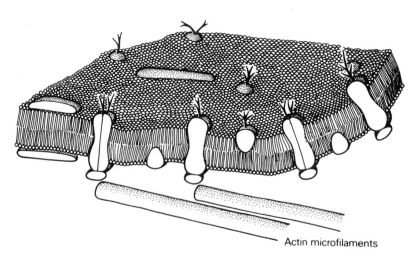

Fig. 3.7   The fluid mosaic model for membrane structure: a schematic three dimensional and cross-sectional view. The membrane is represented by a lipid bilayer composed mainly of phospholipids. The integral proteins are those which penetrate partially or wholly through the bilayer. Peripheral proteins are shown in contact with either the bilayer or the integral proteins. These lack a hydrophobic extension into the lipid bilayer. The extensions (⅃) on the upper end of the proteins are intended to represent carbohydrate chains. Some proteins are also thought to be associated with the cytoskeletal system of the cell, particularly the microfilaments. This may limit their lateral mobility in the lipid layer.

Actin microfilaments

Let us now summarize the evidence for this generalized concept of membrane structure. In part, the model was based on thermodynamic considerations and the realization that stable associations between lipids and proteins would only exist if hydrophobic and hydrophilic interactions were maximized. Protein molecules were visualized as having a strongly hydrophobic portion interacting with the lipid 'tails' of the bilayer, and hydrophilic regions exposed to the aqueous surroundings and phospholipid head groups. As pointed out earlier, cytochrome b_5 fits this concept ideally. Similarly, glycophorin, a molecule which actually spans the erythrocyte membrane, has a hydrophilic segment protruding from each surface and a hydrophobic segment in between (Fig. 3.8). The —NH$_2$ terminus, which faces out from the erythrocyte, bears sixteen separate carbohydrate chains, which are themselves extremely hydrophilic groups (see Furthmayr, 1977). Indeed, it appears that many integral membrane proteins are glycoproteins. These polar groups, protruding through the membrane, and the hydrophobic groups within seem to lock the proteins in position. There is now overwhelming evidence that such proteins are so stabilized in this orientation that they cannot invert their positions (i.e. 'flip-flop') within the plane of the membrane. The phospholipid molecules are also believed to be partitioned asymmetrically such that one face of the bilayer differs in composition from the other (see Rothman and Lenard, 1977, for a detailed review).

A second line of evidence favouring the fluid mosaic model has been the demonstration that proteins can migrate laterally in the plane of the membrane. This was shown most dramatically by Frye and Edidin (1970) who followed the fate of surface proteins of two different types

Fig. 3.8 The amino acid sequence of human glycophorin, the major glycoprotein of erythrocytes. The darkened residues represent residues 73 to 95 which do not contain any charged side chains, but have instead a high number of hydrophobic residues. The carbohydrate chains are shown by the

symbol /CHO\ and are found

on the amino terminal segment. The identity of amino acids at positions 1 and 5 are in some doubt and may vary from individual to individual.

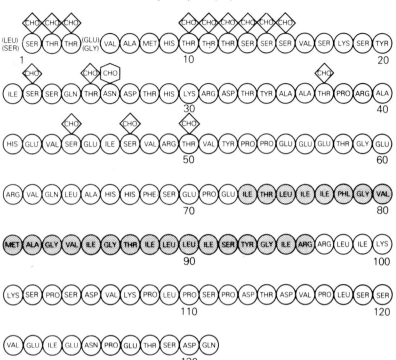

Human Erythrocyte Glycophorin A

Fig. 3.9 Diagram illustrating the mixing of fluorescently tagged cell surface antigens of mouse and human cells after cell fusion induced by Sendai virus. Total mixing of surface antigens (□, ▲) was seen after 40 min. (From Frye and Edidin, 1970).

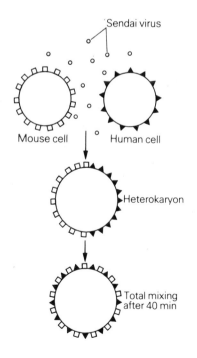

of cell when they were fused into a heterokaryon (Fig. 3.9). They tagged the surface of a mouse cell with specific antibodies bearing a green fluorescent dye, and a human cell with an antibody–red dye combination. After cell fusion (which can be promoted by means of Sendai virus), there was a rapid mixing of the two types of surface marker, suggesting that the membranes were indeed fluid-like, and that proteins were able to diffuse laterally over the entire surface of the hybrid cell.

A third line of evidence consistent with the fluid mosaic model has been derived from the use of plant *lectins*, proteins which are characterized by their abilities to bind to specific carbohydrate groups. Although their function in plants is still uncertain, they can attach selectively to various glycoproteins on cell surfaces. For example, concanavalin A from jack beans is highly specific for α-mannosyl and α-glucosyl residues in oligosaccharide linkages, while wheat germ agglutinin binds preferentially to *N*-acetyl *D*-glucosamine residues. The properties and specificities of lectins have been reviewed by Liener (1976) and are discussed further in Chapter 13. Moreover, because these lectins have more than one combining site (i.e. they are multivalent), they are able to cross-link surface glycoproteins (Fig. 3.10). By employing lectins tagged with fluorescent dyes or electron dense markers, it has been shown that patches of surface glycoproteins quickly form on certain types of cell after they are exposed to the lectin. Usually this is followed by agglutination, or clumping, as multiple cross-links are gradually formed between

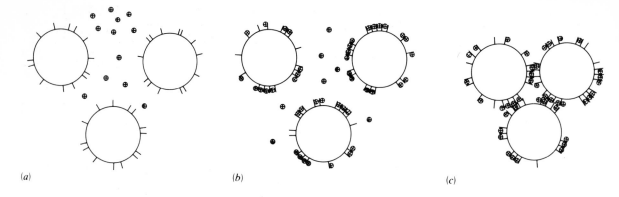

(a)　　　　　　　　　　　　　(b)　　　　　　　　　　　　　(c)

Fig. 3.10　　Diagram illustrating one proposed mechanism to explain agglutination of naked cells by lectins. (a) Cells with glycoproteins (|) arranged randomly over their surfaces are mixed with a multivalent lectin (⊕) capable of binding to specific carbohydrate groups. In (b), patches of glycoprotein begin to develop on surface membranes due to cross-links forming between adjacent carbohydrate chains. Patching is possible, presumably because glycoproteins are laterally mobile in the lipid bilayer. In (c), clumping of cells begins as cross-links are formed between patches of glycoproteins on different cells. The model suggests that patching and hence lateral movement of glycoproteins occurs prior to agglutination. Such lateral movement of lectin binding sites has provided evidence for the fluid mosaic model of membrane structure.

adjacent cells. In certain cells, such as lymphocytes, a large patch or 'cap' is formed on one pole of the cell which is eventually drawn into the cell by endocytosis. These experiments clearly demonstrated that surface glycoproteins are able to diffuse laterally in the plane of the membrane. Moreover, experiments with lectins have also shown that membranes are asymmetric structures, since they usually bind to only the outer surface of the plasmalemma and the internal, luminal face of the endoplasmic reticulum.

Other evidence for the fluid mosaic model has been derived by the technique of freeze fracture (Figs. 3.11 and 3.12). This procedure allows the microscopist to view membranes in both surface and cross-sectional view from material which has not been chemically fixed, and therefore remains chemically unchanged (see p. 22). Freeze fracture studies of a variety of biological membranes have shown that the bilayer is impregnated with a series of globules or particles, normally 6–18 nm in diameter, which are most probably multi-sub-unit aggregates of membrane proteins (Fig. 3.11). It is now known that the fracturing procedure splits membranes in half, exposing the internal faces, rather than the true surface of the membrane (Branton, 1966). These aggregates have been shown to undergo considerable rearrangement in response to such treatments as lectin or antibody binding, again implying that they are mobile.

Although the fluid mosaic model is now assumed to apply to membrane structures of all types, it is generally agreed that proteins cannot migrate laterally in an altogether unrestricted manner. Different parts of the surface of a cell, for example, undoubtedly have different functions and the plasmalemma is not thought to have a uniform structure throughout. The endoplasmic reticulum, though part of a structural continuum, is also non-uniform since it is clearly differentiated into regions which bear ribosomes and others which do not. Some proteins may be rigidly fixed in position by overlying peripheral proteins and others more loosely tethered to cytoplasmic structures, such as the actin-containing microfilaments (Fig. 3.7) or other networks of structural proteins. In addition, a membrane can no longer be regarded as an entirely random structure. Domains seem to exist which are probably determined thermodynamically by particular protein–protein and protein–lipid associations.

Fig. 3.11 Freeze-fractured plasma membrane of a radish root cell. The particles that are seen on the membrane face are located *in vivo* within the continuum of the bilayer membrane. They have been exposed by splitting of the membrane during the freeze-fracture process. The protrusions on the membrane face represent cross-fractured plasmodesmata. In the upper left corner some cellulose fibrils may be recognized in the cross-fractured cell wall. × 60 000. Reproduced from Straehelin and Probine (1970).

Membrane biosynthesis

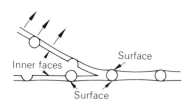

Fig. 3.12 Membrane model proposed to explain the images seen in freeze-etched preparations of membranes. Reproduced from Branton (1966).

Most of the important questions regarding the synthesis of biological membranes have still to be answered. These include the elucidation of the site of membrane synthesis in the cell, the mechanism of the control of synthesis, and the sequence of assembly of the various membrane components.

With a simple model, such as that proposed by Danielli and Davson, membrane synthesis could be envisaged as occurring by self-assembly depending on the associative properties of the component molecules; presumably a phospholipid bilayer would be formed first, followed by the addition of proteins. However, the more complex membrane models, including the fluid-mosaic systems, require a more ordered sequence of events. For example, how can integral proteins, particularly those that span the thickness of the bilayer and have exposed hydrophilic groups on either side, be inserted in such a manner that they always have one particular orientation in the membrane?

There are generally considered to be two alternatives to the problem of the mechanism of membrane synthesis. Either the various

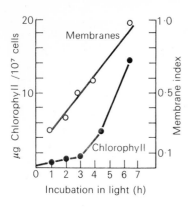

Fig. 3.13 Synthesis of chlorophyll and increase in chloroplast membranes in a mutant of *Chlamydomonas reinhardi* transferred to light after a period of several generations in the dark. Reproduced from Ohad, Siekevitz and Palade (1967).

membrane components – protein, lipid, glycoprotein – are assembled simultaneously in a one-step process or there is a sequential, multi-step assembly mechanism. These possibilities have been investigated by studies of the rates of synthesis of membranes and their components in various plant and animal cells often with conflicting results or interpretations.

Some interesting evidence regarding this problem has come from a study of the biogenesis of chloroplast membranes in a mutant of *Chlamydomonas reinhardi* which is unable to synthesize chlorophyll in the dark (Ohad, Siekevitz and Palade, 1967). The synthesis of plastid membranes is dependent on chlorophyll synthesis and so when grown in the dark for several generations, the cells contained a plastid with only a trace of membranes. On illumination, there is a rapid synthesis of photosynthetic membranes which is accompanied by the synthesis of chlorophyll, lipids and proteins (Fig. 3.13). de Petrocellis, Siekevitz and Palade (1970) followed the changes in chemical composition of thylakoid membranes during the greening of this mutant. During this period, it was shown that the ratios of chlorophyll to a cytochrome designated 554 and to carotenoids, and the relative concentrations of the individual carotenoids were continuously changing; this was consistent with a multistep assembly process. Thorne and Boardman (1971) came to a similar conclusion concerning the assembly of the photosynthetic membranes in etiolted pea seedling exposed to illumination. They found that the appearance of chlorophyll *b* and of energy transfer from carotenoids to chlorophyll *a* preceded by several hours the appearance of Hill activity (see p. 337) and the formation of grana. These observations were not consistent with a single-step process.

The process of protein synthesis takes place on the ribosomes (see Ch. 5). Therefore, the precursors of the plasmalemma and various organelle membranes are synthesized away from their final location. Recent experiments with cultured mammalian cells have suggested that, whereas peripheral proteins of the plasmalemma may be formed on free polyribosomes in the cytoplasm and only later associate with the membrane, integral proteins are assembled on polysomes attached to the rough endoplasmic reticulum. Indeed, the current view is that they constitute a special class of secreted protein (see Fig. 3.14). Most

Fig. 3.14 Proposed model for the synthesis of integral membrane glycoproteins such as the spike glycoprotein of vesicular stomatitis virus. As the nascent chain of the polypeptide emerges from the groove of the large ribosomal sub-unit, the ribosome can attach firmly to the membrane and the hydrophobic signal sequence penetrate the bilayer. As it emerges in the lumen, the signal sequence is clipped off and blocks of carbohydrate (▲) are added. An intermediate hydrophobic sequence then bonds the glycoprotein firmly within the membrane leaving the −COOH terminus free in the cytoplasm. In the case of secreted proteins, the entire polypeptide passes into the lumen of the ER.

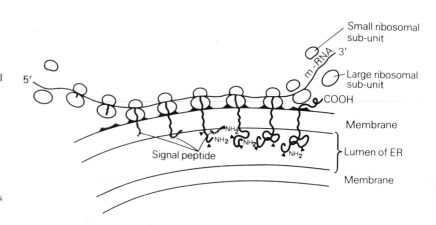

information has come from experiments designed to follow the assembly of coated viruses, such as vesicular stomatitis virus (Rothman and Lodish, 1978; Lodish, 1979).

The advantage of using a virus to study membrane assembly is that host protein synthesis is shut down soon after infection and only viral proteins, which are few in number, are formed. In addition, the viral coat consists of the host plasmalemma into which a single species of glycoprotein, the G–protein (of molecular weight 63 000), is inserted. Assembly of membrane begins in the ER (Fig. 3.14). The mRNA of the virus codes for a precursor peptide which has an extension of about sixteen amino acids on its amino terminus. This peptide extension is not found on the mature glycoprotein of the viral membrane, and is usually known according to Blobel's designation as a 'signal' sequence (see Blobel, 1979). Signal sequences are not restricted to virus proteins but are typical features of the amino termini of precursor chains of most secreted proteins of mammalian cells, including pre-insulins, pre-collagens and pre-immunoglobins, and are noted whenever the mRNAs of such proteins are translated in a cell free system which lacks microsomes (i.e. vesicles of ER) (see p. 192). If such membranes are included in the incubation mixture, the proteins formed are rapidly sequestered in the vesicles and the signal sequence removed. There is no one particular signal sequence seemingly common to a group of related proteins. However, the peptide is always rich in hydrophobic amino acids and, as it emerges from the groove on the large ribosomal sub-unit (see p. 195), it appears to facilitate the attachment of the ribosome and its associated message to the membrane of the rough ER (Fig. 3.14). There are also special accommodation sites for the ribosome on the rough ER. However, the polysome does not attach unless nascent polypeptide chains with signal sequences are present. Thus, the signal sequence is able to form what is assumed to be a firm hydrophobic association with the phospholipid of the membrane. Although the mechanism whereby the polypeptide threads through the bilayer is unclear, soon after its amino terminus emerges into the lumen of the ER it is rapidly processed in a series of post-translational events. The signal sequence, for example, is usually clipped off by means of a specific endopeptidase and carbodydrate groups added at appropriate positions on the chain. However, whereas a secretory protein would be released in soluble form within the lumen, the integral glycoprotein of the virus is partially retained in the membrane. The glycosylated amino terminus is thus exposed within the lumen of the ER while the carboxyl end of about thirty amino acids is retained on the cytoplasmic side. As with glycophorin (Fig. 3.8), a linking, hydrophobic region spans the membrane itself.

The membrane glycoprotein is then transferred to the cell surface. This involves sequential movement through the smooth ER and Golgi apparatus and ultimate delivery of a segment of membrane to the plasmalemma (see p. 497). As the Golgi-derived vesicles fuse with the existing surface membrane, the amino termini of the glycoproteins with their attached carbodydrate become exposed on the outside of the cell (Fig. 3.15). The viral nucleic acid and its associated protein (the

Fig. 3.15 Scheme for the budding of an enveloped virus particle, such as vesicular stomatitis virus, from a eukaryotic cell. In (a), the spike glycoprotein of the viral membrane moves to the surface from the Golgi bodies in the form of vesicles and in (b) becomes associated with the plasmalemma. A second viral protein (M protein), formed on free polyribosomes, attaches to the inner surface of the membrane (b) and seems to help organize the G protein into patches (c) which exclude host proteins but not phospholipids. The M protein is thus an example of a peripheral protein. Finally, (d) the nucleocapsid binds beneath the patches and the whole particle can bud from the surface as an infective virus.

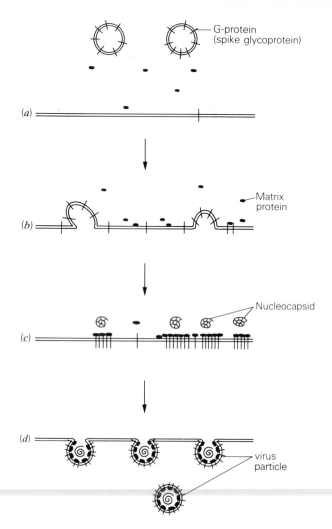

nucleocapsid) assemble beneath areas of host plasmalemma in which the glycoprotein has become concentrated and intact, infective particles bud away from the cell.

If this model of membrane assembly holds true for all eukaryotic cell types, it implies that membrane synthesis is a discontinuous process. The lipid bilayer is assembled first, although this may be immediately modified by either the attachment of peripheral proteins or by integral proteins flowing back into it. Further integral proteins are then introduced directly from attached polysomes.

An alternative approach to the assembly of integral proteins is the trigger hypothesis (Wickner, 1979). This model emphasizes the ability of a membrane lipid bilayer to trigger the folding of a polypeptide chain into a conformation that spans the membrane. The leader sequence of the peptide may again be important, its interaction with the membrane causing the conformational change which allows the protein to cross the membrane.

It is unclear what controls the flow of these constituents to the Golgi apparatus and hence to the surface, or how the initial phospholipid bilayer is assembled. Nor do we know how the distinctive compositions of the sequentially related membrane systems (the ER, Golgi and plasmalemma) are maintained. The synthesis of the membrane systems of the various organelles is even more obscure than the origins of the plasmalemma, and what is presently known is dealt with in the relevant chapters. Even in the chloroplast and mitochondrion, the majority of proteins are probably coded for in the nucleus, so that special recognition signals on such polypeptides must allow for their entry into the organelle and their appropriate orientation and positioning within the membrane system itself.

Membrane transport

The most fundamental property of cellular membranes, and the one to which most of their functions are ultimately linked, is their ability to act as selective permeability barriers, regulating the nature and amount of the substances that move across them. Apart from maintaining a distinct cellular environment in response to external changes, membranes allow intracellular compartmentation to occur which produces a number of specific metabolic microenvironments while also permitting the movement of important metabolites from one compartment to another. The control of metabolism is thus closely linked to the transport properties of membranes. In the remainder of this chapter, the mechanisms by which substances cross membranes will be discussed in general terms, while more specific transport processes involving ions, metabolites and macromolecules are described in later sections in relation to particular organelles.

The forces responsible for the movement of solutes across membranes depend on the nature of the solute. Uncharged solutes respond to the gradient of chemical potential (or activity), while charged solutes are influenced by an additional component resulting from the difference in electrical potential. The former is related to the concentration of the solute and the latter to its net charge; together they make up the electrochemical potential of a charged solute.

Solutes move across a membrane down a chemical or electrochemical gradient, resulting in a loss of free energy by the solute (Fig. 3.16). Such *passive* or *downhill* transport is brought about by diffusion. Some substances are relatively soluble in the lipid phase of the membrane, but this is not the case with polar solutes for which membranes usually show low permeabilities. Movement of charged solutes through membranes may be aided either by the presence of aqueous pores, or by combination with specific *carrier* molecules which are soluble in the membrane. The latter process is known as *facilitated diffusion*. Carrier-mediated transport is generally considered to involve three steps: solute recognition in which the solute binds to the carrier, transport of this complex across the membrane, and release of the solute on the other side of the membrane. The carrier can then move back across the membrane to accept another solute molecule (Fig. 3.17). Such facilitated diffusion may be contrasted with *exchange diffusion* in which there is an equal movement of the solute in the

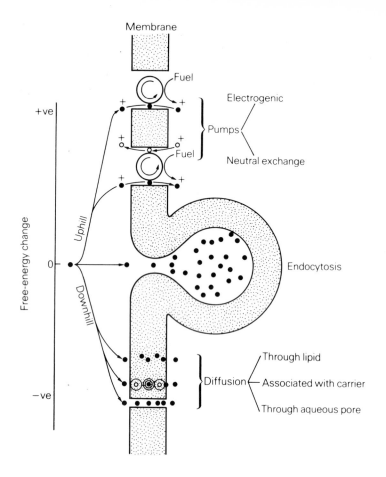

Fig. 3.16 Diagrammatic representation of the types of active (uphill) and passive (downhill) transport that may occur across cellular membranes. Redrawn after Clarkson (1977).

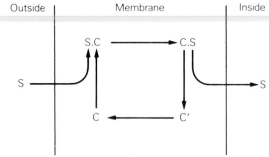

Fig. 3.17 Simplified scheme for carrier-mediated transport. S is the solute, C the carrier, C. S the carrier/solute complex, and C' the mobile carrier precursor which moves back across the membrane and is reconverted to C.

opposite direction, thus involving no net flux. This again is usually attributed to a carrier system. Facilitated movement by carriers is also involved in active transport and is discussed below.

In the case of an ion in an ideal solution, the electrochemical potential, $\bar{\mu}_j$, is given by the following equation:

$$\bar{\mu}_j = \mu_j^* + RT \ln c_j + z_j F \Psi$$

where μ_j^* is the chemical potential of the ion in its standard state, R the gas constant, T the absolute temperature, c_j the concentration of the ion j, z_j the valency, F the Faraday constant, and Ψ the electrical potential of the medium which is determined by its total electrolyte composition and does not depend on the ion j in particular. The driving force on the ion will thus depend on the gradient of electrochemical potential across the membrane which in turn depends on both the concentration gradient and the electrical potential gradient. Since c_j and Ψ can influence $\bar{\mu}_j$ independently, it follows that the demonstration of the accumulation of an ion (against a concentration gradient) does not establish that uphill or active transport is occurring. For example, a solute may be precipitated on entering a cell and so may not move against an overall electrochemical potential gradient. Even without such immobilization, providing a sufficiently large electrical potential difference is maintained, a solute may be passively transported even against a concentration gradient.

An ion will be in equilibrium across a membrane when its electrochemical potential is the same on both sides of a membrane; no change in free energy will therefore occur if ions move from one side to another providing the net flux is zero. This equilibrium state, when the tendency of an ion to move down its chemical potential gradient in one direction is balanced by its tendency to move in the opposite direction down its electrical potential gradient, is described by the *Nernst equation*:

$$\Psi^i - \Psi^0 = \frac{RT}{z_j F} \ln\left(\frac{c_j^0}{c_j^i}\right)$$

where Ψ^i, Ψ^0, c_j^i and c_j^0 are the electrical potentials and ion concentrations inside and outside the membrane and the other symbols are as described above. Thus, if the concentration of an ion is known on either side of a membrane, it is possible to calculate the electrical potential difference required to maintain an equilibrium state across the membrane. The calculated Nernst potential ($E_{N_j} = \Psi^i - \Psi^0$) may then be compared with the measured potential difference across the membrane. If the calculated and measured values are the same, then an equilibrium state exists. If there is a marked difference between the two values, then the system is not in equilibrium and energy must be expended to maintain this situation, i.e. it indicates that some form of active transport is involved.

Active or *uphill transport* may be defined for most practical purposes as a process by which a solute is moved against a gradient of electrochemical potential. Movement is therefore dependent on the decrease in free energy of a metabolic process to drive the solute in a thermodynamically uphill direction: mechanisms which perform this active transport are known as *pumps*. Pumps are envisaged as special types of carriers which are coupled to the utilization of metabolic energy, permitting them to work against chemical or electrochemical gradients. This allows them to absorb solutes from the environment which may be present at very low levels, and to maintain a constant internal ionic environment which may be very different from that outside.

The best characterized pumps are those concerned with the transport of ions of which there are two basic types, *neutral* and *electrogenic*. With neutral ion pumps, there is no net charge transfer across the membrane. This can be accomplished by a one-to-one coupled exchange of ions of equal and like charge in opposite directions which is known as *antiport*. Alternatively two ions of opposite charge may be moved in the same direction in a process known as *symport*. The terminology antiport and symport was developed by Mitchell for translocation across mitochondrial membranes (see Ch. 7) in which the whole macromolecular system catalysing transport (and including the carrier) is known as a *porter*. Neutral pumps are involved in the exchange of intracellular Na^+ for extracellular K^+, and in the active influx of Cl^- which is coupled to passive cation influx. With electrogenic pumps there is a net transfer of charge across the membrane. This may be achieved by the independent, unidirectional transport of an ion (*uniport* in the terminology mentioned above), or by an exchange pump which has a stoichiometry other than one-to-one. Such movement provides the driving force for the passive transfer of an ion of the same charge in the opposite direction or of an opposite charge in the same direction. It should be noted however that the different types of pump are postulated to explain the observed membrane fluxes and potentials in living cells although in most cases little is known concerning the molecular mechanisms involved. Thus the details of how transport occurs remain uncertain. It should also be noted that the terminology uniport, symport and antiport can also describe passive movements since they refer to a type of mediated transport irrespective of whether it is energized.

Having established that pumps are mechanisms that use metabolic energy to drive uphill transport, the means by which this energy is supplied must now be considered. Firstly it is usual to distinguish between *primary* and *secondary active transport* processes. In the former, energy is supplied directly by a chemical reaction involving the transport system. In secondary transport, the uphill movement of one solute is coupled to and driven by the flow of a second solute down its gradient of electrochemical potential; this process is termed *co-transport* or *counter-transport* depending on whether the movement is in the same direction or opposite direction to the solute moving downhill (see below).

The two major energy sources for active transport are ATP hydrolysis and electron flow linked to certain membrane-associated redox reactions. Both ATP and electron flow are generated in respiration and photosynthesis and full descriptions of these processes are included in Chapters 7 and 8. The most thoroughly understood ATP-dependent transport process is the Na^+/K^+ pump of mammalian plasma membranes, particularly red blood cells. The pump is driven by a membrane-bound ATP-ase and moves Na^+ out of and K^+ into the cell (Fig. 3.18). Both the ATP-ase and the pump are dependent on ATP, require Na^+ and K^+ together for maximal activity, and are specifically inhibited by the cardiac glycoside ouabain. Transport is believed to involve a phosphorylated enzyme intermediate in which Na^+ efflux is associated with the phosphorylation step and K^+ influx with the subsequent hydrolysis. The pump regulates the internal ionic environment and is fundamental to many physiological processes, e.g.

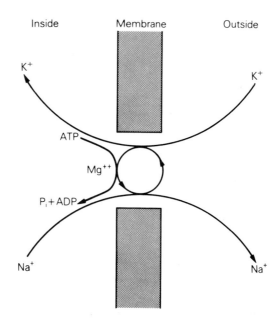

Fig. 3.18 A simplified model of the ATP-dependent, Na^+/K^+ transport system found in animal cell membranes.

Inside   Membrane   Outside

K^+

K^+

ATP

Mg^{++}

$P_i + ADP$

Na^+

Na^+

nerve conduction, secretion. Furthermore, since a major part of cellular metabolism is involved in supplying energy for this active transport, the pump provides a pacemaker for metabolism. For example, in some cells it can regulate O_2 consumption by mitochondria through the production of ADP and P_i from the hydrolysis of ATP (see p. 295). There are other transport systems which also appear to depend on ATP including cation transport in giant algae, although evidence for higher plants is less clear cut. Higher plants certainly contain cation-stimulated ATP-ase activity but correlations with transport are not as strong as in other cells (see Hodges, 1976; Lüttge and Higinbotham, 1979).

The alternative to ATP hydrolysis is to utilize electron flow to produce a potential gradient down which ions can move across membranes. Any discussion of such a mechanism must include the *chemiosmotic hypothesis* which is discussed in detail in relation to phosphorylation and transport in mitochondria and chloroplasts in Chapters 7 and 8. Very simply, electron carriers in mitochondrial and chloroplast membranes are considered to have a vectorial arrangement which allows them to generate a gradient of protons across the membrane as a result of electron flow (see Fig. 7.18). This produces a *proton motive force* which may be used to drive ATP synthesis or solute transport (see p. 309). Transport of cations and anions are thus secondary, counter-transport systems in the above terminology. The processes are reversible so that ATP hydrolysis may be used to generate a proton gradient. However, it is not easy to see how electron flow in these organelles can be linked to transport at the plasmalemma or tonoplast. Nevertheless, using similar concepts to those involved in the chemiosmotic hypothesis, it may be postulated that ATP-ase at the plasmalemma could drive an electrogenic efflux pump producing a proton gradient which in turn could be linked to K^+ influx and to a

Fig. 3.19 Diagrammatic representation of the possible linkage between an ATP-dependent proton pump at the plasmalemma and potassium and anion transport into the cell.

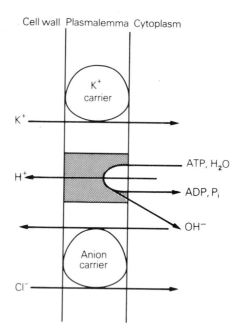

second anion carrier system (Fig. 3.19). Direct evidence for such a scheme is however lacking at the present time.

Co-transport is the third mechanism which can supply energy for active transport. This is achieved indirectly by coupling the energetically uphill movement of one solute to the downhill movement of another (see Baker, 1978). The best known example is the sodium co-transport of amino acids and sugars in various animal cells. An electrochemical gradient for Na^+ is maintained by an ATP-dependent sodium efflux pump as described earlier. The resulting downhill movement of sodium into the cell is coupled to the influx of an organic solute by a facilitated diffusion carrier which has specific binding sites for sodium and the organic solute.

In many microbial systems, protons take the place of sodium which indicates that proton co-transport may be a more fundamental process. As described earlier, a proton gradient could be created by proton extrusion linked to a membrane ATP-ase or electron flow. Uptake of organic solutes could be coupled to the resulting inward flow of protons. A similar system may be involved in sugar loading in the phloem of higher plants and is consistent with the high pH of phloem sap. Experiments have shown that loading is greater with an external medium of pH 5 rather than pH 8 and is inhibited by the uncoupling agent CCCP (see p. 295). A possible scheme for co-transport involving a proton-pumping ATP-ase at the plasmalemma of sieve elements is shown in Fig. 3.20.

Finally, we must briefly consider the macromolecular mechanisms involved in the movement of solutes across membranes by facilitated diffusion or active transport. This can be achieved by carriers or by pores. Carriers are believed to be proteins which are capable of binding specific solutes and transporting them across the membrane (Fig. 3.17). Various mechanisms have been proposed to account for

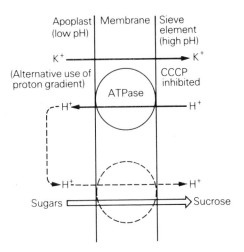

Fig. 3.20 A scheme for a proton-pumping ATP-ase at the plasmalemma of a sieve element. Sugars are co-transported down the resultant proton gradient by combination with a protonated carrier. Redrawn after Baker (1978).

movement across the membrane, including diffusion of the carrier complex or its rotation in the plane of the membrane which brings the binding sites alternately to the inner and outer surfaces. However, little is known of the nature of the carrier or the means by which translocation is coupled to the energy source.

Alternatively transient aqueous pores have been proposed to account for the high permeability of membranes to water and certain small molecules, although their existence is by no means universally accepted. The estimated size of these pores, some 0·4 nm radius, and their possible transient nature, means that they would not be readily visualized by electron microscopy. Pores could be formed by the integral proteins which span the membrane in the fluid mosaic model (Fig. 3.7) and an interesting model for membrane transport has been suggested by Singer (1974). It is proposed that the membrane is bridged by aggregates of integral proteins which form water-filled pores. The pore is initially closed to the diffusion of molecules except water. A specific site for a solute exists on the surface of a pore and faces one side of the membrane when the protein is in a particular conformational state. Rearrangement of the proteins, perhaps triggered by the energy-yielding step in active transport, translocates the binding site and solute from one side of the membrane to the other (Fig. 3.21). This model can be extended to involve the peripheral membrane proteins (see p. 139), some of which may act as specific binding proteins for the solute. Again a conformational change in the integral proteins moves the solute from one side to the other.

Fig. 3.21 A scheme for the translocation event in active transport. Sub-unit aggregates of specific integral proteins span the membrane forming a water-filled pore. The pore is initially closed to the diffusion of molecules other than water. A binding protein with an active binding site for a solute X attaches specifically to the exposed surface of the integral protein. Some energy-yielding step results in a rearrangement of the sub-units of this structure, opening the pore and releasing X to the other side of the membrane. Redrawn after Singer (1974).

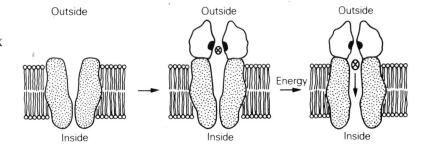

The study of *ionophores* has provided a clue to the molecular basis of carrier transport of ions across membranes. These are antibiotic molecules produced by bacteria and fungi which greatly increase the permeability of artificial lipid bilayers and biological membranes to ions, in some cases in a selective manner. Ionophores are effective even when present at very low concentrations (10^{-8}–10^{-3} mol m^{-3}), and it has been speculated that molecules with similar properties maybe the carriers in biological membranes. There appear to be two major ways in which ionophores act. They either form a complex with the ions which diffuses through the lipid phase of the membrane in a facilitated manner, or they induce transient pores in the membrane through which ions enter.

Valinomycin and monactin are two ionophores which form lipid-soluble complexes with cations allowing the rapid transport of the ions across membranes. They are both macrocyclic structures (Fig. 3.22). The chemical groups on the outside of the ring are non-polar, accounting for the lipid-solubility, while the oxygen atoms on the inside of the ring provide a polar environment with sites for the electrostatic binding of cations. Thus an ion in this exchange site could cross the membrane surrounded by a hydrophobic layer.

The pore-forming ionophores are different in structure, as illustrated by nystatin (Fig. 3.22), a polyene which shows considerable anion selectivity. This molecule is also macrocyclic, but polar groups project from both the outer and the inner faces of the ring. Therefore it is unlikely that the observed ionophoric properties of nystatin are due to the formation of ion-carrier complexes, and it has been suggested that nystatin opens up pores in the membrane. For example, the phospholipid bilayer is normally impermeable to urea (radius 0·23 nm), glycerol (radius 0·27 nm) and glucose (radius 0·40 nm), whereas on addition of nystatin both urea and glycerol can cross the membrane, but not glucose. This suggests the formation of a pore smaller than 0·40 nm. One possibility is that the nystatin interacts with the cholesterol in the lipid bilayer, thereby inducing the formation of lipid micelles (see Ch. 2) which, when closely packed together, produce hydrophilic pores with a radius of approximately 0·4 nm.

It is interesting to note that an artificial, lipid bilayer membrane with ionophores of the above types would resemble the properties of a biological membrane, containing both mobile cation-selective carriers and anion-selective water-filled pore. The properties and uses of ionophores are discussed further in Chapter 7.

A third possible mechanism for solute transport which is not as widely discussed involves membrane vesiculation in the process of *endocytosis*, a term which includes both phagocytosis and pinocytosis (Fig. 3.16). In this process solutes and macromolecules may be drawn into cells by means of invagination of the plasmalemma. The resulting vesicles later break down and discharge their contents into the cytoplasm or fuse with other structures such as the vacuole. Certainly microscopic studies suggest that endocytosis is a widely occurring phenomenon in living cells: it can be readily demonstrated by using electron dense tracers in conjunction with electron microscopy, and

Fig. 3.22 Structural formulae of (a) the carrier-type ionophores, valinomycin and monactin, and (b) the pore-forming ionophore, nystatin.

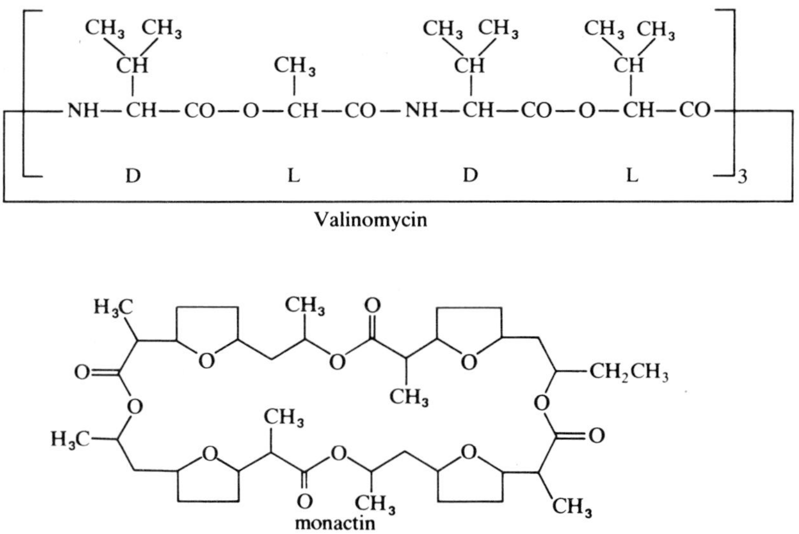

(a) The cyclic dodecadepsipeptide valinoymycin and the macrotetrolide monactin

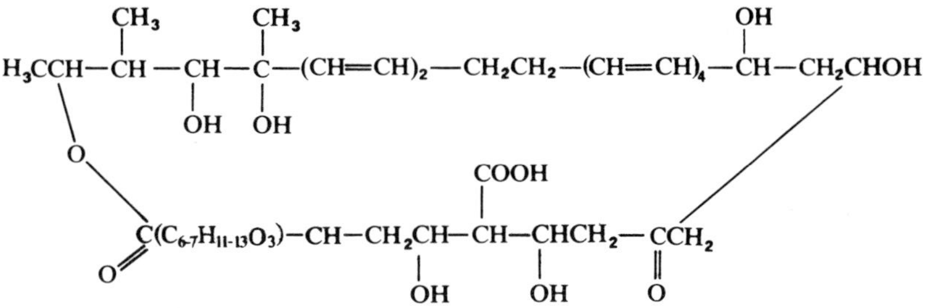

(b) The approximate structural formula for nystatin

cytochemical staining indicates that ATP-ase is associated with the vesicles in some cells (see Baker and Hall, 1973). However, the contribution that such a process makes to the total solute transport is not clear. Furthermore such bulk transport may be considered to lack selectivity although, if solutes become specifically bound to the membrane surface and so become concentrated in comparison to the external medium, considerable selectivity would be introduced.

Further reading

Bretscher, M. and Raff, M. (1975) Mammalian plasma membranes. *Nature, Lond.*, **258**, 43.

Clarkson, D. A. (1977) Membrane structure and transport. In *Molecular Biology of Plant Cells,* Ed. H. Smith. Blackwell, Oxford.

Davis, B. D. and Tai, P-C. (1980) The mechanism of protein secretion across membranes. *Nature, Lond.*, **283**, 433.

Hall, J. L. and Baker, D. A. (1977) *Cell Membranes and Ion Transport.* Longman, London.

Jain, M. K. and Wagner, R. C. (1980) *Introduction to Biological Membranes.* Wiley.

Lodish, H. F. and Rothman, J. E. (1979) The assembly of cell membranes. *Sci. Amer.*, January 1979, p. 48.

Lüttge, U. and Higinbotham, N. (1979) *Transport in Plants.* Springer-Verlag, New York.

Morré, D. J. (1975) Membrane Biogenesis. *Ann. Rev. Plant Physiol.*, **26**, 441.

Rothman, J. E. and Lenard, J. (1977) Membrane asymmetry. The nature of membrane asymmetry provides clues to the puzzle of how membranes are assembled. *Science N.Y.*, **195**, 743.

Sabatini, D. D. and Kreibich, G. (1976) Functional specialization of membrane-bound ribosomes in eukaryotic cells. In *The Enzymes of Biological Membranes*, Ed. A. Martonosi. Plenum Publishing Co., New York. p. 531.

Singer, S. J. (1974) The molecular organization of membranes. *Ann. Rev. Biochem.*, **43**, 805.

Spanswick, R. M. (1981) Electrogenic ion pumps. *Ann. Rev. Plant Physiol.*, **32**, 267.

Stekhoven, F. S. and Bonting, S. L. (1981) Transport adenosine triphosphatases: properties and functions. *Physiol. Rev.*, **61**, 1.

Waksman, A., Hubert, P., Crémel, G., Rendon, A. and Burgun, C. (1980) Translocation of proteins through biological membranes. A critical review. *Biochim. Biophys. Acta*, **604**, 249.

Wickner, W. (1979) The assembly of proteins into biological membranes. *Ann. Rev. Biochem.*, **48**, 23.

Literature cited

Baker, D. A. (1978) Proton co-transport of organic solutes by plant cells. *New Phytol.*, **81**, 485.

Baker, D. A. and Hall, J. L. (1973) Pinocytosis, ATP-ase and ion uptake by plant cells. *New Phytol.*, **72**, 1281.

Blobel, G. (1979) Mechanism for the transfer of newly synthesized proteins across intracellular membranes. In *From Gene to Protein: Information Transfer in Normal and Abnormal Cells*. Eds. T. R. Russel, K. Brew, J. Schultz and H. Faber. Miami Winter Symposium Vol. 16, Academic Press, New York.

Branton, D. (1966) Fracture faces of frozen membranes. *Proc. natl. Acad. Sci. U.S.A.*, **55**, 1048.

Collander, R. (1954) The permeability of *Nitella* cells to nonelectrolytes. *Physiol. Plant.*, **7**, 420.

Danielli, J. F. and Davson, H. (1935) A contribution to the theory of permeability of thin membranes. *J. cell. comp. Physiol.*, **5**, 495.

De Petrocellis B., Siekevitz, P. and Palade, G. E. (1970) Changes in chemical composition of thylakoid membranes during greening of the y-l mutant of *Chlamydomonas reinhardi*. *J. Cell Biol.*, **44**, 618.

Frye, L. D. and Edidin, M. (1970) The rapid intermixing of cell surface antigens after formation of mouse-human heterokaryons. *J. Cell Sci.*, **7**, 319.

Furthmayr, H. (1977) Structural analysis of a membrane glycoprotein. *J. Supra. Molec. Struct.*, **7**, 121.

Hodges, T. K. (1976) ATP-ases associated with membranes in plant cells. In *Transport in Plants. II*. Encyclopedia of Plant Physiology. Eds. U. Lüttge and M. G. Pitman. Springer-Verlag, Berlin.

Korn, E. D. (1969) Cell membranes: structure and synthesis. *Ann. Rev. Biochem.* **38**, 263.

Liener, I. E. (1976) Phytohemagglutinins (Phytolectins). *Ann. Rev. Plant Physiol.*, **27**, 291.

Lodish, H. F. (1979) Biochemical and genetic studies on the biosynthesis of the Vesicular stomatitis virus (VSV) glycoprotein. In *From Gene to Protein: Information Transfer in Normal and Abnormal Cells*. Eds. T. R. Russel, K. Brew, J. Schultz and H. Faber. Miami Winter Symposium. Vol. 16. Academic Press, New York.

Nicolson, G. L. and Singer, S. J. (1974) The distribution and asymmetry of mammalian cell surface saccharides utilizing ferritin-conjugated plant agglutinins as specific saccharide stains. *J. Cell Biol.*, **60**, 236.

Ohad, I., Siekevitz, P. and Palade, G. E. (1967) Biogenesis of chloroplast membranes. *J. Cell. Biol.*, **35**, 553.

Robertson, J. D. (1959) The ultrastructure of cell membranes and their derivatives. *Biochem. Soc. Symp.*, **16**, 3.

Robertson, J. D. (1967) The organization of cellular membranes. In *Molecular Organization and Biological Function*. Ed. J. M. Allen, Harper and Row, New York.

Rothman, J. E. and Lenard, J. (1977) Membrane assymetry. *Science*, **195**, 743.

Rothman, J. E. and Lodish, H. F. (1978) Synchronized transmembrane insertion and glycosylation of a nascent membrane protein. *Nature, Lond.*, **269**, 775.

Singer, S. J. and Nicolson, G. L. (1972) The fluid mosaic model of the structure of cell membranes. *Science N.Y.*, **175**, 720.

Strittmatter, P., Rogers, M. J. and Spatz, L. (1972) The binding of cytochrome b_5 to liver membranes. *J. Biol. Chem.*, **247**, 7188.

Thorne, S. W. and Boardman, N. K. (1971) Formation of chlorophyll b, and the fluorescence properties and photochemical activities of isolated plastids from greening pea seedlings. *Plant Physiol.*, **47**, 252.

Vanderkooi, G. (1974) Organisation of proteins in membranes with special reference to the cytochrome oxidase system. *Biochim. Biophys. Acta*, **344**, 307.

4   The nucleus

The cell nucleus was first described by Robert Brown in 1831 when examining the stamen filments of *Tradescantia*. The nucleus is the largest of the cellular organelles and nuclei are now known to be present in all cells, with a few notable exceptions such as the mature sieve tubes of higher plants and mammalian erythrocytes, although even in such cells nuclei are present during the early stages of development. In the bacteria and the blue-green algae there is no clearly defined nucleus, although localized regions of DNA, known as nucleoids, can usually be recognized by various staining techniques. The presence or absence of a distinct membrane-bound nucleus is usually taken as the basic distinguishing feature between eukaryotic and prokaryotic cells (see p. 10). Although large, the nucleus has proved to be a difficult organelle to study. Its size means that it is difficult to isolate undamaged, since techniques used to break open cells are liable to disrupt the nuclei themselves. With the light microscope much has been learnt of the structure and behaviour of the nucleus and chromosomes, but fine structural observations with the electron microscope have so far proved to be relatively disappointing. The various regions within the nucleus are not separated by membranes and the closely integrated macromolecular structures have presented considerable problems with regard to fixation and staining for electron microscopy.

During interphase or the non-dividing stage of the cell cycle, the nucleus is usually roughly spherical in shape and about 10 μm in diameter (Fig. 4.1), although there may be considerable variation in both the shape and size between different cells and species. In meristematic cells the nucleus is usually spherical and may form as much as 75% of the cell volume, while in differentiated cells which have a large central vacuole it has a more flattened shape and lies against the cell wall. In elongated cells the nucleus may become cylindrical in shape while in others it appears highly lobed. Irregularities in the nuclear outline increase the relative surface area through which molecules may pass and may be an indication of increased activity between nucleus and cytoplasm.

With both light and electron microscopy, the nucleus can be seen to consist of a number of distinct components. The nuclear contents are separated from the cytoplasm by a *nuclear envelope* or membrane which surrounds the nuclear sap (Fig. 4.1). Within this sap there are regions known as *chromatin* which are heavily stained by basic dyes

Fig. 4.1   Electron micrograph of nucleus with a prominent nucleolus in a cell from the root meristem of *Suaeda maritima*. × 13 500.

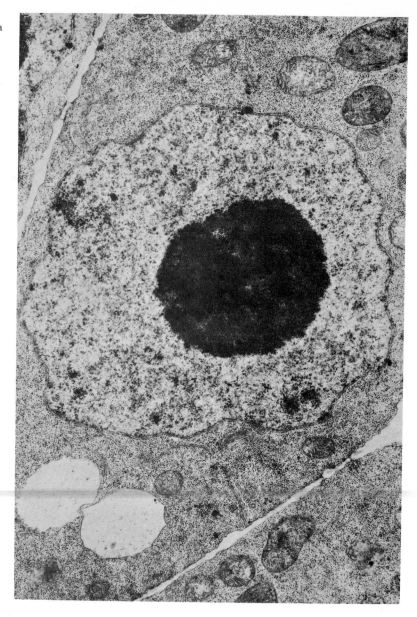

and by Feulgen's technique which is specific for DNA. The chromatin consists of condensed regions known as *heterochromatin* which are usually most heavily stained, and lighter regions called *euchromatin*. In addition, the nucleus contains one or more spherical bodies known as *nucleoli* (Fig. 4.1). When cell division occurs, the nuclear envelope and nucleoli disappear and the chromatin becomes organized into the darkly staining *chromosomes*.

Chemically, nuclei consist chiefly of nucleic acids and proteins, with other compounds such as lipids present only in very small amounts. Table 4.1 shows the proportions of DNA, RNA and protein in isolated nuclei from various higher plant species. There is considerable variation in composition between the different nuclei, although in all cases

Table 4.1 Amounts of DNA, RNA and protein in plant nuclei.

Dry mass $(10^{-12} g)$	DNA $(10^{-12} g)$	% DNA	% RNA	% Protein	Plant
–	–	3	9	88	Yeast
150	13	8	17	75	Tobacco cell culture
70	10	14	12	74	Pea seedling
–	–	16	8	76	Pea stem
84	21	25	21	54	Pea root tip
81	16	20	19	61	Pea root tip
76	31	41	12	47	Pea epicotyl
–	–	7	3	90	Pea embryos
129	33	25	11	63	Pea embryos
–	–	17	4	79	Wheat embryos
–	–	5	4	53	Wheat embryos
152	60	39	9	52	Broad bean root tip
230	77	33	6	61	Onion root tip

For sources see Lynden (1968). Variation in the DNA content of the nucleus within a species can result from polyploid cell lines in certain organs.

protein forms by far the largest part (up to 90%) of the mass of the nucleus and there is usually more DNA present than RNA. The protein component is made up of two major groups, the basic proteins and the non-basic or acidic proteins, the latter making up the largest part of the interphase nucleus. The most well known of the basic proteins are the protamines which are found in fish spermatozoa, and the histones, which are found in the nuclei of higher plants and animals. These proteins have a high content of basic amino acids, particularly arginine and lysine, which gives them very basic isoelectric points of pH 10–11. There are five main classes of histone, designated H1, H2A, H2B, H3 and H4 (Isenberg, 1979), although other nomenclatures have also been used. They complex with DNA by ionic bonding to form the nucleo-proteins, and the possible function of these complexes, particularly nucleo-histones, in the control of gene activity will be discussed later. The non-basic or acidic proteins which constitute up to 70% of the nuclear proteins do not bind with DNA, although they may be associated with the DNA/histone complex. They form the proteins of the nuclear sap, ribosomal proteins, and part of the chromosomal protein.

The complex structure of the nucleus suggests that its rôle in cell function and metabolism may also be complex. It has been known for many years that the nucleus is the major control centre of living cells, and is essential for their survival. Cells from which the nucleus has been removed by microsurgery may continue to metabolize for some time but eventually die unless the nucleus is replaced. The rôle of the nucleus as a source of information was demonstrated by Hämmerling in a classic series of experiments using the large single-celled, mushroom-shaped alga, *Acetabularia*. The various species of this plant differ in the shape of the cap, and Hämmerling showed by a series of grafting and transplant experiments between decapitated plants, that the regenerated cap is always characteristic of the species that contributed the nucleus, clearly demonstrating the importance of the nucleus in the control of cell development (Fig. 4.2). The nucleus contains the

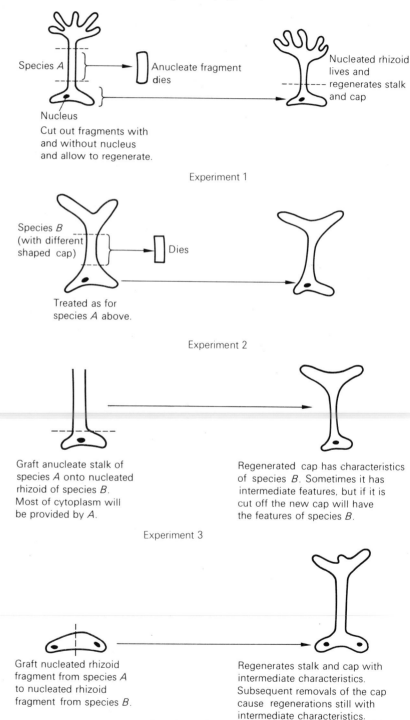

Species A

Nucleus

Cut out fragments with
and without nucleus
and allow to regenerate.

Anucleate fragment
dies

Nucleated rhizoid
lives and
regenerates stalk
and cap

Experiment 1

Species B
(with different
shaped cap)

Dies

Treated as for
species A above.

Experiment 2

Graft anucleate stalk of
species A onto nucleated
rhizoid of species B.
Most of cytoplasm will
be provided by A.

Regenerated cap has characteristics
of species B. Sometimes it has
intermediate features, but if it is
cut off the new cap will have
the features of species B.

Experiment 3

Graft nucleated rhizoid
fragment from species A
to nucleated rhizoid
fragment from species B.

Regenerates stalk and cap with
intermediate characteristics.
Subsequent removals of the cap
cause regenerations still with
intermediate characteristics.

Experiment 4

Fig. 4.2   Examples of grafting
experiments carried out with the large
single-celled alga *Acetabularia*.

160

chromosomes which in turn contain the bulk of the genetic material, DNA, and so clearly has a central rôle in reproduction and in the regulation of cellular activities. At the molecular level these functions involve two basic processes both of which occur in the nucleus. The first is the biosynthesis of DNA which ensures the duplication and continuity of the genetic apparatus. The second concerns the biosynthesis of various RNA molecules which are responsible for the synthesis of enzymes and other proteins and so for the expression of genetic information and the control of cellular function. These processes represent what is called the central dogma of molecular biology and are summarized in Fig. 4.3. This concept involves the duplication of the genetic information contained in DNA, the transfer of this information to RNA, which is called *transcription*, and the expression of this information in the form of protein synthesis, which is called *translation* (see Ch. 5). It must be mentioned, however, that the central dogma is not absolute since in some systems reverse transcription has been demonstrated in which RNA transcribes DNA. It is perhaps more correct to say that information flows from nucleic acids to protein. Thus the nucleus is clearly the major site in the cell both for the storage of genetic information and for the regulation of cellular function. However, it must be remembered that some degree of autonomy exists outside the nucleus as we will see in the chapters concerned with chloroplasts and mitochondria. The rest of this chapter will be concerned with the fine structure of the nucleus in relation to the biochemical activities described above.

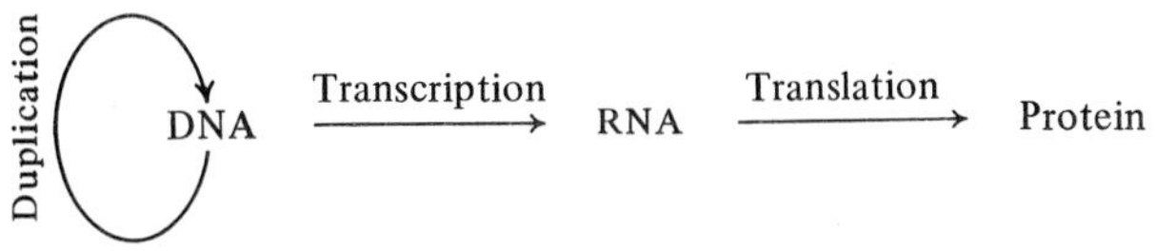

Fig. 4.3   The central dogma of molecular biology.

$$\text{Duplication} \left(\text{DNA} \xrightarrow{\text{Transcription}} \text{RNA} \xrightarrow{\text{Translation}} \text{Protein} \right.$$

Structure of the nucleus

The nuclear envelope

Although it is well below the resolution of the light microscope, the presence of a discrete envelope or membrane around the nucleus was inferred before the use of the electron microscope by various physiological experiments which showed, for example, that isolated nuclei possessed osmotic properties and that dyes injected into the cytoplasm did not penetrate the nucleus. Electron microscopy has shown that the nucleus is, in fact, surrounded by an envelope consisting of two unit membranes, each about 8 nm wide, and separated by a gap called the *perinuclear space* which may vary from 10 to 60 nm in width (Fig. 4.4). The membrane is pierced by pores which vary in diameter between species from 40 to 110 nm and occupy up to 8% of the area of the nuclear envelope. These pores are clearly seen in thin sections (Fig. 4.4) and, more strikingly, in surface views of the nuclear envelope obtained by the freeze-etch technique (Fig. 4.5).

The fine structure of the pore complex has been the subject of considerable discussion, although there now appears to be some

Fig. 4.4   Electron micrographs of
portions of nuclei from plant cells
showing the nuclear envelope,
perinuclear space and pores. (*a*) From
Suaeda maritima roots, fixed in
glutaraldehyde and OsO$_4$. × 54 000.
N, nucleus; V, vacuole; nuclear pores
are marked by arrows. (*b*) High power
micrograph of nuclear envelope of
onion (*Allium cepa*) root tip cells. In
the interpore regions, the inner
membrane is attached to the peripheral
condensed chromatin (*c*). The arrows
denote the annular granules lying in the
pore margin. Note the peripheral
granules which project from the pore
wall into the lumen, and the centrally
placed electron-opaque elements. The
outer nuclear membrane bears
ribosomes. × 150 000. Reproduced
from Franke (1974).

(*a*)
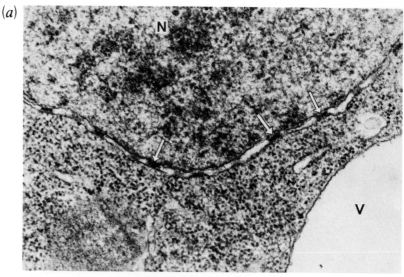

(*b*)
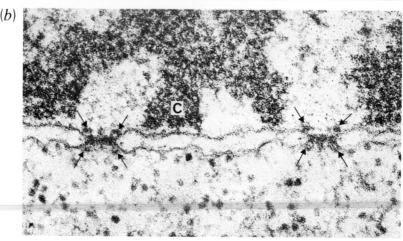

general agreement about certain basic features (see Roberts and
Northcote, 1970; Franke, 1974). The ringlike material lying at the
inner and outer pore margins is known as the annulus and is made up
of distinct globular sub-units, known as the annular granules, each
about 18 nm in diameter. There are usually eight granules attached to
the membrane in a symmetrical arrangement (Fig. 4.6). The pores also
appear to contain an inner series of electron-dense clumps, the
peripheral granules, attached to the membrane in the equatorial plane.
The centre of most pores is occupied by a central granule, usually
about 15 nm in diameter, which may be connected to the pore wall by
fibrillar material. The remainder of the pore often appears to be filled
with an amorphous material with an increased electron density. The
method of operation and significance of the pores is not fully understood. Certainly the presence of the central granule and ground
substance in the pores suggests that they do not allow free communication between nucleus and cytoplasm. It has been suggested that the
pores may open and close to control and passage of substances from

Fig. 4.5 Freeze-etch preparation of nucleus of pea root cell showing nuclear pores. ONM (outer nuclear membrane); INM (inner nuclear membrane). × 58 200. Provided by K. Roberts and reproduced from Northcote (1968).

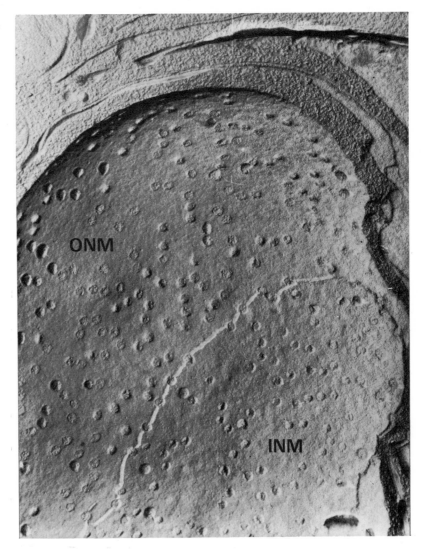

the nucleus but as yet, there is no unequivocal evidence that large molecules such as RNA pass through the pores. Franke (1974) has emphasized that the various globular components described above do not always have a compact form but appear as aggregates of fine filaments. Furthermore, the nuclear pore complex is, after the ribosomes and polysomes, the most RNA-rich structure known and it has been suggested that the pore complex material consists largely of ribonucleoproteins in various stages of coiling (Fig. 4.6). One possibility is that the ribonucleoprotein in the nuclear sap uncoils when it comes into contact with the nuclear membrane and passes as a thread through the pore complex. The central granule may represent RNA molecules in the process of passing from nucleus to cytoplasm, with recoiling and combination with proteins occurring in the cytoplasm.

In higher plant cells the nuclear envelope breaks down at the beginning of cell division and the fragments cannot be distinguished

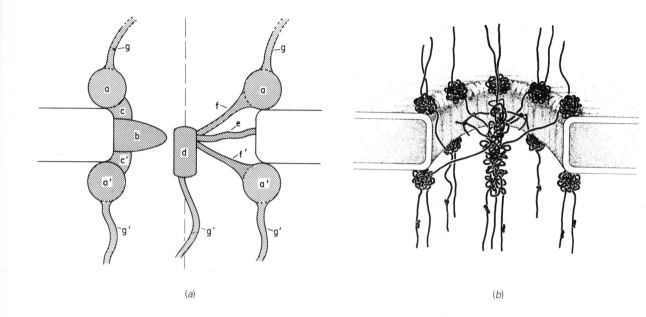

(a)   (b)

Fig. 4.6   (a) Schematic diagram of
the individual components of the
nuclear pore complex: a, annular
granules; b, peripheral granules; c,
material between annular and
peripheral granules; d, central granule;
e, equatorial filaments of the pore
interior; f, inner pore filaments
connecting annular and central
granules; g, axial filaments. The prime
sign denotes components of the
nucleoplasmic side. (b) Diagrammatic
view of the universal principles of the
nuclear pore complex architecture.
Embedded in diffuse material, eight
regularly spaced annular granules lie
upon either pore margin. A central
granule or rod is located in the
innermost part of the pore and is often
attached by filaments to the pore wall
and/or annular granules. The diagram
represents the fibrillar aspects of the
complex, where the granules are seen
as coils of filaments. The peripheral
elements are not included. Reproduced
from Franke (1974).

from strands of endoplasmic reticulum (ER). At the end of cell
division of the envelope is reformed from pieces of the ER which may
or may not have been part of the original nuclear envelope. This
identity of the ER with the nuclear envelope is clearly seen in many
cells where the outer nuclear membrane is continuous with the ER
giving an open communication of the perinuclear space with the ER
cisternae. The significance of the relationship between the ER and
nuclear membrane is discussed in relation to the unit membrane and
endomembrane concepts in Chapters 3 and 11.

The chromosomes

The importance of chromosomes as the major DNA containing
structures in cells and so as carriers of the great majority of hereditary
information has been known for some considerable time. In the
non-dividing or interphase nucleus the chromatin is scattered throughout the nuclear matrix although, of course, it must have an extremely
important function in the control of cellular metabolism. When the
process of nuclear division or *mitosis* begins, the chromatin becomes
organized into the chromosomes which are formed largely form the
euchromatin. The more condensed heterochromatin is usually found
in small segments along the length of the chromosomes; it stains
differently from euchromatin and is also known to synthesize DNA at
a different time. The nuclei of eukaryotes contain a specific number of

chromosomes and during mitosis each chromosome replicates longitudinally, the separate halves moving to opposite poles of the cell to form two identical daughter nuclei. The details of cell division and chromosome behaviour are fully described in textbooks of cytology and genetics and will not be discussed further here. We shall be concerned with the structure of chromosomes in relation to the structure and synthesis of nucleic acids.

Under the light microscope chromosomes show considerable variation between species in both length and thickness. During the process of nuclear division each chromosome is duplicated and becomes divided longitudinally into two halves known as *chromatids* which are joined at a region called the *centromere*. This duplication is associated with DNA synthesis which occurs during interphase before the appearance of the chromosomes in their condensed form. In some chromosomes the chomatids appear to consist of a number of threads called *chromonemata* which have a number of bead-like nodules called *chromomeres* arranged along them.

The fine structure of chromosomes has proved difficult to study with the electron microscope due to the complex coiling of the macromolecular fibrils of which they are composed. In fact, the DNA in chromosomes is so tightly packed that, if uncoiled, it would be many times longer than the chromosome itself; packing ratios (the length of DNA: the length of chromatin fibrils) range from 25:1 to 150:1. Electron microscopic studies have been employed in an attempt to determine how the DNA is folded and coiled to produce the compact chromosome, and ideas in this rapidly developing field have been reviewed by Ris (1978). It should be noted, however, that the relationship between the structures observed by light microscopy as discussed above and the fibrils seen by electron microscopy is by no means clear.

Electron microscopy has revealed globular repeating units in chromatin, each about 10 nm in diameter, which have been called *nucleosomes* (Fig. 4.7). These are thought to contain a piece of DNA (about 200 base pairs) wound around the outside of a core of histone consisting of two molecules each of histones H2A, H2B, H3 and H4. A chain of these nucleosomes forms a fibril which is approximately 10 nm in diameter; the H1 histone does not form an integral part of the nucleosome but is associated with the outside of the chromatin and may perform a cross-linking function, controlling the tightness of the coiling. The 10 nm fibril in turn forms a 20–30 nm fibril which may be produced by helical coiling of the 10 nm fibril or by close packing of groups of nucleosomes. However, it is not clear how these fibrils are organized to form the chromonemata and chromomeres although this presumably could involve further supercoiling. There is also some evidence that inactive chromatin and transcribed chromatin are organized differently: the former appears to be condensed into nucleosomes while the latter may occur as extended thin fibrils. These observations are however not fully in agreement with certain biochemical findings and, in addition, the factors which control these transitions are largely unknown.

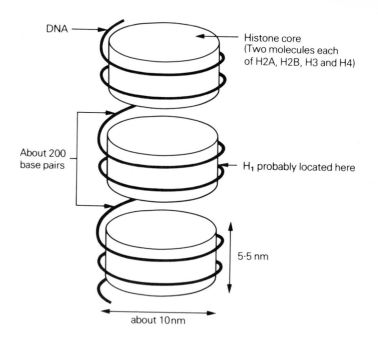

Fig. 4.7   Diagram showing possible arrangement of DNA and histones to form a chain of nucleosomes.

DNA

Histone core
(Two molecules each
of H2A, H2B, H3 and H4)

About 200
base pairs

H_1 probably located here

5·5 nm

about 10nm

Microtubules and chromosome movement

The general morphology and composition of microtubules will be described in Chapter 6 and one potential function of these structures, their rôle in cell wall formation, will be discussed in Chapters 10 and 11. In addition, the microtubules appear to play an important part in the mechanism of chromosome movement during mitosis.

The separation of the daughter chromatids during mitosis is closely related to the structure of the mitotic spindle which is formed at right angles to the plane of cell division and has a fibrous appearance when examined by light microscopy. The chromosomes become arranged along the equator of the spindle during the metaphase of mitosis. The centromeres then divide and the chromatids move to opposite poles of the spindle during anaphase. The spindle fibres are now thought to consist of bundles of microtubules. The spindle region contains numerous microtubules when examined by electron microscopy (Fig. 4.8) and a number of different types have now been distinguished. The *continuous microtubules* run from pole to pole, the *chromosomal microtubules* extend from the poles to the centromere region of the chromosomes, while, during anaphase the *interzonal* microtubules appear in the region between the separating chromatids.

During anaphase the separating chromatids move at a rate of about one μm per minute and usually the spindle itself elongates. A number of models involving the microtubules have been proposed to explain this movement. One possibility is that the chromosomal microtubules contract, but no change in the thickness of these structures has been observed. Inoue (1964) has proposed that the microtubules may shorten or lengthen, without a corresponding change in thickness, by the loss or gain of sub-units which are in a state of dynamic equilibrium

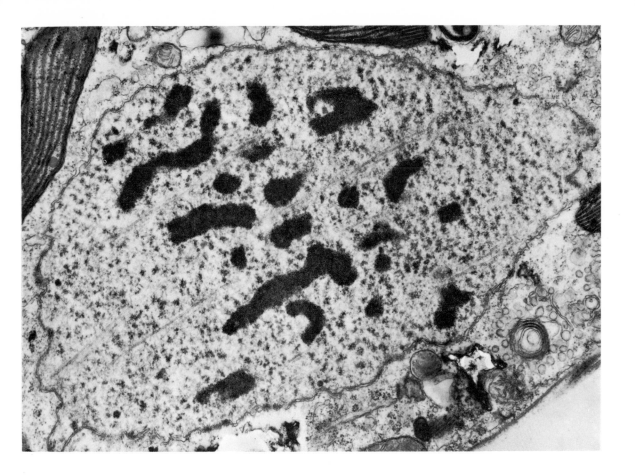

Fig. 4.8   Bundles of microtubules running between the chromosomes in dividing nucleus of *Euglena gracilis*. × 20 600. Reproduced from Leedale (1970).

with the microtubule polymer. However, there is no obvious structure at the poles in higher plants to which the chromosomal microtubules are anchored, while disruption of the fibres close to the poles by ultraviolet microbeams has no effect on chromosome movement (Bajer and Mole–Bajer, 1972). McIntosh, Hepler and van Wie (1969) have suggested an alternative model which involves the very thin bridges which interconnect the microtubules. These are thought to have a mechanochemical function similar to muscle cross-bridges and are able to push one tubule past another. The bridges may contain enzymes which are capable of releasing energy from compounds such as ATP and of converting some of this energy into mechanical work.

The Nucleolus

The nucleolus is roughly spherical in shape and is the most obvious structure in the nucleus of non-dividing cells (Fig. 4.1). Athough it is not bounded by a membrane, it is very clearly defined when observed with the light or electron microscope. Its major rôle in cell metabolism is concerned with the synthesis of ribosomal RNA and so it is very intimately associated with the process of protein synthesis (see Ch. 5).

Fig. 4.9 Normal interphase
appearance of nucleolus of *Spirogyra
sp.* The chromosome (nucleolar
organizer) is the most lightly staining
area and is seen in longitudinal section
(double arrows) and transverse section
(single arrows). F, fibrillar region; *g*,
granular zone; V, nucleolar vacuole.
× 33 000. Reproduced from
Jordan (1971).

Under the electron microscope a number of distinct regions can be
recognized within the nucleolus (Fig. 4.9).

1. A granular region found usually at the periphery of the nucleolus
   which is composed of densely staining granules of 15–20 nm in
   diameter which are thought to be the precursors of the cytoplasmic ribosomes;

2. A fibrillar region composed of fibres of about 10 nm diameter

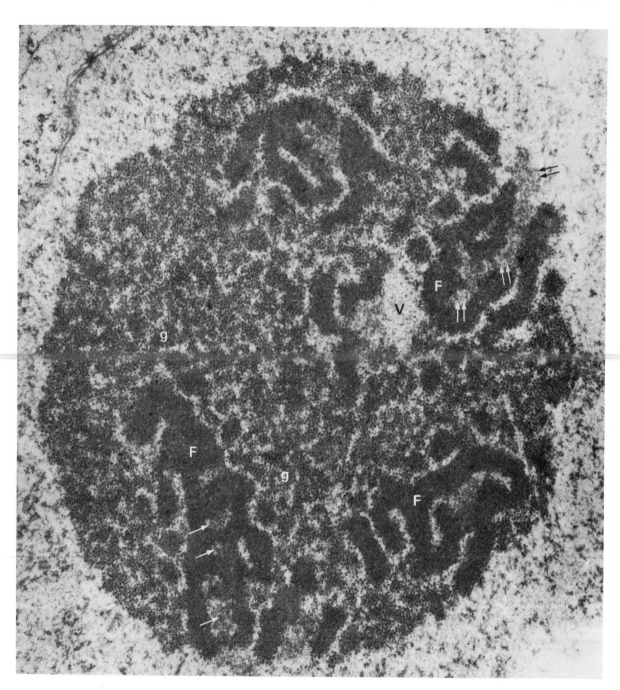

Fig. 4.10   Nucleolus of artichoke
tuber tissue, *Helianthus tuberosus*.
Nuo, nucleolus; c, chromosome; arrow,
the nucleolar organizer. × 46 000.
Reproduced from Jordan (1971).

which are in turn made up of 2 nm diameter fibrils. The fibrils
are thought to consist of ribonucleoprotein and to be the
precursors of the granular region;

3. The fibrils are embedded in amorphous protein, and
4. The nucleolus also contains chromatin which is found both at the

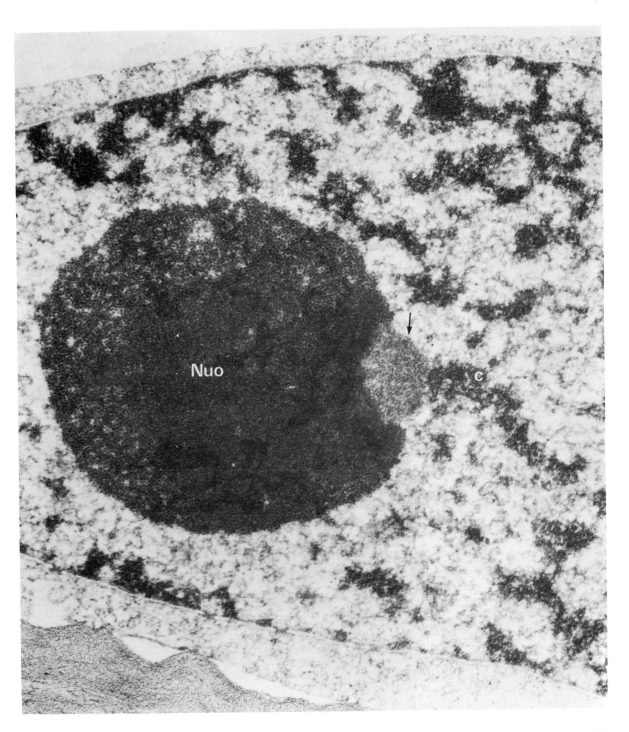

periphery and well inside the nucleolus. DNA makes up 7% of the mass of isolated nucleoli from *Pisum sativum* and 37% of *Vicia faba* nucleoli, although these large variations may represent the problem of preparing nucleoli free of contaminating chromatin (see Lord and Lafontaine, 1969). The DNA is probably composed of several types including the ribosomal RNA cistrons and perhaps that of the nucleolar organizer which will be discussed below.

During the process of cell division the nucleoli follow a series of cyclic changes. In the early stages of mitosis (prophase) the various regions of the nucleolus become less distinct, the fibrillar and granular regions merge with the surrounding nucleoplasm, and the nucleolus finally disappears (Lafontaine, 1974). In higher plants, this usually occurs before the breakdown of the nuclear membrane. Normally, nucleoli are not observed during metaphase and anaphase but reappear again at the end of mitosis (telophase). The reappearance of the nucleolus is closely associated with special regions of certain chromosomes which are known as the *nucleolar organizers* and contain the DNA template for the synthesis of the nucleolar structures (Fig. 4.10). The reassembling nucleolus is usually first seen as a region of fibrillo-granular material adhering to the decondensing chromosomes and it is possible that the nucleolus is permanently attached to the nucleolar organizer which forms part of the nucleolar chromatin material. In an electron microscope study of roots of *Plantago ovata*, Hyde (1967) described heterochromatin intrusions into the nucleolus which may represent a permanent attachment from the nucleolar organizing chromosome in the interphase nucleolus. There may be more than one nucleolar organizer but the developing nucleoli usually eventually fuse to form a single nucleolus. In some cells, however, there may be more than one nucleolus present at the end of the cell division, although the nucleoli frequently coalesce as the cell matures.

Functions of the nucleus

As discussed earlier the nucleus is the major site of DNA in the eukaryote cell and so forms the functional genetic apparatus responsible for the preservation, replication and expression of genetic information. At the molecular level this function involves the biosynthesis of both DNA and of the various RNA molecules which are reponsible for the translation of genetic information into various protein molecules (see Ch. 5). The chromosomes direct the synthesis of messenger RNA while the major function of the nucleolus is the synthesis of ribosomal RNA. The nuclear envelope probably has an essential rôle in the regulation of the movement of molecules between nucleus and cytoplasm but it is not clear at present how this is achieved. The other major component of the nucleus is the nuclear sap. Studies of nuclei isolated from animal tissues suggest that the nuclear sap may contain a variety of enzymes including a number associated with glycolysis and the TCA cycle, although these findings may well be due to cytoplasmic contamination. However, it appears that in animal cells the nucleus is the major site of NAD production due to the presence of the enzyme NAD pyrophosphorylase, which catalyses the reversible reaction:

$$\text{Nicotinamide mononucleotide} + \text{ATP} \rightleftharpoons \text{NAD}^+ + \text{pyrophosphate}$$

This may enable the nucleus to exert some control over the oxidative processes of the cell, although there is also increasing evidence of a linkage between NAD metabolism and DNA synthesis in some animal cells and slime moulds. NAD is the substrate for the synthesis of an unusual polynucleotide known as poly-(ADP-ribose). This is a polymer of ADP units each of which is attached to an extra ribose molecule. NAD is split to give nicotinamide and an ADP-ribose unit which is added to the end of the poly-(ADP-ribose) chain. Poly-(ADP-ribose) inhibits DNA replication which suggests that NAD accumulation and DNA synthesis may be inversely correlated.

Evidence for DNA as the genetic material

Before considering the steps involved in the duplication of DNA and in the DNA-directed synthesis of RNA it is perhaps worth summarizing the evidence that has led to the conclusion that DNA is the genetic material in the vast majority of cells, except certain viruses where the genetic information is contained in RNA. The evidence, both direct and indirect, has come from several lines of approach.

1. DNA is metabolically very stable and the amount of DNA per cell in a given species of higher organism remains fairly constant and unaffected by changes in the environment or nutrition. This stability would be expected of the genetic material.
2. In higher organisms, experiments have shown that the chromosomes contain the genes and that the DNA of the cell is very largely found in the chromosomes.
3. The composition of the bases in DNA (see p. 70 is very specifically related to the species. The ratio of the four purine and pyrimidine bases found in DNA is constant for all tissues of the same organism but differs from that of other species. This would be expected if all cells of an organism contain the same genes. This observation is supported by experiments using the technique of molecular hybridization (see p. 75 which shows that there is greater homology between the base sequences of DNA from closely related species than between DNA from more widely separated organisms.
4. There is a direct correlation between the absorption of ultraviolet light by DNA and the rate of mutation. The rate of mutation is highest at the wavelength of maximum absorption, suggesting that DNA is the hereditary material and when irradiated undergoes chemical changes which cause mutations.

5. Similar interpretations can be advanced with the use of some chemical mutagens. For example, the incorporation of base analogues such as the 5-bromouracil into DNA results in an increase in the mutation rate.

6. The best direct evidence for the rôle of DNA as the genetic material has come from the study of transformation in bacteria. Transformation is a phenomenon in which a certain genetic characteristic, such as antibiotic resistance, may be transferred to a non-resistant strain of the same or related species by the addition of extracts from the resistant strain. The transformation is permanent and the transforming principle has been identified as DNA.

7. Further evidence has come from a study of bacteriophages which are viruses which attack bacteria and are composed only of DNA and protein. Again, it has been thoroughly established that it is the DNA and not the protein which is injected into the bacteria and is responsible for the replication and multiplication of the virus.

8. Finally, it has been shown that in the process of sexual reproduction in bacteria where the genome of one bacterium is transferred to another, there is a correlation between the number of genes and the amount of DNA transferred at any given time after mating has begun.

Replication of DNA and the chromosomes

The DNA of the chromosomes must be duplicated before cell division to ensure that the daughter nuclei each receive the full complement of genetic information. With the Watson–Crick model for the structure of DNA (see Ch. 2), it is considered that the two strands of the double-helical DNA model are complementary. Replication of DNA occurs by a separation of the strands together with the synthesis of complementary strands to form two daughter double helices, each identical to the parent DNA and containing one strand from the parent (Fig. 4.11). This is known as *semi-conservative replication*. The first experimental evidence for this hypothesis was obtained by Meselson and Stahl in 1958 using *E. coli* cells. The bacteria were grown for several generations using ammonium chloride labelled with the ^{15}N isotope rather than the usual ^{14}N as the only nitrogen source. The bacterial DNA thus became labelled with ^{15}N and could be separated from the lighter, normal ^{14}N DNA by centrifugation in a gradient of caesium chloride. The cells labelled with the heavy DNA were then transferred to a medium containing normal ammonium chloride and grown for several generations. At various intervals the cells were harvested and the DNA extracted and separated by centrifugation. After one generation the isolated DNA was of one type and of density midway between that of the light and heavy types of DNA which is consistent with the semi-conservative hypothesis as the DNA would be expected to contain one heavy (^{15}N) and one light (^{14}N) strand. After two generations, two types of DNA could be isolated, one of medium

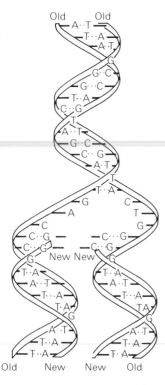

Fig. 4.11   The replication of DNA. Reproduced from James D. Watson, *Molecular Biology of the Gene*, Second edition, copyright © 1970 by J. D. Watson, W. A. Benjamin, Inc., Menlo Park, California.

density and of light (^{14}N) density. These results are illustrated in Fig. 4.12 and are clearly consistent with the replication of DNA predicted by the Watson–Crick hypothesis.

Fig. 4.12 Schematic representation of the Meselson–Stahl experiment which demonstrated the semi-conservative representation of DNA. Redrawn from Lehninger (1970).

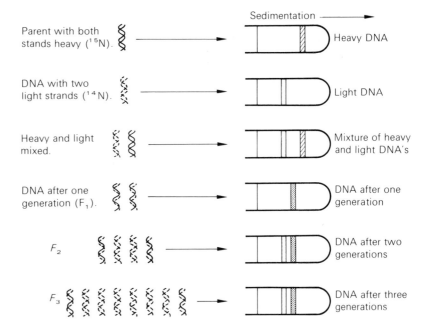

Since this pioneer work with bacterial systems, semi-conservative replication of DNA has been demonstrated in both mammalian and higher plant cells. For example, Filner (1965) studied cultured tobacco cells which multiplied exponentially with a generation time of 2 days and, using ^{15}N-nitrate as the nitrogen source, demonstrated a semi-conservative replication of DNA.

At about the same time as the experiments of Meselson and Stahl, Kornberg and his co-workers isolated an enzyme from *E. coli* which was thought to be responsible for the replication of DNA and was known as DNA polymerase. This enzyme catalyses the synthesis of DNA from the four deoxyribonucleoside triphosphates in the presence of magnesium ions and either native or single-stranded DNA as a template.

n (Deoxyribonucleoside triphosphate)

$$\xrightarrow[\text{Mg}^{++}]{\text{DNA}} \text{(Deoxyribonucleoside monophosphate)}n + n \text{ pyrophosphate}$$

Replication of DNA is initiated by the formation of a 'nick' or single-stranded break in the DNA by the action of a DNA endonuclease. This allows DNA polymerase to bind and catalyse the synthesis of two new strands of DNA. Synthesis occurs by the addition of mononucleotide units to the free 3′ end of a DNA chain (see p. 70) with replication occurring in the 5′ to 3′ direction. Since the two strands of DNA are anti-parallel, this means that the enzyme must work in two physical directions (Fig. 4.13). For this to occur, one of

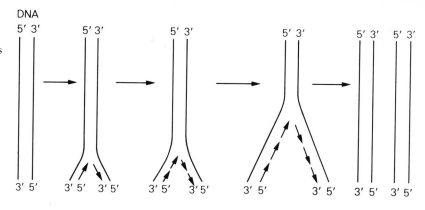

Fig. 4.13 DNA replication. Both strands are synthesized in the 5' to 3' direction. One is copied continuously and the other as a number of fragments Redrawn after Bryant (1976).

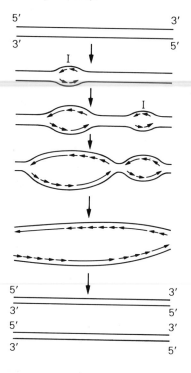

Fig. 4.14 DNA replication in a eukaryotic chromosome. The synthesis of new strands occurs in both directions from each initiation point (I). Redrawn after Bryant (1976).

the strands must be synthesized discontinuously in sections (known as *Okazaki fragments* after the discoverer of short pieces of newly synthesized DNA in bacteria) to be later joined by DNA ligase activity to make a complete strand (Fig. 4.13).

Since this original discovery of DNA polymerase activity, now known as DNA polymerase I, two more such enzymes, DNA polymerases II and III, have been detected in *E. coli* and other bacteria which are able to synthesize DNA in a semi-conservative manner. It now appears that polymerase I is not responsible for DNA replication *in vivo*; its replication rate *in vitro* is much slower than the rate *in vivo*, while mutants have been found with very low levels of the enzyme but which nevertheless replicate and grow at the normal rate. Other studies with *E. coli* mutants suggest that polymerase III is involved in replication while polymerase I (and perhaps II) are important in gap-filling and DNA repair.

Eukaryotic cells also contain more than one DNA polymerase. Apart from those associated with mitochondria and chloroplasts (see Chs. 7 and 8), mammalian and higher plant cells appear to contain two major nuclear DNA polymerases (see Stevens and Bryant, 1978; Stevens, Bryant and Wyvill, 1978). DNA polymerase–α is a high molecular weight ($1-2 \times 10^5$), freely soluble enzyme, while DNA polymerase–β has a lower molecular weight (about 5×10^4) and is bound to the chromatin. The activity of the soluble enzyme shows a closer correlation with DNA synthesis than the chromatin-bound polymerase and so is believed to be involved in DNA replication; the latter may be involved in DNA repair. Lower eukaryotes (e.g. algae, fungi) apparently lack a chromatin-bound enzyme but possess two or three soluble, high molecular weight DNA polymerases.

Eukaryotic chromosomes contain very much more DNA than those of prokaryotes and DNA replication occurs at a much faster rate (see Bryant, 1976). This is only possible because DNA synthesis is initiated at many points (perhaps up to a 1 000) along each chromosome (Fig. 4.14). Two replication forks travelling away from each other are produced at each initiation point and replication continues until the forks merge with those from adjacent initiation points. DNA synthesis is thus continuous in both strands. DNA ligase activity, which is necessary to join the newly synthesized fragments together, has been detected in higher plants and has been shown to have an absolute requirement for ATP and magnesium ions.

How is the synthesis of DNA related to the replication of the chromosomes? Evidence that chromosomes replicate in a semi-conservative manner has been obtained by Taylor and his co-workers in a series of autoradiographic studies of plant chromosomes (see Lewis and John, 1963; John and Lewis 1969). Roots were grown in a medium containing thymidine labelled with tritium. Thymidine is a precursor of DNA and so is incorporated into the chromosomes which can then be identified by autoradiography. After a period in the radioactive solution the roots were transferred to a non-labelled medium containing the drug colchicine. Colchicine not only contracts the chromosomes making them more clearly visible, but also prevents cell division though not chromosome duplication and so results in the retention of the chromosomes of two divisions within the original cell. The results of these observations are summarized in Fig. 4.15. Chromosomes duplicating during the period of division in labelled precursor showed both chromatids to be equally labelled throughout. After one replication in the unlabelled medium, one chromatid was labelled while its sister was completely unlabelled. After another division, half the chromosomes were unlabelled while the remainder showed one labelled and one unlabelled sister chromatid. One interpretation of these observations is that the interphase chromosomes already consist of two components, behaving as though they had two units of DNA along their length, although these are not visible by the usual staining methods. Each of these components behaves as a DNA template and so when duplication occurred in the presence of labelled thymidine, each chromatid contains an original non-labelled strand, although the whole chromatid appears to be labelled by autoradiography. When division occurs in the non-labelled medium, the labelled and non-labelled strands each build a non-labelled strand, resulting in one labelled and one non-labelled sister chromatid. This same general pattern of labelling has now been described in a variety of higher plant species, although exceptions have also been reported. However John and Lewis (1969) conclude that 'there seems little doubt that the mechanism of chromosome duplication, like that of DNA replication, is semi-conservative'.

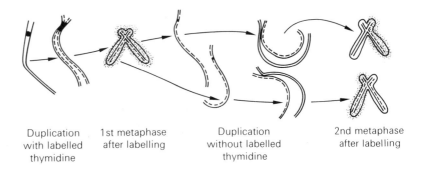

Fig. 4.15 Diagrammatic representation of the distribution of DNA sub-units during chromosome replication. Broken lines represent labelled sub-units and unbroken lines represent grains in the autoradiograms. Reproduced from Taylor (1963).

Duplication with labelled thymidine   1st metaphase after labelling   Duplication without labelled thymidine   2nd metaphase after labelling

Synthesis of RNA

The DNA in the nucleus serves as a template for the synthesis of RNA molecules which thus have a base sequence complementary to that of the DNA. The genetic information in the DNA is thus mobilized and amplified many-fold. The enzyme responsible for this synthesis is known as RNA polymerase or DNA-dependent ribonucleoside triphosphate: RNA nucleotidyl transferase and catalyses the following reaction:

n(ribonucleoside–5'-triphosphate)

$$\xrightarrow[\text{Mg}^{++}\text{ or Mn}^{++}]{\text{DNA template}}$$ (ribonucleoside–5'-monophosphate)n + n pyrophosphate

The basic reaction is therefore similar to that catalysed by DNA polymerase (see above). It requires a DNA template, Mg^{++} or Mn^{++} ions, and the four ribonucleoside triphosphates. The template specificity of the enzyme varies with the source and with the state of purification, some prefering single-stranded DNA while others are more active with double-stranded DNA. However, some strand separation must take place for initiation, although this may be quite restricted as the RNA polymerase moves along the duplex structure to produce a complementary copy (Fig. 4.16). The antibiotic actinomycin −D inhibits the action of RNA polymerase by forming non-ionic complexes with DNA.

Fig. 4.16   A schematic mechanism for the action of RNA polymerase. The attachment of the enzyme RNA polymerase to a DNA molecule opens up a short section of the double helix, thereby allowing free bases on one of the DNA strands to base pair with the ribonucleoside-(P) ~ (P) ~ (P) precursors. As RNA polymerase moves along the DNA template, the growing RNA strand peels off, allowing hydrogen bonds to form between the two complementary DNA strands. Reproduced from James D. Watson, *Molecular Biology of the Gene,* Second edition, copyright © 1970 by J. D. Watson, W. A. Benjamin, Inc., Menlo Park, California.

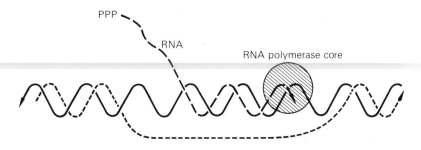

Eukaryotic cells contain multiple forms of RNA polymerase which differ in their properties, intracellular location and function. Apart from the enzymes found in mitochondria and chloroplasts, there appear to be three forms associated with the nucleus (see Duda, 1976; Key, 1976; Jendrisak, 1980) which are known as RNA polymerases I, II and III. These differ not only in their chromatographic properties, but also in their sensitivity to α-amanitin, a toxin isolated from mushrooms. RNA polymerase I is resistant while polymerase II is considerably more sensitive than III. RNA polymerase I is localized in the nucleolus and primarily transcribes ribosomal RNA. RNA polymerase II is probably responsible for the synthesis of messenger RNA, while RNA polymerase III appears to synthesize transfer RNA and perhaps 5 S ribosomal RNA.

The concept of messenger RNA (mRNA) dates from 1961 when Jacob and Monod proposed the existence of short-lived RNA molecules with base sequences complementary to DNA, which carry information for the assembly of amino acids into polypeptides from the DNA of the structural genes to the sites of protein synthesis in the cytoplasm, the ribosomes. The general features of mRNA have been described in Chapter 2. The demonstration of specific mRNA is however difficult to achieve since the cell must contain a large number of different mRNAs which carry the information for the synthesis of a wide range of proteins although they make up only a small percentage of the total cellular RNA.

Recently this problem has been made more approachable by two findings. Firstly many, but by no means all, mRNAs have been shown to contain sequences of poly-(adenylic acid) (poly-(A)) at the 3' end of the molecule, usually from 50 to 250 nucleotides in length. The function of this poly-(A) is not clear; it is not translated although it has been suggested that it may increase the stability of mRNA in the cytoplasm by slowing down its degradation by ribonuclease. It is also interesting to note that the large poly-(A) sequences are added after transcription by RNA polymerase, although short sequences of oligo-(A) may be transcribed. However, the presence of poly-(A) provides a valuable means of purification by affinity chromatography (see p. 40). For example, poly-(A) will bind to columns of poly-(uridylic acid) attached to sepharose while other types of RNA pass straight through. Secondly the development of *in vitro* (cell-free) assay systems for protein synthesis has made it possible to demonstrate that a mRNA fraction can direct the synthesis of a clearly definable protein. The most successful system isolated from plants is that prepared from wheatgerm. Very simply, dry wheat embryos are ground thoroughly and the homogenate centrifuged at about 20 000 g to remove the larger organelles (e.g. cell walls, nuclei, mitochondria). The supernatant containing the ribosomes and other factors needed for protein synthesis may then be used as a 'translation' system for various mRNAs. Technical advances such as these have lead to the identification of a number of mRNAs from higher plants including those for phenylalanine ammonia lyase, flavone synthase, α-amylase, leghaemoglobin from root nodules, and a variety of seed storage proteins (see Leaver, 1980).

When cells are incubated with radioactive precursors of RNA, up to 90% of the RNA that becomes rapidly labelled is located in the nucleus. Much of this becomes widely dispersed in separation systems such as sucrose gradients and polyacrylamide gels (see Ch. 2) with molecular weights ranging from about 300 000 to several million. This is generally known as *heterogeneous nuclear (Hn) RNA* although it has also been described as polydisperse, messenger-like and DRNA (see Key, 1976). Much of this RNA is rapidly broken down in the nucleus while a small proportion (up to 10%) is transferred to the cytoplasm where it becomes associated with polyribosomes. It is thus assumed to be a precursor of mRNA and has been shown to contain poly-(A) sequences (see above). However, the exact relationship between Hn RNA and mRNA is not clear although a comparison of nuclear and cytoplasmic poly-(A)-containing RNA suggests that the nuclear RNA is reduced in size before entering the cytoplasm (Grierson, 1977). Why only a small proportion of the HnRNA appears in the cytoplasm is also unclear. It may be that it is more efficient to transcribe large parts of the genome at the same time and later to break down this polygenic RNA into individual RNAs than to activate single genes in a controlled manner. The possible rôle of HnRNA in gene regulation is discussed later in this chapter.

Synthesis of ribosomal RNA

As described in Chapter 5, the RNA of cytoplasmic ribosomes of higher plants consists of two major and two minor components with sedimentation coefficients of 25 S, 18 S, 5·8 S and 5 S, and there is now a wide variety of evidence which suggests that the nucleolus is the site of synthesis of this RNA. Inactivation of the nucleolus by micro-beam ultra-violet light inhibits the formation of new cytoplasmic RNA, while mutants of the toad *Xenopus* which lack nucleoli are unable to synthesize ribosomal RNA (rRNA). Isolated nucleoli have been shown to synthesize rRNA and this finding has been supported by a variety of autoradiographic studies on intact cells. For example, Brady and Clutter (1972) isolated radioactive (H^3)rRNA from bean hypocotyls and hybridized it to complementary DNA in the polytene chromosomes observed in cytological preparations. Autoradiography showed that the ribosomal genes were localized in the nucleolar organizer region in four of the giant chromosomes.

In most plants, between 0·02 and 1·0% of the total nuclear DNA hybridizes with rRNA. This DNA has a higher content of guanine and cytosine than the bulk of the nuclear DNA and occurs in reiterated sequences. In fact, there are from several hundred to several thousand genes for rRNA, the number varying between species. This multiplicity of genes is presumably necessary to meet the heavy demand for ribosomes in the cytoplasm. The two large rRNAs hybridize separately to DNA and there appear to be equal numbers of genes for each of these molecules. However, hybridization evidence shows that these genes are transcribed together, the best described eukaryotic systems being those of *Xenopus* and certain mammalian cells. Here the genes for the two large rRNAs alternate with each other and are separated by spacer sequences that are not transcribed. In addition there are extra sequences of transcribed DNA which are interspersed between the rDNA genes and subsequently discarded during maturation. Thus rRNA is initially transcribed as a single large precursor molecule that undergoes post-transcriptional cleavage and modification to give the 28 S and 18 S rRNAs of the ribosomal sub-units (see Ch. 5). The precursor molecule is said to be *polycistronic*; that is, it is a single transcription product of more than one gene (or cistron). There is considerable diversity in the size of the rRNA precursors from different species. Those from birds and mammals are generally the largest (e.g. a molecular weight of about 4.2×10^6 for Hela cells) while higher plant precursors range from $2.3 - 2.9 \times 10^6$. Polycistronic precursors are not normally found in prokaryotes; the 23 S and 16 S rRNAs appear to be transcribed as a single strand but are immediately separated to give individual precursors which do however contain some excess sequences. After transcription the RNA is modified by the methylation of certain bases (see p. 76). The rRNA is processed as a ribonucleoprotein since proteins appear to be added during transcription. rRNA is never found in the cytoplasm as free RNA.

Although not as fully characterized as some animal systems, higher plants appear to show a similar sequence of events for rRNA synthesis. When plant cells are incubated with radioactively labelled precursors of RNA such as ^{32}P-orthophosphate or ^{3}H-uridine for various periods of time, the distribution of radioactivity shows that certain high molecular weight molecules are synthesized before the 25 S and 18 S rRNAS (see Grierson, 1976, 1977; Key, 1976). For example, if artichoke tuber tissue is fed with ^{32}P-orthophosphate and the rapidly labelled RNA fractionated by gel electrophoresis, peaks of radioactivity are found in RNA with molecular weights of $2.3, 1.4$ and 0.9×10^6 (Fig. 4.17). The 25 S and 18 S rRNAs have molecular weights 1.3 and 0.7×10^6 and have a much lower level of radioactivity in the early stages of the experiment. The rapidly labelled molecules do not accumulate, have similar base compositions to rRNA, and compete with 25 S and 18 S rRNA in DNA hybridization studies. These

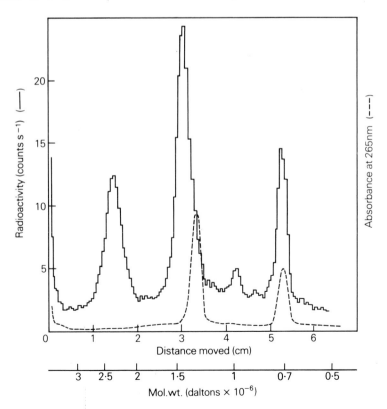

Fig. 4.17 Rapidly-labelled RNA from artichoke (*Helianthus tuberosus*) tuber tissue. Cultured tuber explants were incubated in ^{32}P-orthophosphate for 15 min, followed by non-radioactive phosphate for 45 min. Extracted RNA was fractionated by electrophoresis in polyacrylamide gels. Three peaks of radioactivity were observed (molecular weights of 2.3, 1.4 and 0.9×10^{-6}) which had only a short life and did not normally accumulate. The mature 25 S and 18 S rRNAs have molecular weights of 1.3 and 0.7×10^{-6}. Reproduced from Rogers *et al.* (1970).

findings clearly indicate that the high molecular weight, rapidly labelled molecules are precursors of rRNA.

A generalized scheme for the synthesis of rRNA in higher plants based on findings using both plant and animal tissues is shown in Fig. 4.18. It should be noted that some of the detail is uncertain while results obtained with some plants are not consistent with the scheme. For example, with carrot, an extra, larger precursor is detected (Leaver and Key, 1970) and it is not clear whether the two largest precursors are synthesized separately or if one is converted to the other.

Fig. 4.18   The synthesis and processing of the precursor to rRNA. Redrawn after Grierson (1977).

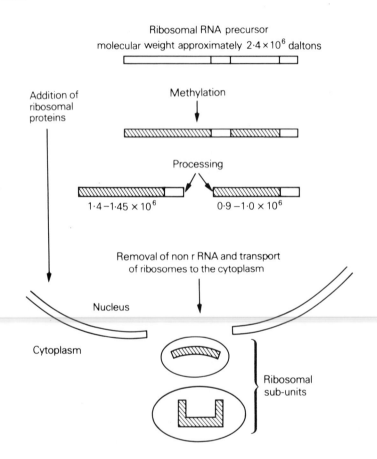

The visualization of rRNA transcription was first strikingly demonstrated in amphibian egg cells by Miller and Beatty (1969). Nucleoli were isolated, the nucleolar DNA material separated and then dispersed for examination by electron microscopy. This material consisted of thin fibres about 10–30 nm diameter which were periodically coated with a matrix of hairlike fibrils (Fig. 4.19). Enzymic digestion experiments suggested that the axial fibres consisted of DNA coated with protein while the hairlike fibrils consisted of RNA and protein. Each matrix region contained approximately 100 hair-like fibrils which were connected to the axial fibre at one end and increased in length from the narrow to the wide end of the region. Each matrix-covered

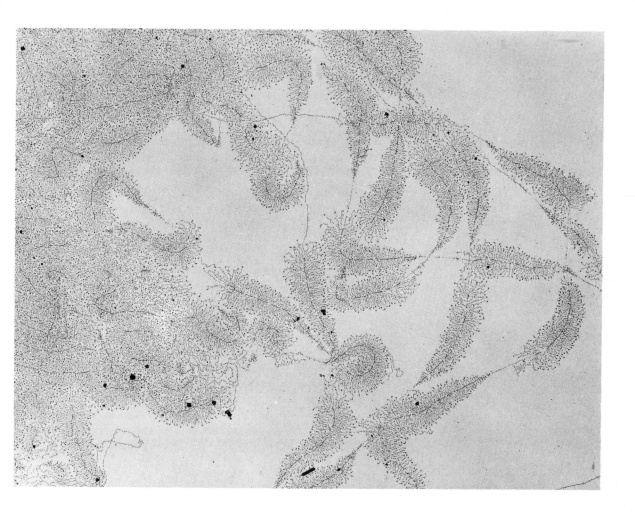

Fig. 4.19 Electron micrograph of nucleolar genes from an amphibian egg cell. These genes code for rRNA and are identifiable because each is transcribing for about 100 fibrils of RNA as the RNA polymerase moves along the gene. The segments between the fibrillar regions are presumably genes that are inactive at this time of high rRNA synthesis. × 16 000. Reproduced from Miller and Beatty (1969). Copyright © 1969 by the American Association for the Advancement of Science.

DNA segment is believed to be a gene transcribing rRNA and each hairlike fibril a growing rRNA precursor molecule. Thus many precursor molecules are synthesized simultaneously as RNA polymerase molecules move along each gene. The segments between the matrix regions are presumably genes that are inactive at the time of high rRNA synthesis. Similarly striking micrographs have been obtained from the green alga *Acetabularia* (Trendelenburg *et al.*, 1974).

Synthesis of 5 S ribosomal RNA and transfer RNA

The 5 S RNA of the large sub-unit of eukaryote ribosomes (see Ch. 5) appears to be synthesized outside the nucleolus and so separately from the two large rRNAs. Similarly tRNA is transcribed outside the nucleolus since it is still produced normally in mutants of *Xenopus* which lack nucleoli or in cells in which the nucleoli have been inactivated by ultra-violet light. Both molecules appear to be transcribed as larger molecules which are trimmed during maturation. tRNA undergoes extensive modification after transcription including methylation and the formation of pseudouridine (see p. 76). However, very few details are known of the synthesis of these molecules in higher plants. One of the major difficulties is in distinguishing between the various types of RNA which are in the 5 S size range.

Control of gene expression

Apart from its obvious role in the conservation and duplication of genetic information, the nucleus is very much involved in the regulation of expression of this information. The sequence of events leading to the expression of gene activity has been summarized by Siebert (1968) as follows:

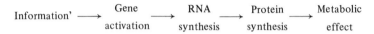

The stimulus leading to gene activation may be a small molecule such as an amino acid or a larger molecule, such as a hormone. A wide range of studies have shown that the effects of plant hormones such as gibberellic acid and auxins on processes such as enzyme induction or cell elongation are frequently associated with a marked effect on nucleic acid metabolism (Key, 1969; Jacobsen, 1977). Generally, an enhancement of RNA synthesis is observed in hormone-sensitive systems. There are several possible sites at which hormones may exercise their effect. They may act at the level of transcription in the regulation of mRNA synthesis, at the level of translation with an effect on tRNA or on the mRNA-ribosome interaction, or at some intermediate level such as an effect on the movement of RNA between nucleus and cytoplasm. These possibilities have been investigated by studies on isolated nuclei and chromatin, where a direct effect of the hormone has been strongly suggested by an increased RNA synthesis in the presence of the hormone. For example, Johri and Varner (1968) showed that pea nuclei incubated in the presence of 10^{-8} M gibberellic acid incorporated up to 80% more labelled nucleotide into RNA than those incubated without added hormone. It was necessary to add the hormone during the isolation of the nuclei, the response decreasing as the hormone addition was delayed, suggesting that some hormone-sensitive factor might be lost from the nuclei during isolation. These workers also demonstrated that gibberellic acid affected that nature as well as the quantity of RNA synthesized. One interpretation of the results is that the hormone regulates the synthesis of specific mRNA molecules which are then responsible for the synthesis of the proteins necessary for a certain physiological response.

However, there are many problems associated with the interpretation of such experiments. For example, with intact tissues, artifacts may arise due to effects on the uptake of labelled precursors or to differences in turnover rates or precursor pools. There are also considerable difficulties associated with the identification of transcription products since tRNA and, in particular, rRNA as well as mRNA synthesis frequently increases in response to hormone treatment. Nevertheless, using *in vitro* translation systems (see p. 177) and immunoprecipitation techniques to detect mRNA activity, evidence for the hormonal or environmental control of transcription has been produced for several enzymes from higher plants.

Perhaps the most widely studied example concerns the effect of gibberellic acid on the synthesis and secretion of α-amylase and other hydrolases in barley aleurone layers. The enzymes are secreted into the endosperm from the surrounding aleurone tissue and are responsible for the mobilization of reserves during seed germination. During the first 12 hours of gibberellic acid treatment, there is a lag phase of several hours followed by a steady increase in α-amylase synthesis. Inhibitors of transcription, such as actinomycin D and α-amanitin (see p. 176), are effective during the first period but not when added after 12 hours. Using a cell-free translation system, it has been shown that mRNA for α-amylase begins to increase after several hours of gibberellic acid treatment and correlates well with α-amylase synthesis over the first 12 hours (see Jacobsen, 1977). However, it should be noted that this increase in mRNA could result from enhanced synthesis, reduced breakdown, or activation of inactive forms, although the first possibility seems the most likely. Furthermore, the ineffectiveness of transcription inhibitors after 12 hours, even though gibberellic acid is still required for continued synthesis of the enzyme, suggests that regulation may switch to the level of translation at this stage.

A second system of hormonal control of mRNA synthesis is provided by the large increase in cellulase synthesis observed in auxin-treated pea epicotyls (Verma et al., 1975). However, in this system, the increase in translatable mRNA for cellulase begins immediately on auxin treatment whereas the increase in cellulase activity is not detectable for another 12–24 hours. Thus some additional control, perhaps at the level of translation, must be operating in this tissue.

A third example of mRNA enchancement concerns not hormones but light, which is clearly a critical factor in the control of plant development. Irradiation of dark-grown cell suspension cultures of parsley with ultra-violet light leads to the induction of various enzymes involved in the synthesis of flavonoid glycosides, including phenylalanine ammonia-lysase (see p. 468) and flavone synthase (Schroder et al., 1979). Increases in mRNA for the two enzymes were detectable after 2 hours and the levels increased rapidly for several more hours; the peak for phenylalanine ammonia–lyase mRNA was reached before that for flavone synthase mRNA. The apparent half lives of the enzymes (see p. 111) were 7–10 hours for phenylalanine ammonia–lyase and 5–7 hours for flavone synthase. Using these data, further calculations indicated that the light-induced changes in the two enzymes resulted from changes in mRNA activity.

Thus, although some change in RNA metabolism would seem to be the inevitable consequence of all hormone treatments in the long term, there is no clear evidence of more specific responses involving RNA transcription. However, as mentioned above, other post-transcriptional sites of control must also be considered. Furthermore it must be remembered that the regulation of nucleic acid metabolism is not the only mechanism of hormone action. Other metabolic changes may be induced before an effect on nucleic acids is observed. Thus, with auxin treatment, the fastest response involves the secretion of protons into the cell wall compartment and a subsequent effect on wall plasticity (see Ch. 10). These responses may be observed within a few minutes and are therefore too rapid to stem from changes in nucleic acid or protein synthesis. Presumably both short-term and longer-term effects are involved in the overall regulation of plant growth by hormones.

Gene activation is usually thought to involve some mechanism of derepression and repression. Repression occurs when the accumulation of a substance results in the inactivation of the templates necessary for its synthesis. In bacterial systems this is thought to involve highly specific repressor molecules for single genes or small groups of genes. This seems less likely to operate in higher cells, since a much greater number and variety of these molecules would be required: the evidence available at present suggests that repression occurs at a more general level with a much lower degree of specificity. However, the concept of repressors could explain the action of gene-activating molecules such as certain hormones which may act by binding with the repressors and so freeing the genes for transcription. In eukaryotes, the mechanism of repression may be considered at both the chromosomal and molecular levels.

There is good evidence for suggesting that regulation of gene activity may be achieved by the condensation of the chromosomes causing the repression of large blocks of genes. For example, in the 'lampbrush' chromosomes of amphibian oocytes, side loops are formed from the main axis of the chromosomes during certain stages of development in which the DNA is in a stretched and despiralized state. These loops are formed in a specific sequence and disappear when development of the oocytes is complete. It is thought that RNA synthesis occurs in sequence in the stretched loops of chromosomes while the rest is in the condensed state. This concepts is supported by studies of the giant chromosomes found in the salivary glands of the larvae of certain insects. These *polytene chromosomes* usually consist of over 1 000 identical chromosomes arranged side-by-side and show a striking pattern of transverse bands (Fig. 4.20). At certain times some of these bands show a more disorganized arrangement to form 'puffs', which are thought to result from an uncoiling for the DNA in these regions. Autoradiographic studies using radioactive precursors of RNA have demonstrated that the puffs are sites of intense RNA synthesis (Fig. 4.16). It is also interesting that the pattern of puffing along the chromosome varies at different stages of development of the larvae and under different physiological conditions. Puffing is therefore considered to indicate specific gene activation, with different genes showing enhanced activity at different stages of development. Further evidence has come from the work of Frenster, Allfrey and Mirsky (1963) who showed that when thymus nuclei were disrupted by sonication it was possible to separate 'condensed' and 'extended' chromatin material; it was the 'extended' chromatin that showed the

Fig. 4.20 Salivary gland *(a)* chromosomes from *Drosophila* after feeding larva with tritiated cytidine and examination by autoradiography. The top photograph is taken with the microscope focused on the chromosome and the lower when focused on the silver grains. The arrows indicate the puff regions which are more heavily labelled with radioactivity suggesting that they are sites of intense RNA synthesis. Scale = 10 μm. Reproduced from Rudkin and Woods (1959).

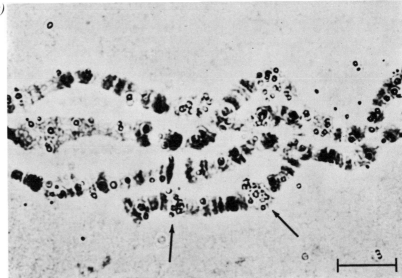

(b)

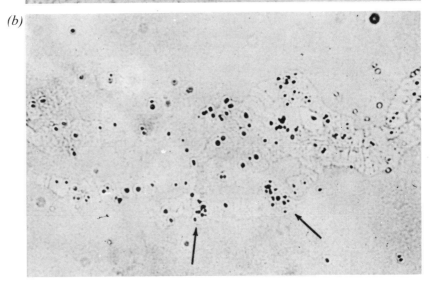

highest incorporation rates when nuclei were incubated with RNA precursors. In plants, variable degrees of chromatin condensation are observed in ultrathin sections of nuclei (Nagl, 1976). Inactivated nuclei, such as those in maturing sieve elements, display extremely condensed chromatin, while chromatin in interphase (functioning) nuclei may show spherical puff-like segments. The application of cytochemical techniques for the detection of nucleic acids and protein suggests that these spherical segments consist of a network of DNA fibrils in an amorphous matrix containing both RNA and protein (Lafontaine, 1974); they are considered to be micropuffs which are actively involved in RNA synthesis, and may be compared with the puffs of the polytene chromosomes described above. Thus at the cytological level, the activation of genes may involve a change in the DNA fibrils from a condensed to an extended form.

At the molecular level, a number of hypotheses have been suggested to explain the repression and depression of gene activity, including the action of histones, which as we have seen earlier readily form complexes with DNA and of specific RNA molecules which act as repressors or activators. It is the action of histones that has been the most widely discussed, particularly in relation to plant systems, since Huang and Bonner (1962) published the first experimental evidence for this concept when they showed that RNA-polymerase activity associated with pea seedling chromatin was inhibited by histones. Further evidence came from the studies of Bonner, Huang and Gilden (1963) who reported that chromatin isolated from pea seedlings could direct RNA and protein synthesis in the presence of RNA-polymerase and ribosomes from *E. coli*. They claimed that chromatin isolated from the cotyledons could direct the synthesis of a globulin which is synthesized by the tissue *in vivo*, although chromatin from the buds and roots which contain very low levels of this protein could not support its synthesis. However, if histone was removed from the bud chromatin, the DNA was able to direct globulin synthesis. This suggests that histones are effective inhibitors of DNA-directed RNA synthesis although it does not demonstrate that histones are highly selective regulators of mRNA synthesis.

However, a number of other observations suggest that the explanation is not so simple. For example, if DNA-protein complexes are broken into small fragments by sonication, the effectiveness of the complex as a template for RNA synthesis is increased although the ratio of DNA to histone remains unaltered. Another problem concerns the range of histones available to act as repressors. Although the cells of higher plants and animals contain many thousands of genes, the number of different histones is very much lower and this argues against their rôle as specific gene repressors. Furthermore, the histones vary little in their general properties between different tissues in the same organism, between different metabolic states, or between species. Thus it is now widely believed that histones are involved primarily in the organization of chromatin structure and play only a rather general, non-specific rôle in the control of gene expression. Although DNA–histone complexes are relatively inactive in RNA synthesis when compared with purified DNA, histones are unlikely to control the transcription of individual genes.

In contrast to histones, the acidic proteins of chromatin (see p. 159) show much greater heterogeneity. They occur in greater numbers (up to several hundred although still probably too few to act as specific repressors) and show a wider range of molecular weights. Furthermore, they show species and tissue specificity and vary with changes in gene activity and in response to hormone treatment (see Jacobsen, 1977). There is certainly some evidence that acidic proteins can neutralize the repressive effects of histones (see above) and so facilitate transcription (Paul, 1972). However, it must be remembered that it is difficult to separate cause and effect: the changes observed in acidic proteins could be a consequence and not the cause of changes in gene activity. Nevertheless it seems likely that changes and modifications in chromatin proteins must play a central role in the control of gene expression.

Finally a brief discussion must be included concerning the possible role of HnRNA (see p. 177) and repetitive sequences of DNA in the regulation of gene activity in eukaryotes. Eukaryote DNA is of several types including *single-copy* or *unique* DNA which is assumed to contain the *structural genes*, i.e. those genes which specify the amino acid sequences or structure of polypeptides. Eukaryotes also contain *repetitive* DNA in which certain sequences of bases are repeated many times. Highly repetitive DNA (5–10% of the total) contains short sequences of bases repeated from 10^5 to 10^7 times per genome: it may differ in base composition from the bulk of the DNA and is often synonymous with satellite DNA (see p. 74). It occurs at specific sites on the chromosomes and may be involved in chromosome movement. In contrast, moderately repetivitve DNA contains sequences that are repeated from several to many thousands of times, and is interspersed with single-copy DNA in an ordered arrangement along the chromosomes. Several models have proposed a regulatory role for this repetitive DNA and one of these is outlined below.

Britten and Davidson (1969) have proposed that the eukaryotic genome contains a large number of sensor genes which can bind with various signals, such as hormones or enzyme substrates, leading to the activation of an adjacent integrator gene and the transcription of activator RNA (Fig. 4.21). This RNA is recognized by a receptor gene located elsewhere in the genome, causing an adjacent structural (called producer) gene to produce a mRNA molecule. Britten and Davidson envisage a genome containing batteries of structural genes, each controlled by multiple integrator and receptor genes. The integrators can thus induce transcription of many structural genes in response to a single molecular event at the sensor site. Alternatively,

Fig. 4.21 Diagram illustrating a simple example of the integrative system proposed by Britten and Davidson (1969). In (*a*), there are multiple receptor genes and, in (*b*), multiple integrator genes. Stimulation of the sensor genes causes the initiation of transcription of their integrator genes. Activator RNAs diffuse (dotted lines) from their sites of synthesis to the receptor genes, leading to the active transcription of the producer genes. A genome is considered to contain overlapping batteries of structural genes, each controlled by multiple integrator and receptor genes. Redrawn after Britten and Davidson (1969).

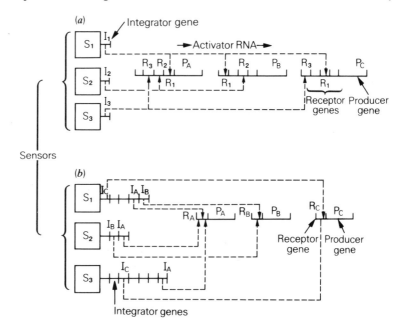

multiple receptors mean that different sensors (and so different signals) may control the transcription of a given structual gene. In this model, regulation is accomplished by the specific RNA molecules and not proteins, and so overcomes the difficulty raised above by the lack of sufficient proteins to act as specific gene regulators. It also accounts for the presence of repeated sequences which are considered to be regulatory in function, and for the rapid turnover of much of the HnRNA.

This model has recently been extended by Davidson and Britten (1979) to account for the observations that coding sequences for genes not expressed in a cell may appear in its nuclear RNA, and that the levels of certain repetitive sequences in nuclear RNA vary between tissues. The structural genes are considered to transcribe continuously at a basic rate to yield a so-called constitutive transcript (CT) containing the mRNA precursor and short repeat sequences. In addition, cell-specific and regulatory regions of the genome are transcribed to give RNA known as the integrating regulatory transcript (IRT). These regions have the same logical function as the integrator genes of the earlier model and are under the control of the sensor genes. The CT and IRT together make up the HnRNA. Thus, as in the earlier model, a battery of structural genes is under the control of a group of repetitive sequences. Gene expression is regulated by RNA–RNA duplexes formed between the repeated sequence regions of CTs and complementary sequences on certain IRTs. These duplexes are considered to be essential for the survival and processing of the cell-specific sets of mRNA and so for eventual translation of mRNA in the cytoplasm. The above description provides only a brief outline of a complex model and the reader is referred to the original papers for further details. Various predictions follow from the model and now require experimental testing. Control in the second model essentially takes place post-transcriptionally. Transcriptional control is indicated if the basic rate of transcription is exceeded as, for example, in the lampbrush chromosomes where increased RNA synthesis occurs in the side loops. Post-transcriptional control can also occur in the cytoplasm as we shall see in the next chapter.

Further reading

Bryant, J. A. (1976) *Molecular Aspects of Gene Expression in Plants*. Academic Press, London.

Bryant, J. A. (1980) Biochemical aspects of DNA replication with particular reference to plants. *Biol. Rev.*, **55**, 237.

Duda, C. T. (1976) Plant RNA polymerases. *Ann. Rev. Plant Physiol.*, **27**, 119.

Flavell, R. (1980) The molecular characterisation and organisation of plant chromosomal DNA sequences. *Ann. Rev. Plant Physiol.*, **31**, 569.

Franke, W. W. (1974) Structure, biochemistry and functions of the nuclear envelope. *Int. Rev. Ctyol.*, Suppl. **4**, 71.

Goodenough, U. (1978) *Genetics.* 2nd edn. Holt, Rinehart and Winston.

Grierson, D. (1977) The nucleus and the organisation and transcription of nuclear DNA. In *The Molecular Biology of Plant Cells.* Ed. H. Smith. Blackwell. Oxford.

Isenberg, I. (1979) Histones. *Ann. Rev. Biochem.*, **48**, 159.

Jacobsen, J. V. (1977) Regulation of ribonucleic acid metabolism by plant hormones. *Ann. Rev. Plant Physiol.*, **28**, 537.

Key, J. L. (1976) Nucleic acid metabolism: In *Plant Biochemistry*. Eds. J. Bonner and J. E. Varner. Academic Press, New York.

Kornberg, R. D. (1977) Structure of chromatin. *Ann. Rev. Biochem.*, **46**, 931.

Lafontaine, J. G. (1974) The nucleus. In *Dynamic Aspects of Plant Cell Ultrastructure*. Ed. A. W. Robards. Mcgraw-Hill, London p.l.

Laskey, R. A. and Earnshaw, W. C. (1980) Nucleosome assembly. *Nature Lond.*, **286**, 763.

Leaver, C. J. (1980) *Genome Expression and Organization in Plants*. NATO, ASI 29. Plenum Press, New York.

Marcus, A. (1971) Enzyme induction in plants. *Ann. Rev. Plant Physiol*, **22**, 313.

McGhee, J. D. and Felsenfeld, G. (1980) Nucleosome structure. *Ann. Rev. Biochem.*, **49**, 1115.

Nagl, W, (1976) Nuclear organization. *Ann. Rev. Plant Physiol.*, **27**, 39.

Literature cited

Bajer, A. S. and Mole-Bajer, J. (1972) Spindle dynamics and chromosome movement. *Int. Rev. Cytol.*, Suppl. **3**.

Bonner, J., Huang, R. C. and Gilden, R. V. (1963) Chromosomally directed protein synthesis. *Proc. natl. Acad. Sci. U.S.A.*, **50**, 893.

Brady, T. and Clutter, M. E. (1972) Cytolocalization of ribosomal cistrons in plant polytene chromosomes. *J. Cell Biol.*, **53**, 827.

Britten, R. J. and Davidson, E. H. (1969) Gene regulation in higher cells: a theory. *Science, N. Y.*, **165**, 349.

Davidson, E. H. and Britten, R. J. (1979) Regulation of gene expression: possible role of repetitive sequences. *Science, N. Y.*, **204**, 1052.

Filner, P. (1965) Semi-conservative replication of DNA in a higher plant cell. *Expl. Cell Res.*, **39**, 33.

Frenster, J. H., Allfrey, V. G. and Mirsky, A. E. (1963) Repressed and active chromatin isolated from interphase lymphocytes. *Proc. natl. Acad. Sci. U.S.A.*, **50**, 1026.

Grierson, D. (1976) RNA structure and metabolism. In *Molecular Aspects of Gene Expression in Plants*. Ed. J. A. Bryant. Academic Press, London.

Huang, R. C. and Bonner, J. (1962) Histone, a suppressor of chromosomal RNA syntheis. *Proc. natl. Acad. Sci., U.S.A.*, **48**, 1216.

Hyde, B. B. (1967) Changes in nucleolar ultrastructure associated with differentiation in the root tip. *J. Ultrastruct. Res.*, **18**, 25.

Inoue, S. (1964) Organization and function of the mitotic spindle. In *Primitive Motile Systems in Cell Biology*. Eds. R. D. Allen and N. Kam. Academic Press, New York, p. 549.

Jendrisak, J. (1980) Purification, structures and functions of RNA polymerases from higher plants. In *Genome Organization and Expression in Plants*. Ed. C. J. Leaver. Plenum Press, New York.

Johri, M. M. and Varner, J. E. (1968) Enhancement of RNA synthesis in isolated pea nuclei by gibberellic acid. *Proc. natl. Acad. Sci. U.S.A.*, **59**, 269.

John, B. and Lewis, K. R. (1969) The chromosome cycle. *Protoplasmatologia*, **4B**, 1.

Jordan, E. G. (1971) *The Nucleolus*. Oxford Biology Readers. Oxford University Press.

Key, J. L. (1969) Hormones and nucleic acid metabolism. *Ann. Rev. Plant Physiol.*, **20**, 449.

Leaver, C. J. and Key, J. L. (1970) Ribosomal RNA synthesis in plants. *J. molec. Biol.*, **49**, 671.

Leedale, G. F. (1970) Phylogenetic aspects of nuclear cytology in the algae. *Ann. N. Y. Acad. Sci.*, **175**, 429.

Lehninger, A. L. (1970) *Biochemistry*. Worth Publishing Inc., New York.

Lewis, K. R. and John, B. (1963) *Chromosome Marker*, Little, Brown, Boston.

Lord, A. and Lafontaine, J. G. (1969) The organization of the nucleolus in meristematic plant cells. *J. Cell Biol.*, **40**, 633.

Lynden, R. F. (1968) Structure, function, development of the nucleus. In *Plant Cell Organelles*. Ed. J. B. Pridham. Academic Press, London and New York.

McIntosh, J. R., Hepler, P. K. and van Wie, G. (1969) Model for mitosis. *Nature, Lond.*, **224**, 659.

Miller, O. L. and Beatty, B. R. (1969) Visualization of nucleolar genes. *Science, N. Y.*, **164**, 955.

Northcote, D. H. (1968) Structure and function of plant-cell membranes. *Br. med. Bull.*, **24**, 107.

Paul, J. (1972) General theory of chromosome structure and gene activation in eukaryotes. *Nature, Lond.,* **238,** 444.

Ris, H. (1978) Higher order structure in chromosomes. *Ninth Int. Cong. Electron Micros.* Ed. J. M. Strugess. **3,** 545. Micros. Soc. Canada, Toronto.

Roberts, K. and Northcote, D. H. (1970) Structure of the nuclear pore in higher plants. *Nature, Lond.,* **228,** 385.

Rogers, M. E., Loening, U. E. and Fraser, R. S. S. (1970) Ribosomal RNA precursors in plants. *J. molec. Biol.,* **49,** 681.

Rudkin, G. T. and Woods, P. S. (1959) Incorporation of H^3-cytidine and H^3-thymidine into giant chromosomes of *Drosophila* during puff formation. *Proc. natl. Acad. Sci. U.S.A.,* **45,** 997.

Schroder, J., Kreuzaler, F., Schafer, E. and Hahlbrock, K. (1979). Concomitant induction of phenylalanine ammonia-lyase and flavone synthase in mRNAs in irradiated plant cells. *J. biol. Chem.,* **254,** 57.

Siebert, G. (1968) The nucleus. In *Comprehensive Biochemistry.* Eds. M. Florkin and E. H. Stoty. **23, 1.**

Stevens, C. and Bryant, J. A. (1978) Partial purification and characteristics of the soluble DNA polymerase. (polymerase-α) from seedlings of *Pisum sativum L. Planta,* **138,** 127.

Stevens, C., Bryant, J. A. and Wyvill, P. C. (1978) Chromatin-bound DNA polymerase from higher plants. *Planta,* **143,** 113.

Taylor, J. H. (1963) The replication and organization of DNA in chromosomes. In *Molecular Genetics.* Part I.E. J. H. Taylor, Academic Press, New York.

Trendelenburg, M. F., Spring, H., Scheer, U. and Franke, W. W. (1974) Morphology of nucleolar cistrons in a plant cell, *Acetabularia meditterranea. Proc. natl. Acad. Sci.,* **71,** 3626.

Verma, D. P. S., MacLachlan, G. A., Byrne, H. and Ewings, D. (1975) Regulation and in vitro translation of messenger ribonucleic acid for cellulase from auxin-treated pea epicotyls. *J. biol. Chem.,* **250,** 1019.

Watson, J. D. (1970) *The Molecular Biology of the Gene.* W. A. Benjamin, New York.

5 Ribosomes

Ribosomes are small subcellular particles that have been found in every plant tissue so far examined and are known to be the sites of protein synthesis where amino acids are assembled in a specified sequence to produce polypeptide chains. Some thirty years ago Brachet and Caspersson independently carried out experiments which led to our present concept of the relationship between ribosomes and protein synthesis. They found a correlation between the ability of cells to synthesize protein and the presence of a material with a strong affinity for basic dyes which was shown to be RNA. We now know that over nine-tenths of the RNA of plant cells is ribosomal and that this is found largely in the cytoplasm. Ribosomes were first noted in plant cells by Robinson and Brown in 1953 when studying bean roots with the electron microscope and shortly afterwards Palade described similar structures in animal cells. Since then improvements in the techniques of electron microscopy and cell fractionation have played a large part in the elucidation of the structure, properties and function of these cellular particles.

In sections of cells viewed under the electron microscope, ribosomes appear as oblate spheroids, 17–23 nm in diameter that are clearly far below the resolution of the light microscope. They are seen after fixation with osmium (Fig. 5.1) but not when potassium permanganate is used since this fixative destroys nucleic acids. Ribosomes occur both free in the cytoplasm and attached to the outer (cytoplasmic) surface of the membranes of the endoplasmic reticulum (ER) (Fig. 5.1). The ER is then known as 'rough-surfaced' as opposed to the 'smooth-surfaced' state when ribosomes are not attached; the significance of this attachment is not clear although we shall see later that there is evidence that different types of protein are specifically synthesized on either the free or the membrane-bound ribosomes. In addition, in many cells, clumps or strings of ribosomes have been observed which are joined by a thin thread about 1 nm wide (Fig. 5.2). The thread is apparently RNA as it is destroyed by treatment with the enzyme ribonuclease. These aggregates of ribosomes are known as *polyribosomes* or *polysomes* and their significance in protein synthesis will be discussed later.

Soon after they were recognised by electron microscopy, techniques were developed for the isolation of ribosomes. These methods depend on the fact that ribosomes are small and so can be separated from

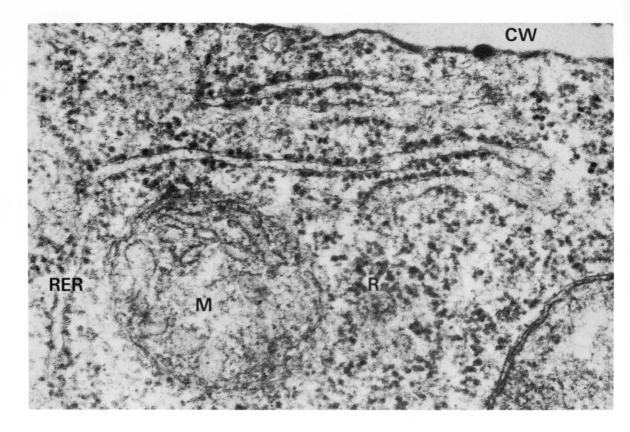

Fig. 5.1 Electron micrograph of a cell from a root of *Suaeda maritima* fixed with glutaraldehyde and osmium tetroxide and showing ribosomes free and attached to the ER. RER, rough ER; R, ribosomes; M, mitochondrion; CW, cell wall. ×85 700.

other cellular organelles by centrifugation at high speeds. Very simply, cells are broken up, centrifuged to remove cell walls, nuclei, mitochondria and other structures larger than ribosomes and then the resulting supernatant is centrifuged more rapidly (100 000 g for 2 hr) to sediment the ribosomes. Frequently such preparations consist of ribosomes attached to fragments of endoplasmic reticulum and are known as a *microsomal fraction*. To free the ribosomes, the microsomal pellets are resuspended in detergents such as deoxycholate, which solubilize the membranes, and a pure ribosome pellet is obtained by recentrifugation. Ribosomes have been shown to consist of approximately equal parts of RNA and protein and can catalyse the incorporation of radioactive amino acids into proteins provided ATP, Mg^{++} and other cofactors are present. If ribosomal preparations are further subjected to sucrose-gradient centrifugation (see p. 33) a number of sedimentation bands are normally observed migrating from the top of the tube (Fig. 5.3). The slowest moving band consists of single ribosomes while the more rapidly sedimenting bands contain clusters of between two and six ribosomes which are thought to correspond to the polyribosomes seen in thin sections of cells by electron microscopy. Aggregates of ribosomes will, of course, sediment more rapidly than single ribosomal particles. The isolated polysomes can be broken down by a brief treatment with the enzyme ribonuclease to give a corresponding increase in the number of single particles which are known as *monosomes* (Fig. 5.3). The term monosome distinguishes the single

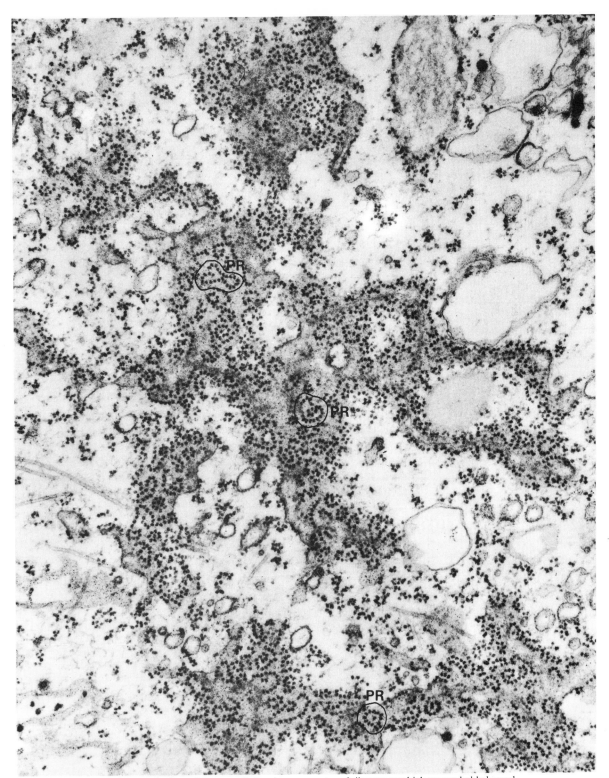

Fig. 5.2   Electron micrograph of a radish root cell showing aggregates of ribosomes which are probably bound together by messenger RNA to form polyribosomes. ×41 500. PR (polyribosomes). Provided by E. Newcomb and H. Bonnett, Department of Botany, University of Wisconsin.

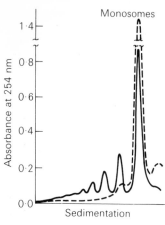

Fig. 5.3 Sucrose density gradient centrifugation of polyribosomal preparation from radish leaves. —control; ----polysomes treated with RNA-ase. Reproduced from Pearson (1969).

Properties and Structure of ribosomes

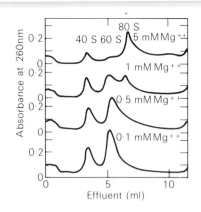

Fig. 5.4 Density gradient centrifugation of ribosomal preparations from imbibed rice embryos showing the effect of changes in the magnesium ion concentration of the sucrose gradient medium on the dissociation of ribosomes into sub-units. Reproduced from App, Bulis and McCarthy (1971).

194

units derived from the breakdown of isolated polysomes from the naturally occurring monomer, the ribosome.

Ribosomes may be grouped into two major classes based on their sedimentation coefficients (see p. 34). The cytoplasmic ribosomes of eukaryotes have a sedimentation coefficient of about 80 S while those isolated from prokaryotic cells and from chloroplasts and mitochondria of higher organisms have a value of about 70 S (Table 5.1). As we shall see later, these classes are further distinguished by differences in the molecular weights of their component RNA molecules and in their sensitivity to antibiotics. It must be noted however that exceptions occur, particularly in relation to mitochondrial ribosomes, which have been reported to have sedimentation coefficients ranging from 55 (mammalian) to 78 S. However, mitochondrial protein synthesis is inhibited by the same antibiotics that inhibit chloroplast and prokaryote ribosomal activity, indicating that mitochondrial ribosomes are of the 70 S type.

Cytoplasmic ribosomes from plant cells are similar in many respects to those from animal cells. Both are about 20 nm in diameter, contain 40–50% RNA and 50–60% protein, have a molecular weight of about 4×10^6 and a sedimentation constant of about 80 S. These ribosomes are composed of two sub-units which require Mg^{++} ions for structural cohesion. If Mg^{++} is removed from isolated cytoplasmic ribosomes by dialysis the 80 S particle dissociates into two sub-units with sedimentation constants of 60 S and 40 S (Fig. 5.4). The S values are not additive since the rate of sedimentation is also influenced by the shape of the particle (see p. 34). If the Mg^{++} concentration is restored to 1×10^{-3} M they will reassociate. If the Mg^{++} concentration is increased still further, two ribosomes associate to form a dimer with a sedimentation constant of about 120 S (Fig. 5.5). The 70 S ribosomes of prokaryotes are also made up of two sub-units with sedimentation coefficients of 50 S and 30 S and usually contain about twice as much RNA as protein. Nonomura, Blobel and Sabatini (1971) have carried out a

Table 5.1 Comparison of S values of cytoplasmic, chloroplast and mitochondrial ribosomes from various species.

Organism	Cytoplasmic	Chloroplast	Mitochondrial
E. coli	70	—	—
Oscillatoria	70	—	—
Chlamydomonas reinhardi	80	68	—
Neurospora crassa	77	—	73
Saccharomyces cerevisiae	80	—	74/75
Candida krusei	—	—	76
Pisum sativum	80	70	—
Spinacea oleracea	78	67/70	—
Nicotiana tabacum	80	70	—
Zea mays	—	—	77/78

Source: Boulter, Ellis and Yarwood, 1972 and Bryant, 1976

Fig. 5.5 Diagrammatic representation of the effect of magnesium ions on the sub-unit structure of ribosomes.

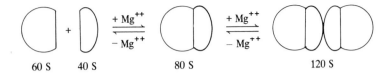

60 S   40 S   80 S   120 S

Fig. 5.6 Three-dimensional model of rat liver ribosome. Reproduced from Nonomura, Blobel and Sabatini (1971).

thorough electron microscopal study of rat liver ribosomes after negative staining, looking at many images, and have built up an interesting three-dimensional model of ribosomal structure (Fig. 5.6). The small sub-unit appears to lie to one side of the large sub-unit and has a characteristic cleft which divides the sub-unit into two unequal regions. The model includes a tunnel which runs between the two sub-units directly under the cleft in the small sub-unit and is thought to be occupied by the messenger RNA strand in polysomal ribosomes. A somewhat similar channel appears to exist in the prokaryotic ribosome and explains why part of a mRNA molecule is protected from nuclease attack (see Stöffler and Wittman, 1977).

Ribosomal RNA and protein

If the RNA is extracted from the 80 S ribosomes of plants and animals it is seen to consist of four components which may be separated on the basis of their sedimentation coefficients. There is one molecule of 25–28 S RNA which is hydrogen-bonded to one molecule of 5·8 S RNA, one molecule of 18 S RNA and one molecule of 5 S RNA (Fig. 5.7). The sedimentation coefficient of the largest RNA molecule is about 25 S in plants and is usually somewhat higher in animals. In

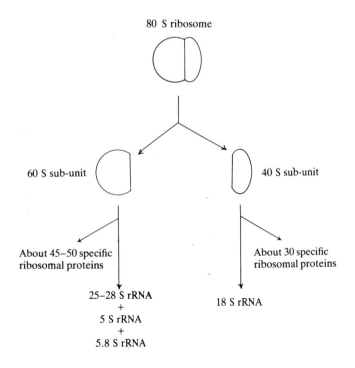

80 S ribosome

60 S sub-unit

40 S sub-unit

About 45–50 specific ribosomal proteins

About 30 specific ribosomal proteins

25–28 S rRNA
+
5 S rRNA
+
5.8 S rRNA

18 S rRNA

Fig. 5.7 The structure of an 80 S eukaryotic ribosome.

contrast, the 70 S ribosomes of prokaryotes and chloroplasts contain three molecules of RNA with sedimentation coefficients of 23 S, 16 S and 5 S. The 25–28 S and 18 S molecules of the 80 S ribosomes constitute approximately two-thirds and one-third respectively of the total ribosomal RNA and are derived from the 60 S and 40 S sub-units respectively. They appear to be considerably folded to form tight packages at the centres of the sub-units. The 5·8 S RNA of eukaryotes is believed to be analogous to the 5 S RNA of prokaryotes, while the 5 S RNA of the large sub-unit of 80 S ribosomes is specific to eukaryotes.

The proteins of ribosomes are difficult to study because of their insolubility in dilute aqueous media and their tendency to precipitate when RNA is removed by ribonuclease activity. However, the proteins can be extracted in soluble form by treatment with strong salt or urea solutions and then separated by gel electrophoresis. Treatment of ribosomes with increasing concentrations of monovalent cations results in the release of some of the proteins, known as *split proteins*, leaving protein-depleted core particles which have no activity for protein synthesis. More drastic treatment with urea is needed to remove the remaining *core proteins* from the core particles. This selective release presumably reflects differences in binding strengths and in the location of proteins within the sub-units. It is interesting to note that functional sub-units can be reconstituted by mixing together a large RNA molecule and the corresponding proteins under suitable conditions, showing that ribosomes are self-assembling particles which depend on the specific association of protein and RNA molecules to form supra-molecular structures. This technique allows the investigation of the function of ribosomal proteins by omitting or modifying specific proteins and studying the effect on the ability of the ribosomes to synthesise proteins. For example, bacteria may become resistant to the antibiotic streptomycin which is an inhibitor of protein synthesis. By substituting in turn single proteins purified from streptomycin-resistant strains into the sub-units of the sensitive strains, it was possible to demonstrate that the property of streptomycin resistance was due to a single protein of the 30 S sub-unit. However, as we shall see below, this approach is limited since particular functions may involve contributions from several molecules rather than individual proteins.

Ribosomal proteins are generally small, extremely basic, and consist of a mixture of polypeptides, most of which have molecular weights in the range 10 000 to 30 000 (see Brimacombe, Stöffler and Wittman, 1978; Wool, 1979). The average molecular weights of the proteins from the ribosomes of rat liver and *E. coli* are about 21 000 and 18 000 respectively. Eukaryotic ribosomes contain about 80 proteins of which 45–50 are associated with the large sub-unit and about 30 with the small sub-unit. In contrast, *E. coli* ribosomes contain a total of 55 proteins. Each of these proteins appears to be unique and the complete amino acid sequence has been determined for most of the proteins of *E. coli* ribosomes. Studies on their topography (i.e. arrangement within the sub-units) is also well advanced with the technique of immunological electron microscopy providing an exciting new approach. In this procedure, antibodies are raised against individual ribosomal proteins and are used to form dimers with the ribosomal sub-units. The bound antibody can be visualized by electron microscopy after negative staining and the antibody attachment site taken as the point where a given ribosomal protein has an antigenic determinant exposed on the ribosomal surface. Antibody binding sites are mapped and topographical models built up of the location of proteins on the sub-units.

Thus there is increasing knowledge available on the composition, structure and arrangement of the proteins and RNA in the ribosome; it will presumably be the first organelle for which the total sequences of its components will be known. However, there is still a great deal to be learnt about the relationship between structure and function in this organelle and about the dynamics of the ribosomal sub-units in the process of protein synthesis. The two most thoroughly studied ribosomes in this repect are those from *E. coli* and rat liver. Although they differ in size, their basic function in the mechanism of protein synthesis appears to be essentially the same. However, as mentioned above, 70 S and 80 S ribosomes contain very different numbers of proteins and it is interesting to speculate on the role of these extra proteins in eukaryotes. The greater complexity perhaps supports the concept that greater control of gene expression exists at the ribosomal level in higher cells. In addition, the extra proteins could be involved in the interaction of ribosomes with receptors in the endoplasmic reticulum or in the control of membrane-associated protein synthesis. In eukaryotes, ribosomal proteins are synthesized in the cytoplasm before transportation to the nucleolus where they are assembled into sub-units (see Ch. 4). The 5 S RNA which is synthesized at a different site to the large RNA molecules is then incorporated and the complex transported to the cytoplasm. Again, some of the extra proteins may be involved in these processes. Another interesting feature of eukaryotes is that proteins for both 80 S and 70 S (mitochondrial and plastid) ribosomes are synthesized in the cytoplasm and then the two types are later partitioned; but we know very little of how this is accomplished.

The individual proteins are however no longer considered to act as separate and distinct units of function within the ribosome. In addition to producing the compact shape of the ribosome, the interaction between proteins and RNA produces functional sites involving both types of molecule; it is these functional domains, rather than individual molecules, that are currently of interest in assigning functions to parts of the ribosome (Kurland, 1977). For example, in the prokaryotic ribosome, the mRNA binding site appears to be restricted to the 30 S sub-unit and to involve the 16 S RNA and at least three proteins at a well-defined site. Again, the interaction between tRNA and the ribosome appears to involve the 5 S RNA and proteins associated with both the large and small sub-units. It is beyond the scope of this text to discuss these interactions in greater detail since much of this research involves prokaryotic systems. The reader is referred to the extensive reviews quoted at the end of the chapter for further information.

Synthesis of ribosomes

As with other aspects of ribosomal metabolism, evidence for the site and mechanism of synthesis of ribosomes in higher cells has come largely from studies using animal tissues. Although few details from higher plant cells are known, the evidence available suggests that the mechanism for the biogenesis of ribosomes is similar in plants and animals.

Evidence for the function of the nucleolus in ribosomal RNA synthesis is discussed fully in Chapter 4 and is summarized only briefly here. The relationship was first suggested by cytochemical observations which showed that the nucleolus had strong basophilic staining properties which were shown to be due to RNA rather than DNA. This evidence is supported by the finding that nucleolar and cytoplasmic RNA, but not chromosomal RNA, have the same base composition, and that the appearance of new RNA in the cytoplasm is prevented if the nucleolus is specifically destroyed by micro-beam ultra-violet irradiation.

In mammalian cells ribosomal RNA first appears in the nucleolus as a large molecule with a sedimentation coefficient of 45 S (see p. 178). This is probably the precursor of both the larger RNA molecules and is split to give 32 S and 18 S molecules. The 18 S RNA is moved rapidly into the cytoplasm to form the small sub-unit of the ribosome. The 32 S molecule is degraded to a 28 S molecule and reaches the cytoplasm about 40 min later. The synthesis of ribosomal proteins occurs on the cytoplasmic ribosomes, although the RNA precursor rapidly gains proteins and the processing of the RNA as described above takes place in ribonucleoprotein particles before the two ribosomal sub-units are moved to the cytoplasm. There is now evidence that a similar process may operate for ribosome biogenesis in higher plant cells. Certainly the structure of the nucleolus and the molecular weights of the RNA molecules are similar and there is evidence of a high molecular weight RNA precursor molecule in the nuclear fraction from higher plant tissues which may be the equivalent of the ribosomal precursor molecule found in animal cells (see p. 179).

Ribosomes and protein synthesis

Plant cells contain a vast number of different protein molecules which are largely composed from just twenty amino acids, each protein having a specific sequence of amino acids in its polypeptide chains (see Ch. 2). These sequences are precisely predetermined and errors appear to occur only very rarely. Thus any proposed mechanism for protein synthesis must account not only for the polymerization of amino acids but must also explain how they enter the polypeptide chains in the required order. Our present concept of the mechanism of protein synthesis is based on studies of a wide range of biological species although the greater part and most of the critical experiments have been carried out with two cell-free systems, one from *E. coli* and the other from reticulocytes or immature red cells. So far there have been fewer observations reported from studies with higher plants, but these suggest that in general similar biochemical steps are in operation.

The ultimate information for protein synthesis is contained in the nucleotide sequence of chromosomal DNA which forms the structural genes and specifies the amino acid sequence of proteins (see Chs. 2 and 4). Nuclear control is not direct, but is mediated through a special type of RNA molecule known as messenger RNA (mRNA) which carries information from the chromosomes to the sites of protein synthesis in the cytoplasm, the ribosomes. Information is copied onto mRNA which has a base sequence complementary to that of a strand of chromosomal DNA which specifies the amino acid sequence of one or more polypeptides (see p. 177). From work on bacterial systems it is known that units of three nucleotides which are known as *codons* contain the information for individual amino acids. Permutation of the four bases of DNA and mRNA shows that there are sixty-four different possible codons which are clearly plenty to code for the twenty amino acids. In association with the ribosomes, the mRNA is able to act as a template for protein synthesis in the cytoplasm. The polymerization of amino acids does not take place directly but utilizes another special RNA molecule known as transfer RNA (tRNA). These molecules serve as intermediates between the nucleotide sequence in mRNA and the amino acids. There is no specific affinity between amino acids and the bases in mRNA and so the amino acids are not recognized by either the ribosomes or the mRNA; transfer RNAs act as adaptors enabling the amino acids to line up on the mRNA templates. The tRNA molecules attach to amino acids through a covalent bond and there are specific tRNA molecules for specific amino acids. Part of the tRNA molecule which is known as the *anticodon* contains a triplet of bases which hybridizes with the complementary bases on the mRNA attached to the ribosome. The repeated interaction of codon and anticodon allows the ordered polymerization of amino acids into polypeptide chains. Transfer RNA molecules therefore play the rôle of bilingual adaptors to which amino acids are attached so that they can be fitted into the polypeptide chain as directed by the nucleotide triplet language of mRNA.

The various stages in the synthesis of proteins will now be discussed more fully. They involve one of the central dogmas of molecular biology, which postulates that genetic information flows from nucleic acids to protein. The first step is known as transcription and does not involve a change of code since DNA and mRNA are complementary. The second step involves a change of code from nucleotide sequences to amino acid sequences and is called translation.

Evidence for ribosomes as sites of protein synthesis

The first direct evidence relating ribosomes and protein synthesis was described by Zamecnik and coworkers in a series of investigations from 1952 onwards. Radioactive amino acids were injected into rats and at various time intervals after the injection the liver was removed, homogenized and centrifuged to separate the various cell organelles, which were studied for evidence of incorporation of amino acids into proteins. If examined shortly after injection, only the microsomal fraction contained labelled protein whereas at later times labelled protein was found in all cell fractions. These results suggested that proteins were first synthesized in the microsomes and then transferred to other parts of the cell. It was later found that isolated microsomal fractions, which were supplemented with ATP, catalysed the incorporation of amino acids into proteins in the presence of the unsedimentable, solube fraction which provided essential cofactors.

Since this pioneer work, cell-free systems capable of catalysing amino acid incorporation into protein have been isolated from a variety of species and have been shown to require the following components: ribosomes, the twenty common amino acids (one or more labelled with radioactivity to measure incorporation), tRNAs, mRNA, an ATP generating system, GTP, Mg^{++} and K^{+} ions and a supernatant enzyme fraction. The importance of ribosomes themselves in such a system, rather than any contaminant of the ribosomal fraction, can be tested in a number of ways: by omission of the ATP generating system (as ribosomes have no capacity to produce ATP), by varying the Mg^{++} concentration which will affect ribosome integrity, and by addition of the enzyme ribonuclease which will attack both mRNA and the ribosomes themselves. In addition, when using preparations from higher cells, possible contamination by bacteria should always be considered and care taken to use sterile material or to keep infection to very low levels. This is particularly important when studying amino acid incorporation by isolated cell organelles such as mitochondria or chloroplasts, which resemble bacteria more closely in size than do ribosomes and so are more likely to sediment with the bacteria in the centrifuge.

Cell-free sytems of cytoplasmic ribosomes capable of incorporating amino acids into protein have now been reported from a wide variety of higher plant tissues. In general, the activity of the cell-free systems for amino acid incorporation is very much less than the *in vivo* rates of incorporation by whole cells or tissues. This is perhaps not surprising since the two systems are hardly comparable. With intact tissues there are clearly permeability problems and any supplied precursor may be considerably diluted by the endogenous supply. With the *in vitro* cell-free systems considerable disruption and loss of activity may occur during preparation. The isolated systems may contain impurities such as proteases or ribonucleases which will seriously affect the incorporation rates. In addition, there is the problem of supplying the various components at their optimal levels.

However, although the use of cell-free systems has some disadvantages, it is clearly an extremely useful method for studying the various stages of amino acid incorporation into proteins and has enabled the construction of a detailed scheme for protein synthesis. Although the evidence has come largely from bacterial and mammalian systems, a similar sequence of reactions is known to operate in higher plants. The process is broadly divided into four major stages, each with a requirement for specific enzymes and cofactors. These are:

1. activation of amino acids with the formation of specific aminoacyl-tRNA molecules.
2. the binding of these aminoacyl-tRNA molecules to the mRNA-ribosome complex and the initiation of the polypeptide chain.
3. elongation of the polypeptide chain.
4. the termination and release of the chain from the ribosome complex.

These stages will now be described in more detail with particular emphasis on protein synthesis on the 70 S bacterial ribosome (usually from *E. coli*) which is the most thoroughly investigated system.

Amino acid activation and transfer RNA

As explained previously, the polymerization of amino acids does not occur directly, but through the use of tRNA which acts as an intermediate in the arrangement of activated amino acids in order on the mRNA template. This process involves the formation of an aminoacyl-tRNA complex which takes place in two steps. Firstly, the amino acid is activated by ATP and the activated amino acids are then esterified to specific tRNA molecules:

Amino acid + ATP + enzyme $\rightleftharpoons$

(amino acid–AMP–enzyme) + pyrophosphate

(Amino acid–AMP–enzyme) + tRNA $\rightleftharpoons$

aminoacyl–tRNA + AMP + enzyme

The enzymes which catalyse both of these reactions are known as aminoacyl-tRNA synthetases and each is highly specific for one amino acid. Each cell therefore requires at least twenty different activating enzymes and most of these have now been isolated and purified from bacterial cells. All have an absolute requirement for magnesium ions and are highly specific for both the amino acid and the corresponding tRNA. The enzymes must therefore have at least two highly specific binding sites for these molecules and a third for binding ATP. In eukaryotic cells, multiple forms of these enzymes exist but their presence is associated with protein synthesis in cell organelles which appear to have their own complement of these enzymes, often with different properties to those found in the cytoplasm.

Fewer aminoacyl-tRNA synthetases have been purified from plant tissues compared with other organisms which is partly due to their greater inherent instability (see Lea and Norris, 1977). However, the enzymes that have been isolated from higher plant tissues appear to resemble those from other sources in having precise substrate requirements. For example, Attwood and Cocking (1965) have isolated a highly purified preparation of alanyl-tRNA synthetase from tomato roots which was highly specific for alanine. In contrast, Peterson and Fowden (1963) have shown that an amino acid analogue, azetidine-2-carboxylic acid (Fig. 5.14), may be activated by the proline-activating enzyme. Azetidine-2-carboxylic acid occurs naturally at high concentrations in some plants, whereas it is toxic to others at lower levels, although this toxicity may be overcome by the addition of proline. Peterson and Fowden studied prolyl-tRNA synthetase from various plants and found that the enzyme from plants which contain azetidine-2-carboxylic acid (e.g. lily-of-the-valley) does not activate this compound whereas the enzymes from susceptible plants (e.g. mung bean) activate the analogue almost as effectively as proline. Similarly, Lea and Fowden (1972) studied the amino acid substrate specificity of glutamyl-tRNA synthetase from three higher plant species. In two of these, *Hemerocallis fulva* leaves and *Caesalpina bonduc* seeds, γ-substituted analogues of glutamic acid occur naturally but they are not found in the third tissue used, *Phaseolus vulgaris* seed. Lea and Fowden found that the enzyme from *Phaseolus* was activated by a range of γ-substituted glutamic acids. In contrast, the analogue-producing plants have developed a discriminatory mechanism based on altered enzyme specificity which prevents these plants incorporating their own natural products into protein.

Little is known, however, about the range of aminoacyl-tRNA synthetases which may occur in those tissues containing a number of tRNA species capable of binding with a specific amino acid. Some interesting observations relating to the problem have come from Kanabus and Cherry (1971) working on soybean seedlings which contain six leucine-accepting tRNAs (tRNA^{leu}). They found that leucyl-tRNA synthetase activity from soybean cotyledons could be fractionated into three components. One of these exclusively acylates two of the six tRNAs^{leu} whilst the remaining two components charge the other four tRNAs equally well. The hypocotyls from the same seedlings contained only these last two components and so lacked the capacity to acylate all of the tRNA^{leu} species. These aminoacyl-tRNA synthetases therefore appear to show organ specificity, a property which could be associated with the synthesis of organ-specific proteins and so be closely related to the problem of differentiation in plants.

The highly specific relationship between the aminoacyl-tRNA synthetases, the amino acids and the tRNAs is closely related to the structure and nucleotide sequence of tRNA. The different tRNAs from all species appear to have a number of common characteristics. Each contains about eighty nucleotides, has a molecular weight of about 25 000 and sedimentation coefficient of 4 S. They contain a high proportion of guanine and cytosine, up to 10% of some unusual bases which are largely the methylated forms of normal bases, and an unusual nucleoside known as pseudouridine (see p. 76). Interchain base pairing occurs and if models are constructed to allow the maximum possible pairing, folding of the polynucleotide chain produces the so-called clover leaf pattern of layout (Fig. 5.8) in the 100 or so cases of tRNA where the base sequence is known, even though these sequences show considerable differences. This model has five special regions.

1. All tRNA molecules have the same terminal sequence of bases (CCA) at one end of the chain. The last residue, adenylic acid, is

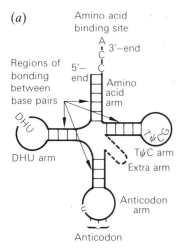

(a)

Amino acid
binding site

Regions of
bonding
between
base pairs

Amino
acid
arm

DHU arm

DHU

TψC arm

Extra arm

Anticodon
arm

Anticodon

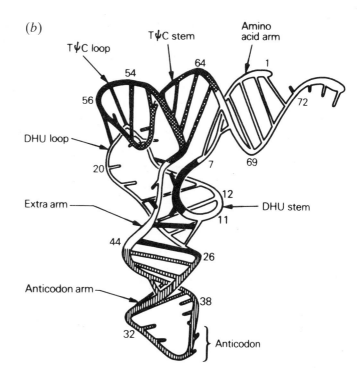

(b)

TψC loop

TψC stem

Amino
acid arm

DHU loop

Extra arm

Anticodon arm

DHU stem

Anticodon

Fig. 5.8   *(a)* Diagrammatic representation of the structure of transfer RNA showing the characteristic clover leaf pattern of layout (Ψ, pseudouridine; DHU, dihydrouridine). *(b)* Schematic diagram showing the tertiary structure of yeast tRNA^{phe}. The bases are numbered from the 5′ end. The ribose phosphate backbone is shown as a continuous cylinder with ladder rungs to indicate hydrogen-bonded base pairs. Unpaired bases are indicated by shortened rods. The black rods indicate tertiary structure hydrogen bonds. Redrawn after Quigley *et al.* (1975).

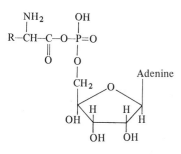

Fig. 5.9   The structure of aminoacyl adenylic acid.

the amino acid attachment site. The structure of aminoacyl adenylic acid is shown in Fig. 5.9.

2. All tRNAs have another part of the chain in common which consists of seven unpaired bases including pseudouridine and is known as the TψC arm.

3. There is one nucleotide triplet in the chain which is different in all tRNAs examined. This is the codon recognition site or *anticodon* and is complementary to the corresponding codon triplet in mRNA.

4. The DHU arm shows some variability containing from seven to eleven nucleotides.

5. The extra arm is highly variable. It contains three to five nucleotides in most tRNAs, but can have from thirteen to twenty-one nucleotides in some cases.

The tertiary structure of tRNA has been studied by X-ray diffraction (see Ch. 2) and a three-dimensional model for one tRNA (yeast tRNA^{phe}) has been produced (Fig. 5.8). This has been described as an L-shaped model in which the amino acid and TψC stems form one arm and the DHU stem and anticodon stem form the other. This means that the anticodon is about 8 nm away from the amino acid at the other end of the molecule. The reason for this is not clear but is presumably related to the mechanism involved in the transfer of amino acids to the growing polypeptide chain. The model may well apply to all tRNAs since many of the hydrogen bonds in the molecule involve the invariant nucleotides.

Since all tRNAs possess the same two-dimensional clover leaf pattern, certain highly specific interactions, such as that between the synthetase and tRNA, must presumably involve subtle features of the three-dimensional structure. Little detail of these recognition sites is available at present and will presumably await, for example, crystallization of a tRNA bound to a synthetase and analysis of such structures by X-ray diffraction.

The problem of specificity is complicated somewhat by the occurrence of more than one specific tRNA molecule for each amino acid; these are known as *isoacceptor tRNAs*. This is a direct consequence of the degeneracy of the genetic code (see below) which has multiple code words for most of the amino acids. This means that more than one tRNA is required for most amino acids and the number of tRNAs would therefore be expected to equal the number of codons. However, this is not so since it was found that some tRNAs can recognize more than one codon. This is explained by the *wobble hypothesis* which suggests that base pairing between the third base in the mRNA codon and the corresponding first base in the anticodon may be atypical. Thus, in addition to A = U and G = C (see p. 72), U or G in the first anticodon base can correspond to A or G or U or C respectively in the mRNA codon. Similarly inosine, a modified adenine base found in many eukaryotic anticodons, may pair with U, C or A. Thus for some amino acids which are specified by a number of codons, there may be fewer tRNAs than there are code words. In contrast, there are other cases where there are more isoacceptor species of tRNA than are necessary to recognize all the code words. Thus degeneracy of the genetic code cannot fully explain the multiplicity of tRNAs found in many tissues and other physiological rôles may be involved. For example, different tRNAs may have different intracellular localizations relating to protein synthesis in the cytoplasm, mitochondria and chloroplasts. There is also some evidence that isoaccepting tRNA species may be involved in development since different complements of isoacceptor tRNAs have been reported in different tissues, while patterns of tRNA have been shown to change in response to hormone treatment (see Lea and Norris, 1977). However, not all reports have been positive and interpretation is complicated by the problems associated with the reliable isolation and identification of the different tRNA species.

It is interesting that certain *cytokinins* have been found in tRNA from a wide range of organisms. There are many natural or synthetic cytokinins which are characterized by their ability to promote cell division but also have other effects on plant growth and development. The most effective cytokinins are N^6-substituted adenine derivatives and a number have been isolated from tRNA after hydrolysis. The cytokinin isopentenyl adenine and other substituted adenines always occur immediately adjacent to the 3′ end of the anticodon and appear to be necessary for the specific tRNA to function efficiently in protein synthesis. However, initial optimism that cytokinins may function by incorporation into tRNA has not been substantiated, since exogenous cytokinins do not appear in tRNA to any significant extent. Cytokinins have not only been shown to have multiple effects on the synthesis of proteins and nucleic acids, but also on the levels and synthesis of other hormones. However, there is little evidence that these regulatory rôles are exerted through a structural involvement in tRNA.

Initiation of the polypeptide chain

Complete polypeptide chains contain an amino acid with a free carboxyl group at one end and one with a free amino group at the other. Synthesis always begins with the amino-terminal amino acid and ends with the carboxyl-terminal. In order to ensure that polypeptide synthesis proceeds in the right direction, i.e. from the terminal-NH_2 group to the $-COOH$ end, initiation of the chain in bacterial systems always involves a special amino acid, N-formylmethionine (Fig. 5.10). The formyl group is added to the methionine after it has become attached to the tRNA. The enzyme that catalyses this reaction is not able to formylate free methionine or all methionine tRNAs; there are two types of methionine tRNA but only one of these is involved in the formylation reaction.

Fig. 5.10   The structure of N-formylmethionine and methionine.

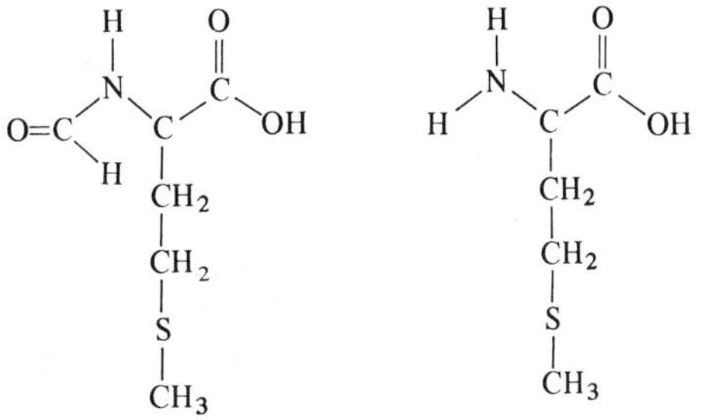

N-formylmethionine   Methionine

In the complex N-formylmethionyl-tRNA (fmet-tRNA$_f$), the $-NH_2$ group is protected by the formyl group leaving only the carboxyl group free to react with the $-NH_2$ group of the next amino acid. The N-formyl group does not, however, appear in the completed protein since it is later removed by a special hydrolytic enzyme. In most cases the terminal methionine is also removed by a second enzyme.

In bacterial cells, the ribosomes are constantly dissociating into sub-units and reforming, and it is now known that this dissociation is essential for protein synthesis to begin. The 30 S sub-unit binds both the mRNA and the fmet-tRNA to form an *initiation complex* which then binds with the 50 S sub-unit to form the functional 70 S ribosome (Fig. 5.11). The sub-units, fmet-tRNA and mRNA alone, however, are not enough to initiate polypeptide synthesis. There is an additional requirement for GTP (an energy-rich nucleotide) and three different protein *initiation factors* (called IF-1, IF-2 and IF-3 with molecular weights of 9 000, 70 000 and 21 000 respectively) which are not normally attached to the ribosomes. IF-3 is involved with the dissociation of the ribosomes into sub-units and with the binding of mRNA to the 30 S sub-unit. IF-1 and IF-2 appear to be involved in the binding of fmet-tRNA and GTP to form the initiation complex and in the binding

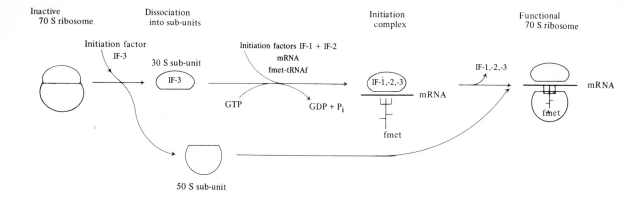

Inactive
70 S ribosome

Dissociation
into sub-units

Initiation factor
IF-3

30 S sub-unit

IF-3

Initiation factors IF-1 + IF-2
mRNA
fmet-tRNAf

GTP

GDP + P$_i$

Initiation
complex

IF-1,-2,-3

mRNA

fmet

IF-1,-2,-3

Functional
70 S ribosome

mRNA

fmet

50 S sub-unit

Fig. 5.11   Formation of the initiation complex and functional 70 S ribosome.

of the complex to the 50 S sub-unit. However, their roles are cooperative and complex and many of the details are ill-defined, particularly for IF-1 which usually binds more weakly than the other initiation factors and may also require their presence for binding. Furthermore, there is still some uncertainty concerning the order in which the various components needed for protein synthesis bind to the sub-units. The initiation factors are released when the initiation complex associates with the 50 S sub-unit.

In eukaryotic cells, initiation of 80 S ribosome protein synthesis involves an unformylated methionyl-tRNA. In higher plants, both bean and wheat tissues have been shown to contain two major and one minor species of tRNA^{met}. The minor species can be formylated and is presumed to be the initiator of chloroplast protein synthesis (see p. 216). One of the major species of tRNA^{met} acts as an initiator in the cytoplasm, while the other donates methionine into internal positions in the growing polypeptide chain. In the case of the initiator tRNA$_f^{met}$, it is not clear how the terminal $-NH_2$ group is protected to ensure that polypeptide synthesis occurs in the right direction (see above); the specific utilization of tRNA$_f^{met}$ in initiation results entirely from the selective recognition of this tRNA by the initiation factors.

Other differences from prokaryotes exist. Mammalian protein synthesis has been shown to require at least seven initiation factors (eIF-1, etc, although this is a rapidly developing and changing field), whereas only two initiation factors have been clearly identified to date in higher plants. In addition, initiation in eukaryotic systems requires both ATP and GTP.

Elongation

The 70 S ribosome contains two sites at which the aminoacyl-tRNAs bind in order that a peptide bond may be formed between the amino acids. These are the aminoacyl and peptidyl sites and are designated 'A' and 'P' sites respectively. Once the initiation complex has been formed, the fmet-tRNA must be attached to the peptidyl site before the second aminoacyl-tRNA can bind to the aminoacyl site of the complete ribosome at the next codon of the mRNA (Fig. 5.12). Additional cytoplasmic factors are required, however, for this binding

Fig. 5.12   Elongation of the
polypeptide chain.

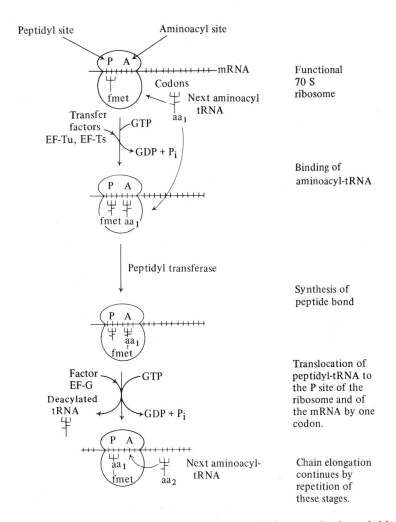

to occur. These are GTP and two proteins which occur in the soluble
phase of the cell and are known as *transfer* or *elongation factors*
(designated EF-Tu and EF-Ts). Factor EF-Tu forms a stable complex
with GTP and aminoacyl-tRNA and this ternary complex interacts
with the ribosome resulting in the transfer of the aminoacyl-tRNA to
the 'A' site. This involves the hydrolysis of GTP and the release of an
EF-Tu · GDP complex. The second factor EF-Ts reconverts this
complex to EF-Tu · GTP which can then interact with another molecule of aminoacyl-tRNA, allowing EF-Tu to function catalytically in
the binding cycle. These reactions may be summarized as follows:

Aminoacyl – tRNA · EF-Tu · GTP + ribosome · mRNA $\longrightarrow$

aminoacyl – tRNA · ribosome · mRNA + EF-Tu · GDP + P_i.

EF-Tu · GDP + EF-Ts $\rightleftharpoons$ EF-Tu · EF-Ts + GDP

EF-Tu · EF-Ts + GTP $\rightleftharpoons$ EF-Tu · GTP + EF-Ts

Aminoacyl-tRNA + EF-Tu · GTP $\rightleftharpoons$ Aminoacyl–tRNA · EF-Tu · GTP

The function of GTP hydrolysis appears to be to release EF-Tu
from the ribosome and thus allow the aminoacyl–tRNA to participate
in peptide bond formation.

The next stage in the elongation process involves the synthesis of a peptide bond by a reaction between the free amino group of the incoming amino acid and the carboxyl group of the first amino acid which is esterified to tRNA. The enzyme which catalyses this reaction is called peptidyl transferase (or peptide synthetase) and is an integral part of the large sub-unit. It contains two substrate binding sites (the A', acceptor site, and P', donor site) and a catalytic centre.

After the peptide bond has been formed, the peptidyl-tRNA, which has just been lengthened by the addition of one amino acid, must be moved from its attachment to the 'A' site on the ribosome to the 'P' site to make way for the next aminoacyl-tRNA. This movement of the ribosome relative to the mRNA is termed translocation and again requires the hydrolysis of GTP and another protein elongation factor called EF-G. At the same time there is a translocation of mRNA which moves three nucleotides (one codon) over the ribosomal surface. One molecule of GTP is hydrolysed for every translocation step. The function of EF-G is not clear since, although it is generally regarded as the 'translocase', recent work suggests that it may be primarily involved in the release of the deacylated tRNA from the 'P' site of the ribosome.

This sequence of events involved in elongation must take place very rapidly since it has been calculated that, in *E. coli* growing under optimal conditions, a polypeptide chain of about forty amino acids can be produced in 20 seconds. In eukaryotes, the process of chain elongation appears to be essentially similar to that in prokaryotes and requires GTP and potassium and magnesium ions. Two protein elongation factors have been isolated from various tissues, including wheat embryos, and are designated EF-1 and EF-2. EF-1 appears to be equivalent to EF-Tu; it forms a ternary complex with GTP and aminoacyl-tRNA and is responsible for the binding of the aminoacyl-tRNA to the ribosomes. There appears to be no factor corresponding to the EF-Ts of prokaryotes. The second elongation factor, EF-2, is involved in the process of translocation and is equivalent to EF-G. The elongation factors are exchangeable between eukaryotes but not between eukaryotes and prokaryotes.

Termination and release of the polypeptide chain

The termination of the polypeptide chain and the release of the completed chain from the ribosome is controlled by two factors. The termination of the chain is indicated by three special termination codons in the mRNA which are UAG, UAA and UGA. These are the so-called *nonsense codons* which do not encode for any amino acid. The chain is, however, still bound to the tRNA which is in turn attached to the mRNA. In prokaryotes, the chain is released from the ribosome under the direction of three distinct proteins which are called *release factors* and are designated RF-1, RF-2 and RF-3. These are bound to the ribosome and control the hydrolysis of the ester link between tRNA and the polypeptide chain. RF-1 recognizes UAA or UAG, while RF-2 recognizes UAA or UGA. RF-3 has no release activity itself, but facilitates the binding of the other factors to the ribosomes and so affects the rate of release of the polypeptide chain. GTP hydrolysis occurs and this appears to be required for the binding and dissociation of the release factors rather than peptidyl–tRNA hydrolysis. Little is known about chain termination in plants although, in animals, there appears to be a single release factor (RF) which recognizes all three termination codons.

Once the chain has been terminated and released, the ribosome separates from the mRNA and is ready to enter a new cycle of polypeptide synthesis. To do this, it must first dissociate into its two sub-units, a step which is promoted by initiation factor IF-3 as previously described. The cycle of events involved in protein synthesis on the 70 S bacterial ribosome is illustrated in Fig. 5.13. Although not as fully described, protein synthesis in plants appears to be essentially very similar. The chief differences are in the initiator tRNA employed in the recognition of mRNA during initiation (see p. 206), in the numbers of soluble protein factors involved at various stages and, perhaps, in the involvement of GTP and ATP in different steps. Important differences may also exist in the control of protein synthesis at the level of translation (see Bray, 1976; Revel and Groner, 1978).

The above account is, of necessity, a brief one. For full details of protein synthesis in both prokaryotes and eukaryotes, the reader is referred to the extensive reviews quoted at the end of the chapter.

The genetic code

As we have seen in the previous sections, protein synthesis on the ribosomes involves both tRNA and mRNA. The genetic information for the synthesis of proteins contained in DNA is carried to the ribosomes by mRNA. Since there is no specific affinity between amino acids and the nucleotides of RNA, tRNAs act as adaptors, enabling the amino acids to attach to mRNA (see p. 201). What then is the correlation between the nucleotide sequence in DNA (and mRNA) and the amino acids in the polypeptide chains? The variable parts of the nucleic acids are the bases. These consist of adenine (A), guanine (G), cytosine (C) and thymine (T) in DNA and A,G,C and uracil (U) in RNA. Since there are four bases in the nucleic acids and about twenty amino acids in the proteins, there is clearly not a one-to-one relationship. It is now known that groups of three nucleotides in DNA which are known as *codons* correspond to specific amino acids.

The details of the methods used to elucidate the *genetic code*, i.e. the base sequences that correspond to specific amino acids, will be discussed only briefly here. The problem has been studied mainly with bacteria and viruses and the techniques are thoroughly described in texts on genetics and molecular biology. One method which has been widely used involves the incubation of ribosomes with synthetic mRNA of known base composition and ratio. For example, if co-polymers of, say, A and U are prepared, these will contain eight possible triplets (AAA, AAU, AUA, UAA, AUU, UAA, UUA,

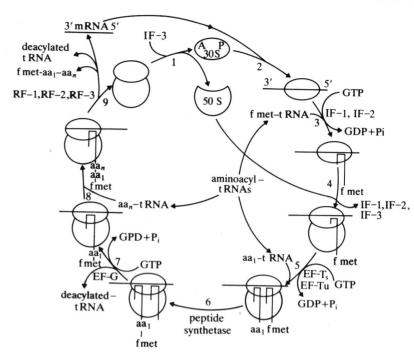

Fig. 5.13 Mechanism of protein synthesis on the 70 S microbial ribosome. Modified after Boulter, Ellis and Yarwood (1972).

Initiation

1,2: Dissociation of monomer into native sub-units requires initiation factor IF-3, which also promotes binding of mRNA to 30 S sub-unit. 'A' and 'P' indicate relative positions of aminoacyl and peptidyl sites respectively.

3: Formation of initiation complex requires factors IF-1 and IF-2, GTP and fmet-tRNA. The latter is shown entering directly into the 'P' site, alternatively it may enter the 'A' site and then move to the 'P' site.

4: Functional ribosome formed by addition of 50 S sub-unit; initiation factors released.

Chain elongation

5: Elongation factors EF-Ts and EF-Tu and GTP required for binding of aminoacyl-tRNA in 'A' site.

6: Peptide bond formation requires peptide synthetase (peptidyl-transferase) thought to be a function of the ribosome.

7: GTP and transfer factor *G* are required for displacement of discharged tRNA and translocation of peptidyl-tRNA to the 'P' site. Although it has been considered as the 'translocase', factor EF-G appears to promote the release (uncoupled from translocation) of the deacylated tRNA molecule from the 'A' site, while translocation *per se* seems to be a function of the 50 S sub-unit.

8: Chain elongation proceeds by repetition of steps 5, 6 and 7.

Chain termination

9: Requires release factors RF-1, RF-2 and RF-3. A specific peptidyl-tRNA hydrolase may also be required to give the free polypeptide. Ribosome may be released in form of sub-units, but in absence of IF-3 these will spontaneously recombine.

UUU). If the ratio of the two bases is known, the relative abundance of the possible codons can be calculated. This can then be compared with the amount of incorporation of various radioactive amino acids into protein using these synthetic polynucleotides as mRNA. The value of this procedure was greatly enhanced by the development of techniques for the synthesis of co-polymers of known sequences of repeated bases. These procedures gave considerable information regarding the base composition of the triplets coding for specific amino acids, but told little of the sequence of bases in these triplets. However a method was developed which enabled this sequence to be studied directly. It was found that, in the absence of GTP, ribosomes did not synthesize proteins but were still able to bind mRNA and aminoacyl-tRNA. The shortest oligonucleotide which supported the binding of tRNA was a trinucleotide. For example, the trinucleotide AAA was found to support the binding of lysine-tRNA. The synthesis of a large number of these trinucleotides greatly helped to determine the base sequence of the codons for certain amino acids.

The use of these methods and others established the codons of all of the amino acids (Table 5.2). Of the sixty-four possible permutations allowed by a triplet code, sixty-one have been assigned to specific amino acids; the remaining three, the nonsense triplets, function as signals for the termination of polypeptide synthesis (see p. 208). The main features of the code are as follows.

Table 5.2   The genetic code

Alanine	GCU	GCC	Leucine	UUA	UUG
	GCA	GCG		CUU	CUC
Arginine	CGU	CGC		CUA	CUG
	CGA	CGG	Lysine	AAA	AAG
	AGA	AGG	Methionine	AUG	
Asparagine	AAU	AAC	Phenylalanine	UUU	UUC
Aspartic acid	GAU	GAC	Proline	CCU	CCC
				CCA	CCG
Cysteine	UGU	UGC	Serine	UCU	UCC
Glutamic acid	GAG	GAA		UCA	UCG
				AGU	AGC
Glutamine	CAA	CAG	Threonine	ACU	ACC
Glycine	GGU	GGC		ACA	ACG
	GGA	GGG	Tyrosine	UAU	UAC
Histidine	CAU	CAC	Tryptophan	UGG	
Isoleucine	AUU	AUC	Valine	GUU	GUC
	AUA			GUA	GUG
			Nonsense codons	UAA	UAG
				UGA	

1. The code is highly *degenerate*. This means that all of the amino acids except tryptophan and methionine are coded by more than one codon. Thus leucine is coded by six different triplets and tyrosine by two triplets. In most cases the variation in the codon for a specific amino acid involves only the third base in the sequence.

2. The code is not overlapping and there is no punctuation or spacing between codons. There must therefore be a fixed starting signal at the beginning of mRNA which ensures that its nucleotide sequence is correctly read. There is only one codon, AUG, for methionine and, in bacteria, this also codes for the initiating amino acid N-formyl methionine. However, it appears that the initiator codon is not at the 5' terminus of mRNA but is set some number of nucleotides in, producing a long initiator region. In prokaryotes, the initiation signal consists of the codon

Fig. 5.14 Structures of proline and azetidine-2-carboxylic acid.

$$H_2C\text{---}CH_2$$
$$H_2C\diagdown_N\diagup CH\cdot COOH$$
$$H$$

Proline

$$CH_2$$
$$H_2C\diagup\diagdown CH\cdot COOH$$
$$N$$
$$H$$

Azetidine-2-carboxylic acid

AUG which is preceded on its 5' side by a sequence of purine nucleotides. This region base pairs to a complementary sequence near the 3' end of the 16 S rRNA (see p. 196). Most eukaryotic mRNA possesses a capping structure preceding the first AUG which takes the form $m^7 G^{5'}$ ppp5' $N_m Y_m \ldots$ (i.e. the terminal 7-methyl guanosine is linked by a pyrophosphate bridge to another methylated nucleoside). The cap probably has at least two functions. 7-Methyl guanosine is recognized during initiation while the cap may protect the mRNA from degradation by nucleases.

3. The code is widely believed to be universal. At least, it has been shown to be identical for the amino acids from viruses, bacteria, the amphibian *Xenopus* and certain mammals. However there is now evidence that mitochondrial genetic systems (see p. 322) differ from the standard code in certain respects. For example, UGA codes for tryptophan instead of 'stop' (see p. 208) in yeast mitochondria. The implications of this finding in relation to the evolution of mitochondria are far from clear.

Polysomes

In early studies with the electron microscope, it was noticed that ribosomes frequently occurred in groups joined by a thin thread although the significance of this arrangement in relation to the mechanism of protein synthesis was not at first appreciated. The function of these polysomes was first suggested by differential centrifugation experiments carried out by Warner, Rich and Hall (1962). They found that if reticulocytes were incubated with radioactive amino acids and then lysed and centrifuged, the maximum labelling of nascent protein was associated not with the single ribosome 80 S fraction but with a more rapidly sedimenting 170 S fraction. Electron microscopy showed that most of the ribosomes in this fraction were present in groups of five. The filament joining the ribosomes was known to be RNA from studies using RNAase and this led to the idea that polysomes consisted of a number of ribosomes passing along a strand of mRNA, each at a different stage in the assembly of the same polypeptide chain (Fig. 5.15). The number of ribosomes that can be associated together in a polysome is clearly determined by the length of the mRNA. For example, the polypeptide chains in haemoglobin contain about 150 amino acids which would require a mRNA chain of about 450 nucleotides. Nucleotides are spaced 0·34 nm apart, giving a mRNA molecule about 100 nm long which could accommodate a number of ribosomes of approximately 20 nm diameter. In some bacterial and animal cells, polysomes containing up to fifty units have been reported. Although the polysome structure represents a more

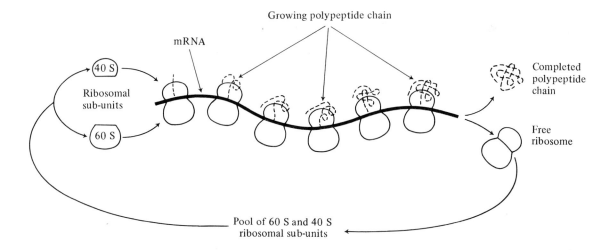

Growing polypeptide chain

mRNA

40 S

Ribosomal sub-units

60 S

Completed polypeptide chain

Free ribosome

Pool of 60 S and 40 S ribosomal sub-units

Fig. 5.15 Diagrammatic representation of polyribosomal function. The individual ribosomes function independently, each synthesizing a polypeptide chain as it moves along the mRNA.

efficient use of mRNA, one ribosome alone acting on mRNA can produce a single polypeptide chain. Evidence from bacteria suggests that the two sub-units of the ribosome dissociate at the termination of the polypeptide chain to join a pool of sub-units and then randomly associate at chain initiation. It is possible to envisage a ribosome–polysome cycle with a continuous dissociation and association of sub-units (Fig. 5.15).

Polysomes occur in higher plant cells and are found both free in the cytoplasm and attached to membranes of the endoplasmic reticulum. Studies with animal cells suggest that this attachment is by the large sub-unit, leaving the smaller sub-unit free to dissociate. It is also believed that certain proteins are specifically synthesized on the ribosomes and polysomes attached to the ER. For example, with animal cells, there is good evidence that ribosomes synthesizing a wide range of secretory proteins are located almost exclusively on membranes. The mechanism by which these proteins are synthesized and exported is discussed in Chapter 3. Intracellular proteins, however, appear to be synthesized on both membrane-bound and free ribosomes, although different kinds of protein may be produced by the two types of ribosome. The most detailed investigation on plants involves legume storage protein synthesis. In bean cotyledons, just prior to the stage of storage protein synthesis, there is a marked increase in the amount of ER and in the proportion of membrane-bound ribosomes. These observations were extended by Bailey, Cobb and Boulter (1970) who showed that if slices of bean (*Vicia faba*) cotyledons were incubated with radioactive leucine, 80% of the amino acid incorporation into protein was into globulin protein which is stored in particles called protein bodies. Autoradiographic studies suggested that this incorporation is initially associated with the rough ER and is then transferred to the protein bodies. This concept has gained support from Bollini and Chrispeels (1979) who investigated the biosynthesis of polypeptides by free and membrane-bound polysomes isolated from *Phaseolus vulgaris* which stores vicilin and phytohaemoglutinin as the major reserve proteins (50% and 10% respectively of the total protein of mature seeds). The two classes of polysome were shown to make different sets of polypeptides. Four polypeptides, similar in size to the four polypeptides of vicilin, were made by membrane-bound polysomes and not by free polysomes. The membranes were identified as ER. Free polysomes only made polypeptides which did not bind to antibodies specific for vicilin. Thus the ER was clearly identified as the site of vicilin synthesis in developing bean cotyledons. It should be mentioned however that not all reports have found such a clear distinction between free and bound ribosomes in relation to reserve protein synthesis. Furthermore, the exact pathway taken by the storage protein in moving from the ER to the protein bodies has not been established.

The importance of polysomes in protein synthesis has been demonstrated by a variety of different experiments. If radioactive amino acids are fed to plant tissues, there is a preferential incorporation into the polysome fraction, demonstrated when polysomes are separated from free ribosomes by centrifugation. Increased protein synthesis, for example in germinating seeds, is generally accompanied by an increased polysome content, while polysome levels tend to fall during senescence and when ripening seeds begin to dry out. The polysome content of cells may also be influenced by plant growth substances. For example, during the lag period before the appearance of enzymes in gibberellic acid-treated barley aleurone layers (see p. 183), there is an increase in polysome formation and in membranes of the endoplasmic reticulum. Addition of abscisic acid reduces the percentage of polysomes and inhibits further α-amylase synthesis, although the total number of ribosomes per cell is not affected (Evins and Varner, 1972). The removal of gibberellic acid during the midcourse of α-amylase production also arrested the synthesis of the enzyme and decreased the percentage of polysomes. Another example is provided by Short, Tepfer and Foskett (1974) who showed that cytokinin treatment can evoke a rapid and specific increase in the polysome content of cultured soybean cells. Similarly, there is evidence that light, acting through phytochrome, can regulate protein synthesis through modification of ribosomal properties, including the formation of polysomes (Smith, Billett and Giles, 1977). Thus an effect on polysome formation may provide an important means by which gene expression can be regulated at the translational level. There are a variety of other ways by which this might occur including the availability and binding of tRNA and initiation factors and the regulation of polypeptide chain elongation and termination. Phosphorylation and dephosphorylation (see p. 111) of ribosome proteins and other necessary components may be involved in some of these processes. In some animal systems there is good evidence that post-transcriptional control of both the rate and quality of protein synthesis occurs and that some of the mechanisms mentioned above are involved. However, much less is known about such control processes in higher plants.

The extent to which polysomes occur and may be isolated from plant cells depends very much on the physiological state of the tissue. Wilting, anaerobiosis, or addition of the uncoupling agent dinitrophenol causes a complete loss of polysomes, whereas treatment with actinomycin D (an inhibitor of RNA polymerase) may have little effect on the polysome content. Such experiments suggest that a continuous supply of energy, but not a continuous synthesis of mRNA, is needed to maintain polysome structure. Lin and Key (1967) studied the recovery of soybean roots after a period of anaerobiosis and found that polysomes reformed rapidly in the first 30 min of the aeration period and that this was followed by a much lower rate of recovery. Actinomycin D had little effect on the first rapid stage of recovery, although the inhibitor had a pronounced effect on the second stage. This experiment suggested that at least some of the mRNA in the cells survives unfavourable conditions and can be utilized to form polysomes as soon as energy production is resumed.

Perhaps the most striking example of mRNA survival is the so-called *long-lived* or *preformed mRNA* that is reported to occur in dry seeds. Although both RNA and protein synthesis are initiated very early in seed germination, making it very difficult to determine which comes first, there is increasing support for the concept that intact mRNA is present in dry, ungerminated seeds (see Payne, 1976). However, although this preformed mRNA is translated during early germination, it appears that its function its rapidly taken over by newly synthesized mRNA which encodes for the same polypeptides (Caers, Peumans and Carlier, 1979). Thus the role of this mRNA in seed germination remains uncertain at the present time.

Protein synthesis and ribosomes in cell organelles

It has frequently been demonstrated in a variety of higher cells that protein synthesis occurs at sites apart from the cytoplasmic ribosomes. With higher plants, the presence of ribosomes and the capacity to incorporate amino acids into protein has been reported for both chloroplasts and mitochondria. Probably the most important criterion in these studies is the isolation of the organelle fractions free from contamination by other organelles or by bacteria. The ability of chloroplasts and mitochondria to synthesize protein is perhaps not surprising, since it is well established that they contain DNA and so have some degree of genetic autonomy (see Chs. 7 and 8). Evolution presumably would preserve the machinery to process the information contained in the DNA. It has also been reported that cell wall fractions from barley shoots contain ribosomes, although there is no evidence that they are active or perform any function (Jervis and Hallaway, 1970). The characteristics of these ribosomes are identical to the cytoplasmic ribosomes and they may well originate in the cytoplasm and become incorporated into the cell wall during cell plate formation (see Ch. 10). In contrast, the ribosomes isolated from chloroplasts and mitochondria are distinct from the cytoplasmic ribosomes and appear to resemble bacteria ribosomes in many respects. The characteristics of ribosomes and protein synthesis in higher plant mitochondria are not as well understood as the chloroplast systems and so it is the chloroplast ribosomes that will be discussed more fully here.

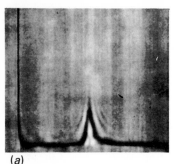

(a)

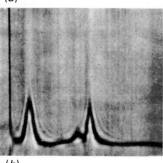

(b)

Fig. 5.16   Ultracentrifuge diagrams of leaf and chloroplast extracts from spinach. (a) Chloroplast ribosomes; (b) leaf extract. Reproduced from Lyttleton (1962).

Ribosomes and protein synthesis in chloroplasts

The first report that photosynthetic tissue contains two classes of ribosomes came from Lyttleton in 1962. He found that two classes of ribosomes could be separated by ultracentrifugation from a general leaf extract of spinach, whereas only the smaller class was released from a chloroplast preparation from the same tissue (Fig. 5.16). The chloroplast ribosomes had a sedimentation coefficient of about 70 S, similar to bacterial ribosomes. Since this first report, two size classes of ribosomes have been demonstrated in leaf and algal extracts from a variety of species, while other experiments have revealed further similarities between chloroplast and bacterial ribosomes. The major characteristics in which ribosomes from chloroplasts resemble those from bacteria but differ from cytoplasmic ribosomes are as follows.

1. Chloroplast ribosomes have a sedimentation coefficient of about 70 S as compared to the 80 S cytoplasmic ribosomes (Table 5.1).
2. The two major RNA components from chloroplasts are comparable in size to bacterial and blue-green algal RNAs and smaller than the cytoplasmic RNAs. Loening and Ingle (1967) examined RNA sizes from seven higher plant species by gel electrophoresis. They found that the green tissues were enriched in RNAs with mobilities corresponding to 23 S and 16 S RNA (see p. 196), whereas non-green tissues contained only 25 S and 18 S RNA.
3. The 70 S ribosomes require higher Mg^{++} concentrations for stability. Boardman, Franki and Wildman (1966) found that brief dialysis of tobacco extracts caused almost complete dissociation of the chloroplast ribosomes but not of the cytoplasmic ones.
4. Chloroplasts ribosomes from *Euglena* and higher plants have a requirement for *N*-formylmethionyl-tRNA in chain initiation (see Bryant, 1976); this resembles prokaryotic systems but not 80 S ribosomes.
5. The two classes of ribosomes are further distinguished by their response to antibiotics such as chloramphenicol. Ellis (1969) examined the effect of the various isomers of chloramphenicol on amino acid incorporation by cytoplasmic and chloroplast ribosomes from tobacco. Only the naturally occurring D-*threo*-isomer inhibited incorporation and this was only effective against the chloroplast system (Table 5.3). There is a similar response of higher cell ribosomes to spectinomycin, lincomycin and erythromycin, all of which inhibit bacterial ribosomal activity (Ellis, 1970). These antibiotics had no appreciable effect on amino acid incorporation by cytoplasmic ribosomes from tobacco leaves, beetroot discs or rat liver but they did inhibit incorporation by bean and tobacco chloroplasts (with the exception that erythromycin did not inhibit incorporation by tobacco chloroplasts). These experiments, in which a range of structurally unrelated antibiotics were used, clearly demonstrate the similarity between bacterial and chloroplast ribosomes, and further

Table 5.3   Effect of chloroamphenicol (CAM) isomers and cycloheximide on radioactive leucine incorporation into protein by chloroplast and cytoplasmic ribosomes from tobacco. The results are expressed as a percentage of the number of counts incorporated by the complete assay mixture which contained L-leucine-^{14}C, ATP, an ATP generating system, GTP and potassium and magnesium ions in a final volume of 0·5 ml. In some experiments, chloroplasts were treated with the detergent Triton X-100 to remove any bacterial contamination.

| | L-Leucine-^{14}C incorporation into protein (%) | | |
Assay mixture	Intact chloroplasts	Chloroplasts treated with Triton X-100	Cytoplasmic ribosomes
Complete	100·0	100·0	100·0
Zero time	1·5	0·5	11·0
−ATP and GTP	5·0	3·0	7·0
Complete, Triton X-100 insoluble	5·0		
+Ribonuclease (1 μg)	20·0	2·5	18·0
+D-threo CAM (150 μg)	23·0	23·0	95·0
+L-threo CAM (150 μg)	90·0	98·0	93·0
+D-erythro CAM (150 μg)	90·0	92·0	100·0
+Cycloheximide (150 μg)	96·0	99·0	76·0
+Cycloheximide (12 μg)	90·0	95·0	86·0

Reproduced from Ellis (1969).

distinguishes them from the cytoplasmic ribosomes of higher cells.

The mechanism of protein synthesis in chloroplasts appears to be similar to that previously described for prokaryotic cells. Isolated chloroplasts have been shown to contain aminoacyl-tRNAs, aminoacyl-tRNA synthetases, transformylase activity and two species of methionyl-tRNA, one of which can be formylated. The tRNAs^{met} from the cytoplasm of higher plants cannot be formylated. Isolated chloroplasts require magnesium ions, GTP and an ATP-generating system in order to incorporate amino acids into protein. The ATP may be generated in the light in the absence of an enzymic ATP-generating system. For example, Raminez, Del Campo and Arnon (1968) showed that isolated, whole spinach chloroplasts were able to incorporate amino acids into protein solely at the expense of light energy. The rate of incorporation on the light was about twenty times that in the dark and was considerably greater than in the presence of an enzymic ATP-generating system. Polysomes have been isolated from chloroplasts and electron microscopy has shown that these may be attached to the thylakoid membranes. In tobacco leaves, the 70 S ribosomes are as much as twenty times more active in amino acid incorporation than the 80 S particles.

Function of the chloroplast ribosomes

Having established that chloroplasts are capable of incorporating amino acids into protein and that they contain ribosomes distinct from those in the cytoplasm, we must then consider the extent to which chloroplast proteins are made on these 70 S ribosomes. Much of the information available at present has been obtained from inhibitor studies in which the proteins made in the chloroplasts are identified by analysis of the effect of 70 S ribosome inhibitors on the proteins made in greening cells and tissues. This approach is illustrated by the results of Ellis and Hartley (1971) who studied the effect of the antibiotic lincomycin, which specifically inhibits protein synthesis on chloroplast ribosomes (see above), on the development of certain photosynthetic enzymes when pea apices were transferred from the dark to the light (Table 5.4). Only the development of ribulose bisphosphate carboxylase and of the protein known as Fraction 1 protein was inhibited by lincomycin.

Fraction 1 protein is a part of the protein complement of chloroplasts which sediments as a single fraction in the ultracentrifuge, but is present in much larger amounts than any other and constitutes about half of the total soluble leaf protein. It has a molecular weight of about 500 000, consists of large and small sub-units, and contains the fixation enzyme ribulose bisphosphate carboxylase (see Ch. 8). Fraction 1 protein and ribulose bisphosphate carboxylase are considered to be the same protein. Further examination of the formation of this protein revealed that the synthesis of the two sub-units shows differing responses to 70 S and 80 S ribosome inhibitors. It appears that although the large sub-unit is synthesized in the chloroplasts, the small one is made in the cytoplasm.

This conclusion is supported by the results of another approach to this problem which aims to identify the products of protein synthesis by isolated chloroplasts. For example, Blair and Ellis (1973) showed that isolated chloroplasts from *Pisum* incorporated labelled amino

Table 5.4   Effect of lincomycin (LM) on light-induced increases in photosynthetic enzymes in pea apices. Detached pea apices were illuminated in the presence and absence (water) of lincomycin and then homogenized and assayed for five enzymes. The specific activities were compared with those from apices kept in the dark

	Specific activity ($\mu mol/min/mg\ protein$) 48 h light			
	Dark	Water	LM 2 $\mu g/ml$	LM 1 $\mu g/ml$
Ribosephosphate isomerase	2·4	7·5	10·6	9·2
Phosphoribulokinase	0·02	0·31	0·31	0·32
Ribulose bisphosphate carboxylase	0·015	0·1	0·022	0·025
Fraction I protein	1·0	4·6	1·0	1·2
Phosphoglycerate kinase	1·4	3·4	3·8	3·0
Triosephosphate dehydrogenase (NADP)	0·02	0·12	0·1	0·11
Triosephosphate dehydrogenase (NAD)	0·22	0·21	0·21	0·22

Reproduced from Ellis and Hartley (1971). Copyright © 1971 by the American Association for the Advancement of Science.

acids into only one soluble polypeptide which was identified as the large sub-unit of Fraction 1 protein. These findings are summarized in the model shown in Fig. 5.17. Experiments with both *Chlamydomonas* and *Pisum* have identified a polypeptide of molecular weight about 20 000 known as P20, which is immunoprecipitated by antibodies prepared against purified ribulose bisphosphate carboxylase (Dobberstein, Blobel and Chua, 1977; Highfield and Ellis, 1978). P20 is believed to be a higher molecular weight precursor of the small sub-unit, which is translated in the cytoplasm and is cleaved to its final size (MW about 14 000) within or while moving into the chloroplast.

Fig. 5.17 Model for Fraction −1-protein (ribulose bisphosphate carboxylase) synthesis in higher plants. Redrawn after Highfield and Ellis (1978).

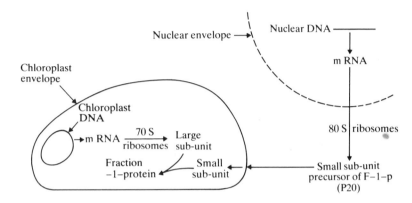

The model raises the interesting question of how such a large molecule is transported across the chloroplast envelope. Highfield and Ellis (1978) demonstrated that the processing of P20 to the small sub-unit requires the presence of chloroplast envelopes in *Pisum*, and they went on to propose that the transport of P20 into chloroplasts may involve combination with a specific carrier in the chloroplast envelope. Protease activity in the envelope then removes part of the polypeptide which produces a conformational change in the membrane, leading to the transport of the small sub-unit into the chloroplast (see below). A similar model may also be applied to mitochondria. A number of proteins are now known to be synthesized within these organelles and in some cases are made up of sub-units which are synthesized in separate cellular compartments (see Schatz, 1976). For example, three of the seven sub-units of cytochrome *c* oxidase are synthesized within the mitochondria, the remainder outside. Furthermore, it is possible that the cytoplasmic sub-units may well exert an important controlling effect on the synthesis of the organelle-derived components. The transport of proteins into mitochondria and chloroplasts has been reviewed by Chua and Schmidt (1979). This does not appear to involve a mechanism similar to the signal hypothesis (see Ch. 3). The proteins are made on cytoplasmic ribosomes which are not bound to the organelle membranes. Movement into the organelles is a post-translational and not a co-translational event. A mechanism such as that described by the trigger hypothesis (see Ch. 3) may be involved.

Thus the experiments using 70 S ribosome inhibitors and isolated chloroplasts indicate that the bulk of chloroplast protein is synthesized in the cytoplasm. Apart from the large sub-unit of Fraction 1 protein, there is evidence that some chloroplast ribosomal proteins and certain proteins of the chloroplast envelope and thylakoids are synthesized within the organelle. The interactions and control processes between the chloroplast and the nucleocytoplasmic system may well be very complex and present a fascinating topic for future research.

Evolution of ribosomes

The marked similarities between the ribosomes of bacteria and chloroplasts and their differences from the cytoplasmic ribosomes of higher cells have revived the old concept that chloroplasts evolved from symbiotic photosynthetic prokaryotic cells, although this does not explain how the 80 S cytoplasmic ribosomes evolved. Alternatively, endosymbiosis may not have been involved. The present day organelles may have arisen by the enclosing of some of the DNA and ribosomes of the cell within membranes. These ribosomes then retained some of their primitive characteristics while cytoplasmic particles evolved into the 80 S type.

Analysis of ribosomes and of ribosomal RNA has shown that there is a general increase in the size of ribosomes during the course of evolution. Loening (1968) determined the molecular weights of the two largest ribosomal RNA components from a wide range of species using electrophoretic methods (Table 5.5). It was found that bacteria, blue-green algae and higher plant chloroplasts have 23 S and 16 S rRNAs (see p. 196) with molecular weights of 1·1 and $0·56 \times 10^{-6}$ respectively. The 25 S and 18 S components of the cytoplasmic ribosomes from higher plants, ferns, other algae and fungi have molecular weights of $1·3 \times 10^{-6}$ and $0·7 \times 10^6$. The $0·7 \times 10^{-6}$ component is also common to all animals but the larger component (with sedimentation coefficient of about 28 S) has a molecular weight of $1·4 \times 10^6$ in sea urchins and $1·75 \times 10^6$ in mammals.

These and other determinations have been incorporated into a general scheme for the molecular evolution of rRNA by Noll (1970) (Fig 5.18). The two major rRNA components are similar in size in prokaryotes and chloroplasts and are smaller than those from higher

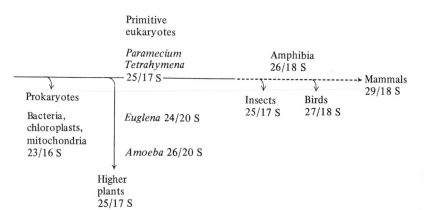

Fig. 5.18   Biochemical evolution of ribosomal RNA. The numbers refer to the sedimentation values of the large and small components of ribosomal RNA characteristic of each class of organism. Reproduced from Noll (1970).

Table 5.5 The molecular weights of ribosomal RNA from various species

Species		Molecular weight ($\times 10^{-6}$)		Markers	No. of determinations
Animals					
	HeLa	1·75	0·70	Standard	
	Rat (liver)	coinc.	coinc.	HeLa	(2)
	Mouse (liver)	1·71	0·70	*E. coli*	(1)
	Rabbit (reticulocytes)	1·72	0·70	*E. coli*	(1)
	Chick (liver)	1·58	coinc.	HeLa	(2)
	Xenopus (tadpole)	1·51	conic.	HeLa	(1)
	(liver, ovary)	1·54	0·69	*E. coli*	(2)
	Drosophila	1·40	0·73	*E. coli*	(2)
		1·41	0·73	*Xemopus*	(1)
	Arbacia	1·40	0·68	*E. coli*	(2)
Plants and protozoa					
Protozoa	*Amoeba*	1·53 ⎤ unstable	0·89	HeLa and *E. coli*	(4)
	Euglene	(1·3?) ⎦	0·85	Own chloroplast, approx.	(3)
	Tetrahymena	1·30	0·69	HeLa and *E. coli*	(2)
	Paramecium	1·31	0·69	*E. coli*	(1)
Higher plants	Pea, bean, radish, corn	1·27–1·31	0·70–0·71	HeLa and *E. coli*	(6)
Algae	*Chlorella*	1·28	0·69	*E. coli*	(1)
	Chlamydomonae	1·30	0·69	*E. coli*	(1)
Fern	*Dryopteris*	1·34	0·72	*E. coli*	(2)
Fungi	*Aspergillus*	1·30	0·73	*E. coli*	(1)
	Botrylis	1·30	0·68	*E. coli*	(1)
	Chaetomium	1·30	0·71	*E. coli*	(1)
	Rhyzopus	1·28	0·72	*E. coli*	(2)
	Saccharomyces	1·30	0·72	*E. coli*	(2)
Prokaryotic cells					
Bacteria	*E. coli*	1·07	0·56	Standard	
	Rhodopseudomonas	1·08	0·59	Pea	(1)
Actinomycetes	*Streptomyces*	1·11	0·56	HeLa	(2)
	Other species	1·13	0·56	HeLa	(2)
Blue-green algae	*Anabaena*	1·07	0·55	Pea	(1)
	Nostoc	coinc.	coinc.	*E. coli*	(1)
	Oscillatoria	1·07	0·56	Pea	(1)
Higher plant chloroplasts		1·07–1·11	0·56	Pea and *E. coli*	(many)

Reproduced from Loening (1968).

organisms. Apart from a few exceptions such as *Amoeba*, there is an increase in size of the RNA of the larger ribosomal sub-unit as evolution progresses from plants and lower animals, to amphibia and mammals. The functional advantage of this increase in size is unknown.

Further reading

Boulter, D. (1977) Protein synthesis in the cytoplasm. In *The Molecular Biology of Plant Cells*. Ed. H. Smith. Blackwell Scientific Publications, Oxford, p. 256.

Boulter, D., Ellis, R. J., and Yarwood, A. (1972). Biochemistry of protein synthesis in plants. *Biol. Rev.*, **47**, 113.

Bray, C. M. (1976) Protein synthesis. In *Molecular Aspects of Gene Expression in Plants*. Ed. J. A. Bryant. Academic Press, London. p. 109.

Brimacombe, R., Stöffler, G. and Wittmann, H. G. (1978) Ribosome structure. *Ann. Rev. Biochem.*, **47**, 217.

Bryant, J. A. (1976) *Molecular Aspects of Gene Expression in Plants*. Academic Press, London.

Chua, N-H., and Schmidt, G. W. (1979) Transport of proteins into mitochondria and chloroplasts. *J. Cell Biol.*, **81**, 461.

Ellis, R. J. (1977) Protein synthesis by isolated chloroplasts. *Biochim. Biophys. Acta.*, **463**, 185.

Ellis, R. J. (1981) Chloroplast proteins: synthesis, transport and assembly. *Ann. Rev. Plant Physiol.*, **32**, 111.

Kurland, C. G. (1977) Structure and function of the bacterial ribosome. *Ann. Rev. Biochem.*, **46**, 173.

Leaver, C. J. (1980) *Genome Expression and Organization in Higher Plants.* NATO, AS129, Plenum Press, New York.

Noll, H. (1970) Organelle integration and the evolution of ribosome structure and function. *Sym. Soc. exp. Biol.*, **24**, 419.

Payne, P. I. (1976) The long-lived messenger ribonucleic acid of flowering-plant seeds. *Biol. Rev.*, **51**, 329.

Revel, M., and Groner, Y. (1978) Post-transcriptional and translational controls of gene expression in eukaryotes. *Ann. Rev. Biochem.*, **47**, 1079.

Stöffler, G. and Wittmann, H. G. (1977) Primary structure and three-dimensional arrangement of proteins within the *Escherichia coli* ribosome. In *Molecular Mechanisms of Protein Biosynthesis.* Eds. H. Weissbach and S. Pestka. Academic Press, New York.

Stutz, E. (1976) Ribosomes. In *Plant Biochemistry.* Eds. J. Bonner and J. E. Varner. Academic Press, New York, p. 15.

Weissbach, H. and Pestka, S. (1977) *Molecular Mechanisms of Protein Biosynthesis.* Academic Press, New York.

Wool, I. G. (1979) The structure and function of eukaryotic ribosomes. *Ann. Rev. Biochem.*, **48**, 719.

Literature cited

App, A. A., Bulis, M. G. and McCarthy, W. J. (1971) Dissociation of ribosomes and seed germination. *Plant Physiol.*, **47**, 81.

Attwood, M. M. and Cocking, E. C. (1965) The purification and properties of the alanyl-transfer ribonucleic acid synthetse of tomato roots. *Biochem. J.*, **96**, 616.

Bailey, C. J., Cobb, A. and Boulter, D. (1970) A cotyledon slice system for the electron autoradiographic study of the synthesis and intracellular transport of seed storage protein of *Vicia faba. Planta*, **95**, 103.

Blair, G. E. and Ellis, R. J. (1973) Protein synthesis in chloroplasts. I. Light-driven synthesis of the large sub-unit of fraction I protein by isolated pea chloroplasts. *Biochim Biophys. Acta*, **319**, 223.

Boardman, N. K., Franki, R. I. B. and Wildman, S. G. (1966) Protein synthesis by cell-free extracts of tobacco leaves. III. Comparison of the physical properties and protein-synthesising activities of 70 S chloroplast and 80 S cytoplasmic ribosomes. *J. molec. Biol.*, **17**, 470.

Bollini. R. and Chrispeels, M. J. (1979) The rough endoplasmic reticulum is the site of reserve-protein synthesis in developing *Phaeolus vulgaris* cotyledons. *Planta*, **146**, 487.

Caers, L. T. Peumans, W. J. and Carlier, A. R. (1979) Preformed and newly synthesized messenger RNA in germinating wheat embryos. *Planta*, **144**, 491.

Dobberstein, B., Blobel, J. and Chua, N-H. (1977) *In vitro* synthesis and processing of a putative precursor for the small subunit of ribulose-1, 5-biphosphte carboxylase of *Chlamydomonas reinhardtii. Proc. natl. Acad. Sci. U.S.A.*, **74**, 1082.

Ellis, R. J. (1969) Chloroplast ribosomes: stereo-specificity of inhibition by chloramphenicaol. *Science, N.Y.*, **163**, 477.

Ellis, R. J. (1970) Further similarities between chloroplast and bacterial ribosomes. *Planta*, **91**, 329.

Ellis, R. J. and Hartley, M. R. (1971) Sites of synthesis of chloroplast proteins. *Nature, Lond.*, **233**, 193.

Evins, W. H. and Varner, J. E. (1972) Hormonal control of polyribosome formation in barley aleurone layers. *Plant Physiol.*, **49**, 348.

Highfield, P. E. and Ellis, R. J. (1978) Synthesis and transport of the small subunit of chloroplast ribulose bisphosphate carboxylase. *Nature, Lond.*, **271**, 420.

Jervis, L. and Hallaway, M. (1970) Isolation of ribosomes from cell-wall preparations of barley (*Hordeum vulgare*). *Biochem. J.*, **117**, 505.

Kanabus, J. and Cherry, J. H. (1971) Isolation of organ-specific leucyl-tRNA synthetase from soybean seedlings. *Proc. natl. Acad. Sci., U.S.A.*, **68**, 873.

Lea, P. J. and Fowden, L. (1972) Stereospecificity of glutamyl-tRNA synthetase isolated from higher plants. *Phytochemistry*, **11**, 2129.

Lea, P. J. and Norris, R. D. (1977) tRNA and aminoacyl-tRNA synthestases from higher plants. In *Progress in Phytochemistry*, Vol. 4. Eds. L. Reinhold, J. B. Harborne and T. Swain. Pergamon Press, Oxford.

Lin, C.Y. and Key, J. L. (1967) Dissociation and reassembly of polyribsomes in relation to protein synthesis in the soybean root. *J. molec. Biol.*, **26**, 237.

Loening, U. E. and Ingle, J. (1967) Diversity of RNA components in green plant tissues. *Nature, Lond.*, **215**, 363.

Loening, U. E. (1968) Molecular weights of ribosomal RNA. *J. molec. Biol.*, **38**, 355.

Lyttleton, J. W. (1962) Isolation of ribosomes from spinach chloroplasts. *Expl. Cell. Res.*, **26**, 312.

Nonomura, Y., Blobel, G. and Sabatini, D. (1971) Structure of liver ribosomes studies by negative staining, *J. molec. Biol.* **60**, 303.

Pearson, J. A. (1969) Isolation and characterization of polysomes and polysomal RNA from radish leaves. *Expl. Cell Res.*, **57**, 235.

Peterson, P. J. and Fowden, L. (1963) Different specificities of proline-activating enzymes from some plant species. *Nature, Lond.*, **200**, 148.

Quigley, G. J., Wang, A. H. J., Seeman, N. C., Suddath, F. L., Rich, A., Sussman, J. L. and Kim, S. H. (1975). Hydrogen bonding in yeast phenylalanine transfer RNA. *Proc. natl. Acad. Sci. USA.*, **72**, 4866.

Raminez, J. M., Del Campo, F. F. and Arnon, D. I. (1968) Photosynthetic phosphory-lation as energy source for protein synthesis and carbon dioxide assimilation by chloroplasts. *Proc. natl. Acad. Sci., U.S.A.*, **59**, 606.

Schatz, G. (1976) The biogenesis of mitochondria–a review. In *Genetics, Biogenesis and Bioenergetics of Mitochondria*. Eds. W. Bandlow, R. J. Schweyen, D. Y. Thomas, K. Wolf, and F. Kaudewitz. de Gruyter, Berlin.

Short, K. C., Tefler, D. A. and Fosket, D. E. (1974) Regulation of polyribosome formation and cell division in cultured soybean cells by cytokinin. *J. Cell Sci.*, **15**, 75.

Smith, H., Billett, E. E. and Giles, A. B. (1977) The photocontrol of gene expression in higher plants. In *Regulation of Enzyme Synthesis and Activity in Higher Plants*. Ed. H. Smith. Academic Press, London

Warner, J. R., Rich, H. and Hall, C. E. (1962) Electron microscope studies of ribosomal clusters synthesizing haemoglobin. *Science, N.Y.*, **138**, 1399.

6

The soluble phase of the cell

The soluble phase of the cell may be defined as the supernatant fraction remaining after sedimentation of all particulate matter in the ultracentrifuge. Cytologically, it is represented by the continuous phase of the cytoplasm in which the various organelles (and we include here the free ribosomes) and the membrane sheets of the endoplasmic reticulum are suspended. It is bounded by the plasmalemma at the exterior of the cell and separated from the vacuole by a second limiting membrane, the tonoplast. These definitions, though useful, may not necessarily, of course, define the same material. However, if cell homogenates are centrifuged at around 100 000 g, the method normally employed to separate the soluble phase from particulate substances, a fairly flat plateau of sedimentable material is reached after about 20 min. Such a sedimentation discontinuity is illustrated in Fig. 6.1 for an homogenate from rat liver. During this time, all of the cell organelles and most of the membrane fragments and ribosomes have settled to the bottom of the centrifuge tube. The rest of the materials from the cell continue to sediment at only a very slow rate. Of course, all molecules denser than the suspending medium will continue to settle

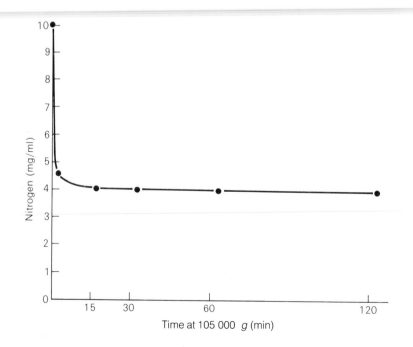

Fig. 6.1 Sedimentation of rat liver brei nitrogen as a function of time during centrifugation in a Spinco No. 40 angle head rotor. (Reproduced from Anderson and Green, 1967.)

and enrich the pellet at rates roughly related to their molecular dimensions, and some protein complexes of high molecular weight may sediment relatively quickly because of their large size. Such large complexes are probably few, however, and because we can observe a relatively sharp break between the readily sedimentable, 'particulate' fraction and most of the substances which are in solution in the higher speed supernatant fraction, we shall assume that our operational definition of the soluble phase (based on centrifugation) has a reasonable basis.

Electron micrographs of cells stained and fixed with potassium permanganate usually show the continuous phase of the cytoplasm, which we believe corresponds with the high speed supernatant fraction, to have a fairly uniform though granular appearance (Fig. 6.2). Permanganate destroys nucleic acid (see p. 20) and hence the ribosomes, so that the matrix, when cleared of small particles such as these, can be seen to extend as a continuum throughout the non-vacuolated portions of the cell, bathing the different organelles. Soluble molecules may well be able to diffuse freely within this matrix. However, the apparent homogeneity that we observe in Fig. 6.2 belies

Fig. 6.2   Electron micrograph of a section of a root cap cell from corn root tip fixed in KMnO$_4$. Note the absence of free ribosomes but the good preservation of membrane structure. The ground plasm shows a fairly uniform appearance throughout the cell. ×19 000. V, vacuole; GB, Golgi body; M, mitochondrion; ER, rough endoplasmic reticulum; W, cell wall. Provided by Dr B. E. Juniper, University of Oxford.

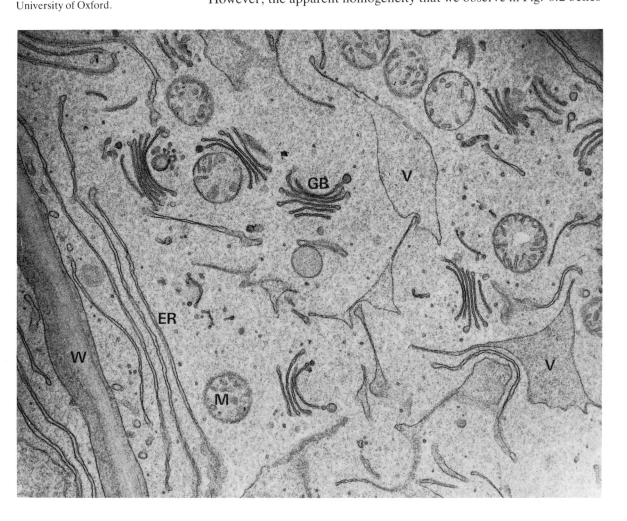

a considerable amount of submicroscopic organization which is not revealed by this method of specimen preparation.

Certainly, living cytoplasm does not behave like a simple colloidal solution of proteins and lipids in water. Nor is it easy to understand how the metabolic pathways, which we believe are localized in this so-called soluble phase, function efficiently in such a random environment in which free diffusion of substrates, products and enzymes is presumably possible. Biochemistry has more and more become a study of multienzyme systems or compartments of enzyme activity in which the product of one reaction can immediately become the substrate for a next (see Ginsburg and Stadtman, 1970). It will not be surprising to discover that all of the enzyme systems of the soluble phase are organized in such a manner.

Structural considerations

The viscosity and apparent elasticity of cytoplasm was a theme of great interest to early cell biologists. Viscosity was normally measured in two ways: by following the rate of fall of a body of known density and size through a unit distance or by measuring the amplitude of Brownian movement of small particles within the cell. By measurements of this kind, the continuous phase of the cytoplasm was shown to possess a viscosity considerably greater than that of water. Moreover, this was not uniform but changed markedly in response to environmental stimuli such as light or hydrostatic pressure. Further, when the plasmalemma of a cell is torn open, the contents flow out as a shreddy, granular mass which does not mix easily with water (Plowe, 1931). This cohesiveness suggests that the cytoplasm is held together in a structural framework of some kind, possibly by a system of interlocking protein fibres, as in a thixotrophic gel. Some clue as to the underlying basis of this gel-like structure was provided by Loewy in 1952 who prepared extracts of the slime mould *Physarum polycephalum* and showed that the viscosity of the preparation decreased in a reversible manner when ATP was added. He concluded that, like muscle, the system was capable of undergoing a structural change under the influence of ATP and that hydrolysis of ATP provided the necessary energy for the contraction to occur. As we shall discuss later, it is likely that a protein very similar to muscle actin in responsible for the gelling of cytoplasm and its movement. In addition, the process may also involve additional proteins which were previously thought to be confined to muscle, but which may be universal in all eukaryotic cells.

Paradoxically, in addition to having a gel-like structure, the cytoplasm of living plant cells can frequently be seen to be in motion. At its lowest level this consists of a Brownian movement of particles which is, of course, a non-vital, random process. Other types of movement range from a simple churning to a well defined protoplasmic streaming. The latter can occur at a high rate, with particles moving several millimetres in a minute, and such processes require the input of cellular energy and are sensitive to metabolic poisons and to anaerobiosis. Streaming is a particularly confusing phenomenon to observe, as particles may be moving at quite different speeds or in opposite directions within the same protoplasmic strand. The strands themselves are also continually changing in form and location. In leaf cells of many aquatic plants such as *Elodea* or *Nitella*, there is a pronounced rotating belt around the margins of the cells, and the cytoplasm in these regions shows a positive birefringence when viewed microscopically by polarizing light. This indicates a high degree of molecular orientation, presumably of protein molecules, within the streaming path, and recently protein fibrils about 5 nm in diameter, known as microfilaments, have been observed in these regions by electron microscopy. However, as yet, no completely satisfactory theory has been proposed to account for their rôle in cytoplasmic streaming. Indeed, we know very little about the nature of structural proteins in the cytoplasm of plants or of the systems that might control their assembly and orientation. Nevertheless, the discovery of microtubules, microfilaments, and other fibrillar structures has given us some clue as to how an intracellular micro-architecture and contractile machinery might be maintained and organized within the cytoplasmic ground substance.

Microtubules and microfilaments

Two major types of slender, proteinaceous threads in the cytoplasm of cells have been described, namely *microtubules* and *microfilaments*. The former were first described in detail in plant cells by Ledbetter and Porter (1963) after the introduction of glutaraldehyde as a tissue fixative. They are usually destroyed by osmium or permanganate and quickly depolymerize when the cells are subjected to low temperature. As osmium tetroxide was a popular fixative and low temperatures a norm in tissue preservation for electron microscopy, it is not surprising that the existence of microtubules had been in doubt for some years after they had been first noted in plant cells by Manton. At best they seemed ephemeral structures, at worst fixation artefacts. It is now clear that certain microtubule assemblies are more stable than others.

Structure and function of microtubules

Microtubules constitute a class of morphologically and chemically related fibres which appear to be common to all eukaryotic cells in which they have been properly sought. There are also reports of similar structures occurring in prokaryotes (Margalis, To and Chase, 1978). Microtubules consist of long tubes of outer diameter 25 nm which have 5 nm thick walls and a hollow core about 15 nm across (Figs. 6.3 and 6.4). The walls can be seen to consist of globular sub-units of about 5 nm diameter aligned in thirteen longitudinal strands known as protofilaments (Figs. 6.4 and 6.5). The tubules themselves vary in length from a fraction of a micrometer up to millimeters.

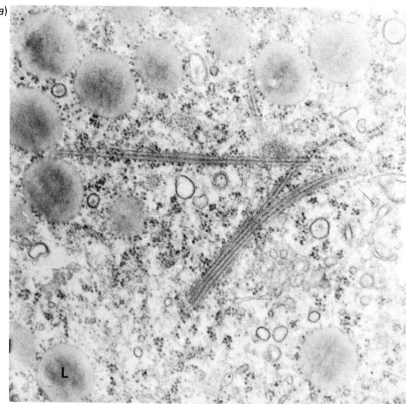
(a)

Fig. 6.3a, b Electron micrograph of cytoplasmic microtubules in the spore of the plasmodial mold *Didymium iridis*. (a) in longitudinal section, ×38 000 and (b) in transverse section, ×74 000. In these cells the microtubules are closely associated in bundles. The apparent association of ribosomes with the microtubules seen in 6.3(a) may be an artefact of fixation. The large spherical structures (L) are lipid droplets. A mitochondrion (M) and the spore wall (W) are also shown in Fig. 6.3(b). Provided by Dr H. Aldrich, University of Florida.

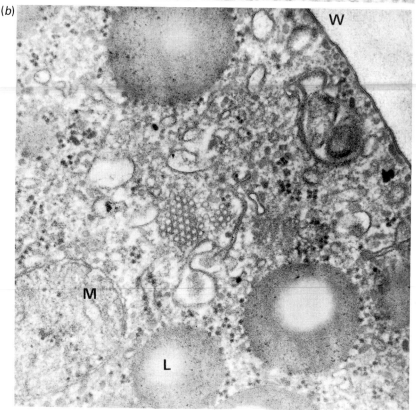
(b)

Fig 6.4   Cross-section of a microtubule in a cell from the root tip of *Juniperus chinensis* showing the sub-unit structure by negative staining. Each tubule is apparently composed of thirteen filamentous sub-units with a centre to centre spacing of 4·5 nm. An electron lucent core approximately 10 nm in diameter is present in the centre of each tubule. ×1 300 000. Reproduced from Ledbetter (1965), Brookhaven National Laboratory.

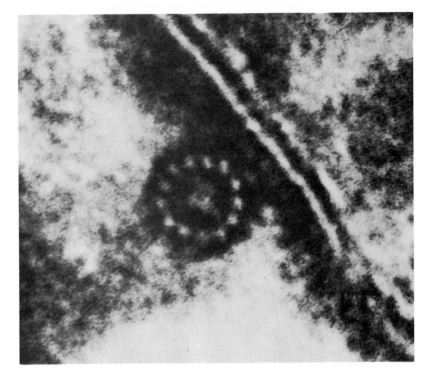

Fig. 6.5   Diagrammatic view of a cross-section (*a*) and of a surface lattice (*b*) of a cytoplasmic microtubule. In (*a*), the arrangement of the 13 protofilaments is shown. In (*b*), the probable location of the α- and β-subunits has been illustrated using stippled and clear ellipsoids. The bar represents 25 nm, the approximate diameter of the microtubule (from Snyder and McIntoch, 1976).

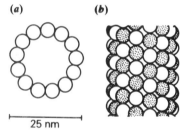

(a)   *(b)*

25 nm

Although cytoplasmic microtubules can occur in groups of loosely associated bundles, there are examples of precise associations of several tubules into more or less permanent structures. The axonemata of cilia and flagella, for example, have a characteristic 9 + 2 arrangement of modified tubules (Fig. 6.6). On occasions, thin thread-like bridges have also been observed between cytoplasmic microtubules. As these structures are easily destroyed by fixation and only visible at the highest resolution of the electron microscope, they may be of more general occurrence than hitherto suspected. Bridges were observed by Hepler and Newcomb in the plate region of dividing plant cells (see Newcomb, 1969) and they may help to order arrays of microtubules into correct patterns or, in some undefined way, assist in tubule associated motility (see p. 167).

The protein chemistry of microtubules has been studied most intensively in animal cells but there is no reason to suppose that they are very different in plants, as they are obviously similar in both appearance and properties, and cross-react immunologically with the

Fig 6.6   Electron micrograph of a cross-section of a flagellum from *Chlamydomonas reinhardi* showing nine pairs of tubules around the circumference of the flagellum and one pair placed centrally. ×170 000. Provided by Dr H. Aldrich, Unversity of Florida.

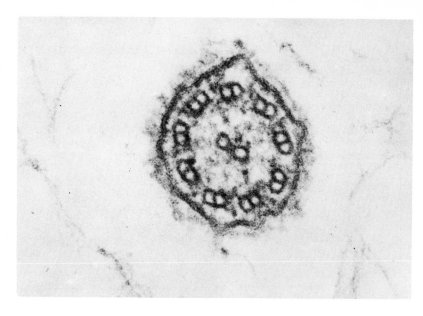

animal protein. For example, antibodies prepared in rabbits against microtubular protein from bovine brain bind to microtubules of carrot cells (Lloyd *et al.*, 1979).

Colchicine, a drug which inhibits mitosis by interfering with microtubule assembly, proved to be a valuable aid in the first isolation of the polypeptide components since it binds tightly to one of the sub-unit proteins preventing polymerization. After adding radioactive colchicine to a brain extract, it was a reasonably simple matter to purify the labelled complex. More recently, by carefully controlling the conditions for tissue disruption, preparations of assembled microtubules have been isolated from tissues such as brain, but not yet from plants. Each microtubule sub-unit has a sedimentation coefficient of about 6 S and consists of two similar but not identical polypeptides of molecular weight about 57 000, known as α- and β-tubulins. Each of the globular structures seen in negatively stained preparations of microtubules (Fig. 6.4) consists of one of these tubulin molecules. There are two distinct sites on the tubulin dimer to which the nucleotide GTP can bind, and a third occupied by Mg^{++}. There are also reports that tubulins may be phosphorylated on certain of their serine residues and that this may be a consequence of a protein kinase associated with the tubulin preparations. Indeed, it is now clear that only about 80 to 90% of a microtubule consists of tubulin and that a number of other proteins including ones of very high molecular weight (>250 000) form part of the structure (Scherline and Schiavone, 1977). The filamentous bridges between associated tubules, for example, are probably not made up of tubulin, but of such associated proteins.

It is apparent from their lability that microtubules readily undergo reversible assembly and disassembly and that a delicate equilibrium exists between sub-units and polymer. Colchicine probably influences this equilibrium by binding to the growing end of microtubules as well as to the 6 S dimers and thus prevents tubule assembly. It has also become evident that the same monomeric units can be assembled into different forms depending upon local conditions of pH, cation concentration and temperature and this may account for the diversity of fibril dimensions that has been noted in different organisms.

Assuming that microtubules are widespread, what rôle do they play in the cytoplasm of the cells in which they are found? Some possible functions have been suggested, but not, of course, proven by observing their intracellular locations and by studying those cellular processes that are disrupted by colchicine. Microtubules are found, for example, as components of cilia and flagella, in extending nerve processes, in pseudopodia and in the cytoplasm adjacent to growing plant cell walls. They compose the spindle fibres of dividing cells (see p. 166), probably accounting for the anisotropy of the division figure. Moreover, they are often observed in the cytoplasm in regions where there is directed movement of materials or organelles. Changes in cell shape or symmetry are often accompanied by the appearance of microtubules, although here they may be involved in the maintenance of shape rather than playing a directive rôle in the morphogenic process. Presumably, therefore, they serve as cytoskeletal agents of some kind, maintaining local areas of stress and rigidity and defining cytoplasmic channels. Such a skeletal function is illustrated in Fig. 6.7 which is a micrograph of a thin section derived from a wall-less mutant of *Chlamydomonas reinhardi*. Cells of this organism have a series of conspicuous ridges radiating out from their posterior end which result from struts of microtubules located just below the plasmalemma.

Polymerization of tubulin. Weisenberg first described conditions which promoted the *in vitro* polymerization of brain tubulin into microtubule-like structures (see Snyder and McIntosh, 1976). Since that time, considerable progress has been made in our understanding of the assembly process. It has become clear the polymerization is promoted at slightly acidic pH and in presence of monovalent ions close to physiological ionic strength. Mg^{++} and GTP, at molar concentrations roughly equivalent to that of the tubulin monomer, also increase polymerization rates. Some of the GTP is hydrolysed during assembly but it is not clear what the function of the nucleotide might be. Although polymerization will proceed spontaneously from solutions of tubulin sub-units, it occurs most rapidly if the preparation is first seeded with tubule fragments from which assembly appears to occur undirectionally. For a complete discussion of this topic, the reader is referred to the review by Snyder and McIntosh (1976).

It is not clear what controls microtubule assembly and disassembly *in vivo*. There is known to be a soluble pool of tubulin, probably comprised mainly of 6 S dimers which can be drawn upon as required. Although GTP and local divalent cation concentrations are undoubtedly important in promoting polymerization, it is also evident that microtubules do not form randomly but only at specific locations in the cell; these are known as nucleating centres or microtubule organizing centres (MTOC). In the case of the mitotic spindle, for example, they appear to emanate from the poles of the mitotic

Fig. 6.7 Electron micrograph of a thin section of *Chlamydomonas reinhardi* showing microtubules beneath one of the surface ridges. This particular cell is from a mutant strain that lacks a cell wall. but which can grow normally on a supplemented medium. ×125 000. Provided by Dr H. Aldrich, University of Florida.

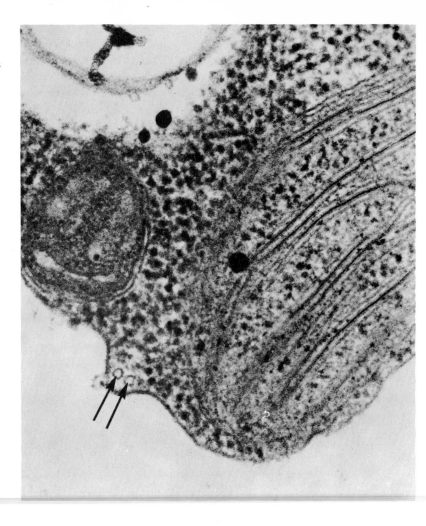

apparatus (see p. 166). In other instances the growth foci are less well defined, although there is now evidence that nucleating centres for cytoplasmic microtubules occur along the edges of cells in *Azolla* root tips (Gunning, Hardham and Hughes, 1978).

It is encouraging that antibodies prepared against mammalian tubulin react with plant microtubules (Lloyd *et al.*, 1979, 1980). Not only does this indicate that the protein structures have been highly conserved and are probably very similar, but it has allowed microtubule distribution to be studied in intact cells using indirect immunofluorescence techniques. In these procedures, the cell wall is first partially digested with cellulases to allow the ready penetration of antibodies. After brief fixation, the cells are exposed to antibody raised in rabbits against highly pure tubulins from mammalian brain. The preparation is then washed thoroughly to remove any non-specifically bound antibody and finally stained with a second antibody, usually prepared from goats and directed against rabbit γ-globulins, to which a fluorescent dye, fluoroscein, has been covalently attached. When illuminated by ultraviolet light and viewed under a fluorescent microscope, structures which have bound the two antibodies fluoresce brightly. Thus in Fig. 6.8, in which cultured carrot cells were used, it is clear that a tubulin network spirals around the periphery of the cytoplasm of these elongated cells, again suggestive of a skeletal or morphological rôle. If, as is most likely, this tubulin is in the form of microtubules, it is also clear that these structures branch and form an interconnected cytoskeleton the entire length of the cell. The method of staining used in Fig. 6.8 is known as indirect immunofluorescence because two kinds of antibody are employed, with the second, relatively less specific species, bearing the stain. Its advantage over direct staining by a single antibody preparation is that several fluorescent goat immunoglobulins can bind to a single rabbit antibody

Fig. 6.8   Fluorescence patterns reflecting the distribution of tubulin in cultured carrot cells. (a) elongated cell (×300), unstained and photographed under interference contrast optics; (b) a cell (×740) visualized by fluorescein immunofluorescence after successive treatment with anti-tubulin antibody and fluorescein labeled goat antibody directed against rabbit immunoglobulins. This cell had been converted to a protoplast by cellulase treatment (60 min) which dissolves the wall; (c) and (d) (×740) were stained in a manner identical to b but were treated for only 30 min with cellulase and had not rounded up. Note that the fibrils of tubulin run around the cell in transverse interconnected hoops. Controls using non-immune rabbit serum did not fluoresce (from Lloyd et al., 1979).

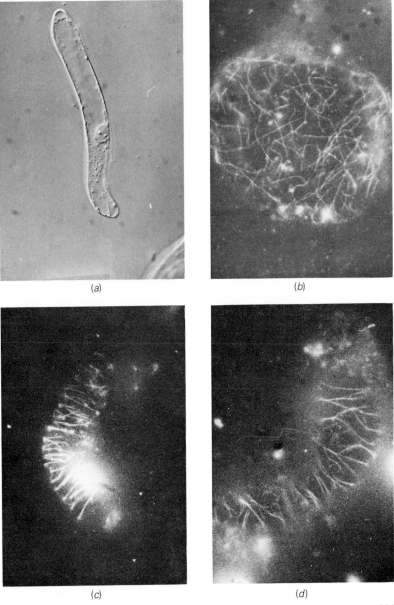

(a)

(b)

(c)

(d)

molecule thus amplifying the fluorescence. This technique is likely to have great impact in studying the distribution of molecules within plant cells, particularly in those cases where antibody against the usually more easily purifiable, homologous animal protein has been prepared.

Microfilaments

Microfilaments are much narrower than microtubules, being only about 5–6 nm in diameter. They often occur in bundles and have been noted at a number of cellular locations, including within, or close to, the path of streaming cytoplasm in plant cells (Fig. 6.9). Just as colchicine provided a tool to investigate microtubules, another drug, cytochalasin B, appears to have a specific and reversible effect on microfilaments. The use of this drug has demonstrated a clear correlation between the integrity of the microfilament systems and the contractility of cytoplasm. Microfilaments have been noted, for example, just below the cleavage furrow of mammalian cells during cytokinesis and in a number of other instances when cell invagination occurs. Invariably they are found in regions of cytoplasm which are undergoing contraction, and circumstantial evidence suggests that they provide the contractile machinery of the cell. Streaming of cytoplasm in cells, for example, can be explained most simply on the basis of contractility. Cytochalasin B has been supplied to a number of cell types and shown to disrupt microfilament assemblies and simultaneously inhibit any contractile processes, including streaming, going on within the cell. Furthermore, when the drug is removed, filaments reappear and the previously inhibited activity is resumed. An example of the action of cytochalasin B is given in Fig. 6.10, which shows an outer cap cell in a maize (*Zea Mays*) root-tip after treatment with the drug. A massive accumulation of secretory vesicles has occurred deep in the cytoplasm close to their sites of origin, the Golgi bodies (see Ch. 11). Cytochalasin B has inhibited the vectorial transport of the vesicles to the cell surface, thus implicating microfilaments in the process. The drug does not alter microtubule morphology or the processes in which microtubules are involved. Indeed, cytochalasin B appears so specific that any sensitivity to the drug is usually taken to mean that the cell type possesses some type of contractile microfilament system.

Microfilaments, like microtubules, are complex assemblies of different protein monomers. In organisms as diverse as mammals, slime molds, amoebae and probably plants, the major protein component of these structures is a special form of the muscle protein actin known as β-actin. A second isomer, γ-actin, is also found in these non-muscle cells, but usually in much smaller amounts than the β-form. All actins have molecular weights of about 43 000 and very similar amino acid compositions. Actins in non-muscle cells constitute anywhere from 2 to 15% of the total cell protein. Like the tubulins, they have been highly conserved during evolution and the forms derived from widely different groups of organisms differ only by a few amino acids and then only at particular regions in their sequence. Most of the diversity is

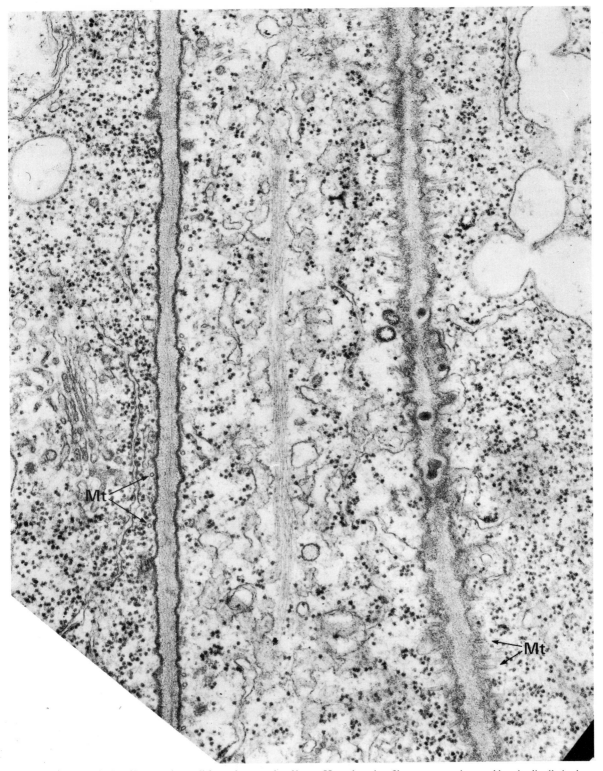

Fig 6.9   Bundle of microfilaments in a cell from the root-tip of bean. Here the microfilaments are orientated longitudinally in the direction of cell growth. Microtubules, by contrast, are orientated transversely and are seen in cross-section on the left of the cell but in oblique section on the right. Magnification ×51 7000 Mt (microtubule). Provided by Dr E. H. Newcomb, University of Wisconsin.

Fig. 6.10   A portion of an outer root cap cell from the root tip of a maize (*Zea mays*) seedling treated with 100 μg/ml cytochalasin B for 2 hours. Note the heavy concentration of secretory vesicles (SV) in the inner portions of the cytoplasm and the lack of secretory vesicles near the cell surface. The results strongly suggest that cytochalasin B prevents the vectorial transport of mature secretory vesicles from the Golgi bodies to the cell surface. For a more normal view of such a cell, the reader is referred to Fig. 11.7. Glutaraldehyde-paraformaldehyde-picric acid-osmium tetroxide fixation. Scale line 1 μm. From Mollenhauer and Morré (1976).

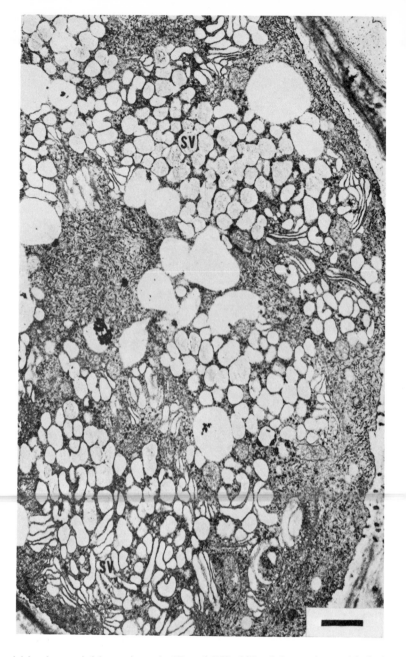

within the variable regions 1–17 and 259–298 of the amino acid chain. Thus, *Physarum* actin differs from mammalian β-action in only seventeen amino acid residues.

That actin itself was a component of microfilaments was demonstrated by Ishikawa, Bischoff and Holtzer (1969), who employed heavy meromyosin, which constitutes the head group of another muscle protein, myosin, to 'stain' or decorate actin assemblies in non-muscle cells. This binding of the heavy meromyosin produces characteristic arrow-like deposits along the microfilaments which are easily identifiable under the electron microscope.

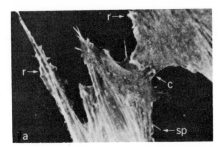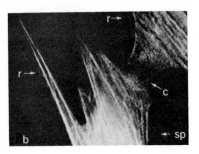

Fig. 6.11 Double fluorescence staining pictures of actin (a) and myosin (b) in rat kidney fibroblasts (×950) grown in culture. Each figure represents the same view of the same cell. In a, the cells were treated first with a rabbit antibody directed against actin and then stained with a fluorescein-conjugate antibody against rabbit immunoglobulins. In b, the same cell had first been treated with the muscle protein heavy meromyosin coupled covalently to biotin. Heavy meromyosin forms a stable complex with myosin and thus mirrors its distribution in the cell. The complex was then stained with a rhodamine-conjugated antibody directed against avidin. Since fluorescein fluoresces with a yellow colour and rhodamine with a red colour, the two dye complexes can be distinguished using suitable filters. The results show that actin and myosin have a fairly similar distribution in the cells, particularly on the stress fibres. Actin staining, however, occurs more intensely at cell ruffles (r) on spicules (sp) projecting from the surface and at points of cell contact (c). Ruffles occur on the moving boundaries of migrating mammalian cells. From Heggeness, Wang and Singer (1977).

7 nm

Fig. 6.12 Actin strands as found in microfilaments, each consisting of polymerized actin monomers. In the contractile elements, additional proteins such as tropomyosin and troponin (which binds Ca^{++}) are expected to be associated with this coil. The filaments themselves would interact with the head groups of parallel strands of myosin in a ratchet-like arrangement to allow side-by-side slippage. Various 'connecting proteins' attaching the strands to membranes and other structures, and to each other, might also be expected to occur.

As with tubulin, it has also been possible to raise antibodies against β-actin, thus allowing the distribution of microfilaments to be studied in whole cells by fluorescence microscopy (Fig. 6.11). In animal cells, the microfilaments exist either as large bundles, often known as stress fibres, located just below the plasma membrane or as an amorphous network. The two forms are both labile and probably interconvertible since rapid transitions between states occur according to the physiological activity of the cell. It is generally agreed that the bundles provide some sort of structural support to the cytoplasm while the amorphous network is involved more directly in cytoplasmic movement and contractile processes.

How the contractility occurs or is controlled is unclear. However, some clues have been provided by the discovery of myosin, tropomyosin and other muscle-like proteins in non-muscle cells, and the demonstration by fluorescent techniques (Fig. 6.11) that these proteins are associated with the microfilaments (see further reading and review by Lazarides and Revel, 1979). Such observations have led to the hypothesis that actin filaments participate in contractility in much the same way as they do in muscle, namely by sliding past myosin filaments in a process dependent upon ATP hydrolysis, Ca^{++} availability and an appropriate geometry of the various protein molecules.

Assembly of actin into microfilaments. A considerable percentage of the actin within a non-muscle cell occurs in unpolymerized monomeric form which presumably can be drawn upon to fabricate microfilaments at appropriate locations. A protein known as profilin, which co-crystallizes with β-actin, inhibits polymerization and is probably responsible for maintaining actin in the monomeric state. Only after its removal can actin be assembled *in vitro* into the twisted helical strands characteristic of the microfilament (Fig. 6.12). The filaments so formed may then be induced to form a gel-like mesh in the presence of yet another protein known as filamin, or actin binding protein, which forms cross-links between strands. Addition of myosin to this complex provides a gel which can contract reversibly upon provision of ATP. The other muscle-like proteins associated with the microfilament presumably assemble with the polymerizing actin molecules, but their actual location on the strand is still not known.

It should be stressed that at the time of writing this chapter, clear-cut evidence for a contractile system involving actin and myosin in higher plant cells has not yet been demonstrated. On the other hand, it seems unlikely that evolution would have by-passed or discarded this system in plants while retaining it in such dissimilar organisms as the slime molds and the mammals. The close similarities in morphology and behaviour of all microfilaments, in whatever cell type they are observed, suggests close structural homology between all these cytoskeletal elements. We predict that the information now available from studying mammalian microfilament systems will be directly applicable to the higher plant.

The cytoskeleton. The partial elucidation of the structures of micro-tubules and actin microfilaments, the description of other, poorly understood, intermediate or 10 nm filaments, plus the immunofluores-cent-evidence that the interior of the cell is criss-crossed with a variety of protein strands, suggests that far from being a homogenous solu-tion, the cytoplasm exists as a meshwork of contractile and structural elements. This has been dramatically demonstrated in mammaliam cells by Porter and his associates (Wolosewick and Porter, 1979) using high voltage electron microscopy in association with stereo-microscopy to probe the three dimensional organization of whole cells after fixation. A second approach used by a number of groups has been to extract cells with non-ionic detergents in such a manner that the more 'soluble' cytoplasmic proteins are removed while the cytoskeleton is retained intact (Webster *et al.*, 1978). Scanning and transmission electron microscopy have revealed an enormously complex system of fibres, and a fine actin lattice extending throughout the cytoplasm. It will not be surprising to find a similar complexity within the soluble phase of plant cells.

In conclusion, therefore, it appears that the continuous phase of the cell, which probably corresponds with the high speed supernatant fraction of homogenates, behaves both as a gel and as a colloid. The sol-gel transitions that are observed, the elastic properties, the move-ment of organelles and the streaming of the cytoplasm itself, can best be explained if the cytoplasm is assumed to have a labile but highly organized micro-architecture. This is probably based on readily assem-bled filamentous proteins. At times, this is manifested by the appear-ance of microtubules and other fine fibres. Such ephemeral structures can probably be polymerized or depolymerized as required at different intracellular loci and fulfil a number of functions. The finer details of the cell micro-architecture may not however be visible at the level of the electron microscope, and awaits further research.

Composition of the soluble phase

The composition of the soluble phase reflects a large part of the cellular biochemistry and so is exceedingly complex. In addition, even under the best conditions of cell fractionation, the more soluble contents of broken organelles will spill out and contaminate the soluble phase, and, in most instances, the water soluble contents of the vacuole will always be recovered in the supernatant solution of centrifuged cell homogenates. In contrast to this, certain soluble enzymes may become adsorbed to membranes, and will be isolated in the particulate fractions.

The contents of a high speed supernatant fraction can be arbitrarily divided into a number of sub-groups: a micromolecular group of substances which includes water, inorganic ions and dissolved gases; a mesomolecular group consisting of metabolites such as lipids, sugars and nucleotides, all of which are essentially molecules of low informational content; and finally free macromolecules consisting chiefly of proteins and RNA with some polysaccharides (Anderson and Green, 1967). Estimates of cytoplasmic pH usually fall within the range 6·8–7·6, although the vacuolar pH may be much lower (see Smith and Raven, 1979). There may also be areas of locally controlled pH or cation concentration in the intact cell. Certainly, the so-called soluble enzymes show a considerable range of pH optima and metal ion requirements which may or may not relate to their normal working conditions.

The monovalent cations are probably free and able to diffuse freely in solution, although some divalent species such as Mg^{++} and Ca^{++} are probably sequestered at exchangeable sites on nucleic acids, nucleotides and acidic polysaccharides. A few are tightly bound to enzymes, although there are relatively few true metalloenzymes.

Nucleotides are found in large numbers in plant cells. Their diversity reflects their many different rôles as cofactors and metabolites. Sugars include glucose, fructose, sucrose, occasionally raffinose and stachyose, but others only rarely. Glycolytic (see p. 240) and other intermediates of carbohydrate metabolism are, of course, ubiquitous. Free amino acids often occur in relatively large amounts and several unusual types not found in proteins and whose function is not clear have been isolated from plant tissues (see p. 79). The commoner amino acids probably serve as precursors of protein, as respiratory substrates and, in the case of the tripeptide glutathione, as a cellular reducing agent. Together with the sugars, the amino acids may act as osmoregulatory substances and there are several reports implicating various low molecular weight metabolites as playing some rôle in cold hardening and frost resistance.

The broad enzymic content of the soluble phase is, of course, reflected by the large numbers of individual proteins that can be isolated. Many of these are enzymes, some of which are discussed in this chapter; others may play a structural rôle within the cytoplasm itself.

Enzyme groups in the soluble phase

We have pointed out that the soluble phase reflects a very large part of the biochemistry of the cell, and so a complete review of the various enzymes and metabolic routes found in this fraction is not feasible in a chapter of this size. We shall concentrate, therefore, on the three major pathways which appear to be characteristic of and largely confined to the soluble phase. These pathways are:

1. the glycolytic or Embden–Meyerhof–Parnas (EMP) pathway (and its reverse known as gluconeogenesis);
2. the pentose phosphate pathway;
3. the system of enzymes responsible for fatty acid synthesis in cells.

The β-carboxylation of pyruvate and phosphoenolpyruvate will also be reviewed as these reactions form important links between the glycolytic and tricarboxylic acid pathways. Another vital process which occurs in the soluble phase is the charging of specific transfer RNAs with their appropriate amino acids, and this topic is more fully discussed in Chapter 5.

Glycolysis

Glycolysis literally means a splitting of sugar. It was a term introduced by Lépine in 1909 to describe the anaerobic fragmentation of glucose to produce carbon dioxide and ethanol in yeast and carbon dioxide and lactic acid in animal muscle. We shall use glycolysis to describe a step-wise degradation of glucose to pyruvate by way of fructose diphosphate. The major features of the pathway are approximately similar in all forms of life and this suggests that the pathway originated in the cell types ancestral to all present day organisms. These cells were most probably living in an atmosphere lacking oxygen and glycolysis provided an efficient and effective mechanism for extracting energy from surrounding nutrient molecules. Because of this, the pathway probably reached an optimal evolutionary state early in biological history. Glycolysis, however, releases only a very small portion of the chemical energy potentially available in a molecule of hexose sugar, and in higher plants and animals the pathway serves primarily to provide pyruvate for oxidation by the tricarboxylic acid cycle (TCA cycle; see p. 289 and as a source of intermediates for biosynthetic processes (Fig. 6.10). The metabolism of glucose to pyruvate involves the consecutive action of ten enzymes, most of which have been thoroughly studied (Table 6.1). All the enzymes except hexokinase (the first priming step in the sequence) are usually considered to be soluble in the sense that they are not associated together in any cell organelle. Furthermore, they show no physical dependence upon each other and are not, as far as can be discerned at present, organized in a multienzyme complex. The plant enzymes have been less well studied than the ones from yeast and mammalian tissues, but, with a few exceptions, seem to display similar properties. Even the control mechanisms, which regulate the flux of material through the pathway, are broadly similar in cell types separated widely by phylogeny. Before looking at these reactions in detail, however, we shall first discuss briefly how glucose and other monosaccharides are made available for metabolism via the glycolytic pathway.

Table 6.1 The reactions, enzymes and standard free energy changes involved in the glycolytic conversion of 1 mole of glucose to 2 moles of pyruvate.

Step	$\Delta G^{\circ\prime}$ (kJ/mole)	Reaction	Enzyme
1	−14.2	Glucose + ATP → Glucose 6-P + ADP	Hexokinase
2	+2.1	Glucose 6-P ⇌ Fructose 6-P	Phosphohexoisomerase
3	−14.2	Fructose 6-P + ATP → Fructose 1,6-BP + ADP	Phosphofructokinase
4	+24.0	Fructose 1,6-P ⇌ Dihydroxyacetone-P + Glyceraldehyde 3-P	Aldolase
5	+7.6	Dihydroxyacetone-P ⇌ Glyceraldehyde 3-P	Triosephosphateisomerase
6	+6.3	Glyceraldehyde 3-P + P_i ⇌ 1,3-BP-Glycerate + NADH + H^+	Glyceraldehyde 3-P dehydrogenase
7	−18.8	1,3-BP-Glycerate + ADP ⇌ 3-P-Glycerate + ATP	Phosphoglycerate kinase
8	+4.4	3-P-Glycerate ⇌ 2-P-Glycerate	Phosphoglyceromutase
9	+1.8	2-P-Glycerate ⇌ P-Enolpyruvate + H_2O	Enolase
10	−31.4	P-Enolpyruvate + ADP → Pyruvate + ATP	Pyruvate kinase

The standard free energy changes refer to pH 7·0 with all other reactants at unit activity.

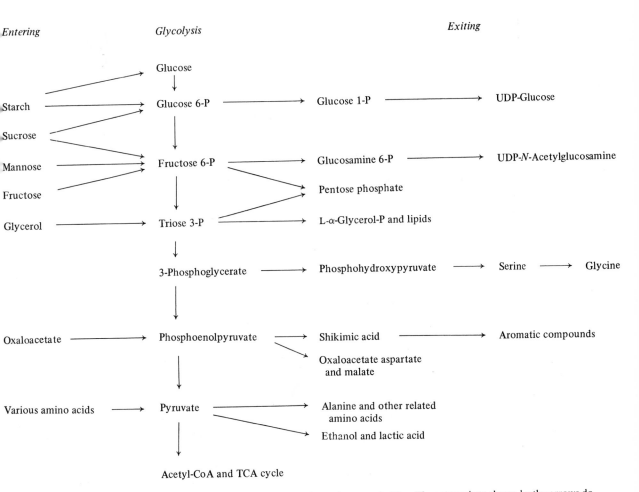

Fig. 6.13 The entry into and exit out of the glycolytic sequence by various metabolites. The conversions shown by the arrows do not necessarily represent single reaction steps, and many intermediate metabolites are not shown.

241

Free glucose is usually considered the natural starting point for glycolysis, though various other metabolites can enter the pathway later in the sequence at the level of fructose 6-phosphate (for example, fructose or mannose) or triose phosphate (for example, glycerol) (Fig. 6.13). Glucose may be transported into the cells directly or be formed intracellularly by breakdown of starch, sucrose, and other complex saccharides. Sucrose is cleaved to free glucose and fructose by invertase, or reversibly to UDP-glucose and fructose by sucrose synthetase (Fig. 6.14). The UDP-glucose formed in this reaction can then conceivably be utilized for biosynthetic processes in the cell (see p. 115) or be converted in a reversible reaction to α-D-glucose 1-phosphate.

Fig. 6.14 Two reactions for the breakdown of sucrose to hexose. In 1, the sucrose is hydrolysed directly to its component monosaccharides in a reaction catalysed by invertase. In reaction 2, UDP-glucose and fructose are produced in a reversible reaction catalysed by sucrose synthetase. The UDP-glucose can then be converted successively to glucose 1-P and glucose 6-P by UDP-glucose pyrophosphorylase and phosphoglucomutase respectively.

$$1.\ \text{Sucrose} \xrightarrow[\ +\ H_2O\ ]{\text{invertase}} \text{glucose} + \text{fructose}$$

$$2.\ \text{Sucrose} + \text{UDP} \xrightleftharpoons[\text{synthetase}]{\text{sucrose}} \text{UDP-glucose} + \text{fructose}$$

Then:

$$\text{UDP-Glucose} + \text{pyrophosphate} \xrightleftharpoons[\text{pyrophosphorylase}]{\text{UDP-glucose}} \text{glucose 1-P} + \text{UTP}$$

Starches, which provide the major storage form of carbohydrate in plants, can be broken down in two main ways: by hydrolytic cleavage and by phosphorolysis (Figs. 6.15 and 6.16). Amylases catalyse the hydrolysis of alternate $\alpha(1 \rightarrow 4)$ glucosidic bonds to yield maltose units, which may in turn be broken down to glucose by the action of the

Fig. 6.15 Two ways in which starch can be converted to hexose.

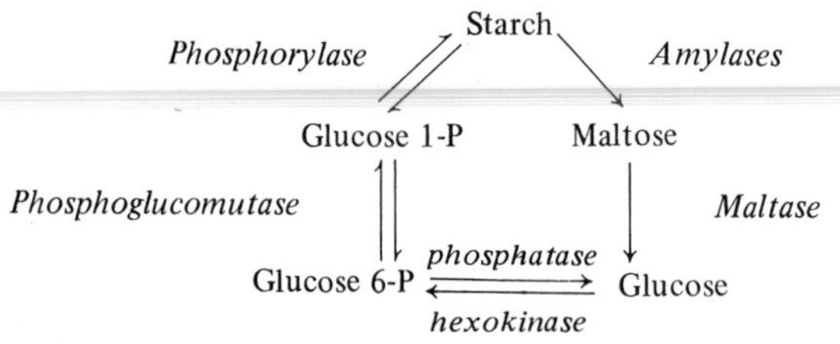

enzyme maltase. β-Amylase, however, has only a limited specificity and restricts its attack to the non-reducing termini of the starch molecules. Consequently, although the linear molecule of amylose can be broken down completely to maltose, amylopectin is only degraded as far as its outer branch points, leaving a so-called limit dextrin (Fig. 6.17). The $\alpha(1 \rightarrow 6)$ glucosidic branch points require a specific debranching enzyme (an $\alpha(1 \rightarrow 6)$ glucohydrolase) before they can be cleaved.

α-Amylase, by contrast, will attack internal $\alpha(1 \rightarrow 4)$ linkages within the core of the molecule as well as the terminal chains. This enzyme will, therefore, degrade amylopectin more or less completely, though small oligosaccharides bearing the $\alpha(1 \rightarrow 6)$ branch point will remain undigested. α-Amylase is usually regarded as the enzyme primarily responsible for starch breakdown in germinating seeds such as peas,

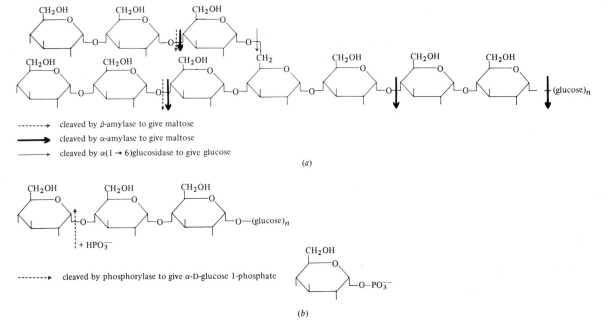

- - - - → cleaved by β-amylase to give maltose
━━━━▶ cleaved by α-amylase to give maltose
────▶ cleaved by α(1 → 6)glucosidase to give glucose

(a)

- - - - ▶ cleaved by phosphorylase to give α-D-glucose 1-phosphate

(b)

Fig. 6.16 Cleavage of amylopectin: (a) by α- and β-amylases and α(1 → 6) glucosidase and (b), by phosphorylase.

corn or barley and it usually shows large increases in specific activity as germination proceeds and as the reserves begin to be mobilized for seedling growth (Fig. 6.18). This increase can be mimicked in barley grains that have had their embryo removed by providing gibberellic acid.

Phosphorylases like β-amylase attack the α(1 → 4) glucan chains from the non-reducing end to give α-D-glucose 1-phosphate, and, ultimately limit dextrins. Although the reaction is reversible, phosphorylase will degrade amylopectin completely in the presence of high levels of inorganic phosphate and a debranching enzyme. The α-D-glucose 1-phosphate that is formed is converted to the 6-phosphate by action of phosphoglucomutase.

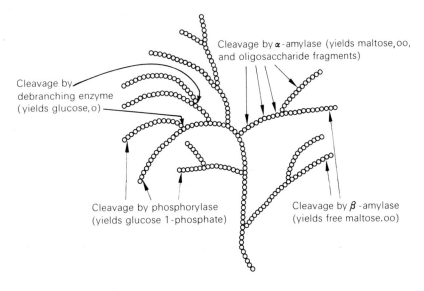

Cleavage by α-amylase (yields maltose, oo, and oligosaccharide fragments)

Cleavage by debranching enzyme (yields glucose, o)

Cleavage by phosphorylase (yields glucose 1-phosphate)

Cleavage by β-amylase (yields free maltose, oo)

Fig. 6.17 Modes of amylopectin breakdown shown diagrammatically. Each glucose molecule in the amylopectin is represented by a 'bead'. α-Amylase attacks the molecule internally and externally; β-amylase and phosphorylase work inwards from the non-reducing ends of the exposed chains; and the α(1 → 6) glucosidase hydrolyses the α(1 → 6) branch points.

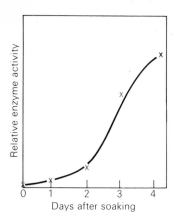

Fig. 6.18 Increase in α-amylase activity during germination of corn (unpublished data from R. M. Roberts). Grain grown without their embryos show no such increase unless they are supplied with gibberellic acid, in which case the α-amylase is induced normally. The enzyme is produced in the aleurone layer and secreted into the endosperm. Maltase and phosphorylase levels also increase during germination.

α-D-Glucose 1-phosphate $\rightleftharpoons$ α-D-Glucose 6-phosphate

Breakdown of starch or sucrose, therefore, ultimately leads to the formation of either free hexose sugars or glucose 6-phosphate. Both types of compound are readily metabolized via the glycolytic pathway.

Priming reactions

Glycolysis can be conveniently divided into two stages: firstly, the so-called priming reactions in which hexose sugars are converted to triose phosphate; secondly, the energy conserving reactions which are associated with oxidation–reduction and which lead to the formation of pyruvate. During the priming stage, there is a net input of cellular energy into the reactants, while in the latter part of the sequence ATP and NADH are generated.

Hexokinase is a membrane-bound enzyme which catalyses the phosphorylation of carbon atom 6 of glucose by ATP in presence of a divalent cation, usually magnesium. The magnesium forms an ATP magnesium complex which is the actual substrate for the enzyme.

α-D-Glucose $\xrightarrow{\text{ATP Mg}^{++}\ \text{ADP}}$ α-D-Glucose 6-phosphate

Glucose 6-phosphate, the product of the reaction, inhibits hexokinase and so will regulate its own biosynthesis from glucose. Hexokinase will also convert D-mannose and D-fructose to their 6-phosphates and is probably responsible for the first step in the metabolism of these sugars. Kinase reactions of this kind result in a large decrease in standard free energy (Table 6.1). They are, therefore, strongly exergonic and effectively irreversible. As we discussed earlier, glucose 6-phosphate may also be formed from glucose 1-phosphate in a reaction catalysed by the enzyme phosphoglucomutase.

Phosphohexoisomerase catalyses the reversible interconversion of fructose 6-phosphate and glucose 6-phosphate:

α-D-Glucose 6-phosphate $\rightleftharpoons$ α-D-Fructose 6-phosphate

A separate, but related, enzyme is responsible for the formation of fructose 6-phosphate from mannose 6-phosphate.

Fructose 6-phosphate is then converted to fructose 1,6-bisphosphate in the presence of ATP and a specific kinase:

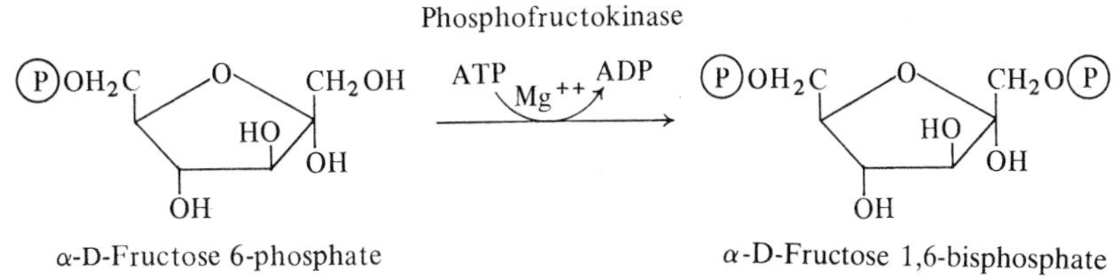

Phosphofructokinase

α-D-Fructose 6-phosphate   α-D-Fructose 1,6-bisphosphate

The properties of this enzyme have been studied in great detail because it is believed to be a main controlling step for the whole glycolytic sequence, and its activity is particularly responsive to changes in metabolite concentration. The enzyme assumes various states of aggregation, and will readily undergo subunit-oligomer equilibrium changes which are dependent upon pH, protein concentration, and the presence of substrate and modifiers. We shall discuss this enzyme in relation to control in a later section. Like the hexokinase reaction, it is essentially irreversible, and can be regarded as the first reaction which is unique to the glycolytic sequence proper. Biochemical pathways frequently begin and end with such irreversible steps which are subject to metabolic control. It is clearly a favourable point at which to exert an effect on the rate of glycolysis without interfering with other reactions which may be common to other phases of carbohydrate metabolism.

Fructose 1,6-bisphosphate is reversibly cleaved to dihydroxyacetone phosphate (from carbon atoms 1,2,3) and glyceraldehyde 3-phosphate (carbon atoms 4,5,6) in the presence of aldolase.

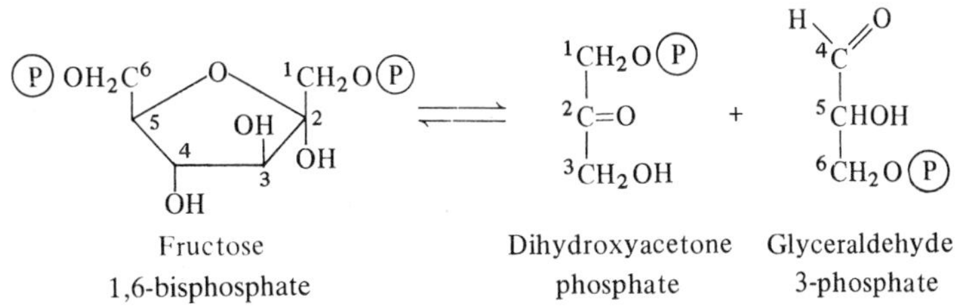

Fructose   Dihydroxyacetone   Glyceraldehyde
1,6-bisphosphate   phosphate   3-phosphate

The standard free energy change at pH $7\cdot0(\Delta G^{\circ\prime})$ of the reaction is relatively high ($+23\cdot9$ kJ/mole) and at high concentrations of fructose 1,6-bisphosphate, the formation of triose phosphate would not be favoured. The reaction, therefore, has a very low equilibrium constant which is related to $\Delta G^{\circ\prime}$, the standard free energy change, in the following manner:

$$\Delta G^{\circ\prime} = -RT \ln K = -RT \ln \frac{[\text{triose-P}]^2}{[\text{F-1,6-P}]}$$

Reactions of this kind in which one molecule of substrate produces two molecules of product, give rise to a square term in the equilibrium constant. The equilibrium position, therefore, will be influenced strongly by the concentration of the reactants. The lower the concentration of fructose 1,6-bisphosphate, the greater is the fraction that is cleaved before equilibrium is attained, and because the concentration of fructose 1,6-bisphosphate in the cell is probably low, the reaction can be considered reversible under normal physiological conditions, and does not represent a barrier to glycolytic carbon flow.

The glyceraldehyde 3-phosphate and dihydroxyacetone phosphate which are produced in equimolar amounts by cleavage of fructose 1,6-bisphosphate can be interconverted in a reversible reaction catalysed by triose phosphate isomerase. This enables all of the dihydroxyacetone phosphate produced by aldolase cleavage to be converted ultimately to 3-phosphoglycerate in the first oxidation stage of glycolysis. It also renders carbon atoms 1, 2 and 3 of the original glucose molecule equivalent to carbons 4, 5 and 6.

$$^1CH_2O\,\text{(P)}\qquad\qquad \xrightarrow[]{\text{Phosphotrioseisomerase}}\qquad\qquad ^1CH_2O\,\text{(P)}$$

$$^2C=O \qquad\qquad\qquad\qquad\qquad\qquad\qquad\qquad ^2CHOH$$

$$^3CH_2OH \qquad\qquad\qquad\qquad\qquad\qquad\qquad ^3C$$

Dihydroxyacetone phosphate

Glyceraldehyde 3-phosphate

The energy conservation stages of glycolysis

The subsequent oxidation of glyceraldehyde 3-phosphate represents the first equation in the energy conservation stage of glycolysis. The reaction is catalysed by glyceraldehyde 3-phosphate dehydrogenase which is NAD dependent.

$$\begin{array}{ccc}
\text{H}\diagdown\;\diagup\text{O} & \xrightarrow[\quad NAD^+\;\;NADH+H^+\quad]{\text{Glyceraldehyde 3-P dehydrogenase}\;\;\;\;\;P_i} & \text{(P)}O\diagdown\;\diagup\text{O} \\
\;\;\;\;C & & \;\;\;\;C \\
CHOH & & CHOH \\
CH_2O\,\text{(P)} & & CH_2O\,\text{(P)}
\end{array}$$

3-Phosphoglyceraldehyde

1,3-Bisphosphoglycerate

Normally, the oxidation of an aldehyde to an acid at neutral pH is highly favoured thermodynamically. However, the dehydrogenation of glyceraldehyde 3-phosphate is coupled with a phosphorylation reaction. Part of the energy of oxidation is made available for the formation of an acyl phosphate, and because the overall free energy change is small, the reaction is reversible. Moreover, the product, 1,3-bisphosphoglycerate, has a high potential for phosphate transfer. The standard free energy of hydrolysis of its acyl phosphate is approximately -49.4 kJ while that of the terminal phosphate group of ATP is only -30.6 kJ (i.e. it is a better phosphate donor than ATP itself).

Triose phosphate dehydrogenase has been investigated in a wide number of plants. It resembles the enzyme of mammalian muscle in its properties (Shulman and Gibbs, 1968). Each of its sub-units (there are at least four) carries a single molecule of tighly bound NAD, and presumably an active site to which substrates can be bound. The NADH probably does not leave the enzyme, but donates its reducing equivalent to an unbound molecule of NAD^+.

A similar enzyme which is specific for NADP is found in chloroplasts. This species of enzyme is believed to be involved in the reduction of 1,3-bisphosphoglycerate produced in photosynthesis (see p. 364).

The high energy acyl phosphate group of 1,3-bisphosphoglycerate is transferred to ADP to yield ATP and 3-phosphoglycerate in a reaction whose equilibrium position is well to the right. The enzyme involved is phosphoglycerate kinase ($\Delta G^{\circ\prime} = -18\cdot8$ kJ).

1,3-Bisphosphoglycerate   3-Phosphoglycerate

In this way, the high phosphate transfer potential of 1,3-phosphoglycerate is converted into a more negotiable form of energy, namely ATP. It provides a good example of *substrate level phosphorylation* (see p. 281).

The phosphoglycerate so formed is always in equilibrium with 2-phosphoglycerate.

3-Phosphoglycerate   2-Phosphoglycerate

The dehydration of 2-phosphoglycerate to yield phosphoenolpyruvate is the second reaction in the glycolytic sequence in which a high energy phosphate group is generated.

2-Phosphoglycerate   Phosphoenolpyruvate

The removal of water from the substrate gives rise to a product with a very high negative free energy of hydrolysis of its phosphate group (-62 kJ) which is much greater than that of the terminal phosphate of ATP. The reaction is reversible, though favouring phosphoenolpyruvate formation. The final step in the glycolytic sequence yields pyruvate and is catalysed by pyruvate kinase. In this reaction, a phosphate group is transferred from phosphoenolpyruvate to ATP.

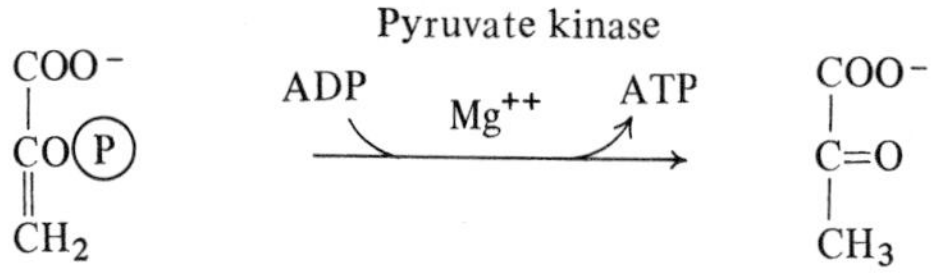

Pyruvate kinase

$$
\begin{array}{c}
\text{COO}^- \\
| \\
\text{CO}\,\textcircled{P} \\
\| \\
\text{CH}_2
\end{array}
\quad
\xrightarrow[\text{Mg}^{++}]{\text{ADP} \qquad \text{ATP}}
\quad
\begin{array}{c}
\text{COO}^- \\
| \\
\text{C=O} \\
| \\
\text{CH}_3
\end{array}
$$

Phosphoenolypruvate   Pyruvate

The reaction is highly exergonic and effectively irreversible. The pyruvate formed can then undergo a number of metabolic fates, including oxidation in the tricarboxylic acid cycle in a sequence of reactions which is confined largely to the mitochondrion.

Glycolysis: the sum

The reactions of the glycolytic sequence are summed up by the following equations:

$$\text{Glucose} + 2\text{ATP} \longrightarrow 2\,\text{triose-phosphate} \tag{1}$$

$$2\,\text{Triose-P} + 4\text{ADP} + 2\text{H}_3\text{PO}_4 + 2\text{NAD}^+ \longrightarrow$$
$$2\,\text{pyruvate} + 4\text{ATP} + 2\text{NADH} + 2\text{H}^+ + 2\text{H}_2\text{O} \tag{2}$$

Equation (1) represents the priming stages; equation (2), the energy conservation reactions. By combining (1) and (2) we get

$$\text{Glucose} + 2\text{ADP} + 2\text{H}_3\text{PO}_4 + 2\text{NAD}^+ \longrightarrow$$
$$2\,\text{pyruvate} + 2\text{ATP} + 2\text{NADH} + 2\text{H}^+ + 2\text{H}_2\text{O}$$

The net change in free energy for the above reaction is $-92\cdot1$ kJ/mole which ensures that the sequence is 'downhill' in the sense that it favours pyruvate formation. However, whereas 2872 kJ are released during the complete oxidation of 1 mole of glucose to CO_2 and water by oxygen, only about 586 kJ, approximately one-fifth of this, has been extracted from the glucose molecules in their conversion to pyruvate. Part of this energy is conserved: two moles of ATP are produced from ADP and inorganic phosphate, for example, with a conservation of about $62\cdot8$ kJ (Fig. 6.19). Two moles of NADH also result, and, depending on their fate, these can represent a further net gain to the cell in terms of available energy (see p. 295). If they are reoxidized by the electron transport chain of the mitochondrion, for example, they are each capable of generating further amounts of ATP (see Ch. 7). Alternatively, reoxidation might occur through the involvement of cytoplasmic NADH oxidase systems or via ethanol or

	Energy balance		Reaction
	Produced	Utilized	
	—	ATP	Glucose → Glucose 6-P
	--	ATP	Fructose 6-P → Fructose 1,6-BP
	2 NADH	—	Glyceraldehyde 3-P → 1, 3-d: P-Glycera
	2 ATP	—	1,3-BP-Glycerate → 3-P-Glycerate
	2ATP	—	P-Enolpyruvate → Pyruvate
Sum 4 ATP + 2 NADH		2 ATP	
	—	2 NADH	Pyruvate → Ethanol or → Lactate
Total*		2 ATP	

Fig. 6.19   A summary of the main bioenergetic steps of glycolysis.

* Total if pyruvate is broken down anaerobically to either lactate or ethanol.

lactate production (Fig. 6.19). In these cases, no ATP is formed and most of the energy is released as heat.

Reoxidation of NADH

There is thought to be only a limited amount of NAD available for oxidation reactions within the cell. Consequently, there must be some means of regenerating it from its reduced form since, otherwise, glycolysis would come to a halt. At present, we can only speculate as to how this reoxidation is accomplished, though there are a number of possibilities.

1. NADH can penetrate the mitochondria and be reoxidized by the electron transport chain. This process would be coupled to the formation of ATP. Isolated plant mitochondria, unlike those of animal cells, are able to oxidize exogenous NADH, and two molecules of ATP are probably produced for each NADH reoxidized (i.e. two coupling sites) (see p. 297).
2. NADH may be used to drive reduction reactions in the cytoplasm, such as the formation of malate from oxaloacetate or glycerophosphate from dihydroxyacetone phosphate. The reduced compounds then pass into the mitochondria and are themselves reoxidized by the electron transport chain. This is one means whereby glycolytic NADH is reoxidized in animal cells.
3. Soluble oxidase systems in the cytoplasm may facilitate the oxidation of NADH and provide a direct pathway to oxygen. Alternatively, this may occur indirectly by the reducing equivalents being first transferred to NADP by the action of transhydrogenase:

$$NADH + NADP^+ \rightleftharpoons NAD^+ + NADPH$$

The following series of reactions provide examples of how NADH or NADPH may be oxidized in the cytoplasm. These

(a)

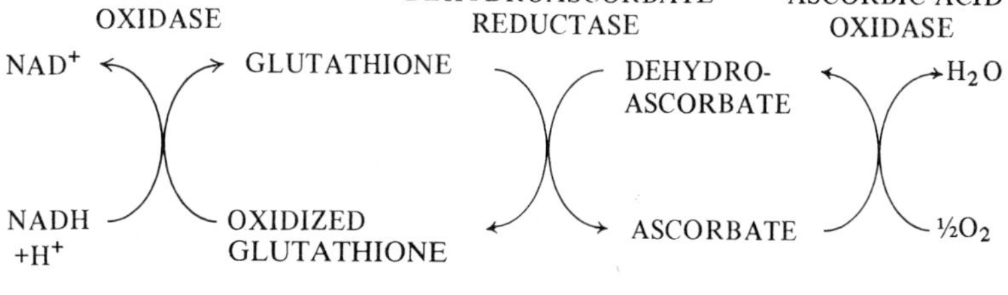

NADH OXIDASE	DEHYDROASCORBATE REDUCTASE	ASCORBIC ACID OXIDASE

NAD$^+$ ← / → GLUTATHIONE ╳ DEHYDRO-ASCORBATE ← / → H$_2$O

NADH +H$^+$ — OXIDIZED GLUTATHIONE ╳ ASCORBATE ╳ ½O$_2$

(b)

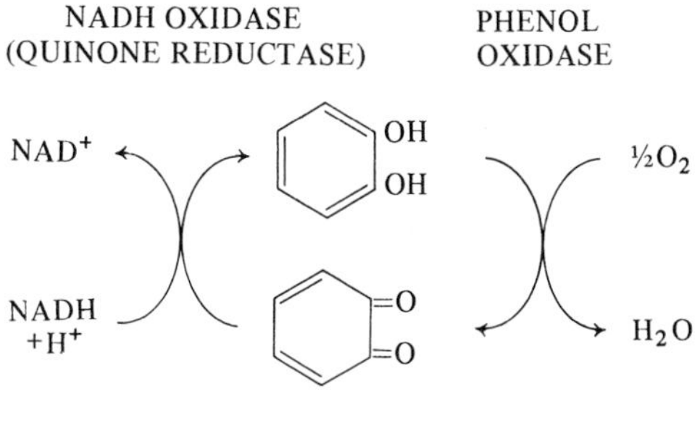

NADH OXIDASE (QUINONE REDUCTASE)	PHENOL OXIDASE

NAD$^+$ ← / → [benzene ring with OH, OH] ╳ ½O$_2$

NADH +H$^+$ [benzene ring with =O, =O] ← / → H$_2$O

reactions are not coupled to ATP production. In scheme (a), glutathione is itself an important reducing agent in the cell and probably participates as a reductant in numerous other reactions.

4. NADH is reoxidized by the formation of either lactic acid (a) or ethanol (b) from pyruvate. This probably occurs predominantly under anaerobic conditions or when oxygen is limited and allows the rapid regeneration of the NAD^+ which is required to promote continued glycolytic flow.

(a)
$$\begin{array}{ccc}
COO^- & & COO^- \\
| & \text{Lactate dehydrogenase} & | \\
C{=}O & \longleftrightarrow & CHOH \\
| & \underset{NADH+H^+ \quad NAD^+}{} & | \\
CH_3 & & CH_3 \\
\text{Pyruvate} & & \text{Lactate}
\end{array}$$

(b)
$$\begin{array}{ccccc}
COO^- & & CHO & & CH_2OH \\
| & \underset{CO_2}{\overset{\text{Pyruvate decarboxylase}}{\longrightarrow}} & | & \underset{NADH+H^+ \quad NAD}{\overset{\text{Alcohol dehydrogenase}}{\longleftrightarrow}} & | \\
C{=}O & & CH_3 & & CH_3 \\
| & & & & \\
CH_3 & & & & \\
\text{Pyruvate} & & \text{Acetaldehyde} & & \text{Ethanol}
\end{array}$$

Lactate formation is favoured in mammalian muscle during exercise because this tissue relies heavily on the ATP produced during glycolysis. However, lactate is only occasionally detected in plant tissues (for example in potatoes or in chlorococcal green algae) and is not a major metabolite. On the other hand, ethanol readily accumulates under anaerobic conditions, or when the tricarboxylic acid cycle is inhibited by metabolic poisons. On occasions, carbohydrate may be converted almost quantitatively to carbon dioxide and ethanol (Table 6.2). As the oxygen concentration is lowered, fermentation and respiration may occur simultaneously and ethanol may accumulate temporarily, particularly in deeper lying tissues. Moist seeds, storage roots, seedling tissues and embryonic tissues characteristically show this type of alcoholic fermentation, and both alcohol dehydrogenase and pyruvate decarboxylase are widespread in plants. However, most plant tissues soon die in absence of oxygen and it is unlikely that the alcohol dehydrogenase serves as a major enzyme to oxidize NADH under normal aerobic conditions. Under aerobic conditions, ethanol can be oxidized to acetaldehyde, then converted to acetate (Oppenheim and Castelfranco, 1967), and finally utilized in the tricarboxylic acid cycle.

Table 6.2 Anaerobic carbohydrate utilization in carrot root tissue*†

Expt. no.	Sugar utilization	Alcohol formed	CO_2 formed
1	0·378	0·220	0·115
2	0·628	0·405	0·203

*Taken from James and Ritchie (1955).
† The data are in grams carbon/100 g of fresh weight of carrot root tissue.

Evidence for the glycolytic pathway

Evidence for the occurrence of the EMP pathway in plants is very strong. All the intermediates involved have been isolated. The individual enzymes have all been shown to be present and in many cases they have been purified. Plant extracts will also convert glycolytic intermediates quantitatively to ethanol and carbon dioxide. Each enzyme has been recovered in amounts at least potentially sufficient to satisfy the full needs of the plant for pyruvate formation. Iodoacetate and fluoride, compounds which inhibit triose phosphate dehydrogenase and enolase, respectively, and which, therefore, inhibit glycolysis, also reduce the rate of respiration of plant cells. This indicates that the EMP sequence is required to produce pyruvate for the TCA cycle. Inhibition of the TCA cycle by such compounds as arsenite and malonate also induce an accumulation of glycolytic end products, suggesting an obligatory link between the two pathways. Furthermore, a number of tissues metabolize specifically labelled D-[^{14}C] glucose in a manner consistent with the operation of the EMP sequence. For example, during anaerobiosis carbon atoms 3 and 4 appear in carbon dioxide while carbon atoms 1 and 6 appear in the methyl carbon of ethanol (Table 6.3). There is now little doubt that the EMP sequence is of major importance in all tissues of higher plants.

Table 6.3 Fate of specific carbon atoms of glucose during glycolysis: location of ^{14}C in ethanol produced during arsenite-induced aerobic fermentation in corn root tips.

Predicted Results	Observed Results		
		Percentage of ^{14}C in ethanol	
$\overset{\displaystyle 1\ 2\ 3\ 4\ 5\ 6}{\underset{H}{\overset{O}{\diagdown}}}$ C–C–C–C–C–C Glucose	Substrate	In CH_2OH	In CH_3
	Glucose 1-^{14}C	2·5	97·5
1 2 3 4 5 6	Glucose 2-^{14}C	97·2	2·8
	Glucose 3,4-^{14}C	0	0
C–C–C C–C–C Trioses			
1,6 2,5 3,4			
CH_3–C–COO$^-$ Pyruvate $\underset{O}{\overset{\|}{ }}$			
1,6 2,5 3,4			
$CH_3CH_2OH + CO_2$ Ethyl alcohol + CO_2			

Source: Beevers and Gibbs, 1954.

Gluconeogenesis

Gluconeogenesis and the metabolism of fats and storage protein to carbohydrate

Gluconeogenesis usually refers to the formation of glucose from various non-carbohydrate precursors. In a more narrow sense, it relates to the reversal of the glycolytic or EMP sequence of reactions (Fig. 6.20). It is a well-documented process in animals and is under strict metabolic and hormonal control, particularly in such gluconeogenic tissues as liver and kidney. It is also a process of great importance in certain plant tissues, particularly for the metabolism of fats or storage protein. Thus, during active lipid breakdown, which occurs extensively in the glyoxysomes of seeds such as the castor bean (Ch. 9), or during protein metabolism in protein rich seeds (Stewart and Beevers, 1967), there is active production of TCA cycle intermediates due to the oxidation of fatty acids or to the deamination of amino acids. In order to convert these intermediates to carbohydrate, exit from the TCA cycle must be negotiated and two energetically unfavourable reactions in the glycolytic sequence, the pyruvate to phosphoenolpyruvate and the fructose 6-phosphate to fructose 1,6-phosphate steps, reversed or bypassed. How are these energy barriers circumvented? Acetyl-CoA, the product of fat breakdown, is first converted to oxaloacetate by the glyoxylate pathway, a series of

Fig. 6.20   A summary of the rate limiting reactions of glycolysis and gluconeogenesis.

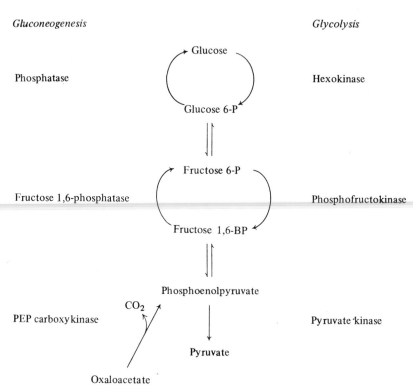

reactions which in seeds are largely restricted to the glyoxysomes (Fig. 6.21) (see Ch. 9). Oxaloacetate is then converted to phosphoenolpyruvate by a reversible decarboxylation reaction, which in plants utilizes ATP and is catalysed by the enzyme phosphoenolpyruvate carboxykinase. We discuss this enzyme in a later section on β-carboxylation.

Oxaloacetate + ATP $\rightleftharpoons$ phosphoenolpyruvate + CO_2 + ADP

Fig. 6.21 The conversion of fats to carbohydrate. A summary of the main steps.

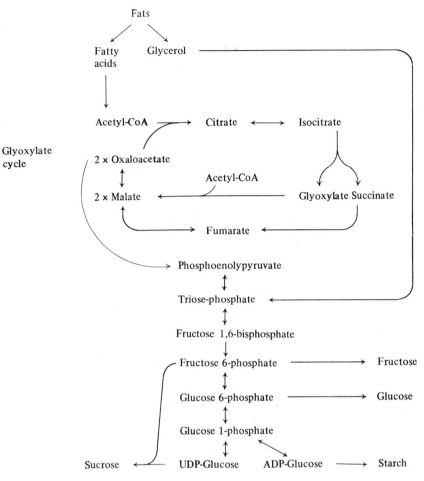

In this way, the energy barrier between pyruvate and phosphoenol-pyruvate is avoided and an easy route from the TCA cycle into the EMP sequence is created. The pathway can then proceed in reverse up to the point of fructose 1,6-bisphosphate formation. Reversal of this step by phosphofructokinase is again thermodynamically unfavourable and it requires a specific phosphatase, fructose 1,6-bisphosphatase. This enzyme, like the kinase catalysing the formation of fructose 1,6-diphosphate is a regulatory enzyme subject to allosteric control. Its properties will be discussed in the next section. The fructose 6-phosphate, once formed is then converted to glucose 1-phosphate, UDP-glucose and hence sucrose (Fig. 6.21). Sucrose may then be transported to other regions of the plant.

Regulation of glycolysis and gluconeogenesis

Energy charge

Krebs suggested that glycolysis and gluconeogenesis are in part regulated by the available ratio of ATP to AMP in the cell. More recently, this idea has been extended by Atkinson (see Atkinson, 1969 and 1966) who used the term 'energy charge' of the cell to describe the concentration of ATP in relation to that of ADP and AMP. He pointed out that adenylates are important effector molecules with regard to enzyme activity (see p. 109). Phosphofructokinase and fructose 1,6-bisphosphatase, for example, which are rate limiting enzymes for glycolysis and gluconeogenesis respectively, are extremely sensitive to control by adenylates (see next sections). In general, pathways responsible for generating ATP (i.e. catabolic processes) are least active at high energy charge (i.e. when ATP levels are high), while those utilizing ATP in biosyntheitc reactions are most active under these conditions.

It should also be noted that ADP and ATP are substrate and product respectively for the glycolytic sequence and as such could influence the overall rate by a mass action effect. Furthermore, the ATP generated in the oxidative part of the sequence can be used to drive the earlier priming stages. This might be expected to accelerate the system to full capacity. This type of control, however, is clearly cruder than the more subtle type of allosteric effects on individual enzymes. Current opinion favours the latter as the most important means whereby glycolysis is regulated.

Regulation of enzyme activities

Phosphofructokinase. Phosphofructokinase has been established as the main controlling step of glycolysis in mammalian systems. It is an allosteric enzyme (see p. 109) characteristically inhibited by ATP and citrate, and hence subject to feedback control by the accumulation of TCA cycle intermediates and decreased energy demand (see Turner and Turner, 1975). This inhibition can be overcome by AMP and in some cases ADP, which in turn are activators of the enzyme. It is a typical allosteric enzyme subject to second site control, exhibiting sigmoid-shaped substrate–saturation curves under certain conditions. Indeed, the mammalian enzyme may have as many as six distinct sites to which various rate-modifying substrates become attached. It is not surprising, therefore, that the formation of fructose-1,6-bisphosphate is a highly complex process and probably subject to many subtle metabolic influences in *vivo*. The plant phosphofructokinase is also subject to allosteric control, though there are often distinct differences between tissues and species as to the nature of the metabolite repression or activation observed. Generally, the enzyme is inhibited by ATP and citrate and in some instances by phosphoenolpyruvate (Dennis and Coultate, 1967; Kelly and Turner, 1968). In contrast to the mammalian enzyme, AMP or ADP seem to exert little influence as activators. Fructose 6-phosphate, however, will overcome the ATP inhibition, so that glycolysis may function even under conditions of high energy charge. This may be particularly important in modulation of glycolytic carbon supply for lipid and protein biosynthesis, when flux through the cycle may be required in the absence of respiratory demand for pyruvate.

Fructose bisphosphatase. In animals and bacteria, this important enzyme of gluconeogenesis is strongly inhibited by AMP, itself a useful indicator of ATP depletion, and hence energy demand within the cells. The gluconeogenic fructose bisphosphatase of plants also appears to be AMP sensitive and is probably the main control point for gluconeogenesis (Kobr and Beevers, 1971).

Pyruvate kinase. In gluconeogenesis, tricarboxylic acid cycle intermediates pass into the EMP sequence via a decarboxylation of oxaloacetate. Phosphoenolpyruvate is, however, readily converted to pyruvate and this step will clearly compete with the reverse common flow towards hexose unless the processes are either modulated or spatially separated within the cell. In liver, pyruvate kinase is stimulated by fructose bisphosphate, a metabolite which indicates a build up of glycolytic intermediates, and inhibited by acetyl-CoA, which probably only accumulates *in vivo* when the TCA cycle is choked. Phosphoenolpyruvate to pyruvate is also a regulatory point in the glycolytic sequence of plant cells, but the effector molecules are not known (Adams and Rowan, 1970; Kobr and Beevers, 1971).

The Pasteur effect

Pasteur noted that the rate of fermentation of sugars by yeast was reduced in the presence of oxygen. Anaerobically, glucose is converted almost quantitatively to ethanol and carbon dioxide, while under aerobic conditions a large proportion becomes cellular material. This oxygen inhibition of the fermentation process and the switch to decreased cellular synthesis under anaerobic conditions is usually known as the *Pasteur effect*. Frequently, there is also a greater rate of carbohydrate breakdown under nitrogen than in air, indicating that the overall rate of glycolysis has been accelerated. More recently, this phenomenon has been shown to be associated with a decrease in ATP, 3-phosphoglycerate and phosphoenolpyruvate, and an increase in ADP, fructose 1,6-bisphosphate and inorganic phosphate (i.e. a drop in energy charge). The Pasteur effect can now be explained, at least in part, in terms of the metabolic control exerted over phosphofructokinase activity. In the absence of air, the continuing energy demand will result in reduced levels of ATP. In turn, this will relieve the ATP inhibition normally exerted over phosphofructokinase (Givan, 1968). In some tissues, this enzyme is also known to respond positively to inorganic phosphate and to increased levels of substrate. Because it is a pacemaker enzyme for the glycolytic sequence, anaerobic conditions are likely to lead to increased activity, promotion of fructose diphosphate breakdown, and hence greater rates of carbohydrate utilization. This also implies that the rest of the enzymes of the pathway are not normally working at full capacity.

The pentose phosphate pathway

Function of the pathway

The presence of pentose sugars, amino sugars, uronic acids and numerous other compounds in living cells which are clearly related to carbohydrates but which are not formed as intermediates of glycolysis, indicates the existence of pathways of carbohydrate interconversion alternative to, or supplementary to, the glycolytic scheme. The most important of these is the pentose phosphate pathway whose primary function appears to be to provide ribose and deoxyribose for nucleic acid biosynthesis, erythrose as a precursor of shikimic acid and aromatic compounds, and reducing power in the form of NADPH. The cycle probably also plays a rôle in the maintenance of reduced sulphydryl compounds since NADPH is the primary reductant for glutathione in cells. The relationship of this pathway to the EMP sequence is shown in Fig. 6.22. The enzymes of this pathway appear to be in the soluble portion of the cytoplasm in so far as we can judge from present analytical data. However, the pathway is in part reproduced in the chloroplasts as part of the Calvin cycle which describes the path of carbon in photosynthesis (Ch. 8).

Fig. 6.22   The metabolic relationship between the pentose phosphate pathway and the glycolytic pathway. Reactions shown by single arrows are considered irreversible. Double arrows indicate reversible steps.

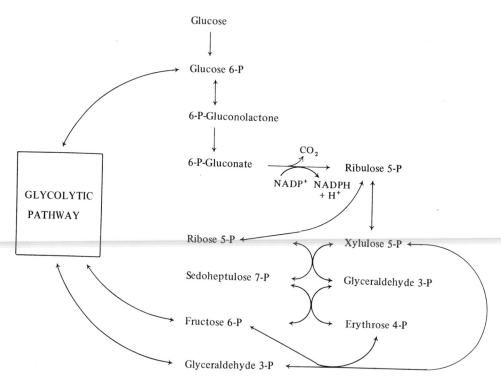

Reactions of the cycle

The early work of Warburg, Dickens and Lipmann and their colleagues, and later by Horecker and Racker, established the presence of enzymes in plant and animal cells which were capable of bringing about the direct oxidation of glucose 6-phosphate to pentose phosphate. The initial reaction is catalysed by glucose 6-phosphate dehydrogenase, called *Zwischenferment* by Warburg and Christian in 1931. It is widespread in plants and it is relatively specific for NADP, rather than NAD. The formation of 6-phosphogluconate is now known to proceed in two steps. Firstly, glucose 6-phosphate is oxidized to the δ-lactone of 6-phosphogluconic acid. This is followed by the hydrolysis of the lactone to 6-phosphogluconate. Although such a lactonization will proceed spontaneously at physiological pH, a lactonase speeds up the reaction.

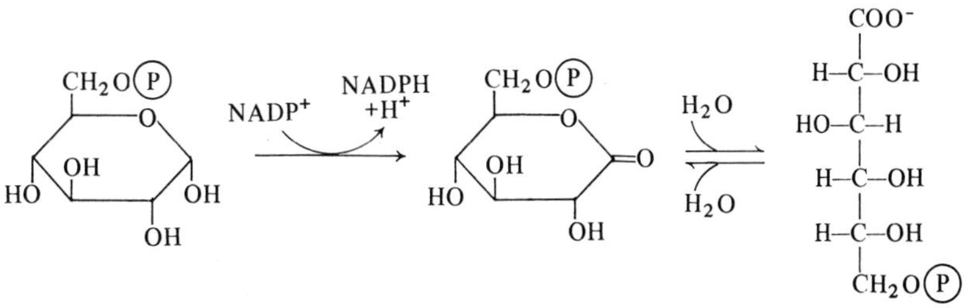

$CH_2O(P)$ NADP$^+$ NADPH +H$^+$ $CH_2O(P)$ H$_2$O   COO$^-$ / H—C—OH / HO—C—H / H—C—OH / H—C—OH / $CH_2O(P)$

D-Glucose 6-phosphate   6-phosphoglucono-δ-lactone   D-6-Phosphogluconate

The next step produces ribulose 5-phosphate and carbon dioxide. The enzyme 6-phosphogluconate dehydrogenase is again NADP specific and carbon atom 1 of the original glucose atom is lost. It should be noted, therefore, that any pentose phosphate formed from glucose 1-^{14}C by these reactions would be unlabelled, while pentose phosphate derived from glucose labelled in the C-2 position would now be labelled in carbon atom 1. The reaction probably proceeds by way of 3-ketogluconate 6-phosphate but this has not been detected as a free intermediate.

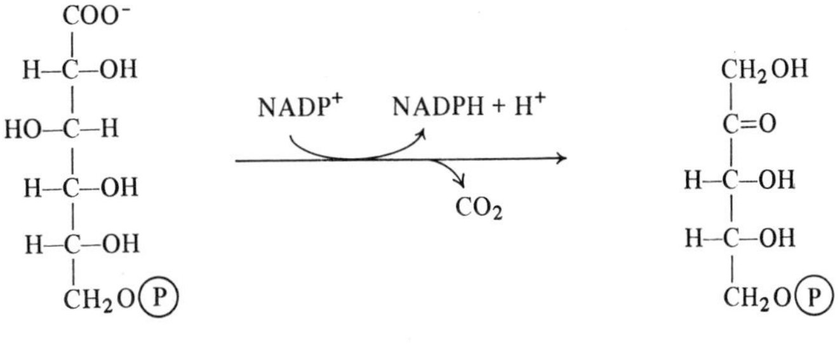

COO$^-$ / H—C—OH / HO—C—H / H—C—OH / H—C—OH / $CH_2O(P)$ NADP$^+$ NADPH + H$^+$ CO$_2$ CH_2OH / C=O / H—C—OH / H—C—OH / $CH_2O(P)$

D-6-Phosphogluconate   D-Ribulose 5-phosphate

Following the two oxidative steps are a series of enzymic interconversions and rearrangements which involve little change in free energy. In these steps, ribulose phosphate may be converted to ribose 5-phosphate by a specific isomerase. An epimerase has also been recognized which catalyses the inversion of the configuration of the

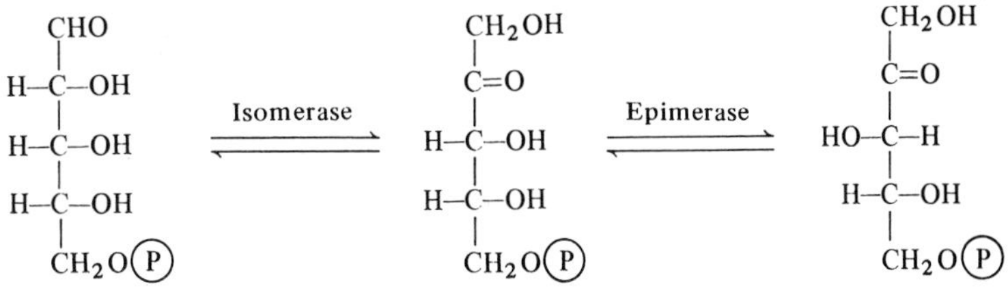

CHO / H—C—OH / H—C—OH / H—C—OH / $CH_2O(P)$ Isomerase CH_2OH / C=O / H—C—OH / H—C—OH / $CH_2O(P)$ Epimerase CH_2OH / C=O / HO—C—H / H—C—OH / $CH_2O(P)$

D-Ribose 5-phosphate   D-Ribulose 5-phosphate   D-Xylulose 5-phosphate hydroxyl group on carbon atom 3 of ribulose 5-phosphate to give xylulose 5-phosphate.

Xylulose 5-phosphate is one known substrate for the enzyme transketolase which catalyses the transfer of the two carbon moiety from a donor ketose to an acceptor aldehyde (in this case ribose 5-phosphate) in the presence of Mg^{++}. Thiamine pyrophosphate (TPP, see p. 121) is a cofactor in the reaction and is found tightly bound to the enzyme, only being removed by treatment with alkali or by prolonged dialysis against EDTA. The cleavage of xylulose 5-phosphate by transketolase leads to the reversible formation of D-glyceraldehyde 3-phosphate and sedoheptulose 7-phosphate. The C—2 moiety is never free but remains attached to the TPP as an active glycoaldehyde group. It is this unit which is transferred to D-ribose 5-phosphate to form the addition product:

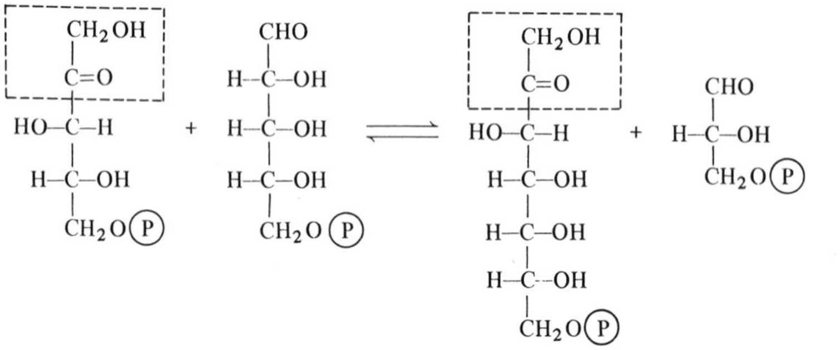

| D-Xylulose 5-phosphate | D-Ribose 5-phosphate | D-Sedoheptulose 7-phosphate | D-Glyceraldehyde 3-phosphate |

A number of other acceptors and donors of the C—2 moiety have been recognized for the transketolase reaction, and are listed in Table 6.4.

The next step in the conversion of pentose 5-phosphate to hexose 6-phosphate is catalysed by the enzyme transaldolase. In this reaction, the upper three carbon moiety of sedoheptulose 7-phosphate is transferred reversibly to glyceraldehyde 3-phosphate. The products are fructose 6-phosphate and erythrose 4-phosphate. Unlike transketolase, no coenzyme is involved in the transfer. The dihydroxyacetone residue becomes attached covalently to the free NH_2 of a lysine on the enzyme forming a Schiff's base, a relatively stable, though transient, intermediate.

Table 6.4   Donors, acceptors and products for the transketolase reactions.*

Donors	Acceptors	Product of C-2 addition
L-Erythulose	Glycoaldehyde	L-Erythulose
D-Xylulose 5-P	D-Glyceraldehyde	D-Xylulose
D-Fructose 6-P	D-Glyceraldehyde 3-P	D-Xylulose 5-P
D-Sedoheptulose 7-P	D-Erythrose 4-P	D-Fructose 6-P
D-Sedoheptulose 7-P	D-Ribose 5-P	D-Sedoheptulose 7-P

* As these reactions are reversible, the product can also act as a C-2 donor.

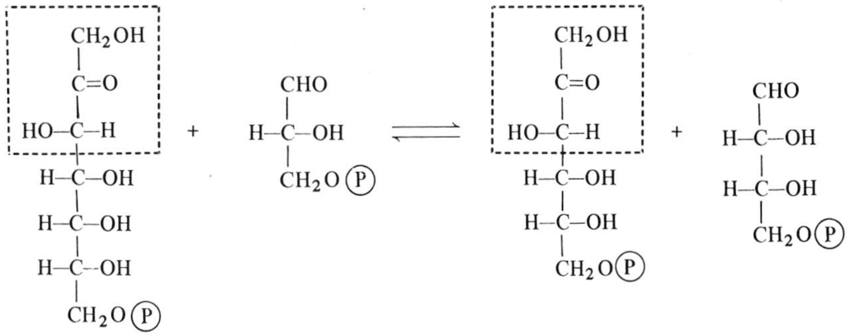

CH2OH   CHO   CH2OH   CHO

(First reaction:)

D-Sedoheptulose 7-phosphate	+	D-Glyceraldehyde 3-phosphate	⇌	D-Fructose 6-phosphate	+	D-Erythrose 4-phosphate

D-Sedoheptulose 7-phosphate:
CH2OH
|
C=O
|
HO–C–H
|
H–C–OH
|
H–C–OH
|
H–C–OH
|
CH2O℗

D-Glyceraldehyde 3-phosphate:
CHO
|
H–C–OH
|
CH2O℗

D-Fructose 6-phosphate:
CH2OH
|
C=O
|
HO–C–H
|
H–C–OH
|
H–C–OH
|
CH2O℗

D-Erythrose 4-phosphate:
CHO
|
H–C–OH
|
H–C–OH
|
CH2O℗

The erythrose 4-phosphate produced in the transaldolase reaction can serve as an acceptor for a second reaction catalysed by transketolase. In this, it accepts a C—2 fragment of xylulose 5-phosphate to form a further molecule of fructose 6-phosphate plus D-glyceraldehyde 3-phosphate.

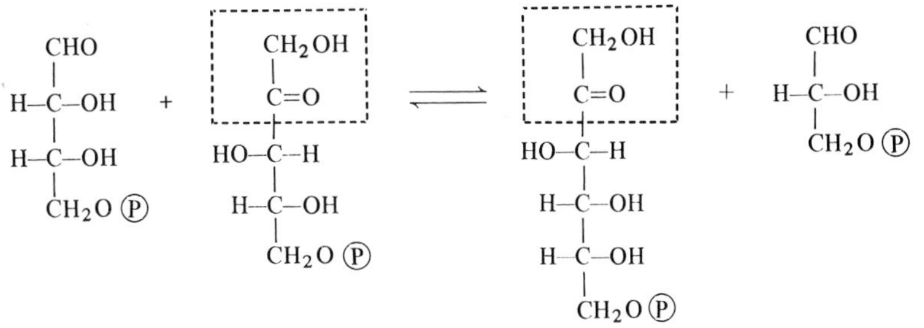

(Second reaction:)

D-Erythrose 4-phosphate	+	D-Xylulose 6-phosphate	⇌	D-Fructose 6-phosphate	+	D-Glyceraldehyde 3-phosphate

D-Erythrose 4-phosphate:
CHO
|
H–C–OH
|
H–C–OH
|
CH2O℗

D-Xylulose 6-phosphate:
CH2OH
|
C=O
|
HO–C–H
|
H–C–OH
|
CH2O℗

D-Fructose 6-phosphate:
CH2OH
|
C=O
|
HO–C–H
|
H–C–OH
|
H–C–OH
|
CH2O℗

D-Glyceraldehyde 3-phosphate:
CHO
|
H–C–OH
|
CH2O℗

The net result of the non-oxidative portion of the cycle is that three moles of pentose phosphate are converted into two moles of hexose phosphate and one of triose phosphate. Figure 6.23 shows schematically the passage of six moles of NADPH and six moles of carbon dioxide (derived exclusively from C—1 of the glucose) are generated. The two moles of D-glyceraldehyde phosphate can be converted to fructose 1,6-phosphate by the combined action of triose phosphate isomerase and aldolase.

Evidence for the cycle

The enzymes and intermediate products of the sequence have been shown to be present in several plant tissues and reside in the soluble phase of the cytoplasm. Furthermore, a variety of tissues will phosphorylate sedoheptulose, erythrose and pentose sugars such as ribose, and metabolize them in a manner consistent with the operation of the pentose phosphate cycle. The labelling pattern in the hexose units recovered is consistent with the scheme shown in Fig. 6.23, and

Fig. 6.23   A schematic representation of the passage of six molecules of hexose phosphate through the pentose phosphate cycle. The numbers refer to the positions of the carbons in the original hexose. Reproduced from Axelrod and Beevers (1956).

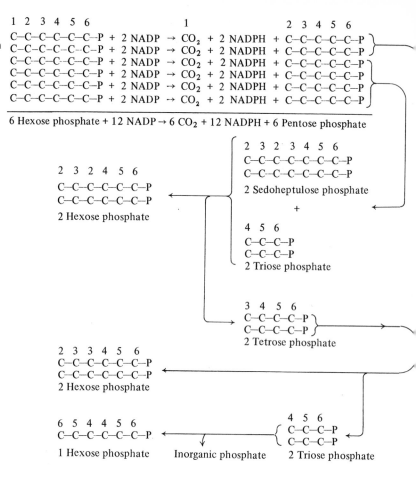

$$
\begin{array}{ccccccccccccc}
1 & 2 & 3 & 4 & 5 & 6 & & & 1 & & & 2 & 3 & 4 & 5 & 6 \\
\end{array}
$$

C–C–C–C–C–C–P + 2 NADP → CO$_2$ + 2 NADPH + C–C–C–C–C–P
C–C–C–C–C–C–P + 2 NADP → CO$_2$ + 2 NADPH + C–C–C–C–C–P
C–C–C–C–C–C–P + 2 NADP → CO$_2$ + 2 NADPH + C–C–C–C–C–P
C–C–C–C–C–C–P + 2 NADP → CO$_2$ + 2 NADPH + C–C–C–C–C–P
C–C–C–C–C–C–P + 2 NADP → CO$_2$ + 2 NADPH + C–C–C–C–C–P
C–C–C–C–C–C–P + 2 NADP → CO$_2$ + 2 NADPH + C–C–C–C–C–P

6 Hexose phosphate + 12 NADP → 6 CO$_2$ + 12 NADPH + 6 Pentose phosphate

2 3 2 3 4 5 6
C–C–C–C–C–C–C–P
C–C–C–C–C–C–C–P
2 Sedoheptulose phosphate

2 3 2 4 5 6
C–C–C–C–C–C–P
C–C–C–C–C–C–P
2 Hexose phosphate

+

4 5 6
C–C–C–P
C–C–C–P
2 Triose phosphate

3 4 5 6
C–C–C–C–P
C–C–C–C–P
2 Tetrose phosphate

2 3 3 4 5 6
C–C–C–C–C–C–P
C–C–C–C–C–C–P
2 Hexose phosphate

6 5 4 4 5 6
C–C–C–C–C–C–P
1 Hexose phosphate

Inorganic phosphate

4 5 6
C–C–C–P
C–C–C–P
2 Triose phosphate

strongly suggests that transaldolase and transketolase have been involved in the interconversion.

Pentose phosphate pathway as a cycle

Although we have presented the pentose phosphate pathway as a cycle, it may not operate as such. It branches from the EMP sequence at two points: at the glucose 6-phosphate oxidation step and at the level of fructose 6-phosphate (Fig. 6.22). While the oxidative mechanism, which proceeds via gluconate 6-phosphate and provides NADPH, is essentially irreversible, the interconversions of fructose 6-phosphate and pentose phosphates involve little change in free energy and are, therefore, potentially reversible. The sequence, therefore, need not operate as a cycle, but could be viewed as two independent mechanisms for the conversion of hexose monophosphate to pentose phosphate. In those tissues where the pathway appears to function in a cyclic manner, this may be related to the requirements of the tissue for NADPH in reduction reactions, and, if under these circumstances more pentose or erythrose is generated than is required in synthetic processes, the excess will be converted back to hexose monophosphate by the non-oxidative pathway. Where NADPH is not required, pentose can conceivably be formed without involving the oxidative steps of the cycle. These facts must be borne in mind in any attempts to assess the flow of carbon through the pathway.

C-6/C-1 Ratios and assessment of carbon flow

In its complete respiration, all of the individual carbon atoms of a molecule of glucose will ultimately appear as carbon dioxide. However, depending upon the particular pathway taken, different carbon atoms will appear in carbon dioxide at different times. In the EMP sequence, the carbon skeleton of the glucose molecule is cleaved and the two triose molecules generated converted to two molecules of pyruvate. No carbon dioxide is given off in these steps. Under aerobic conditions, in actively respiring tissues, pyruvate is decarboxylated and enters the TCA cycle as acetyl-CoA. In this step, the carbon atoms 3 and 4 are lost as carbon dioxide. Carbon atoms 2 and 5, on the other hand, will only begin to appear after one complete circuit of the TCA cycle. Carbon atoms 1 and 6 will be released last of all during turns 3 and 4 and will appear at the same rate (see p. 293).

Clearly, if glucose is oxidized entirely by the combined action of the EMP sequence and the TCA cycle, there will be an anticipated delay, whose extent depends upon respiratory rate, before carbon C-1 and C-6 appear in respired carbon dioxide. On the other hand, if glucose enters the pentose phosphate pathway, carbon dioxide from C-1 is generated after only four enzymatic steps (Fig. 6.22) and it will not be accompanied by carbon dioxide from carbon atom 6. The rate of appearance of C-6 would be determined according to whether the hexose and triose phosphate generated were recycled (Fig. 6.23) or converted to pyruvate. In either event, its appearance would be delayed.

The contribution of the pentose cycle to overall glucose metabolism has often been estimated, therefore, by measuring the changes in the ratio of yields of $^{14}CO_2$ from [1-^{14}C] glucose to those from [6-^{14}C] glucose during incubation of the tissue slices in a respirometer (Bloom and Stettin, 1953). A C-6/C-1 carbon dioxide ratio less than unity indicates the operation of the oxidative portion of the pentose phosphate cycle. A ratio of unity is consistent with, but does not necessarily prove, that all of the carbon is passing through the EMP sequence into the tricarboxylic acid cycle. Various plant materials have been investigated in this way and usually yield ratios less than one. Ratios close to unity were only noted in young meristematic tissues (Table 6.5). More mature parts of plants yielded lower ratios, indicating the participation of the pentose phosphate pathway in their metabolism. Oxidizing agents such as methylene blue or phenazine methosulphate and natural electron acceptors such as nitrate tend to cause a decrease in the ratio when supplied to plant tissues. This is consistent with an increased demand for NADPH (Table 6.6). Ratios also fall during the well-known respiratory rise which occurs when slices of storage tissues such as potato or carrot are washed for several hours. This again can be explained if it is assumed that the pentose

Table 6.5 The Ratio of C-6 and C-1 labelled glucose in aerobic metabolism.

Plant	Part	Age	C-6/C-1
Castor bean	Root	0–1·0 cm from tip	0·98
		1·0–2·0 cm from tip	0·77
		2·0–3·0 cm from tip	0·64
		3·0–4·0 cm from tip	0·50
Castor bean	Cotyledons	4 days old	0·77
		5 days old	0·81
		6 days old	0·54
		7 days old	0·56
Pea	Internode of stem	18 days old	0·39
		23 days old	0·46

Taken from Gibbs and Beevers (1955).

Table 6.6 Effect of exogenous electron carriers on the release of $^{14}CO_2$ from glucose 1,2 or 3,4-^{14}C*

		Conversion of applied glucose into $^{14}CO_2$ (%)			
Additions	Period (h)	1-^{14}C	2-^{14}C	3,4-^{14}C	6-^{14}C
None	0–3	2·0	0·7	5·1	1·8
	3–6	3·6	2·6	10·2	3·2
2 mM Methylene blue	0–3	4·4	1·8	6·1	1·5
	3–6	5·3	3·9	8·9	2·2
1 mM Phenazine methosulphate	0–3	10·8	4·4	3·0	0·8
	3–6	13·3	8·3	7·6	2·3

Reproduced from Butt and Beevers (1961).
* Thirty corn root tips were incubated with 2 μmoles of glucose in 0·067 M phosphate buffer for the times specified.

phosphate pathway and NADPH are required to fulfil a need for new biosynthetic processes. These tissue slices are known, for example, to show increased rates of nucleic acid, fatty acid and protein biosynthesis, as well as well-known respiratory changes.

However, although the C-6/C-1 carbon dioxide ratios have been widely used to assess the contribution of the pentose phosphate pathway, there have been considerable disagreements about the significance of the data and, particularly, as to how the results should be calculated. For example, if materials are drawn out of the TCA cycle or the EMP pathway to form cellular material, much of the radiocarbon will be retained, and will never make a contribution to carbon dioxide even though considerable glycolytic flow may be occurring. On the other hand, carbon dioxide is released from carbon atom 6 of UDP-glucuronic acid during the formation of UDP-xylose (see p. 459). Thus, whenever active cell-wall synthesis is occurring, carbon dioxide originating from C-6 of glucose will probably be released after a relatively short period. This will tend to balance the loss of C-1 in the decarboxylation of gluconate and push the ratio towards unity. Because of such limitations, therefore, it has not been possible to estimate accurately the contribution of the two pathways to glucose metabolism. It has become increasingly clear, however, that the pentose phosphate pathway does play an important rôle in most plant tissues, and in some cases up to about 30% of the glucose utilized may pass by this route. In most mammalian cells, it is equally clear that the C-1 oxidation pathway does not usually operate as a cycle and about 70% of the ribose phosphate presently in nucleic acid or nucleotides arises by the non-oxidative formation of pentose phosphate from fructose 6-phosphate and triose phosphate. C-6/C-1 ratios are usually close to unity, also indicating only a modest or negligible flow of carbon through the pathway. Only in tissues engaged in active synthesis, such as lactating mammary glands, does the pentose phosphate appear to function as a complete cycle. The relatively higher flow in plant cells might be related to the greater ease whereby electrons are transferred from NADPH to oxygen in plant cells, due to the presence of NADPH dehydrogenases in the soluble phase of the cytoplasm, or to a greater reliance on NADPH produced in the pentose phosphate pathway for biosynthetic processes.

Function of NADPH and NADH

In any discussion of the pentose phosphate pathway, the question arises as to why two related coenzymes, NADH and NADPH, should exist side by side in the cell. NAD is usually regarded as the coenzyme of fermentation; it also plays a major rôle in respiration and in the formation of high energy phosphate through oxidative phosphorylation. NADPH is oxidized only sluggishly by molecular oxygen in tissue extracts and this process is not coupled to phosphorylation of ATP. Because it is oxidized relatively slowly, NADPH can be considered to be a storehouse of metabolic hydrogen which is used in reduction reactions essential for the biosynthesis of proteins, lipids and nucleic acids. NADPH, for example, is a cofactor required in the reductive amination of α-ketoglutarate, which gives glutamate. It is also involved in nitrate reduction, in fatty acid biosynthesis, in hydroxylation reactions, in photosynthesis, and in the formation of deoxyribonucleotides from ribonucleotides. NADP is usually present in its reduced form which is of course appropriate to its rôle as a reducing agent in the cell. By contrast, NAD appears to be in more ready equilibrium with molecular oxygen and is usually found in its oxidized form. The enzyme transhydrogenase catalyses the transfer of reducing units between the two coenzymes and may regulate the amount of each reduced coenzyme in the cell (see p. 249).

Control of the pentose phosphate pathway

Little information is as yet available regarding control of the pentose phosphate pathway. However, as pointed out earlier, the pathway fulfils a major rôle in the production of NADPH and there is some evidence to suggest that this process may be regulated by the requirements of the cell for reducing power. Thus, during nitrite or nitrate reduction or in cells undergoing rapid biosynthetic activity, NADPH demand is presumably high. Simultaneously, there is an increased flow of carbon through the pentose phosphate pathway, as indicated by a lowering of the C-6/C-1 carbon dioxide ratio. This demand for NADPH may be sufficient in itself to regulate the flux of carbon into the cycle. There are indications that glucose 6-phosphate dehydrogenase is an allosteric enzyme and subject to various forms of feedback control (see Turner and Turner, 1975).

The biosynthesis of fats

The third major metabolic pathway which is usually regarded as being 'soluble' and therefore associated with the high speed supernatant fraction of the cell is that which leads to the formation of long chain fatty acids. Although other cell fractions do usually show some limited ability to synthesize lipids, it is the soluble system which predominates in most non-green tissues and is responsible for the production of the major portion of the saturated long chain fatty acids, particularly stearic and palmitic acids. Insertion of double bonds, or desaturation as it is called, phospholipid formation, and further elongation of the chain occurs mainly on the microsomes. Chloroplasts, however, contain their own complete synthetase system.

The enzymes responsible for fatty acid synthesis in yeast and animals comprise a large, globular protein body of high molecular weight which contains all of the enzymes (and there are at least seven of them) and all of the other components (cofactors, etc.) which are required for synthesis of palmitic acid. Because of its large size, it can be readily purified by sedimentation in the ultracentrifuge, but any attempt to loosen the complex and free the enzymes leads to complete loss of activity. Only one of the enzymes required for the complete synthesis of palmitate from acetyl-CoA, namely the acetyl-CoA carboxylase, is not tightly bound to the complex, though it is usually associated with it when isolated. By contrast, the group of enzymes responsible for fatty acid synthesis in plant cells are not bound together when isolated, and the complex, if indeed it exists, must dissociate readily during fractionation. Neither does it reassociate upon standing. In these respects, the plant synthetase system resembles that from bacteria, and the general scheme of fatty acid biosynthesis is probably similar to that proposed for the soluble *E. coli* system (see Wakil, 1970). When functioning, the enzymes may associate temporarily with one of the membrane systems of the cell such as the ER or plasmalemma.

The fatty acids of plant cells

Each plant cell contains a complement of fatty acids in its membranes which must be synthesized *in situ*, as there is no evidence that lipid is

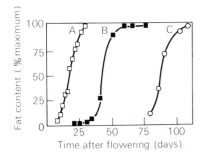

Fig. 6.24   Changes in the triglyceride content of seeds after flowering and pollination. In each case the fat content is expressed relative to the final content of the mature seeds. Curves: A, flax; B, cotton; C, soybean. From data assembled and complied by Butt and Beevers (1965).

transported in plants, except in very specialized cases. However, even though membranes and hence their constituent lipids show a high rate of turnover, most biosynthetic studies have been performed on the plant tissues which accumulate large quantities of lipid as storage material. Maturing seeds of castor bean, cotton, flax and peanut, for example, are favourite subjects for study. In these, the fatty materials are deposited mainly as triglycerides in oil droplets within the cell at a particular developmental stage after pollination (Fig. 6.24). Although the fatty acid composition of these plant oils can be peculiar to the species, genus or even family (Tables 2.5, 2.6), the unsaturated fatty acids, oleic, linoleic and linolenic acids, usually predominate. Nevertheless, we shall first discuss the biosynthesis of the long chain saturated fatty acids, in particular palmitic and stearic acids, which are primary products of the soluble synthetase system, and the likely precursors of the unsaturated derivatives.

Biosynthesis of saturated fatty acids from acetate

In 1953, Newcomb and Stumpf showed that acetate was an effective precursor of higher fatty acids in slices of cotyledons from developing peanuts. This observation has since been extended to a wide number of tissues actively synthesizing fats. Higher homologues are incorporated much more sluggishly into lipids and it is likely that free fatty acids which are intermediate in chain length between acetate and the completed C-16 or C-18 product play little or no rôle as normal intermediates in the reaction sequence. Radioactivity from acetate $1\text{-}^{14}\text{C}$ is incorporated into alternate carbon atoms of the carbon skeletons of newly formed fatty acids. This suggests that fatty acids are produced by the condensation of acetate units. Such a mechanism would also explain the preponderance of even numbered carbon chains among the natural long chain fatty acids.

By the early 1950s, it was known that when fatty acids were broken down, they underwent fission by β-oxidation to yield acetyl-CoA (see Ch. 9). It seemed logical, therefore, to assume that their synthesis occurred by a simple reversal of this catabolic route. Pathways of breakdown and synthesis, however, rarely coincide in biochemistry, and in 1958, Gibson, Titchener and Wakil showed that fatty acid synthesis in partially purified extracts of avian liver required ATP and, unexpectedly, bicarbonate, two compounds not involved in β-oxidation. The ATP underwent dephosphorylation to yield ADP and P_i but ^{14}C from bicarbonate was not incorporated into the final long chain fatty acid. Wakil then demonstrated that bicarbonate was required in the carboxylation of acetyl-CoA to give malonyl-CoA. This compound acted as the C—2 donor for chain elongation. Malonyl-CoA could replace the requirement for acetate, ATP and bicarbonate in the cell free enzyme system.

Not long after the rôle of malonyl-CoA was discovered, the main enzymic features of fatty acid biosynthesis were elucidated. These features appear generally similar in all organisms, though, as stressed earlier, the organization of the synthetase system in plants resembles that from bacteria more closely than that from animals.

The *de novo* synthesis of palmitic acid: overall scheme

Before looking at the individual enzymic steps in detail, it is convenient to summarize our present knowledge of the overall process of fatty acid biosynthesis as it occurs in the cytoplasm of cells.

The formation of one molecule of palmitic acid requires one molecule of acetyl-CoA and seven molecules of malonyl-CoA which condense to form the C-16 fatty acid. Reducing power is furnished largely by NADPH and seven molecules of carbon dioxide are liberated.

$$\text{Acetyl-CoA} + 7 \text{ malonyl-CoA} + 14\text{NADPH} + 14\text{H}^+ \longrightarrow$$

$$\text{CH}_3(\text{CH}_2)_{14}\text{COOH} + 7\text{CO}_2 + 8\text{CoA} + 14\text{NADP}^+ + 6\text{H}_2\text{O}$$

Acetyl-CoA is a primer molecule. Its methyl and carboxyl carbon atoms become C-16 and C-15, respectively, of the final molecule of palmitic acid and chain growth proceeds by successive addition of acetyl units *derived from malonyl-CoA*. The unesterified carboxyl group of malonyl-CoA is lost as carbon dioxide during this condensation, and a β-keto fatty acyl-derivative is formed which then undergoes reduction, dehydration and a further reduction to a saturated fatty acid acyl residue. The cycle is repeated six more times and results in the eventual formation of one molecule of palmitic acid.

The sequence and the individual enzymes have been investigated in both animals and in microorganisms. In plants, however, only the acetyl-CoA carboxylase has been purified and studied in any detail, though there is little doubt that the plant system is basically similar to that in bacteria.

Formation of malonyl-CoA. Acetyl-CoA carboxylase has been purified from wheat germ. It catalyses the carboxylation of acetyl-CoA to give malonyl-CoA:

$$\underset{\text{Acetyl-CoA}}{\overset{\overset{\displaystyle O}{\|}}{\text{CH}_3-\text{C}-\text{SCoA}}} + \text{CO}_2 + \text{ATP} \underset{}{\overset{\text{Acetyl-CoA carboxylase}}{\rightleftharpoons}} \underset{\text{Malonyl-CoA}}{\overset{\overset{\displaystyle \text{HO}_2\text{C}}{|}\quad\overset{\displaystyle O}{\|}}{\text{CH}_2-\text{C}-\text{SCoA}}} + \text{ADP} + \text{P}_i$$

It requires Mg^{++} ions for full activity and like the enzyme from animals, it contains biotin and is, therefore, strongly inhibited by avidin, a protein from egg whites. This compound forms a complex with biotin, rendering the vitamin, or in this case a biotin-enzyme, biologically inert.

Essentially two steps are involved in the carboxylation:

$$\text{Biotin-protein} + \text{ATP} + \text{HCO}_3^- \rightleftharpoons \text{CO}_2\text{–biotin-protein} + \text{ADP} + \text{P}_i$$

$$\text{CO}_2\text{–biotin-protein} + \text{CH}_3 \cdot \text{CO} \cdot \text{SCoA} \rightleftharpoons$$

$$\text{biotin-protein} + {}^-\text{OOC} \cdot \text{CH}_2 \cdot \text{CO} \cdot \text{SCoA}$$

The acetyl CoA carboxylase of plants can be dissociated into the enzymic components biotin carboxylase and transcarboxylase, which catalyse the two reactions shown above, plus a biotinyl binding protein, sometimes known as biotin carboxyl-carrier protein or BCCP (Fig. 6.25a).

(a) *Biotin*

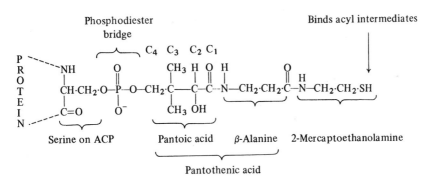

Biotin-CO$_2$ complex

Biotin Lysine of enzyme

Biotinyllysine (biocytin)

(b) *Phosphopantetheine*

Phosphodiester bridge

Binds acyl intermediates

C$_4$ C$_3$ C$_2$ C$_1$

Serine on ACP Pantoic acid β-Alanine 2-Mercaptoethanolamine

Pantothenic acid

Fig. 6.25 Coenzymes involved in fatty acid biosynthesis.

Fatty acid synthetase and ACP. In the second stage of fatty acid biosynthesis, malonyl-CoA reacts with acetyl-CoA and NADPH in a series of condensations and reductions to give the C-16, saturated fatty acid palmitate. There is also an apparent requirement for NADH for palmitate synthesis which probably relates to one or more of the later enzymes in the sequence requiring that coenzyme rather than NADPH (see Stumpf, 1977, in Further Reading).

The acyl intermediates in this cycle of conversions are not free, however. Nor are the thioesters of CoA as in fatty acid oxidation. Throughout, they remain bound as thioesters to a protein of low molecular weight, called the acyl carrier protein (ACP). This serves as an anchor holding the acyl intermediates while the aliphatic chain is

267

built up. Not until the fatty acid is completed is it released from the ACP.

The ACP is a protein of relatively low molecular weight (c. 10 000) and contains a one 4'-phosphopantetheine residue attached to the free hydroxyl group of a serine residue on its peptide chain by means of a phosphate bridge. This cofactor binds the acyl intermediates through a thioester bond (Fig. 6.25). The structure of the 4'-phosphopantetheine moiety is identical to that found in coenzyme A itself (see p. 120). Furthermore, its function is directly analogous to that of CoA in the oxidation of fatty acids. The acyl intermediates remain anchored to the protein while the aliphatic chain is built up two units at a time.

In the first reaction of the sequence, a molecule of acetyl-CoA will react with a molecule of ACP, in a reaction catalysed by a specific transacylase:

$$CH_3 \cdot COS \cdot CoA + HS\text{–}ACP \rightleftharpoons CH_3 \cdot COS \cdot ACP + CoA \cdot SH$$

This acetyl group is then transferred to the condensing enzyme

$$CH_3 \cdot COS \cdot ACP + HS \cdot E_{cond.} \rightleftharpoons HS \cdot ACP + CH_3 \cdot COS \cdot E_{cond.}$$

A molecule of malonyl-CoA can then react with the ACP. The asterisk denotes the position of the radioactive carbon (the carbon that is originally fixed into malonyl CoA):

$$HOO*C \cdot CH_2 \cdot COS \cdot CoA + HS \cdot ACP \rightleftharpoons$$
$$HOO*C \cdot CH_2 \cdot COS \cdot ACP + CoA \cdot SH$$

A separate transacylase (malonyl-CoA transferase) catalyses this reaction.

The condensing enzyme then catalyses the condensation of its attached acyl group (in the first turn of the cycle this is acetyl) with the malonyl-ACP to form the β-ketoacyl-ACP, carbon dioxide, and an uncharged molecule of condensing enzyme:

$$CH_3 \cdot COS \, E_{cond.} + HOO*C \cdot CH_2 \cdot COS \cdot ACP \rightleftharpoons$$
$$CH_3 \cdot CO \cdot CH_2 \cdot COS \cdot ACP + HS \cdot E_{cond.} + *CO_2$$

Note that the carbon dioxide released corresponds to the carbon atom of malonyl-CoA originally fixed in the earlier carboxylation reaction. This explains why no ^{14}C from HCO_3^- is incorporated into long chain fatty acids, even though bicarbonate is an obligatory cofactor in their formation. Furthermore, this decarboxylation renders the reaction strongly exergonic, so that the equilibrium lies well in the direction of synthesis.

The condensing enzyme can elongate all the other intermediates leading to palmitate biosynthesis. These include butyryl-ACP, hexanoyl-ACP and octanoyl-ACP, the products of the first, second and third turns of the cycle, respectively. The rate of the reaction increases with increasing chain length so that these intermediates do not tend to accumulate.

The three reactions which lead to the formation of butyryl-CoA are illustrated below:

$$CH_3 \cdot \overset{O}{\overset{\|}{C}} \cdot CH_2 \cdot COS \cdot ACP + NADPH + H^+ \underset{\text{dehydrogenase}}{\overset{\beta\text{-Ketoacyl}}{\rightleftharpoons}} CH_3 \cdot \overset{OH}{\underset{H}{\overset{|}{C}}} \cdot CH_2 \cdot COS \cdot ACP + NADP^+$$

Acetoacetyl-S-ACP

β-Hydroxybutyryl-S-ACP

$$CH_3 \cdot \overset{OH}{\underset{H}{\overset{|}{C}}} \cdot CH_2 \cdot COS \cdot ACP \underset{\text{dehydrase}}{\overset{\beta\text{-Hydroxyacyl}}{\rightleftharpoons}} CH_3 \cdot CH=CH \cdot COS \cdot ACP + H_2O$$

Crotonyl-S-ACP

$$CH_3 \cdot CH=CH \cdot COS \cdot ACP + NADPH + H^+ \underset{\text{reductase}}{\overset{\text{Enoyl}}{\rightleftharpoons}} CH_3 \cdot CH_2 \cdot CH_2 \cdot COS \cdot ACP + NADP^+$$

Butyryl-S-ACP

The formation of butyryl-S-ACP completes the first of 7 cycles in palmitate biosynthesis. Another malonyl residue can be accepted, and the chain lengthened by repetition of the cycle.

β-ketoacyl reductase, β-hydroxyacyl dehydrase and enoyl reductase have not been purified or characterized in plants, though crude preparations do contain these activities. In bacteria the second of these, the dehydrase, consists of at least three enzymes which have overlapping specificities for different chain lengths. Furthermore, two enoyl reductases are known, one of which utilizes NADPH, the other NADH. The plant system may be similarly complex, but this has been only poorly characterized.

Desaturation, elongation and hydroxylation. It is now clear that the initial product of fatty acid synthesis in plants is palmitoyl-S-ACP. Further two carbon elongation to stearoyl-S-ACP appears to involve an analogous, but distinct enzyme system with different thermal stability, and inhibitor sensitivities (see Packter and Stumpf, 1975). However, malonyl-CoA is again the donor of the C_2 unit and NADPH the sole reductant. The ACP-esters of palmitic and stearic acids serve then as the major precursors of the unsaturated fatty acids.

Biosynthesis of oleic acid. In animal tissues, a membrane bound desaturase converts stearoyl-CoA to oleyl-CoA. It forms part of the 'mixed function' oxygenase system of the cell. The reaction cataylzed is shown below:

$$\text{Stearoyl-CoA} + NADPH + H^+ + O_2 \longrightarrow \text{oleyl-CoA} + NADP^+ + 2H_2O$$

However, recent evidence in plants suggest that desaturation occurs at the level of stearoyl-S-ACP in a reaction catalysed by a soluble cytosolic enzyme (Jaworski and Stumpf, 1974). NADPH can again act as the reductant, although in chloroplasts it may be replaced by ferredoxin, a non-heme iron protein (see p. 346), involved in NADPH formation during photosynthesis.

Biosynthesis of linoleic and linolenic acids. A second and third double bond may be inserted into oleic acid to yield linoleic and linolenic acids respectively (Fig. 6.26). The reactions are catalyzed by enzymes in microsomal particles derived from the endoplasmic reticulum, and appear to be of the 'mixed function' oxygenase type (see previous section). The starting substrate is oleyl-CoA and not in this case oleyl-S-ACP.

Fig. 6.26   The formation of polyenoic acids from derivatives of palmitic and stearic acids. These derivatives are probably either thioesters of ACP or coenzyme A.

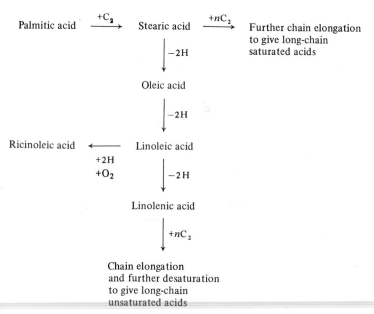

However, an alternative pathway for linolenic acid synthesis may exist (see Stumpf, 1977, in Further Reading). In this, desaturation occurs at the level of lauric acid (C_{12}), while subsequent elongation gives rise to linolenic acid.

$$C\ 12 \longrightarrow 12{:}1 \longrightarrow 12{:}2 \longrightarrow 12{:}3 \longrightarrow$$
$$14{:}3 \longrightarrow 16{:}3 \longrightarrow 18{:}3.$$

All other polyenoic acids are probably formed from monenoic acids by desaturation reactions of the kinds shown in Fig. 6.26 and above.

The saturated fatty acids formed in the cytoplasm may also be elongated. This process is again incompletely understood and is probably carried out in the ill-defined microsomal fraction as well as in a number of individual organelles by C-2 additions from either malonyl- or acetyl-CoA.

Hydroxylation may occasionally be an important process. Ricinoleic acid, for example, constitutes up to 90% of the fatty acids in mature castor bean (see Table 2.6). It is formed by the hydroxylation of oleyl-CoA to ricinoleyl-CoA by a microsomal hydroxylase system requiring NADH and molecular oxygen (Fig. 6.26).

Saturated fatty acids having odd numbers of carbon are known, but are relatively rare. They usually arise from biosynthesis starting with proprionyl-CoA.

Provision of acetyl-CoA. Free acetate, provided in labelling experiments, is converted directly to acetyl-CoA in the presence of ATP, CoA and a specific cytoplasmic synthetase:

$$\text{Acetate + ATP + CoA} \xrightarrow{\text{Mg}^{++}} \text{acetyl-CoA + AMP + pyrophosphate}$$

Fatty acids in seeds, however, are synthesized largely at the expense of sugars translocated from the leaves and the precursor acetyl-CoA is probably formed in the mitochondria by the oxidative decarboxylation of pyruvate, which is itself a product of glycolysis. In plants, acetyl-CoA can probably pass out of the mitochondrion and become available for fatty acid synthesis in the cytoplasm. This is of interest because mammaliam mitochondria are impermeable to acetyl-CoA. In this case, acetate is formed cytoplasmically from citrate in the following reaction:

$$\text{Citrate + ATP + CoA} \longrightarrow \text{acetyl-CoA + oxaloacetate + ADP + P}_i$$

The oxaloacetate is then reduced to malate, passes back into the mitochondrion, and is reconverted back to citrate by the TCA cycle. Such a situation has not been reported for plants.

Formation of glycerides

In the final stages of triglyceride synthesis, the fatty acids become esterified with glycerol. α-Glycerophosphate, the precursor of the glycerol, is formed by reduction of dihydroxyacetone phosphate:

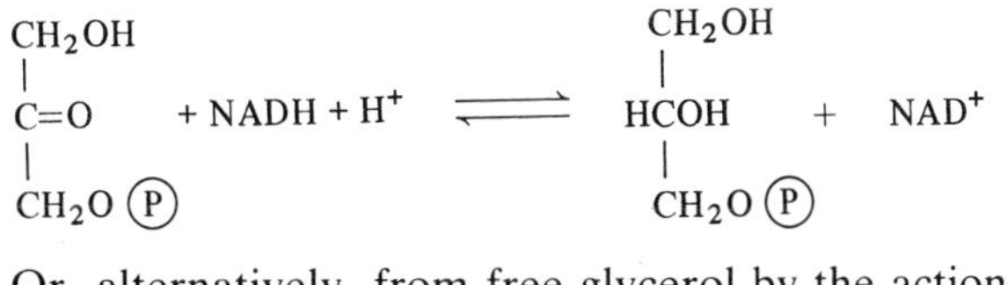

$$
\begin{array}{l}
\text{CH}_2\text{OH} \\
| \\
\text{C=O} \quad + \text{NADH} + \text{H}^+ \\
| \\
\text{CH}_2\text{O} \ \textcircled{P}
\end{array}
\rightleftharpoons
\begin{array}{l}
\text{CH}_2\text{OH} \\
| \\
\text{HCOH} \quad + \quad \text{NAD}^+ \\
| \\
\text{CH}_2\text{O} \ \textcircled{P}
\end{array}
$$

Or, alternatively, from free glycerol by the action of glycerol kinase. The next stage in triacylglycerol formation is the acylation of the free hydroxyl groups of the α-glycerophosphate by two molecules of fatty acyl-CoA to yield L-phosphatidic acid.

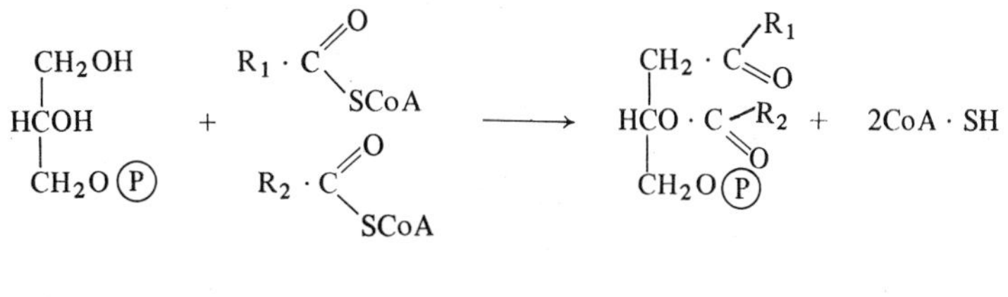

$$
\begin{array}{l}
\text{CH}_2\text{OH} \\
| \\
\text{HCOH} \\
| \\
\text{CH}_2\text{O} \ \textcircled{P}
\end{array}
+
\begin{array}{l}
\text{R}_1 \cdot \text{C} \overset{\text{O}}{\underset{\text{SCoA}}{}} \\
\\
\text{R}_2 \cdot \text{C} \overset{\text{O}}{\underset{\text{SCoA}}{}}
\end{array}
\longrightarrow
\begin{array}{l}
\text{CH}_2 \cdot \text{C} \overset{\text{R}_1}{\underset{\text{O}}{}} \\
| \\
\text{HCO} \cdot \text{C} \overset{\text{R}_2}{\underset{\text{O}}{}} \quad + \quad 2\text{CoA} \cdot \text{SH} \\
| \\
\text{CH}_2\text{O} \ \textcircled{P}
\end{array}
$$

It seems that a saturated acid tends to be transferred preferentially to the 1′ position of the glycerol phosphate, while unsaturated acids come to reside at the 2′ position. The phosphatidic acid may be hydrolysed to a diacyl-glycerol which then reacts with a further molecule of fatty acyl-CoA to yield triacylglycerol, or, alternatively, converted to any of a number of phosphoglycerides.

<table>
<tr><td>

β-Carboxylation and pyruvate metabolism

</td><td>

Pyruvate is a substrate which is common to both EMP pathway and tricarboxylic acid cycle. It can be reduced to lactate and decarboxylated to acetaldehyde. It plays a central rôle in amino acid metabolism as a direct precursor of alanine, isoleucine and valine, and, during protein catabolism, is a product of the breakdown of alanine, cysteine, glycine and serine and a number of other amino acids. Along with phosphoenolpyruvate, it is also involved in β-carboxylation reactions which are of importance in forming links between major biochemical pathways. In this section we shall consider the relative importance of three such carboxylations to the metabolism of plants.

</td></tr>
</table>

The process whereby a new carboxyl group is inserted on a position β to (or once removed from) a pre-existing carbonyl group is known as *β-carboxylation*. In plants, there are a number of such reactions in which carbon dioxide is condensed with either pyruvic acid or phosphoenolpyruvate; a dicarboxylic acid is synthesized in each one. A further β-carboxylation reaction with which we shall not be concerned here, although it occurs in plants, is the carboxylation of α-ketoglutarate to yield isocitrate, catalysed by the isocitric enzyme,

$$\alpha\text{-Ketoglutarate} + CO_2 + NADPH + H^+ \rightleftharpoons \text{isocitrate and } NADP^+$$

This enzyme is NADP specific and is distinct from the mitochondrial NAD requiring isocitric dehydrogenase. Its rôle and possible importance in plant metabolism has not been assessed.

Malic enzyme. The first β-carboxylase which will be discussed is the malic enzyme which catalyses the reversible, reductive carboxylation of pyruvate yielding L-malate:

$$CH_3-\overset{\overset{O}{\|}}{C}-COO^- + H\overset{*}{C}O_3^- + NADPH + H^+ \overset{Mn^{++}}{\rightleftharpoons}$$
$$^-OO\,\overset{*}{C}-CH_2-CHOH-COO^- + NADP^+ + H_2O$$

The asterisk indicates the position of the radioactive carbon atom after provision of $^{14}CO_2$.

The equilibrium position favours oxidative decarboxylation rather than carbon dioxide fixation under low carbon dioxide tensions, and can be pushed in the forward direction only by high concentrations of carbon dioxide. Such concentrations are usually considered to be higher than those normally expected in the living organism, although in deeper lying tissues malate formation may occur by this means. The enzyme is widely distributed and has been isolated from both photosynthetic and non-green tissues. Although it is usually associated with the soluble phase of the cytoplasm, it has been reported to be present in the mitochondria and chloroplasts of some plant species (see p. 370).

Phosphoenolpyruvate carboxylase. This enzyme catalyses the carboxylation of phosphoenolpyruvate to yield oxaloacetate and inorganic phosphate:

$$CH_2=\overset{|}{\underset{O\,\textcircled{P}}{C}}-COO^- + H\overset{*}{C}O_3^- + H_2O \underset{\text{or Mn}^{++}}{\overset{\text{Mg}^{++}}{\rightleftharpoons}} \,^-O\overset{*}{O}C-CH_2-\overset{\overset{O}{\|}}{C}-COO^- + P_i$$

The oxaloacetate formed is very rapidly converted to malate because of the almost universal presence of malic dehydrogenase:

Oxaloacetate + NADH + H$^+$ $\rightleftharpoons$ L-malate + NAD$^+$

The carboxylation is accompanied by a large decrease in free energy and can be regarded as practically irreversible. Moreover, the affinity of the enzyme for carbon dioxide is high, so that it is a highly effective means of fixing carbon dioxide into dicarboxylic acids. Indeed, this enzyme is now regarded as being of prime importance in the photosynthetic fixation of carbon dioxide by certain plants (see Ch. 8).

The enzyme has been shown to be widely distributed in non-green as well as photosynthetic tissues. It is particularly conspicuous in the leaves of Crassulacean plants, such as species of *Bryophyllum, Kalanchoë and Sedum,* and in those plants such as corn and sugar cane which possess the C_4 dicarboxylic acid pathway of photosynthetic carbon dioxide fixation (see p. 364). Although in such photosynthetic plants it is localized in the chloroplasts, it has been found in both cytoplasmic and particulate preparations from non-green tissues.

Phosphoenolpyruvate carboxykinase. This enzyme also catalyses the reversible carboxylation of phosphoenolpyruvate:

$$CH_2=\overset{|}{\underset{O\,\textcircled{P}}{C}}-COO^- + H\overset{*}{C}O_3^- + ADP \overset{\text{Mn}^{++}}{\rightleftharpoons} \,^-O\overset{*}{O}C-CH_2-\overset{\overset{O}{\|}}{C}-COO^- + H_2O + ATP$$

The corresponding enzyme from animal tissues requires GDP or IDP. The free energy change involved is small and so the reaction is freely reversible. Therefore, at low concentrations of carbon dioxide, the enzyme is likely to function as a *decarboxylase.*

Rôles of pyruvate carboxylation

Entry into the tricarboxylic acid cycle

The immediate precursor of pyruvate in the EMP sequence is phosphoenolpyruvate. This yields a molecule each of ATP and pyruvate in presence of ADP and pyruvate kinase. Once produced, pyruvate can undergo a number of reactions including transamination, decarboxylation and reduction. Under normal aerobic conditions, however, most of it is oxidized and fed into the TCA cycle as acetyl-CoA. The further utilization of the acetyl-CoA by the mitochondrion, however, requires catalytic amounts of either oxaloacetate or some other TCA

273

intermediate that is a precursor of oxaloacetate to be present (see p. 294). This is because one molecule of oxaloacetate is required to condense with each molecule of acetyl-CoA to form citrate. In the absence of the oxaloacetate, all available CoA will soon accumulate as acetyl-CoA and pyruvate oxidation and respiratory metabolism will stop.

During normal operation, one molecule of oxaloacetate is generated in the TCA cycle for each pyruvate entering. The TCA cycle, however, is an *amphibolic* pathway, i.e. it functions not only in catabolism to oxidize pyruvate, but also to provide precursors such as amino acids and other compounds for anabolic pathways. Therefore if the cycle is being tapped at any point, such as α-ketoglutarate being aminated to provide glutamate, there must be some other way for providing oxaloacetate if respiration is to continue. We now know that there are special mechanisms by which TCA intermediates can be replenished to allow respiration to continue. These are called *anaplerotic* (or 'filling up') reactions. In plants, the carboxylation of phosphoenolpyruvate by the enzyme phosphoenolpyruvate carboxylase fulfils such a rôle. When the TCA cycle is deficient in oxaloacetate or its dicarboxylic acid precursors, phosphoenolpyruvate may be carboxylated to oxaloacetate.

Entry of TCA cycle intermediates into the EMP pathway

During fat breakdown, acetyl-CoA produced by β-oxidation of fatty acids is converted to malate via the glyoxylate cycle (Ch. 9). Similarly, during catabolism of protein, a number of amino acids give rise to dicarboxylic acids. If these products are to be converted to carbohydrate, there must be some route from the TCA cycle into the EMP sequence. This exit seems to be by way of the phosphoenolpyruvate carboxykinase reaction. In contrast to the reaction catalysed by the malic enzyme, this decarboxylation bypasses the energetically unfavourable pyruvate to phosphoenolpyruvate step. Once formed, the phosphoenolpyruvate can be readily converted to hexose phosphate (see p. 252). For this reason, therefore, it is unlikely that the malic enzyme, which gives rise to pyruvate, plays any rôle in gluconeogenesis.

Crassulacean acid metabolism

Certain succulent plants, particularly some members of the Crassulaceae and Cactaceae, show striking diurnal fluctuations in their content of organic acids, particularly malate (Fig. 6.27). Acid accumulates during the night within the vacuole where it can reach concentrations as high as 0·2 M, while during the subsequent light period there is a massive efflux of acid from the vacuole back into the cytoplasm. There, the malate is decarboxylated and the CO_2 refixed in photosynthesis within the chloroplast. This process of cyclical acidification and deacidification is generally referred to as *Crassulacean acid metabolism* or CAM.

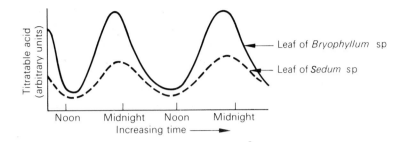

Fig. 6.27 Diurnal fluctuation in acid levels in leaves of Crassulacean plants. Most of the acid is accounted for as malate.

The reactions which are responsible for dark accumulation of malate in CAM plants are the β-carboxylation of phosphoenolpyruvate (PEP), catalysed by PEP carboxylase, and the subsequent reduction of oxaloacetate to malate by malate dehydrogenase. Thus, when $^{14}CO_2$ is supplied for short periods to CAM plants, the C-4 carboxyl group of malate becomes labeled (Sutton and Osmond, 1972). The initial reports of labeling in both C-1 and C-4 were probably the result of partial randomization of the ^{14}C between the C-1 and C-4 positions by the enzyme fumarase, and not, as once believed, to the participation of two separate carboxylation reactions:

$$
\begin{array}{ccccc}
\text{COOH} & \text{*CO}_2 \;\; \text{COOH} & \text{COOH} & \text{COOH} & \text{COOH} \\
| & | & | & | & | \\
\text{C}-\text{O}-\text{\textcircled{P}} \longrightarrow & \text{C}=\text{O} \;\rightleftharpoons\; & \text{CHOH} \;\underset{+\text{H}_2\text{O}}{\overset{-\text{H}_2\text{O}}{\rightleftharpoons}}\; & \text{HC} \;\underset{-\text{H}_2\text{O}}{\overset{+\text{H}_2\text{O}}{\rightleftharpoons}}\; & \text{CH}_2 \\
\| & | & | & \| & | \\
\text{CH}_2 & \text{CH}_2 & \text{CH}_2 & \text{CH} & \text{CHOH} \\
& | & | & | & | \\
& \text{*COOH} & \text{*COOH} & \text{*COOH} & \text{*COOH} \\
\text{Phosphoenol} & \text{Oxaloacetate} & \text{L-Malate} & \text{Fumarate} & \text{L-Malate} \\
\text{pyruvate} & & & &
\end{array}
$$

It is unclear what controls the sudden efflux of malic acid from the vacuole during the light, although it has been proposed that a critical vacuolar concentration is reached which triggers a change in the permeability of the tonoplast. Similarly a mechanism must exist to pump malate from the cytoplasm into the vacuole as soon as it formed during the night.

Until recently, it was thought that the only carboxylase involved in deacidification of malic acid was the NADP-specific malic enzyme. However, it now seems probable that CAM plants can be divided into two groups according to whether decarboxylation occurs by means of the malic enzyme or phosphoenolpyruvate carboxykinase. For example, while the Cactaceae and Crassulaceae make use of the former enzyme, several other CAM species in other families employ the carboxykinase.

The CO_2, once released, is immediately fixed photosynthetically by the ribulose bisphosphate carboxylase in the chloroplasts (p. 361) and channeled into the Calvin cycle of interconversions (Osmond and Allaway, 1974). Thus, CAM is, in many respects, like C_4 photosynthesis (see Ch. 8) in the sense that the same series of carboxylation-decarboxylation reactions are employed. However, it differs in that there is a gradual but massive build-up of malate during the night, whereas C_4 photosynthesis involves only the transient trapping of CO_2 in malate during the light hours. In addition, whereas the phosphoenolpyruvate acceptor in C_4 photosynthesis is generated by the direct phosphorylation of pyruvate in a light-dependent process, in CAM it appears to be formed from stored carbohydrate by way of the glycolytic sequence:

$$\text{Stored carbohydrate} \longrightarrow \text{D-glucose-6-phosphate} \longrightarrow 2\text{PEP} \xrightarrow{2CO_2} 2 \text{ malate}$$

What then is the function of CAM? The process appears to be confined largely to succulent plants which are characterized by a low surface-to-volume ratio, a very high cuticular resistance to water flow and a low stomatal frequency. In short, it occurs in plants which have a highly effective means of restricting water loss in a dry environment. Thus, the functional significance of CAM relates to water economy and the need to restrict gas exchange, a process which in these plants is virtually eliminated during the heat of the day. However, photosynthesis can still proceed at the expense of CO_2 fixed in malate during the preceding night. Under extreme stress, gas exchange may be almost completely eliminated, even during darkness, and the diurnal rhythm in malic acid reflects only the continuous refixation of respiratory CO_2. Of course, under these conditions, though respiratory carbon is conserved, no growth can occur. That CAM is indeed a metabolic response to drought conditions is illustrated by the succulent species *Mesembryanthemum crystalinum*, a plant which normally does not exhibit CAM. However, within two weeks of being subjected to drought conditions it begins to show substantial dark CO_2 fixation and the typical diurnal accumulation and loss of malate (see Winter, 1974). Presumably this change has been made possible by the induction of the appropriate enzymes. The process is reversed once the stress conditions are removed and no longer can night acidification be observed.

Other rôles

β-Carboxylation of phosphoenolpyruvate by the PEP carboxylase enzyme and the decarboxylation of malate by the malic enzyme play major rôles in the photosynthetic assimilation of carbon dioxide in a number of tropical plants including corn and sugar cane. This aspect of β-carboxylation is dealt with in Chapter 8. The carboxylation of phosphoenolpyruvate to a dicarboxylic acid may also play a part in maintaining ionic balance during cation uptake by plant tissues. (Cram, 1975).

Concluding remarks

β-Carboxylation reactions are of considerable importance in governing entry into and exit out of the TCA cycle. The carboxylations provide a link between the EMP sequence and the TCA cycle at points other than the oxidation of pyruvate to acetyl-CoA (Fig. 6.28). Intermediates withdrawn from the cycle are replenished by the oxaloacetate which is formed by carboxylation of PEP. It is interesting that in animals in which PEP carboxylase is absent, this function is fulfilled by the biotinyl-enzyme pyruvate carboxylase which is not found in plants. This enzyme catalyses the following reaction:

$$Pyruvate + CO_2 + ATP \longrightarrow oxaloacetate + ADP + P_i$$

Exit out of the cycle in both plants and animals is governed by the enzyme phosphoenolpyruvate carboxykinase. The malic enzyme probably plays its major rôle in the mobilization of malic acid (Ting and Dugger, 1967), a compound often stored in relatively large amounts in plant tissues. Malate is decarboxylated to pyruvate which is then available for either oxidation or as a precursor of amino acids and other cell materials. NADPH is also generated and the carbon dioxide may in some instances be reassimilated during photosynthesis.

Fig. 6.28 Relationship between the the TCA cycle, glycolysis and β-carboxylation reactions. 1, phosphoenolpyruvate carboxykinase; 2, malic enzyme; 3, phosphoenolpyruvate carboxylase.

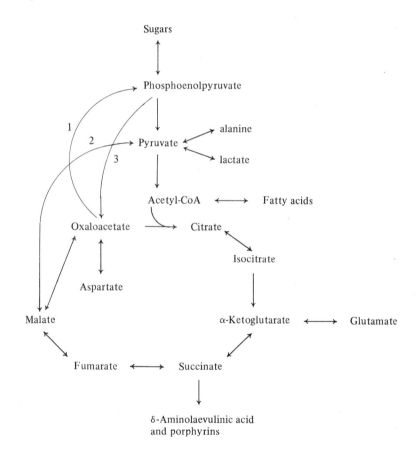

Further reading

The soluble phase of the cell

Anderson, W. G. and Green, J. G. (1967) The soluble phase of the cell. In *Enzyme Cytology*, Ed. D. B. Roodyn. Academic Press, London.
Clarke, M. and Spudich, J. A. (1977) Non-muscle contractile proteins: the role of actin and myosin in cell motility and shape determination. *Ann. Rev. Biochem.*, **46**, 797.

Filner, Ph and Yadav, N. S. (1979) Role of microtubules in intracellular movements. In *Physiology of Movements*. Eds W. Haupt and M. E. Feinleib. Encyclo. Plant Physiology. Vol. 7, p. 95. Springer-Verlag, Berlin.

Helper, P. K. and Palevitz, B. A. (1974) Microtubules and microfilaments. *Ann. Rev. Plant Physiol.*, **25**, 309.

Korn, E. D. (1978) Biochemistry of actomyosin-dependent cell motility. *Proc. natl. Acad. Sci. U.S.A.*, **75**, 588.

Lazarides, E. and Revel, J. P. (1979) The molecular basis of cell movement. *Sci. Amer.*, May 1979, p. 100.

Newcomb, E. H. (1969) Plant microtubules. *Ann. Rev. Plant Physiol.*, **20**, 253.

Parthasarathy, M. V., and Pesacreta, T. C. (1980) Microfilaments in plant vascular cells. *Can. J. Bot.*, **58**, 807.

Snyder, J. A. and McIntosh, J. R. (1976) Biochemistry and physiology of microtubules. *Ann. Rev. Biochem.*, **45**, 703.

Williamson. R. E. (1980) Actin in motile and other processes in plant cells. *Can. J. Bot.*, **58**, 766.

Glycolysis and the pentose phosphate pathway

Ap Rees, T. (1977) Conservation of carbohydrate by the non-photosynthetic cells of higher plants. In: *Intergration of Activity in the Higher Plant*. Symp. Soc. Exp. Biology, Vol. 31. Cambridge University Press, Cambridge, p. 1.

Ap Rees, T. (1980) Integration of pathways of synthesis and degradation of hexose phosphates. In *The Biochemistry of Plants. Vol. 3. Carbohydrates: Structure and Function*. Ed. J. Preiss. Academic Press, New York.

Beevers, H. (1961) *Respiratory Metabolism in Plants*. Row, Peterson, Evanston, Illinois.

Davies, D. D. (1979) The central role of phosphoenolpyruvate in plant metabolism. *Ann. Rev. Plant Physiol.*, **39**, 131.

Gibbs, M. (1966) Carbohydrate metabolism and nutrition. In *Plant Physiology*. Vol. IVB. Ed. F. C. Steward, Academic Press, New York.

Horecker, B. L. (1968) Pentose phosphate pathway, uronic acid pathway, interconversion of sugars. In *Carbohydrate Metabolism and Its Disorders*. Vol. 1. Eds. F. Dickens, P. J. Randle and W. J. Whelan. Academic Press, London.

Scrutton, M. C. and Utter, M. F. (1968) The regulation of glycolysis and gluconeogenisis in animal tissues. *Ann. Rev. Biochem.*, **37**, 249.

Turner, J. F. and Turner, D. H. (1975) The regulation of carbohydrate metabolism. *Ann. Rev. Plant Physiol.*, **26**, 159.

Fatty acid biosynthesis

Bloch, K. and Vance, D. (1977) Control mechanisms in the synthesis of fatty acids. *Ann. Rev. Biochem.*, **46**, 263.

Butt, V. S. and Beevers, H. (1965) The plant lipids. In *Plant Physiology*, Vol. IV B. Ed. F. C. Steward. Academic Press, New York.

Stumpf, P. K. (1970) Fatty acid metabolism in plant tissues. In *Lipid Metabolism*. Ed. S. J. Wakil, Academic Press, New York.

Stumpf, P. K. (1977) Lipid biosynthesis in higher plants. In *International Reviews of Biochemistry*. Ed. Goodwin, T. W. Vol. 14, University Park Press, Baltimore, MD, p. 215.

β-Carboxylation

Kluge, M. (1977) Regulation of carbon dioxide fixation in plants. In *Integration of Activity in the Higher Plant*. Symp. Soc. Exp. Biology, Vol. 31, Cambridge University Press, Cambridge, p. 155.

Osmond, C. B. (1978) Crassulacean acid metabolism: A curiosity in context. *Ann. Rev. Plant. Physiol.*, **29**, 379.

Literature cited

Adams, P. B. and Rowan, K. S. (1970) Glycolytic control of respiration during aging of carrot tissue. *Plant Physiol.*, **45**, 490.

Atkinson, D. E. (1966) Regulation of enzyme activity. *Ann. Rev. Biochem.*, **35**, 85.

Atkinson, D. E. (1969) Gluconeogenesis. *Ann. Rev. Microbiol.*, **23**, 47.

Axelrod, B. and Beevers, H. (1956) Mechanisms of carbohydrate breakdown in plants. *Ann. Rev. Plant Physiol.*, **7**, 267.

Beevers, H. and Gibbs, M (1954) Position of ^{14}C in alcohol and carbon dioxide formed from labelled glucose by corn root tips. *Plant Physiol*, **29**, 318.

Bloom, B. and Stettin, D. (1953) Pathways of glucose catabolism. *J. Amer. Chem. Soc.*, **75**, 5446.

Butt, V. S. and Beevers, H. (1961) The regulation of pathways of glucose catabolism in maize roots. *Biochem. J.*, **80**, 21.

Cram, W. J. (1975) Storage tissues. In *Ion Transport in Plant Cells and Tissues*. Ed. D. A. Baker and J. L. Hall. Elsevier/North Holland, Amsterdam.

Dennis, D. J. and Coultate, T. P. (1967) The regulatory properties of plant phospho-fructokinase during leaf development. *Biochim, Biophys. Acta*, **146**, 129.

Gibbs, M. and Beevers, H. (1955) Glucose dissimilation in the higher plant. Effect of age of tissue. *Plant Physiol.*, **30**, 343.

Ginsburg, A. and Stadtman, E. R. (1970) Multienzyme systems. *Ann. Rev. Biochem.*, **39**, 429.

Givan, C. V. (1968) Short-term changes in hexose phosphates and ATP in intact cells of *Acer pseudoplatanus* L, subject to anoxia. *Plant Physiol.*, **43**, 948.

Gunning, B. E. S., Hardham, A. R. and Hughes, J. E. (1978) Evidence for initiation of microtubules in discrete regions of the cell cortex in *Azolla* root-tip cells, and an hypothesis on the development of cortical arrays of microtubules. *Planta*, **143**, 161.

Heggeness, M. H., Wang, K. and Singer, S. J. (1977) Intracellular distributions of mechanochemical proteins in cultured fibroblasts. *Proc. natl. Acad. Sci., U.S.A.*, **74**, 3833.

Ishikawa, H., Bischoff, R. and Holtzer, H. (1969) Formation of arrowhead complexes with heavy meromyosin in a variety of cell types. *J. Cell Biol.*, **43**, 312.

James, W. O. and Ritchie, A. F. (1955) The anaerobic respiration of carrot tissue. *Proc. Roy. Soc. B.*, **143**, 302.

Jaworski, J. G. and Stumpf, P. K. (1974) Properties of a soluble-steryl-acyl carrier protein desaturase from maturing *Carthamus tinctorius*. *Arch. Biochem. Biophys.*, **162**, 158.

Kelly. G. J. and Turner, J. F. (1968) Inhibition of pea-seed phosphofructokinase by phosphoenolpyruvate. *Biochem. Biophys. Res. Commun.*, **30**, 195.

Kobr, M. J. and Beevers, H. (1971) Gluconeogenesis in the castor bean endosperm. Change in glycolytic intermediates. *Plant Physiol.*, **47**, 48.

Ledbetter, M. C. (1965) Fine structure of the cytoplasm in relation to the plant cell wall. *J. Agric. Fd. Chem.*, **13**, 405.

Ledbetter, M. C. and Porter, K. R. J. (1963) A 'microtubule' in plant cell fine structure. *J. Cell Biol.*, **19**, 239.

Lloyd, C. W., Slabas, A. R., Powell, A. J., and Lowe, S. B. (1980) Microtubules, protoplasts and plant cell shape. An immunofluorescent study. *Planta*, **147**, 500.

Lloyd, C. W., Slabas, A. R., Powell, A. J., MacDonald, G. and Badley, R. A. (1979) *Nature, Lond.*, **279**, 239.

Margulis, L., To, L. and Chase, D. (1978) Microtubules in prokaryotes. *Science, N. Y.*, **200**, 1118.

Mollenhauer, H. H. and Morré, D. J. (1976) Cytochalasin B, but not colchicine, inhibits migration of secretory vesicles in root tips of maize. *Protoplasma*, **87**, 39.

Newcomb, E. H. and Stumpf, P. K. (1953) Fat metabolism in higher plants. Biogenesis of higher fatty acids by slices of peanut cotyledons in vitro. *J. Biol., Chem.*, **200**, 233.

Oppenheim, A. and Castelfranco, P. A. (1967) An acetaldehyde dehydrogenase from germinating seeds. *Plant Physiol.*, **42**, 125.

Osmond, C. B. and Allaway, W. G. (1974) Pathway of CO_2 fixation in the CAM plant *Kalanchoe daigremontiana*. I. Pattern of $^{14}CO_2$ fixation in the light. *Aust. J. Plant Physiol.*, **1**, 503.

Packter, N. M. and Stumpf, P. K. (1975) The effect of cerulenin on the synthesis of medium- and long-chain acids in leaf tissue. *Arch. Biochem. Biophys.*, **167**, 655.

Plowe, J. Q. (1931) Membranes in the plant cell. Morphological membranes at protoplasmic surfaces. *Protoplasma*, **12**, 196.

Scherline, P. and Schiavone, K. (1977) Immunofluorescence localization of proteins of high molecular weight along intracellular microtubules. *Science, N. Y.* **198**, 1038.

Shulman, M. D. and Gibbs, M. (1968) D-Glyceraldehyde 3-phosphate dehydrogenase of higher plants. *Plant Physiol.*, **43**, 1805.

Smith, F.A. and Raven, J. A. (1979) Intracellular pH and its regulation. *Ann. Rev. Plant Physiol.*, **30**, 289.

Snyder, A. A. and McIntosh, J. R. (1976) Biochemistry and physiology of microtubules. *Ann. Rev. Biochem.*, **45**, 702.

Stewart, C. R. and Beevers, H. (1967) Gluconeogenesis from amino acids in germinating castor bean endosperm and its role in transport to the embryo. *Plant Physiol.*, **42**, 1587.

Sutton, B. G. and Osmond, C. B. (1972) Dark fixation of CO_2 by Crassulacean plants: Evidence for a single carboxylation step. *Plant Physiol.*, **50**, 360.

Ting, I. P. and Dugger, W. M. (1967) CO_2 metabolism in corn roots. Kinetics of carboxylation and decarboxylation. *Plant Physiol.*, **42**, 712.

Wakil, S. J. (1970) Fatty acid metabolism in plant tissues. In *Lipid Metabolism*. Ed. S. J. Wakil. Academic Press, New York.

Webster, R. E., Henderson, D., Osborn, M. and Weber, K. (1978) Three-dimensional electron microscopical visualization of the cytoskeleton of animal cells: Immunoferritin identification of actin- and tubulin-containing structures. *Proc. natl. Acad. Sci. U.S.A.*, **75**, 5511.

Winter, K. (1974) Evidence for the significance of Crassulacean acid metabolism as an adaptive mechanism to water stress. *Plant Sci. Lett.*, **34**, 279.

Wolosewick, J. J. and Porter, K. R. (1979) Microtrabecular lattice of the cytoplasmic ground substance. *J. Cell Biol.*, **82**, 114.

7 The mitochondrion

Mitochondria were first described in animal cells in the late nineteenth century and in plants during the early twentieth century. They are rather variable, double-membrane bound particles which have been shown to be sites of gas exchange within the cell. They are the location of a sequence of reactions (known as the *tricarboxylic acid cycle* or *Krebs' cycle*) in which pyruvate, largely produced as a result of glycolysis, is oxidized to carbon dioxide and water. This oxidation of pyruvate is mediated by the production of reduced coenzymes which form the start of a catena of oxidations and reductions, terminating in the reduction of molecular oxygen to water. The catena is known as the *respiratory chain* and at specific points along this chain ATP may be produced from ADP in a process termed *oxidative phosphorylation*. As we shall see later, oxidative phosphorylation bears many similarities to the photosynthetic phosphorylation of the chloroplasts, although in the latter there is no concomitant uptake of oxygen. Both require the transfer of electrons or protons from one carrier to another for ATP to be produced, and contrast with phosphorylation at the substrate level occurring during glycolysis (see p. 247). Mitochondria are thus sites of cellular respiration, which may be defined as the process in which organic substrates are oxidized to yield carbon dioxide and energy with the accompanying reduction of molecular oxygen to water.

Structure

Mitochondria when observed by phase contrast microscopy, appear to be in continuous movement, to change shape rapidly and also to divide. These changes make precise estimates of both size and number difficult. However, on average, mitochondria appear as cylinders about 3–5 μm long and 0·5–1·0 μm in diameter. They vary in number from about 20 to 10^5 per cell; in maize root cap the numbers per cell (100–3 000) are roughly similar to those present in rat liver (500–1 400).

Observations of fixed tissue using an electron microscope also indicate that there are considerable variations in both size and shape. Elongated forms are, however, rarely seen since there is little chance of cutting a complete longitudinal section of a cylinder with a variable plane in the long axis. In cross-section the mitochondria are commonly circular or oval and, following osmium fixation, appear to be bounded by a double membrane envelope. This is seen as two electron dense

Fig. 7.1 Thin sections of plant mitochondria from root tissue. (*a*) from *Suaeda maritima* fixed in glutaraldehyde and osmium ×50 000 (*b*) from *Zea mays* fixed in permanganate ×42 000. Note that permanganate fixation fails, in particular, to preserve the ribosome-like structures.

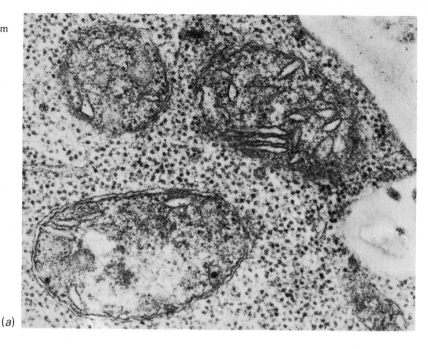

(*a*)

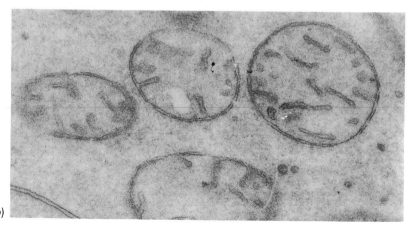

(*b*)

lines each about 5 nm thick, separated by an electron transparent layer about 8·5 nm thick (*the perimitochondrial space*). The innermost, non-membranous, component is known as the *matrix* (Fig. 7.1)

The outer membrane may have a larger average thickness (5·5 nm) than the inner membrane (4·5 nm) and in negatively stained preparations this outer membrane appears to be pitted (Parsons, Bonner and Verboon, 1965). These pits are 2·5–3·0 nm in diameter and spaced rather irregularly (with an average centre to centre spacing of 4·5 nm).

The surface area of the inner membrane is increased by invaginations or *cristae*, which may be plate-like or tubular (Fig. 7.2*a*). In general, the cristae of plant mitochondria are tubular, while those of animal mitochondria assume the more regular plate-like formation. The space within the cristae which is continuous with the perimitochondrial space is known as the *intracristal space* (Fig. 7.2*b*). There is no characteristic number of cristae in the mitochondria of any one species and their frequency tends to increase as the respiratory activity rises.

Fig. 7.2   (*a*) Three dimensional drawings of mitochondria from (1) an animal cell showing the regular plate-like cristae and from (2) a plant cell with less regular and tube-like cristae. (*b*) Mitochondrial membranes and sub-units according to Parsons (1965).

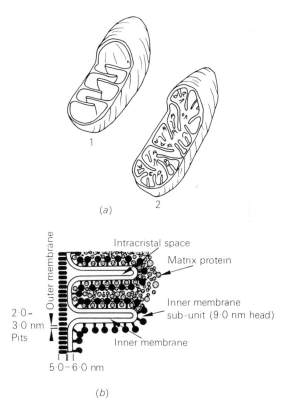

(a)

(b)

When the inner mitochondrial membrane is observed following positive staining with osmium it appears smooth (Fig. 7.1), but after negative staining with phosphotungstate, stalked particles are observed attached to the membrane (Figs. 7.3 and 7.4). These particles were first seen in mammalian mitochondria (Fernandez-Moran, 1962) and have subsequently been demonstrated in mitochondria from a range of higher plant species (Nadakavukaren, 1964; Parsons,

Fig. 7.3   Part of a negatively stained mitochondria from summer squash. The large area (IM) is presumed to be inner membrane and the cristae (C) appear rounded. The inner membrane sub-units (arrow) have a head which is 10 nm in diameter. ×95 000. Reproduced from Parsons, Bonner and Verboon (1965).

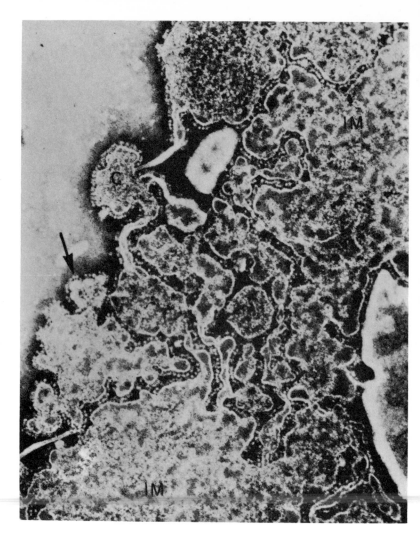

Bonner and Verboon, 1965). The particles vary only slightly in size with the species in question: in onion, the head is about 10 nm in diameter, in *Neurospora crassa*, 8·5 nm, and in beef heart mitochondria, 8–10 nm. In each case the particle is borne on a stalk which is about 4·5–5·0 nm long and 4·0 nm in diameter.

The occurrence of these inner membrane sub-units in material which has not been negatively stained was questioned by Sjöstrand, Andersson-Cedergren and Karlsson (1964), who believe them to be artefacts of lipid or protein material extruded from the membranes during the drying down of unfixed material with phosphotungstate. However, more recently, particles have been observed in mitochondria in which 95% of the lipid has been removed, in mitochondria which were fixed in glutaraldehyde prior to negative staining, and both in isolated mitochondria and in thin sections of neuron mitochondria without negative staining. Parson (1965) claims that any change in the shape of the cristae takes place during swelling of the mitochondrion (necessary for lysis) and not as a result of the negative staining itself.

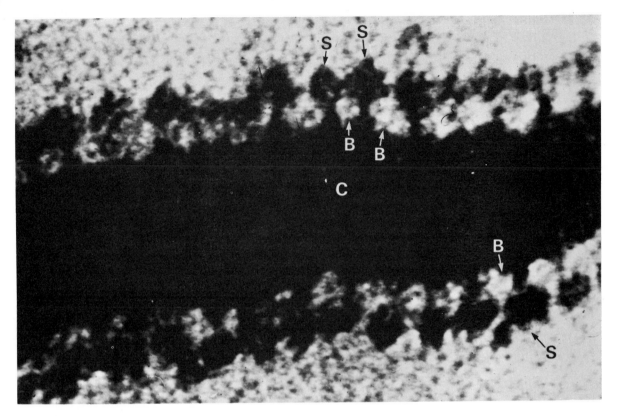

Fig. 7.4   Inner membrane sub-units of bee flight muscle at high magnification showing sub-units (S), and base plate (B): C, Inner membrane. ×1 230 000. Reproduced from Parsons (1965).

The remaining mitochondrial compartment, the matrix, is thought to consist of lipid and protein, and varies considerably in its electron density, particularly in isolated mitochondria (see p. 288). Part of the density of the matrix is accounted for by aggregates of 10 nm diameter particles, while also present are 30–40 nm diameter particles, ribosomes (25 nm in diameter) and DNA filaments (Fig. 7.5). The ribosome-like particles have been observed many times and it has been shown that they may be removed by treatment with ribonuclease. Similarly, the DNA may be removed by treatment with deoxyribonuclease. The 30–40 nm diameter particles which are also characteristic of the mitochondrial matrix may contain metal phosphates.

Isolation

Although the first isolation of a mitochondrial fraction from animal cells was carried out by Bensley in 1934, little progress in understanding the function of the organelles was made until the work of Hogeboom in the late 1940s. The complexity of mitochondria and their intermediate size makes isolation particularly problematical and it was not until 1951 (Millerd, Bonner, Axelrod and Bandurski, 1951) that the first separation of mitochondria from plant tissue was reported.

The inner mitochondrial membrane is semipermeable, so the whole organelle is presumably in osmotic equilibrium with the rest of the cell. It is therefore important to include a non-permeable solute (normally sucrose, but in many cases mannitol or sorbitol) in the isolation

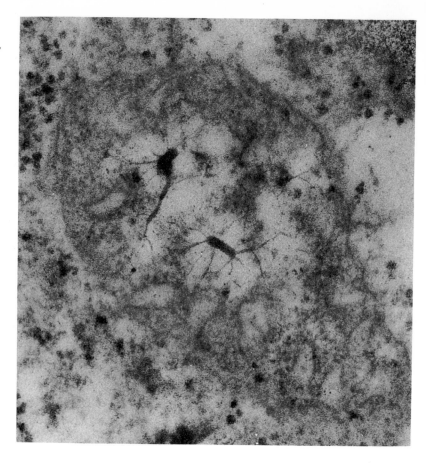

Fig. 7.5 DNA filaments in a mitochondion from *Poelicarpus*. ×108 000. Provided by Dr H. Aldrich, University of Florida.

medium to prevent their swelling and bursting on release from the cytoplasm. Furthermore, it is often necessary to include a chelating agent to remove toxic ions, and often bovine serum albumin to bind fatty acids which might uncouple oxidation from phosphorylation (see p. 295). Mitochondria are generally prepared from dark grown tissues and separated from other cytoplasmic organelles by differential centrifugation (between 600 and 10 000 g; see Fig. 1.21). The whole procedure takes between one and two hours. Since the early 1970s a number of workers have advocated the use of sucrose density gradients to purify such preparations. For example, the mitochondria from potato and mung beans come to lie at the interface between 1·2 and 1·45 M sucrose after centrifugation for 45 min at some 41 000 g on a discontinuous gradient (Douce *et al.*, 1972). However, the value of such a purification, which prolongs the time required for the isolation, depends upon the use to which the preparation is to be put. Where the removal of contaminants is important (for example, in the determination of difference spectra; see p. 298) then density gradient centrifugation may be the answer. However, in other cases, the advantages of a more rapid procedure may outweigh the disadvantages of the presence of contaminants, especially if they do not contribute to the biochemical process being assayed. Consequently, differential centrifugation can produce preparations which are not significantly different from those prepared using a density gradient (see Palmer, 1976; Storey, 1976; Day and Hanson, 1977). Recently, the use of silica-sols has allowed the separation of mitochondria even from broken chloroplasts in preparations from green leaf tissue (see Jackson and Moore, 1979). In contrast to these lengthy procedures, a number of workers in the late 1960s have adopted methods which enable a mitochondrial fraction to be isolated in about 10 min (Palmer, 1967; Sarkissian and Srivastava, 1969). The homogenate is filtered through 50 μm nylon mesh which removes larger particles and other cell debris and then the mitochondria sedimented by centrifugation at about 40 000g for 1–2 min. Starch within this pellet may be removed by turning the centrifuge tube through 90° and recentrifuging. In general, although relatively little work has been carried out on such preparations, the mitochondria do appear to be of higher quality (see p. 296 for criteria of quality) than those prepared by the more conventional procedures, and these techniques will undoubtedly play an important rôle in future mitochondrial research.

Since the 1950s, functional mitochondria have been isolated from a large number of species and variety of tissues. Active preparations have been obtained from green tissues (see Flowers, 1974), roots, storage organs, seeds and seedlings and from very many different species. The mitochondrial fractions thus prepared have rarely been checked for purity by electron microscopy, however and many preparations have presumably been contaminated by other organelles. Parsons, Bonner and Verboon (1965) reported that the chief contaminant of dark grown tissue (dark grown to avoid contamination with chloroplasts) was proplastids. However, even in relatively pure preparations, there may be considerable mitochondrial heterogeneity, possibly arising from the variety of cell types and ages used. For example, three mitochondrial types have been distinguished in 3-day-old etiolated corn shoots. They differ in the number of cristae and in the amount and density of staining of the matrix. Type I mitochondria (Fig. 7.6C) have few cristae and a lightly staining matrix. Type II have more cristae and a matrix which stains more deeply than that of Type I, while Type III mitochondria have many inflated cristae with a very densely staining matrix (Malone, Koeppe and Miller, 1974). Examination of intact tissues suggests that Type I are found in phloem sieve tubes and some adjacent parenchyma cells, while Type II are characteristic of meristematic and undifferentiated cells. Type III are found in all other vacuolated cells. As yet no physiological differences have been described for the different types. Mitochondria do also change their appearance if allowed to swell or contract (Fig. 7.6A & B, p. 318), although this is not thought to involve changes from one mitochondrial type to another.

Biochemical activity

The major biochemical processes occurring within the mitochondria are the tricarboxylic acid cycle and oxidative phosphorylation. The enzymic reactions of the former are believed to occur within the

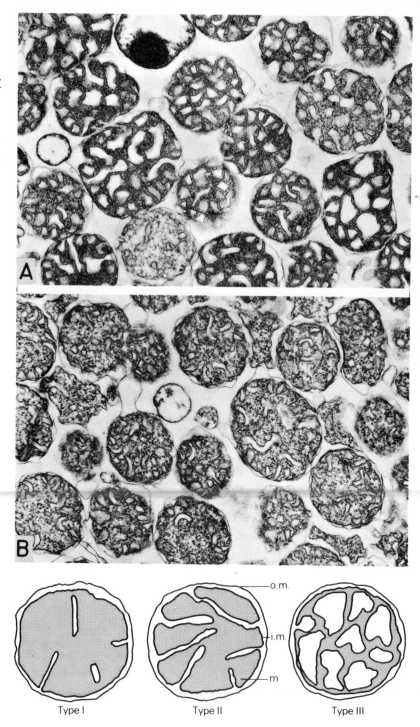

Fig. 7.6   Sweet potato mitochondria fixed with glutaraldehyde in either 0·4 M sucrose (A) or in 0·2 M sucrose (B). Reproduced from Baker *et al.* (1968). C. Three mitochondrial types found in etiolated corn shoots: m, matrix; om, outer membrane; im, inner membrane. Redrawn from Malone *et al.* (1974).

innermost compartment of the mitochondrion, the matrix, although at least one of the enzymes of the cycle is tightly bound to the inner membrane. The enzymes associated with electron transport and oxidative phosphorylation are located in the inner mitochondrial membrane.

Tricarboxylic acid cycle

The pathway by which pyruvate, produced during glycolysis, is oxidized was elucidated using animal tissues during the 1930s. On the basis of this work, Krebs in 1937 proposed the cyclic pathway, now known either as the *citric acid cycle, the tricarboxylic acid cycle* (TCA cycle),' or the *Krebs' cycle*, illustrated in Fig. 7.7.

Mitochondria themselves were first shown to have catalytic activity by Hogeboom in 1946/1947, when he demonstrated that particles isolated from rat liver were the major site of succinic dehydrogenase and of cytochrome oxidase activity. Subsequently, it was shown that, in addition, mitochondria would also catalyse the reaction of pyruvate

Fig. 7.7   The oxidation of pyruvate by the TCA cycle: (The numbers refer to the reactions numbered in the text.)
1. Pyruvate dehydrogenase. 2. Lipoate acetyl transferase. 3. Lipoamide dehydrogenase. 4. Citrate synthetase. 5. Aconitate hydratase (aconitase). 6. Isocitrate dehydrogenase. 7. α-Ketoglutarate dehydrogenase. 8. Succinyl CoA synthetase. 9. Succinate dehydrogenase. 10. Fumarate hydratase (fumarase). 11. Malate dehydrogenase.

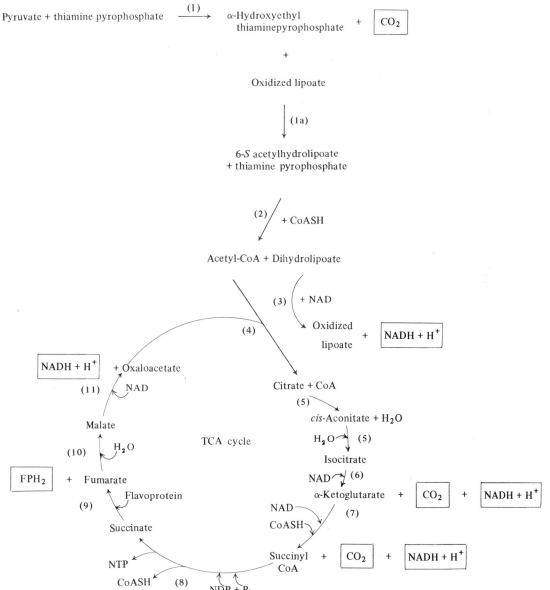

with oxaloacetate, the conversion of α-ketoglutarate to succinate and of succinate to malate. It was apparent, therefore, that animal mitochondria contained at least some of the enzymes concerned in the TCA cycle together with an oxidase system.

The TCA cycle oxidizes acetyl-CoA, transferring hydrogen atoms to NAD^+ and FAD and releasing carbon dioxide. The initial step involves the formation of acetyl CoA from pyruvate in a complex reaction involving three enzymes combined into what is known as the pyruvate dehydrogenase complex. Pyruvate is first decarboxylated in a reaction requiring thiamine pyrophosphate (TPP- see p. 121) and Mg^{++} by pyruvate decarboxylase:

$$
\begin{array}{ccccc}
CH_3 & & & CH_3 & \\
| & & & | & \\
CO & + & TPP \longrightarrow & CHOH & + \quad CO_2 \\
| & & & | & \\
COO^- & & & TPP &
\end{array} \tag{1}
$$

The equation numbers refer to the reactions numbered in Fig. 7.7.

An acetyl group is then transferred from the TPP derivative (α-hydroxyethyl thiaminepyrophosphate) to lipoic acid (which is in an amide linkage with the lysine residues of the enzyme lipoic acetyl transferase):

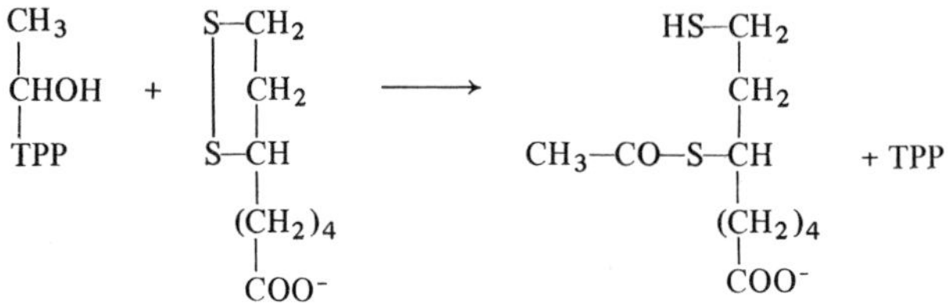

$$
\begin{array}{cccccc}
CH_3 & & S-CH_2 & & HS-CH_2 & \\
| & & | \quad | & & | & \\
CHOH & + & \;\; CH_2 & \longrightarrow & CH_2 & \\
| & & | \quad | & & | & \\
TPP & & S-CH & & CH_3-CO-S-CH & + \; TPP \\
& & | & & | & \\
& & (CH_2)_4 & & (CH_2)_4 & \\
& & | & & | & \\
& & COO^- & & COO^- &
\end{array} \tag{1a}
$$

Lipoate   6-S-Acetylhydrolipoate

Finally the acetyl group is transferred to coenzyme A, producing acetyl coenzyme A and dihydrolipoate (the enzyme is lipoate acetyl transferase):

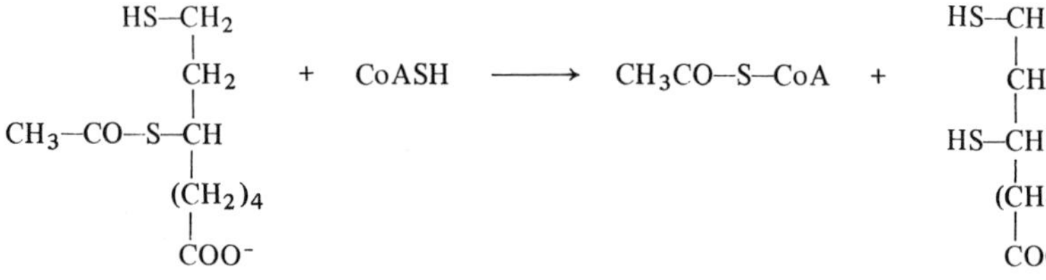

$$
\begin{array}{cccccc}
& HS-CH_2 & & & & HS-CH_2 \\
& | & & & & | \\
& CH_2 & + \; CoASH & \longrightarrow & CH_3CO-S-CoA \; + & CH_2 \\
CH_3-CO-S-CH & & & & & HS-CH_2 \quad (2) \\
| & & & & & | \\
(CH_2)_4 & & & & & (CH_2)_4 \\
| & & & & & | \\
COO^- & & & & & COO^-
\end{array}
$$

The oxidized lipoate is regenerated by dihydrolipoate dehydrogenase, an enzyme which contains tightly bound FAD (E-FAD):

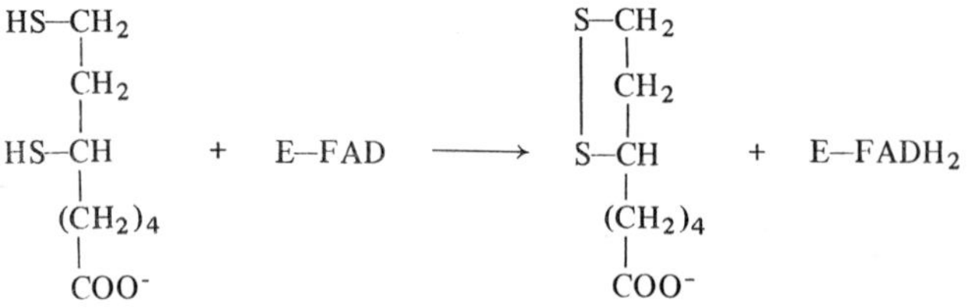

$$
\begin{array}{ccccc}
HS-CH_2 & & & S-CH_2 & \\
| & & & | \quad | & \\
CH_2 & & & CH_2 & \\
| & & & | \quad | & \\
HS-CH & + \; E-FAD & \longrightarrow & S-CH & + \; E-FADH_2 \\
| & & & | & \\
(CH_2)_4 & & & (CH_2)_4 & \\
| & & & | & \\
COO^- & & & COO^- &
\end{array} \tag{3}
$$

Finally the hydrogen atoms are transferred to NAD:

$$E-FADH_2 + NAD^+ \longrightarrow E-FAD + NADH + H^+$$

NADH is a strong competitive inhibitor of NAD for the pyruvate dehydrogenase from potato tubers which suggests that the ratio of NAD to (NAD + NADH) may be a control mechanism at this point. This reaction determines whether the pyruvate produced in glycolysis is converted to acetyl–CoA or to ethanol or lactate.

In the first of the cycle of reactions the acetyl group of acetyl-CoA is transferred to oxaloacetate with the formation of citrate, in a reaction catalysed by the enzyme citrate synthetase:

$$
\begin{array}{l}
COO^- \\
| \\
C{=}O \\
| \\
CH_2 \\
| \\
COO^-
\end{array}
\;+\;
\begin{array}{l}
O \\
\| \\
C{-}SCoA \\
| \\
CH_3
\end{array}
\longrightarrow
\begin{array}{l}
COO^- \\
| \\
CH_2 \\
| \\
{^-}OOC{-}C{-}OH \\
| \\
CH_2 \\
| \\
COO^-
\end{array}
\;+\; CoASH
\qquad (4)
$$

Oxaloacetate   Citrate

Since this enzyme catalyses the entry of acetyl-CoA into both the TCA and the glyoxylate cycles (p. 413), it is a logical point for control by ATP, an end product of the TCA cycle. Citrate synthetase from a number of plant mitochondria has been found to be inhibited by ATP, while that from the glyoxysomes of the castor bean is not.

A reversible dehydration/hydration reaction catalysed by aconitase equilibrates the citrate with *cis*-aconitate and isocitrate:

$$
\begin{array}{l}
COO^- \\
| \\
CH_2 \\
| \\
{^-}OOC{-}C{-}OH \\
| \\
CH_2 \\
| \\
COO^-
\end{array}
\rightleftharpoons
\begin{array}{l}
COO^- \\
| \\
CH \\
\| \\
C{-}COO^- + H_2O \\
| \\
CH_2 \\
| \\
COO^-
\end{array}
\rightleftharpoons
\begin{array}{l}
COO^- \\
| \\
CHOH \\
| \\
CH{-}COO^- \\
| \\
CH_2 \\
| \\
COO^-
\end{array}
\qquad (5)
$$

Citrate   *cis*-Aconitate   Isocitrate

The reductive decarboxylation of the isocitrate to α-ketoglutarate involves an NAD-linked dehydrogenase (isocitrate dehydrogenase) and produces the first of the carbon dioxide released by the cycle. The enzyme requires Mg^{++} or Mn^{++} together with ADP, and is strongly inhibited by ATP:

$$
\begin{array}{l}
COO^- \\
| \\
CHOH \\
| \\
CH{-}COO^- + NAD^+ \\
| \\
CH_2 \\
| \\
COO^-
\end{array}
\longrightarrow
\begin{array}{l}
COO^- \\
| \\
C{=}O \\
| \\
CH_2 + CO_2 + NADH + H^+ \\
| \\
CH_2 \\
| \\
COO^-
\end{array}
\qquad (6)
$$

α-Ketoglutarate

A second decarboxylative dehydrogenase reaction follows, but appears to be of a very different type from that catalysed by isocitrate dehydrogenase. This α-ketoglutarate dehydrogenase is a complex of enzymes somewhat analogous to the pyruvate dehydrogenase complex, and involves TPP and lipoic acid but produces succinyl-CoA instead of acetyl-CoA:

$$
\begin{array}{c}
\text{COO}^- \\
| \\
\text{C=O} \\
| \\
\text{CH}_2 \quad + \text{ NAD}^+ + \text{CoASH} \\
| \\
\text{CH}_2 \\
| \\
\text{COO}^-
\end{array}
\xrightarrow[\text{FAD}]{\text{TPP, lipoate}}
\begin{array}{c}
\text{COO}^- \\
| \\
\text{CH}_2 \\
| \\
\text{CH}_2 \quad + \text{ NADH} + \text{H}^+ + \text{CO}_2 \quad (7)\\
| \\
\text{CO--SCoA}
\end{array}
$$

Succinyl-CoA

The enzyme from cauliflower florets is activated by AMP. The conversion of succinyl-CoA to succinate by succinic thiokinase (succinyl-CoA synthetase) is highly exergonic and the thioester bond energy is utilized to produce a nucleoside triphosphate in a *substrate level phosphorylation*. The reaction appears to involve either guanosine diphosphate or inosine diphosphate, although there are reports of a plant enzyme utilizing adenosine diphosphate.

$$
\begin{array}{c}
\text{COO}^- \\
| \\
\text{CH}_2 \\
| \quad + \text{ NDP} \\
\text{CH}_2 \\
| \\
\text{CO--SCoA}
\end{array}
\longrightarrow
\begin{array}{c}
\text{COO}^- \\
| \\
\text{CH}_2 \quad + \text{ NTP} + \text{CoASH} \quad (8)\\
| \\
\text{CH}_2 \\
| \\
\text{COO}^-
\end{array}
$$

(Nucleoside diphosphate)   Succinate

In any case, the reaction can generate ATP by the action of nucleoside diphosphate kinase:

$$
\text{NTP} + \text{ADP} \xrightleftharpoons{\text{Mg}^{++}} \text{ATP} + \text{NDP}
$$

The production of fumarate from succinate by succinic dehydrogenase is the third dehydrogenation of the cycle. The enzyme succinic dehydrogenase is tightly bound to the mitochondrial membrane and can therefore be used as a marker enzyme for mitochondria (see Table 9.1). In contrast with the other dehydrogenases of the cycle, succinate dehydrogenase does not transfer hydrogen to NAD^+ but to FAD:

$$
\begin{array}{c}
\text{COO}^- \\
| \\
\text{CH}_2 \\
| \quad + \text{ FAD} \\
\text{CH}_2 \\
| \\
\text{COO}^-
\end{array}
\longrightarrow
\begin{array}{c}
\text{COO}^- \\
| \\
\text{CH} \\
\| \quad + \text{ FADH}_2 \quad (9)\\
\text{HC} \\
| \\
\text{COO}^-
\end{array}
$$

Fumarate

The enzyme is, as we have noted earlier (p. 99), competitively inhibited by malonate ($^-\text{OOC--CH}_2\text{--COO}^-$). Fumarate, which is a *trans* isomer (the *cis* isomer, maleic acid, is often used as a buffer) is converted to malate by a reversible hydration/dehydration reaction catalysed by the enzyme fumarase:

$$
\begin{array}{ccc}
\begin{array}{l} COO^- \\ | \\ CH \\ \| \\ HC \\ | \\ COO^- \end{array} & + H_2O \rightleftharpoons & \begin{array}{l} COO^- \\ | \\ HOCH \\ | \\ CH_2 \\ | \\ COO^- \end{array} \\
& & \text{Malate}
\end{array} \qquad (10)
$$

The final reductive step produces oxaloacetate from the malate (malate dehydrogenase) to complete the cycle (Fig. 7.7):

$$
\begin{array}{ccc}
\begin{array}{l} COO^- \\ | \\ HOCH \\ | \\ CH_2 \\ | \\ COO^- \end{array} & + NAD^+ \rightleftharpoons & \begin{array}{l} COO^- \\ | \\ C{=}O \\ | \\ CH_2 \\ | \\ COO^- \end{array} + NADH + H^+ \\
& & \text{Oxaloacetate}
\end{array} \qquad (11)
$$

In sum, one mole of acetyl coenzyme A produces one mole of water, two moles of carbon dioxide, one mole of coenzyme A, one mole of reduced flavoprotein and three moles of NADH.

Although the TCA cycle was first elucidated using animal tissues, there is now no doubt that it also occurs in plants. All the intermediates of the cycle, together with the enzymes, have been isolated from plant tissues, and utilization of specifically labelled pyruvate follows the pattern predicted by the operation of the pathway. The 1-carbon atom appears most rapidly as carbon dioxide, since this involves only a single turn of the cycle. The label in the 2- and 3-carbon atoms takes longer to be released as additional turns of the cycle are necessary (Fig. 7.8). In fact, carbon atoms 2 and 3 are not usually recovered fully as carbon dioxide since a further important function of the cycle, apart from the oxidation of pyruvate, appears to be the provision of carbon skeletons for synthesis. For example, glutamate may be formed from

$$
\begin{array}{l}
1 \quad COO^- \\
\quad\quad | \\
2 \quad CO \\
\quad\quad | \\
3 \quad CH_3
\end{array}
$$

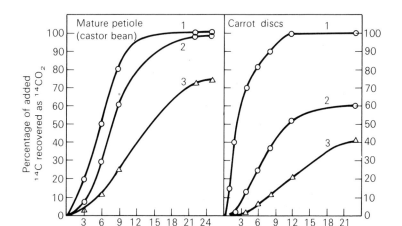

Fig. 7.8 The utilization of pyruvate labelled with ^{14}C in the 1, 2, or 3 positions by castor bean petioles and carrot discs. The tissues were incubated with equal amounts of pyruvate and the amount of label in respired carbon dioxide determined. Reproduced from Beevers (1961).

α-ketoglutarate and aspartate from oxaloacetate by transamination or reductive amination (see Ch. 8).

Inhibitor studies have provided important evidence for the operation of the TCA cycle in animal tissue, although the evidence obtained from their use in plants is at first sight equivocal. For example, malonate at low concentrations (5 mM, pH 7·4) produces approximately 70% inhibition of oxygen uptake in minced pigeon muscle. In plant tissue, however, much higher concentrations (10–50 mM) are needed to produce similar effects and the pH must be low (c. 4·0). At these higher concentrations, it is less certain that malonate is acting as a specific inhibitor of succinic dehydrogenase. However, it has been shown, at least in certain cases, that concentrations of malonate which have little effect on oxygen uptake do inhibit succinic dehydrogenase. The rate of oxygen uptake is maintained by the utilization of supplies of malate from the vacuole, while succinate accumulates in the tissue. Strong inhibition has been obtained with low concentrations at neutral pH values with isolated mitochondria.

The demonstration of TCA cycle activity in the mitochondrial fraction of plants followed the work of Bhagvat and Hill (1951), who showed that plant mitochondria contained both succinic dehydrogenase and cytochrome oxidase activity. Subsequently, all the enzymes of the cycle have been demonstrated in plant mitochondria (Davies and Ellis, 1964). The operation of the TCA cycle is not, however, confirmed merely by the presence of all the constituent enzymes. The sustained oxidation of pyruvate when catalytic amounts of any TCA cycle acid are added has been demonstrated (Fig. 7.9). Furthermore, in a few cases it has been shown that the complete oxidation of individual acids to carbon dioxide and water occurs (Freebairn and Remmert, 1957).

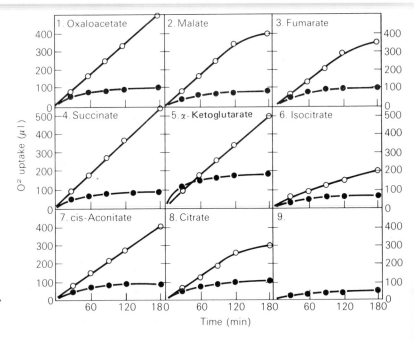

Fig. 7.9   The importance of the components of the TCA cycle in the oxidation of pyruvate by mitochondria from castor bean. The uptake of oxygen by mitochondria utilizing each of the acids alone (black circles) and the acid in the presence of pyruvate (open circles) are shown. Graph 9 shows the O_2 uptake observed in the presence of pyruvate when no second acid was included. From Fig. 7, p. 55 of *Respiratory Metabolism in Plants* by Harry Beevers. (Neal and Beevers, 1960; A. P. Rees and Beevers, 1960). Copyright © 1961 by Harper & Row, Publishers Inc. Reproduced by permission of the publishers.

Oxidative phosphorylation

As has already been indicated, mitochondria catalyse the oxidation of acetyl coenzyme A by the TCA cycle with the net production of one mole of reduced flavoprotein and three moles of NADH. These coenzymes are oxidized by molecular oxygen by way of the respiratory chain, a process which is accompanied by a large free energy change and by the conservation of some of this energy in the form of ATP. The formation of ATP by oxidative phosphorylation occurs only at specific sites in the respiratory chain; these are termed *coupling sites,* since it is here that phosphorylation is coupled to oxidation. The formation of ATP by oxidative phosphorylation only occurs during the passage of electrons or protons down the respiratory chain, although certain chemicals will allow oxidation to occur in the absence of phosphorylation (*uncoupling agents*).

The measurement of the phosphorylative efficiency of the mitochondrion involves the determination of both ATP formation and oxygen uptake. In early work the latter was measured by conventional manometry, but this did not allow the measurement of oxygen uptake over short time periods (e.g. less than 10 min). The use of an oxygen electrode has led to the development of an apparatus for the rapid and reliable recording of oxygen uptake solution.

The commonly used oxygen electrode (Clark, 1956) consists of a platinum cathode connected by a KCl bridge to a Ag/AgCl electrode as a reference anode. The assembly is mounted in plastic and the electrodes separated from the medium (which must be well stirred) by a thin teflon membrane readily permeable to gases but not to other solutes which may poison the platinum cathode. When a polarizing voltage of 0·5–0·8 V is applied across the electrodes, oxygen reacts at the cathode causing a current to flow. This current, which may be recorded, is proportional to the oxygen concentration.

Respiratory control

Oxygen uptake by mitochondrial preparations has been measured polarographically since the mid-1950s (Chance and Williams, 1956) although the method was not applied to plant mitochondria until the early 1960s. Mitochondrial preparations may take up oxygen slowly in the absence of added substrate and this has been termed *state one* by Chance and Williams. This rate is normally considerably increased by the addition of an intermediate of the TCA cycle. Addition of ADP to mitochondria in a medium containing such a substrate, inorganic phosphate and Mg^{++} (a cofactor for oxidative phosphorylation) causes an immediate increase in the rate of oxygen uptake, which then declines when all the ADP has been phosphorylated to ATP (Fig. 7.10). Chance and Williams termed the rapid rate of oxygen uptake in the presence of ADP *state three* and the consecutive slower rate, *state four. State two* is measured in the presence of ADP, but in the absence of a substrate, and *state five* in the presence of both substrate and ADP, but in the absence of oxygen. The control of ADP

Fig. 7.10 A diagrammatic representation of a polarographic (oxygen electrode) trace of the uptake of oxygen by plant mitochondria. M, substrate and ADP mark the addition of mitochondria, a TCA cycle acid and ADP to the assay medium, respectively.

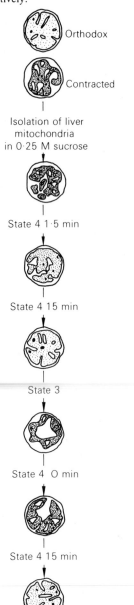

Orthodox

Contracted

|

Isolation of liver mitochondria in 0·25 M sucrose

State 4 1·5 min

State 4 15 min

State 3

State 4 0 min

State 4 15 min

Fig. 7.11 Changes in conformation of rat liver mitochondria during consecutive changes in the metabolic state. Redrawn from Hackenbrock (1966). Copyright © 1966 by the American Association for the Advancement of Science.

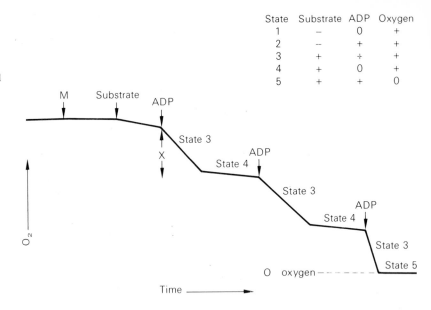

State	Substrate	ADP	Oxygen
1	–	0	+
2	–	+	+
3	+	+	+
4	+	0	+
5	+	+	0

over the rate of oxygen uptake is indicative of the coupling of oxidation to phosphorylation and is characterized by the ratio of state three to state four. This ratio, which is known as the *acceptor control ratio* or the *respiratory control ratio*, is used as a criterion for the preparation of intact mitochondria: for plant mitochondrial preparations the ratio is generally between 2 and 10. Other factors, such as a negligible endogenous respiration rate, a high pyridine nucleotide to cytochrome c ratio and the fact that the addition of exogenous NAD or cytochrome c should not accelerate the oxidation of malate, have also been proposed as important criteria (Chance, Bonner and Storey, 1968). However, as pointed out by Palmer (1976), a high acceptor control ratio is only a valid criterion if all electron flow occurs in a pathway which is obligately coupled to ATP synthesis. Many plants have an alternative pathway which by-passes the synthesis of ATP, allowing a relatively high state 4 rate of respiration which is not indicative of poor quality mitochondria. Furthermore, an effect of NAD on the rate of malate oxidation can be readily explained for plant mitochondria since they have both an externally located malic enzyme (which converts malate and NAD to pyruvate and NADH) and an NADH dehydrogenase (which converts NADH to NAD and feeds electrons into the electron transport chain allowing the uptake of oxygen and the apparent stimulation of malate oxidation). The determination of the level of cytochrome c reductase activity which is very low in intact mitochondria but increases dramatically on osmotic lysis is a more useful measure of any damage.

Plant mitochondria fixed in the various respiratory states appear identical under the electron microscope (Baker *et al.*, 1968) and there are apparently no ultrastructural changes of the type reported for rat liver mitochondria, where changes from state four to state three are accompanied by a decrease in the matrix volume (Fig. 7.11) to a highly condensed state (Hackenbrock, 1966). Mitchondria in state four appear much as those fixed in intact tissues and are called orthodox (Fig. 7.11).

A further feature of the oxygen electrode is its use in calculating directly the amount of ADP phosphorylated per atom of oxygen reduced. Since the oxygen uptake associated with the addition of a given amount of ADP can be calculated (the distance X on Fig. 7.10) the ratio of ADP phosphorylated per atom of oxygen reduced can be determined (ADP/O ratio). This ratio is identical with the P/O ratio (phosphate esterified per atom of oxygen taken up) providing the $1:1$ stoichiometry of the reaction $ADP + P_i = ATP$ is maintained.

The respiratory chain

Most oxidations in a cell end with the reduction of molecular oxygen to water. Generally, however, respiratory intermediates and coenzymes do not react directly with oxygen, but hydrogen atoms or electrons are transferred by way of a series of carriers, which in the case of the mitochondrion forms the so called *respiratory chain*. The final step in the process, the transfer of electrons to oxygen itself, is known as the terminal oxidation and the reaction is catalysed by a *terminal oxidase*. The respiratory chain of plant mitochondria consists chiefly of pyridine nucleotides, flavoproteins, cytochromes, quinones and iron-sulphur proteins.

The *pyridine nucleotide* found in mitochondria, nicotinamide adenine dinucleotide, is the coenzyme of the isocitrate, α-ketoglutarate and malate dehydrogenases of the TCA cycle. The NAD, which is firmly bound to the mitochondrial membrane, is reduced with a characteristic increase in absorption at 340 nm by the addition of two equivalents of hydrogen in the 1 and 4 positions of the nicotinamide ring (see p. 116). There are approximately ten to twenty moles of NAD per mole of cytochrome *c* present in plant mitochondria.

Plant mitochondria will rapidly oxidize externally added NADH and differ in this respect from animal mitochondria which directly utilize externally added NADH slowly, if at all. Respiratory control is generally lower than for other substrates and the maximum ADP/O ratio appears to be two. It is apparent, therefore, that the oxidation of exogenous NADH does not follow the pathway for endogenous NADH: with malate dehydrogenase, for example, which is linked to NAD, ADP/O ratios approach three. The rapid oxidation of exogenous NADH by plant mitochondria also appears to be insensitive to the inhibitors amytal and rotenone (see p. 305) and to be stimulated by divalent metal ions which have no direct effect on the oxidation of endogenous NADH. Exogenous NADH appears to be initially oxidized by a dehydrogenase which is associated with the inner membrane and possibly on its outer surface. Since the oxidations of both exogenous and endogenous NADH are sensitive to antimycin A, it is concluded that subsequently there is a common pathway to oxgyen (Palmer, 1976).

Flavoproteins, of which there are approximately three to six moles in plant mitochondria per mole of cytochrome *c,* are, like the pyridine nucleotides, hydrogen carriers. Their reduction involves the addition of two hydrogen atoms across the quinoid structure in the isoalloxazinc nucleus (see p. 118) and there is an accompanying decrease in the absorbance at about 460 nm with a consequent loss of yellow colour. On oxidation/reduction, these flavoproteins not only exhibit changes in their absorbance but also, when excited with light of 436 nm wavelength, in their fluorescence spectrum (see p. 304; Storey, 1970). Thus individual flavoproteins have been described both by the ratio of change in fluorescence to change in absorbance on oxidation/reduction (*f* for a high ratio and *a* for a low ratio) and by their redox potential (p. 127) which may be high (*h*), medium (*m*) or low (*l*). However, it has not been possible to assign unequivocally either potentials or functions to individual flavoproteins. In animal mitochondria, flavoproteins have been identified associated with succinate dehydrogenase activity, fatty acid oxidation and the NADH-linked dehydrogenases. In plant cells, fatty acid oxidation does not take place within the mitochondria and the flavoproteins of this process are associated with the glyoxysomes (see Ch. 9).

Cytochromes, which are compounds with characteristic absorption spectra in the visible region when reduced, have been known in plants since the work of Keilin in 1925; their importance in plant respiration was established by the work of Bhagvat and Hill. The cytochromes are electron carriers, being oxidized and reduced by the transfer of a single electron. They thus differ from the pyridine nucleotides and flavoproteins both in the nature of the oxidation-reduction and in the number of equivalents carried: i.e. two cycles of reduction and oxidation of a cytochrome must accompany a single oxidation or reduction of a pyridine nucleotide or flavoprotein.

There are four major groups of cytochromes, referred to as A, B, C and D (Fig. 7.12), of which all but D are found in plant mitochondria. All contain iron-porphyrins (see p. 119), and individual compounds within these groups are designated by lower case italic letters. Each cytochrome may be identified by its absorption spectrum both in the oxidized (Fe^{+++}) and reduced (Fe^{++}) form. The cytochromes have three characteristic absorption bands in the visible region, that which occurs at the longest wavelength being designated α and those at the shorter wavelengths β and γ (or Soret) respectively. 'Difference' spectra have been found particularly useful in biological preparations. These are obtained by determining the differences in absorbance in different respiratory states – normally, under aerobic and anaerobic conditions. This is most conveniently carried out using a split-beam spectrophotometer (Fig. 7.13) in which the light from a single monochromator is split and passed through two cells, one of which contains the preparation in the oxidized form and the other in the reduced form. A difference spectrum for a mitochondrial preparation from *Helianthus tuberosum* (Jerusalem Artichoke) is shown in

Fig. 7.12   The porphyrins of the cytochrome classes A, B, C and D.

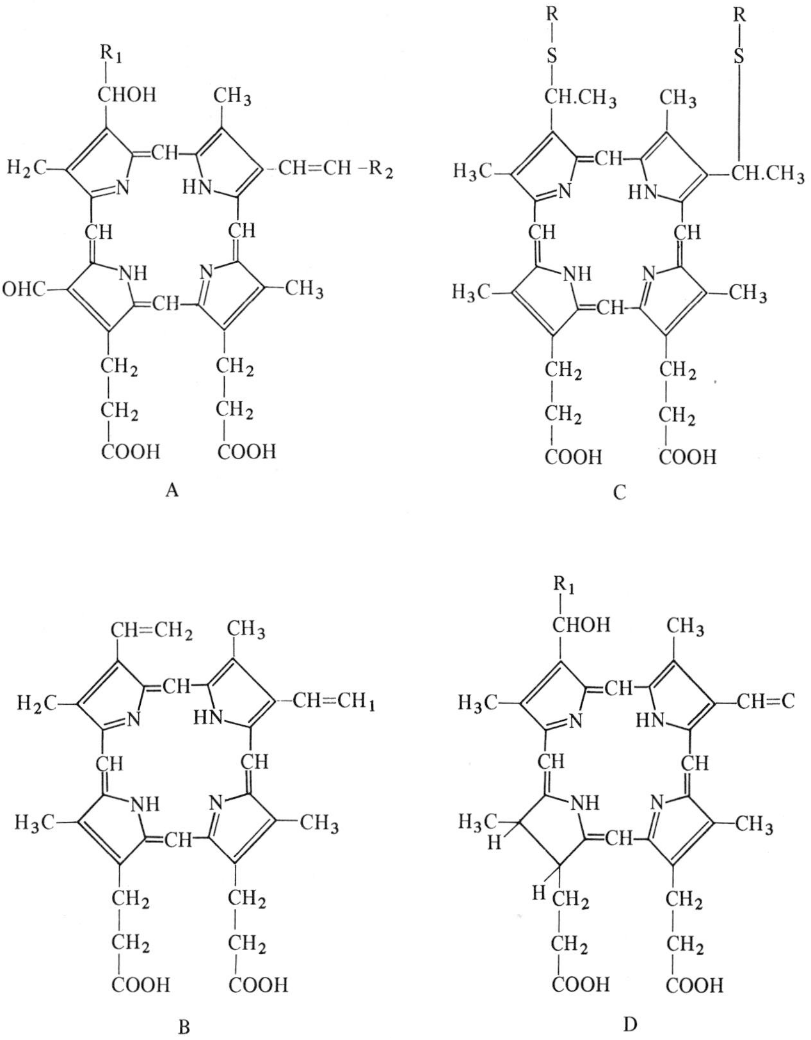

A

C

B

D

Fig. 7.14(a). The peak at 602 nm is the α-band for cytochromes a and a_3, while those at 550 nm and 560 nm (a shoulder) are two α-bands of cytochromes c and b, respectively. The broad peak at 520–530 nm represents the β-bands of b and c. The shorter wavelength spectrum is dominated by the γ or Soret band of cytochrome b (430 nm) with the Soret region of cytochrome oxidase represented as a shoulder around 442 nm.

When the visible spectrum is measured at the temperature of liquid nitrogen (77 K), it is more complex (Fig. 7.14b) and more individual cytochromes are identifiable. The b cytochromes (560 nm shoulder, Fig. 7.14a) are split into three components with peaks at 552, 557 and 561 nm. Recent studies of the b-cytochromes present in plant mitochondria indicate that there may be two further b-type cytochromes. Some confusion exists in the literature over nomenclature since individual cytochromes are identified by the wavelength of their α absorption peak which may be quoted at 77 K or at room temperature

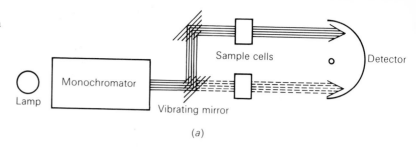

Fig. 7.13   The essential features of a split-beam spectrophotometer (*a*) and a double-beam spectrophotometer (*b*). Redrawn from Chance (1954).

Lamp

Monochromator

Sample cells

Vibrating mirror

Detector

(*a*)

Monochromator set to wavelength *a*

Monochromator set to wavelength *b*

Vibrating mirror

Half silvered mirror

Sample cell

Detector

Lamp

(*b*)

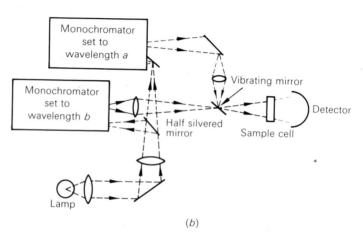

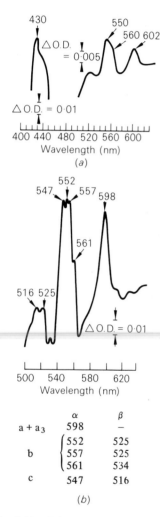

430   550   560 602

$\triangle$O.D. = 0·005

$\triangle$O.D. = 0·01

400 440 480 520 560 600
Wavelength (nm)
(*a*)

552
547   557 598

561

516 525

$\triangle$O.D. = 0·01

500   540   580   620
Wavelength (nm)

	α	β
a + a$_3$	598	–
b	552	525
	557	525
	561	534
c	547	516

(*b*)

Fig. 7.14   Difference spectra of (*a*) Jerusalem artichoke mitochondria at 300 K and (*b*) white potato mitochondria at 77 K. Both spectra were obtained by measurement of the difference in absorption of mitochondria in the presence of oxygen from those in the presence of a powerful reducing agent (dithionite). Redrawn from Lance and Bonner (1968).

In the latter case, which we shall adopt here in accordance with the recommendation of the IUB nomenclature subcommittee, the numerical values are up to 5 nm higher. Thus the cytochrome with α bands at 77 K at 552, 557 and 561 nm are now designated cytochromes b_{557}, b_{560} and b_{566}: the more recent studies indicate two cytochromes known as $b_{557/8}$. Cytochrome oxidase (α = 598 nm) and cytochrome c (α = 547 nm and β = 516 nm) are also visible. The optical separation of cytochromes *a* and a_3 can be achieved after the addition of azide, which only complexes with cytochrome a_3. If mitochondria are washed for a long period, all the cytochrome *c* (α = 547 nm) can be removed and subsequent determination of the difference spectrum indicates that a second cytochrome c (c_1) still remains (α = 549 nm, β = 514 nm, γ = 419 nm). This cytochrome is analogous in solubility (that is, its retention on washing) to the cytochrome c_1 of animal mitochondria, but not identical with it. In general, the cytochromes c are present in the highest amounts with the molar ratio of cytochromes *a* to *c* and *b* to *c* each being about 0·7.

From the amounts of the cytochrome present, expressed per unit weight of mitochondrial protein, and from a knowledge of the state three rate of oxygen uptake, expressed on the same basis, Lance and Bonner (1968) calculated the turnover of the cytochromes. It was assumed in the calculation that four electrons were necessary to reduce one molecule of oxygen and that the oxidation-reduction cycle of each cytochrome involved only one electron. The values for cytochrome c

(60 sec^{-1}) and cytochrome $a + a_3$ (about 30 sec^{-1}) were similar, regardless of the species from which the mitochondria were isolated, and bore no relation to the rate of oxygen uptake by the tissues as a whole. They concluded that respiration rates of tissues depend on the numbers of mitochondria per cell rather than on any great difference between the mitochondria themselves.

Ubiquinone or coenzyme Q_{10} was discovered in large amounts in animal mitochondria and has subsequently been identified in the plant organelle. It is a hydrogen carrier by virtue of its reducibility to a hydroquinone form; reduction is accompanied by a marked decrease in absorption at 270–80 nm and first requires the production of the semiquinone.

Iron-sulphur proteins. Although much of the iron present in the respiratory chain occurs in the cytochromes, it is not restricted to haem groups and a number of non-haem iron proteins have been described. They have a low molecular weight (6 000–12 000) with between two and eight iron atoms per molecule, which are bound to the sulphur atoms of cysteine residues in the protein. These iron-sulphur (Fe-S) proteins (or centres as they are also known) cannot be characterized by their light absorption spectra in the same way as the other components of the electron transport chain, although they do have characteristic electron paramagnetic resonance (EPR) spectra. This technique measures the absorption of energy from an oscillating magnetic field of a fixed frequency as a function of a variable external magnetic field. The spectrometer consists of a source of electromagnetic radiation with a wavelength of about 30 nm, a variable external magnetic field and a suitable detector (see Beinert and Palmer, 1965, for further references and details) and the spectra are normally plotted as the first derivative of the absorption as a function of the magnetic field: they are characterized by a dimensionless constant, the spectroscopic splitting factor, g. Thus, a number of iron-sulphur centres have been described in both plant and animal mitochondria, which are associated with both NADH (prefixed N-) and succinate dehydrogenases (prefix S-). In plant mitochondria, N-1, N-2, N-3, S-1 and S-2 are all paramagnetic in their reduced forms, while S-3 is paramagnetic in its oxidized form (and consequently known as a high potential iron protein or HiPIP (Palmer, 1976; Rich and Bonner, 1978a, b), as is a further iron sulphur protein whose reductants under natural conditions are unknown. The redox potentials of N-1, N-2 and N-3 in mung bean are −260, +60 and

−275 mV respectively and those of S-1, S-2 and S-3, +60, −225 and +65 mV.

Functional complexes. Using preparations from animal mitochondria, the electron transport chain has been fragmented into four functional complexes by the use of detergents and ammonium sulphate fractionation. Each of these complexes (numbered I to IV) consists of two or more components of the chain and a considerable proportion of lipid. They catalyse the reduction of ubiquinone by NADH, the reduction of ubiquinone by succinate, the oxidation of the ubiquinone by cytochrome *c,* and the oxidation of cytochrome *c* by oxygen, respectively (see Nicholls and Elliott, 1974). Most of the ubiquinone and cytochrome *c* appears to be unassociated with these complexes and it has been suggested that they act as mobile electron carriers between the complexes. The complexes generally appear as membrane vesicles in electron micrographs. It is possible to reconstitute a functional electron transport chain by carefully mixing the isolated complexes with various other components, although the reconstituted system cannot be coupled to the phosphorylation of ADP.

Sequence of the respiratory chain

The sequence of reactions in which the pyridine nucleotides, flavoproteins and cytochromes are involved in the respiratory chain may be predicted on thermodynamic grounds, or determined experimentally by the sophisticated spectrophotometric techniques developed particularly by Chance. Since the reactions of the chain are a series of oxidations and reductions, the overall arrangement can be expected to follow from the ease of reduction (or oxidation) of each of the components. Quantitatively, this is represented by the oxidation-reduction potential, which may be equated with the mid-point potential at a defined pH commonly close to 7 (E_0' or E with the defined pH as an added sub-script; see p. 128 and Dutton and Storey, 1971). Arrangement on this basis places pyridine nucleotide ($E_0' = -320$ mV) as the primary acceptor of electrons from the substrate. Subsequently, electrons are passed to the iron-sulphur proteins and the flavoproteins and then to the cytochromes (see Table 7.1, but note that there are some inconsistencies in the data due probably to technical difficulties in making the measurements).

The experimental determination of the kinetics of oxidations and reductions in the cytochrome chain is highly complex, since protons or electrons are transferred from one component to another very rapidly. The techniques employed involve following the oxidation and reduction spectrophotometrically. A small pulse of oxygen is supplied to a mitochondrial suspension, containing substrates under anaerobic conditions. This causes the cytochromes to become first oxidized and then reduced. Changes in the oxidation/reduction state of individual components of the electron transport chain are determined from changes

Table 7.1 Redox potentials and half times for the oxidation of various plant respiratory chain components.*

	α peak 77 K (nm)	$t_{1/2}$ox (m sec)	$E'_{(7.2)}$ (mV)
NAD		260	-320
Fe-S$'$		—	-260 to $+65$
FP		200–500	-155 to $+110$
UQ$_{10}$		350	$+70$ (-12)
$b_{556/557}$	552–554	500	$+75$ to $+100$
$b_{565/566}$	561–56?	35–15	-70 to -80
b_{560}	557	8	$+40$ to $+80$
$b_{557/558}$	553–555	—	-70 to -100
c	547	2·6	$+235$
c_1	549	1·9	$+235$
a	598	2·0	$+190$
a_3		0·8	$+380$

* Data taken from the literature chiefly for potato, mung bean, skunk cabbage and Jerusalem artichoke.

in absorbance. The time sequence of events characterizes the position of individual components in the chain. The time taken for the reactions is too small for complete difference spectra to be determined, so that a separate technique is employed. From the difference spectrum two wavelengths are chosen such that the difference in absorbance on oxidation of the reduced compound is maximal and that other components show no absorbance change at these wavelengths. For example, the wavelength pair 550 and 541 nm may be used to assay cytochrome c (Fig. 7.14(a)): this pair represents a peak and trough for cytochrome c and is insensitive to cytochrome b which shows no significant change in absorption on oxidation/reduction at these wavelengths. The measurements therefore, are carried out in a dual wavelength spectrophotometer (Fig. 7.13(b)). The absorption changes corresponding to the various components are either recorded with a pen on a moving chart if they are slow, or if extremely fast, photographed from an oscilloscope. The oxygen is added with a rapid mixing device and it is often arranged that the mitochondrial sample may be recycled through the spectrophotometer.

Many more studies of the kinetics of electron transport have been made on animal mitochondria than on plant mitochondria and it is only since the late 1960s that the plant respiratory chain has been critically examined. In a series of experiments on mung bean (*Phaseolus aureus*) and skunk cabbage (*Symplocarpus foetidus*), Storey (see Storey, 1976 for refs.) has determined the half time ($t_{\frac{1}{2}}$) for oxidation ($t_{\frac{1}{2}}$ox) and reduction ($t_{\frac{1}{2}}$red) of the chain components. The preparations were uncoupled to avoid the complications of energy conservation and generally succinate was supplied as the substrate. Some of the half times obtained are tabulated in Table 7.1. In mung bean the most rapidly oxidized component was cytochrome oxidase $(a + a_3)$ followed by the cytochromes c. Although cytochrome c_1 has a slightly lower $t_{\frac{1}{2}ox}$ than c in the absence of antimycin A, in the presence of this inhibitor (which isolates the a and c cytochromes from the rest of the respiratory chain) the position is reversed ($t_{\frac{1}{2}ox}$ for c_1 becomes 3·0 msec). It is argued therefore that c lies next to cytochromes a and a_3. Thus cytochrome c_1 which is retained on washing is analogous to the cytochrome c_1 of animal mitochondria, both in solubility and position in the respiratory chain (the $t_{\frac{1}{2}ox}$ values for animal cytochromes c_1 and c are 5·0 and 2·5 msec respectively).

The greatest contrast with the cytochrome chain of animal mitochrondria was for some time thought to lie in the nature of the b cytochromes. Animal mitochondria were thought to have a single cytochrome b with a $t_{\frac{1}{2}ox}$ value of 80 msec. Recently, however, a second cytochrome b has been described and it is possible that there may be a third separate cytochrome b. In plants there are up to five cytochromes b (see Lambowitz and Bonner, 1974) with very different kinetic and redox properties. Although there have been extensive studies over the past few years, there is still considerable confusion over the rôles of these various cytochromes: cytochrome b_{560} has a similar $t_{\frac{1}{2}ox}$ to the other cytochromes and appears to bear a close functional relationship to cytochrome c. However, b_{556} has a rather slow half time for oxidation while that of b_{566} is intermediate between the other two. Thus, while there seems little controversy over the position of b_{560} in the sequence of electron transport from substrate to oxygen, the positions of b_{556} and b_{566} are less certain as is that of the further two cytochromes that have been recently described (see Palmer, 1976).

Although ubiquinone, flavoproteins and pyridine nucleotides may be identified by their absorption spectra, the use of this technique is limited in kinetic studies due to interference by other compounds. It is possible, however, to use the fluorescence of flavoprotein (excited by illumination at 436 nm) and pyridine nucleotides (excited at 336 nm) to help determine their sequence in the respiratory chain, although the sequence of events in this part of the plant electron transport chain remains largely unknown. The mid-point potential of ubiquinone has been determined as $+70$ mV with a further small component having a potential of some -12 mV (Storey, 1973). This mid-point potential and the kinetic measurements are consistent with a role of mediating between the dehydrogenases and the cytochromes as has been proposed for animal mitochondria. Current evidence suggests that a pool of ubiquinone may mediate between the dehydrogenases and a number of separate cytochrome chains. The iron-sulphur proteins are constituents of the succinate and NADH dehydrogenases as are the flavoproteins. Electrons are thought to be transferred from succinate to flavoprotein and the Fe-S centres S-1, S-2 and S-3, and from NADH to flavoprotein and Fe-S centres N-1, N-3 and N-2, respectively, although the exact sequence of events has yet to be elucidated.

Our knowledge of the course of electron transport in plants is summarized in Fig. 7.15. Electron flow generally follows a thermodynamically predictable sequence and is basically similar in plants and animals.

Fig. 7.15 The respiratory chain of plant mitochondria.

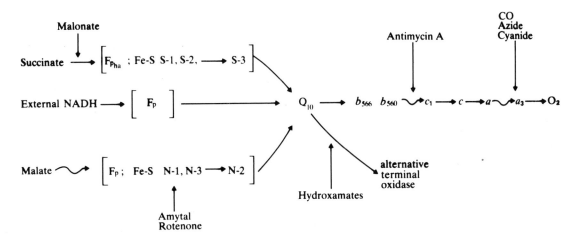

Inhibitors of electron transport. A number of compounds have been found to interact with the respiratory chain consequently to inhibit the reduction of molecular oxygen to water. It has been possible to identify the sites of action of many of these compounds and they have proven useful for indicating the location of coupling sites (see p. 307).

The spectrophotometric identification of an interaction between an inhibitor and an electron carrier relies on changes in the oxidation–reduction state of the respiratory chain. If an inhibitor is added to an aerobic suspension of mitochondria in the presence of substrate, the site of action may be identified spectrophotometrically, since components of the chain on the oxygen side of the inhibition become oxidized while those on the substrate side are reduced. It is therefore possible to identify the crossover from oxidized to reduced forms with the site of action of the inhibitor. For example, by the use of this *crossover theorem* it has been shown that the inhibitors rotenone and amytal inhibit the animal respiratory chain between FP_1 and cytochrome b, while antimycin A blocks the transfer of electrons from cytochromes b to c.

It is a notable feature of the respiration of many plants, however, that many inhibitors of the respiratory chain have relatively little effect on oxygen uptake. Perhaps the most spectacular example is the absence of any marked effect of cyanide on the rate of oxygen uptake by a number of plant tissues and in particular, the tissues of aroid spacides, e.g. *Arum maculatum* and *Symplocarpus foetidus* (skunk cabbage). This absence of any dramatic inhibition by cyanide is not restricted to members of the Araceae, however, and is in fact widespread among flowering plants. It has also been observed in fungi, algae and bacteria although is very rare in animal species (see the extensive review by Henry and Nyms, 1975, for further details). As in animals, the activity of cytochrome oxidase is blocked by cyanide. The

305

continued reduction of molecular oxygen must therefore be mediated by an alternative terminal oxidase, which it may be inferred is either on a separate pathway or one which branches from the main pathway of electron transport. Since cyanide cuts the ADP/O ratios (p. 297) with malate as the substrate from three to one while there is apparently no phosphorylation with succinate as the substrate, it can be argued that only the first phosphorylation site is common to both pathways. The cyanide insensitive respiration is not affected by the inhibitor antimycin A which implies that the branch point is on the substrate side of the cytochromes. It is now generally accepted that this branch point occurs at the level of ubiquinone which may transfer electrons either to cytochrome b or to the alternative terminal oxidase.

Little is known of the nature of this alternative oxidase: its affinity for oxygen is somewhat lower than that of cytochrome oxidase and its action is inhibited by a series of metal chelating compounds, including various substituted hydroxamic acids (e.g. m-chlorobenzhydroxamic acid, m-CLAM and salicylhydroxamic acid, SHAM). It has now been established that it is only the hydroxamic acids whose action is specific and this is not due to their ability to act as metal chelating agents (Rich et al., 1978). The terminal oxidase, the product of whose action appears to be peroxide or superoxide, does not seem to be a b-cytochrome, a flavoprotein, nor an Fe-S protein. Rich and Bonner (1978) have suggested that the alternative terminal oxidase is a quinone, which like most quinones, is autooxidizeable, or some component which is as yet optically and EPR invisible and can accept electrons from a quinone.

The function of the alternative pathway has been the subject of considerable discussion. As far as the members of the Araceae are concerned, it seems relatively clear that the function is thermogenesis. In these species, pollinators are attracted by compounds which are volatalised by elevated temperatures (which may be 20° to 30° above ambient) in the spadix (see Meeuse, 1975). However, this cannot be the function in the majority of species which show cyanide-insensitive respiration and here the function may be related to a requirement for peroxide during ripening (see the review by Solomos, 1977, for details). Alternatively, it may allow electron flow through the TCA cycle in the presence of high levels of ATP (perhaps generated by photosynthesis) in cases where C-skeletons are required for synthetic reactions. Again, the pathway would allow the oxidation of NADH (and the control of cytosolic NADH/NADPH levels) under similar conditions of high ATP levels. The question of the control of which electron pathway is utilized at any one time is discussed below.

Other terminal oxidases. Apart from the cytochrome oxidase system, a number of other terminal oxidases have been demonstrated in plants, although not necessarily associated with the mitochondria. Phenol oxidases, which catalyse the oxidation of phenols to quinones, have been considered to have a possible rôle in respiratory oxygen uptake since the quinones can be reduced by a variety of substances such as ascorbate and NADPH.

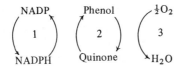

1. Respiratory dehydrogenase
2. Quinone reductase
3. Phenol oxidase

A similar rôle has been cast for ascorbic acid oxidase (ascorbate + O_2 = dehydroascorbate + H_2O), glycolic acid oxidase (glycolate + O_2 = glyoxalate + H_2O_2) and catalase ($H_2O_2 = H_2O + \frac{1}{2}O_2$). Phenoloxidase and ascorbic acid oxidase appear to be components of the soluble phase (see p. 249) and are also found in the cell-wall fraction; it is unlikely that they participate in the uptake of oxygen under normal conditions. Glycolic acid oxidase does play a part in respiratory oxygen uptake in the light (see Ch. 9). These oxidases are, as far as is known, unable to participate in the formation of ATP during oxygen uptake.

The coupling sites

As was indicated at the beginning of the chapter, the primary function of respiration is the conservation of energy as ATP. The breakdown of glucose by glycolysis and the TCA cycle produces ten moles of reduced pyridine nucleotide which on further oxidation to water results in the production of some 2205 kJ (527 kcal). Since this represents 77% of the free energy available in the complete oxidation of glucose, it is clear that this stage in the oxidation can generate the ATP in which the free energy is conserved.

The mitochondrial synthesis of this ATP is obligatorily coupled to the transport of electrons from substrate to water via the respiratory chain (although electron transport can occur in the absence of phosphorylation in the uncoupled state). The stoichiometry of ATP formation with various substrates is well established and this, together with kinetic and inhibitor studies, has allowed the coupling sites on the respiratory chain to be identified.

Mitochondria oxidizing malate synthesize ATP at up to three moles per atom of oxygen reduced (ADP/O of three) while the utilization of succinate results in a ADP/O ratio of only two. From evidence such as this it has been inferred that ATP formation takes place at a number of separate sites along the respiratory chain and that one of these lies between the NAD-linked dehydrogenases and the flavoproteins (not those involved in the pathway from succinate – see Fig. 7.15).

The coupling sites (designated I, II and III and represented by ~ in Fig. 7.15) may be identified more precisely by the crossover theorem. If the action of an inhibitor on the respiratory chain is compared with the state three to state four transition (Fig. 7.10), it can be seen that both result in a relative reduction of the rate of oxygen uptake. Thus it has been possible to identify coupling sites between the flavoproteins (I) between cytochromes b, and c(II), and a and a_3 (III). These sites correspond to large free energy changes, or potential spans (see Table 7.1 and Fig. 7.16) between individual respiratory chain components. Further corroborative evidence has been obtained by fragmenting the respiratory chain with electron donors and inhibitors, and determining the whereabouts of esterification of ADP to ATP.

Cyanide decreases the ATP formation from succinate from two to a maximum of one per atom of oxygen reduced (i.e. there is one coupling site on the pathway to the alternate terminal oxidase). The utilization of exogenous NADH seems to result in a maximum ADP/O ratio of two.

The nature of the coupling reaction has been controversial for many years. The theories are divided chiefly between those which postulate a common chemical intermediate between the reactions of the respiratory chain and of ATP synthesis and those that do not.

There have been many forms of the so-called *chemical coupling hypothesis* and the concept has had many eminent supporters, e.g.

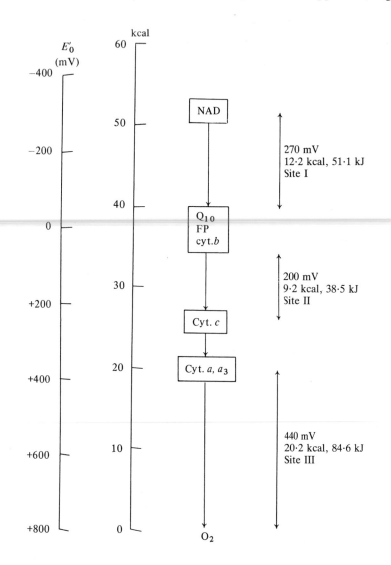

Fig. 7.16   The decline in free energy as electron pairs flow down the respiratory chain to oxygen. There are three segments indicated which yield sufficient energy to generate a molecule of ATP from ADP and phosphate. Redrawn from Lehninger (1970).

$$A_{red} + B_{ox} + I \rightleftharpoons A_{ox} \sim I + B_{red}$$
$$A_{ox} \sim I + X \rightleftharpoons A_{ox} + X \sim I$$
$$X \sim I + P_i \rightleftharpoons X \sim P + I$$
$$X \sim P + ADP \rightleftharpoons X + ATP$$

Fig. 7.17   The chemical coupling hypothesis.

Chance, Ernster, Pressman, Slater. In most formulations, the electron transport carrier (either in the oxidized or the reduced form) reacts with an unknown compound, I, to form an intermediate which then exchanges the respiratory carrier for another compound X. ATP is formed following exchange of I for P_i and phosphorylation of ADP by the intermediate $X \sim P$ so formed (see Fig. 7.17). A phosphorylation intermediate ($X \sim P$) is usually postulated, since the terminal bridging oxygen atom of ATP is derived from ADP and not P_i. The reactions are reversible, since electron transport may be driven in the reverse direction by ATP. The inhibitor oligomycin is thought to prevent the $X \sim I$ to $X \sim P$ step in the reaction which leads to the accumulation of $X \sim I$. Because of the reversibility, the observed inhibition of oxygen uptake must follow. State four respiration also follows the accumulation of $X \sim I$, there being no ADP to accept $\sim P$. Uncoupling agents such as 2,4-dinitrophenol are postulated to cause the hydrolysis of $X \sim I$. One of the chief problems faced by the proponents of this hypothesis is the lack, after many years of research, of evidence for the intermediates X and I and their interaction with the respiratory chain. A central rôle for lipoic acid in the synthesis of ATP has been proposed from work on animal mitochondria and lipoic-acid requiring mutants of *Escherichia coli* (Griffiths, 1976), although the conclusions are controversial (de Chadarevjan *et al.*, 1979). Another suggestion, based on the observation that ATP increases the mid-point potential of b_{566} from animal mitochondria by some 250 mV, that the high E_m form of this cytochrome is a primary intermediate in phosphorylation at site II has been largely discounted through work with plant mitochondria. The shift in E_m on adding ATP appears to be due to reversed electron transport and the essential difference between the plant and animal systems seems to be that the dye, which has to be added to measure E_m, is more effective in short-circuiting reversed electron transport in plant than in animal mitochondria. It must be remembered, however, that the reactions of the electron transport chain are very fast, and it may be that such an intermediate would be both labile and present in low concentration. Furthermore, the intermediates may only be stable in a hydrophobic environment so that isolation in any aqueous medium would result in immediate decomposition.

Chief of the alternative hypotheses is the *chemi-osmotic hypothesis* of Mitchell (e.g. Mitchell, 1968, 1972), which is the subject of an excellent review by Greville (1969). The chief difference from the chemical coupling hypothesis is that there are no common chemical intermediates postulated between the reactions of the respiratory chain and those of ATP synthesis. The operation of the electron transport chain is postulated to generate pH (Δ pH) and electrical potential ($\Delta\Psi$) gradients across the inner membrane which combine to produce an electrochemical gradient of protons ($\Delta \mu_{H^+}$). This gradient constitutes the proton motive force (Δp), which drives the synthesis of ATP through a reversible ATPase (Fig. 7.18). The total protonmotive force (Δp) in mV is given by $\Delta \mu_{H^+}/F = \Delta\Psi - z\Delta pH$, where z is $2\cdot303$ RT/F ($\simeq 60$ at 25°C): R is the gas constant, T the absolute temperature, and F the Faraday.

The inner mitochondrial membrane has been found to be mostly impermeable to protons and yet it remains an experimental fact that protons are ejected during electron transport from substrate to oxygen. The earlier literature suggested that two protons were translocated from the inside to the outside during the passage of electrons through each of the coupling sites (H^+/site $= 2$) and Mitchell has discussed at some length the so-called 'redox loops' which could account for these observations (see Mitchell, 1968, 1976, for example). In essence, it is proposed that the oxidation–reduction components (redox) are arranged in the inner membrane in such a way that successive reactions of the chain involve alternately electrons and hydrogen atoms. If we consider further the transfer of electrons from succinate to oxygen with respect to the spatial arrangement of the components in the inner membrane then we can write for each electron removed:

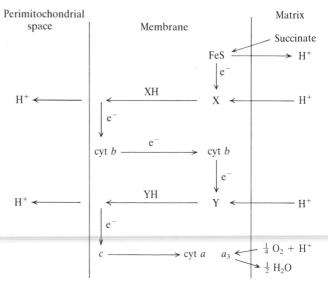

An electron is transferred from succinate to the iron sulphur protein with the release of a proton into the matrix. The next component to be reduced, here designated X, is an hydrogen carrier (for example, a quinone) and so must pick up an electron and a proton. Subsequently, the electron is transferred to a cytochrome b which is located on the perimitochondrial space side of the inner membrane. Since the cytochrome is reduced simply by the addition of an electron, a proton is lost to the outside. This loop achieves the net transfer of one proton from the inside to the outside, although it should be noted that it is not a proton which is translocated across the membrane, but an hydrogen atom. By arranging a second loop involving another hydrogen carrier, Y, which is located on the matrix side of the membrane and cytochrome c, it is possible to transfer a second proton across the membrane and so account for the overall stochiometry. One problem which obviously springs to mind is to identify the components of the electron transport chain functioning between the iron sulphur proteins and the b cytochromes. Recently, Mitchell (see Mitchell, 1976) has

Fig. 7.18   The chemiosmotic hypothesis of oxidative phosphorylation. The figure shows the operation of the proton motive complexes between succinate and oxygen and the ATPase. Two moles of ATP would be produced for each mole of succinate oxidized in accordance with the ADP/O ratio of two for this substrate. Redrawn after Mitchell (1968, 1976).

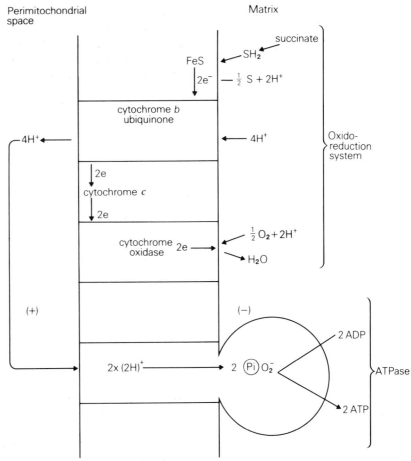

proposed that ubiquinone could function as both X and Y: the redox loops can then be written as:

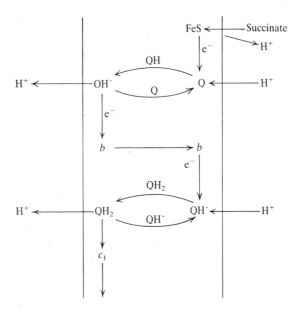

where Q represents ubiquinone, QH_2 the fully reduced ubiquinol and $QH\cdot$ the semi-quinone. Since, in this scheme, $QH\cdot$ moves from left to right in the upper loop and from right to left in the lower loop, Mitchell then proposed that the transmembrane diffusion of $QH\cdot$ be replaced by perpendicular movement of this species, viz.:

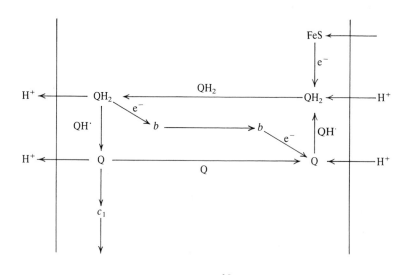

In principle, the scheme can be written in a number of different ways, but it is the concept of a Q-cycle that is important: the details of the cycle are considered in some depth by Mitchell in his paper of 1976. The other crucial factor in the chemi-osmotic chemosmatic hypothesis is that the components of the electron transport system must occupy particular sides of the membrane and this is discussed in more detail below.

Although we have followed Mitchell's scheme based on H^+/site ratio of 2, there is growing evidence that this ratio is 3 (or 4) (see also Ch. 8, p. 354). In fact, a ratio of more than 2 overcomes some thermodynamic objections to the chemiosmotic hypothesis and accounts for proton movement during ATP/ADP exchange across the mitochondrial membrane (see Brand, 1977). This remains a controversial area, and one in which further knowledge of the role of quinones in proton movement is required.

A detailed formulation that can be applied to plants and, incidentally, one that is useful in explaining the interaction of the alternative pathway within the cytochrome chain, has been put forward by Rich and Moore (1976). In this scheme, the precise sequence of events around the b and c cytochromes is deliberately vague since the evidence is still equivocal. The alternative oxidase is shown as being closely associated with the cycle. Reduced quinone may be formed either by a two step reduction or involve the semiquinone species. It is possible that two pools of ubiquinone exist, one of which interacts with the alternative terminal oxidase and the other with the b/c cytochromes.

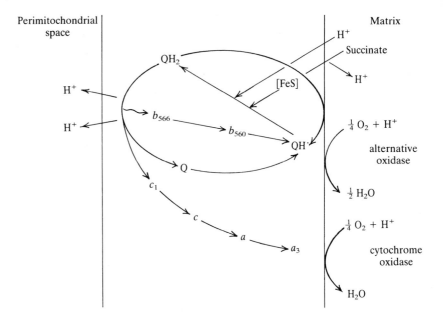

A net return flow of protons is postulated to take place through a reversible ATPase and driven by the protonmotive force, Δp.

The reaction

$$ATP + H_2O \rightleftharpoons ADP + POH$$

may be written as two half-reactions

$$H_2O \rightleftharpoons 2H^+ + O^{--}$$

and

$$ATP + 2H^+ + O^{--} \rightleftharpoons ADP + POH$$

The two half-reactions can be arranged vectorially such that they involve net proton translocation from one side of the membrane to the other:

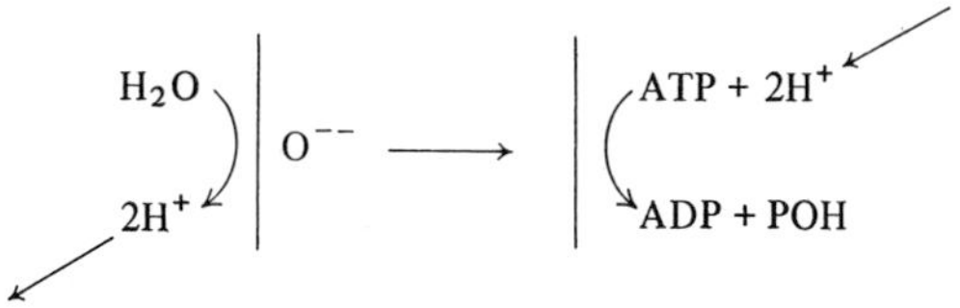

$$
\begin{array}{c|c|c}
\left.\begin{array}{c} H_2O \\ \\ 2H^+ \end{array}\right) & O^{--} \longrightarrow & \left(\begin{array}{c} ATP + 2H^+ \\ \\ ADP + POH \end{array}\right.
\end{array}
$$

In simple terms, since the reactions are reversible, the addition of protons to the left-hand side promotes the formation of ATP. Thus, if the ATPase is strictly coupled to the translocation of $2H^+$ and including proton translocation in the ATPase, we can express this as:

$$ATP + H_2O + 2H_M^+ \rightleftharpoons ADP + POH + 2H_C^+$$

whence, at equilibrium,

$$[ATP] = \frac{1}{k} \cdot \frac{[ADP][POH]}{[H_2O]} \cdot \frac{[H_C^+]^2}{[H_M^+]^2}$$

Thus when the concentration of protons in the matrix is low and that in the perimitochondrial space is high, the synthesis of ATP is promoted.

Mitchell's hypothesis, then, postulates a direct link between phosphorylation and the electrochemical gradient of protons across the inner mitochondrial membrane. When the rate of respiration is high, in the presence of ADP and Pi, then $\Delta \mu_H^+$ is low. For coupled plant mitochondria, the total proton motive force (Δp) in state 4 is about 160 mV and this drops to about 30 mV in the presence of ADP. In both state 3 and state 4, the electrical potential is the major component of Δp, being about four times the value of $z\Delta pH$. During state 4 respiration when $\Delta \mu_H^+$ increases, it should be close to equilibrium with the phosphorylation potential, ΔGp (= $\Delta G^{o\prime}$ + 1.36[ATP]/[ADP][Pi]). Recent data suggests that this requires an H^+/site ratio of at least 3 (Brand, 1977; Moore, 1978).

In summary, the chemi-osmotic hypothesis postulates that the respiratory chain produces an electrochemical potential gradient of protons across the inner membrane and that the return flow of these protons drives ATP production. State four respiration occurs because of the back pressure of protons being prevented from crossing the membrane in the absence of phosphate acceptor. Uncoupling agents promote increased proton conductance in the membrane thus removing the constraint on their rate of production on the C-side of the membrane. This leads to increased oxygen uptake and the bypassing of the ATPase with resultant lack of ATP synthesis.

The chemical coupling and chemi-osmotic hypotheses have been the centre of controversy for a number of years and have resulted in so many publications that it is impossible to review here the experimental evidence in favour of each. At present both hypotheses still have their proponents and antagonists although on balance the chemi-osmotic hypothesis has gained support over recent years. It is now firmly established that the transfer of electrons down the electron transport chain can, reversibly, produce a proton motive force. However, it remains uncertain whether the movement of protons is a part of the synthesis of ATP, although it is clear that the electrochemical gradient of protons is the driving force for ion and metabolite transport.

It should be realized, however, that the two theories outlined above are not the only runners in the field, although they are at the front. The *conformational coupling hypothesis* which was originally put forward by Boyer in 1964 and has lately been extended by Green and his colleagues postulates that conformational changes in enzyme complexes are brought about by electron transport and these in turn induce changes in an ATPase and the synthesis of ATP. The views of a number of the principal workers in the field (Boyer, Chance, Ernster, Mitchell, Racker and Slater) are succinctly expressed in a multi-author review (Boyer et al., 1977).

The ATPase. Sonication of mitochondria produces submitochondrial vesicles or particles, and by a variety of techniques it has been possible to show that these are formed from the inner membrane. This membrane, however, has been turned inside out, so that what was the matrix side of the membrane in the whole mitochondrion becomes the outside of the vesicle. These submitochondrial particles are capable of oxidative phosphorylation although removal of the surface components destroys the phosphorylative ability before affecting the oxidative capacity. The ability to phosphorylate may be restored by a number of so-called *coupling factors*, only one of which, however, has the ability to catalyse the hydrolysis of ATP (F_1 in the terminology of Racker, 1970). The ATPase activity of F_1 is insensitive to oligomycin, although sensitivity may be restored by recombination with either the mitochondrial membrane or with a phospholipid fraction in addition to a particulate hydrophobic protein and an oligomycin-sensitivity-conferring protein, (F_0).

The ATPase (F_1) can be seen under the electron microscope to consist of spherical particles 8·5 nm in diameter. These particles may be identified with the inner membrane sub-units first described by Fernandez-Moran (see p. 283) although they are not seen to be stalked; submitochondrial particles depleted of F_1 are relatively smooth. The stalks of the inner membrane particles appear to be equivalent to the F_0 protein. Mitchell suggests that the F_1 ATPase or inner membrane sub-unit is equivalent to the synthetase part of his reversible ATPase and is drawn as such in Fig. 7.18.

ATP synthesis from glucose. The TCA cycle produces three moles of NADH and one mole of reduced flavoprotein from one mole of acetyl coenzyme A (p. 289). Thus, in theory, the maximum amount of ATP which may be produced by oxidative phosphorylation would result in nine moles from the NADH and two moles from the reduced flavoprotein. Thus in total 12 moles of ATP may be produced from the TCA cycle oxidation of one mole of acetyl coenzyme A, the other mole of ATP being produced in the succinyl coenzyme A synthetase reaction, either directly by the ADP specific enzyme or indirectly by nucleoside diphosphate kinase (GTP + ADP = GDP + ATP). Thus from one mole of pyruvate a total of 15 moles of ATP may be produced and from one mole of glucose 38 moles of ATP. Of these 38 moles, 8 moles arise from glycolysis, two being produced by substrate level phosphorylation (see p. 247). The pentose phosphate pathway may result in up to 36 moles of ATP all being produced by oxidative phosphorylation from the oxidation of one mole of glucose (but see p. 263).

Control of electron flow. The calculation of the extent of ATP synthesis depends upon the assumption that electron transport is tightly coupled to phosphorylation. In many plants which have an alternative terminal oxidase this patently is not the case. Under these circumstances electrons can flow down either pathway. It has been established, using cyanide and hydroxamates, that the total rate is always lower than the sum of the potential rates of the cytochrome and the alternative pathways. The cytochrome pathway appears to be near its maximum at all times and consequently the alternative pathway must be only partially utilized. The regulation of electron flow is postulated to be determined by the degree of reduction of the common component at the branch point (Q) which as a much more positive redox potential than the next component of the alternative pathway. This means that the first component of the alternative pathway can be completely oxidized when the ubiquinone is partially reduced, and only when a greater proportion of the ubiquinone is reduced (by say an inhibition of the cytochrome pathway) are electrons transferred to the alternative pathway.

Inner membrane topography

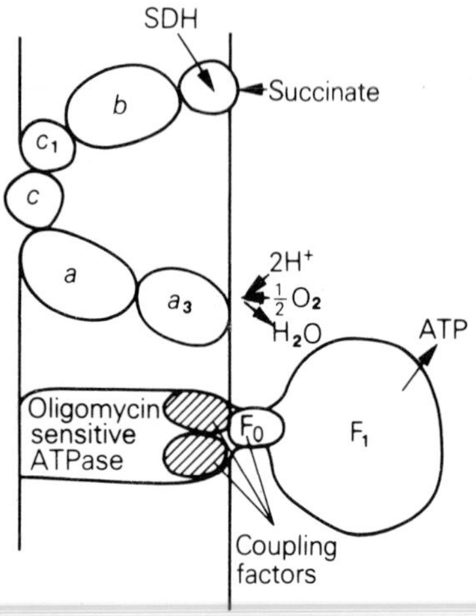

The location of the various components of oxidative phosphorylation within the inner mitochondrial membrane has been studied primarily by the use of histochemistry and of macro-molecular probes (molecules which do not permeate the membrane but which react with membrane components). The probable location of some of the known components is shown in Fig. 7.19; the evidence comes chiefly from work with animal mitochondria. The location of the ATPase on the M-side of the membrane is supported by work on reconstitution of oxidative phosphorylation in submitochondrial particles and by functional tests with an antibody against F_1. Phospholipids are apparently present throughout the membrane, since they are liable to degradation by phospholipase from either side of the membrane. Cytochrome c is located on the side of the perimitochondrial space since specific antibodies against cytochrome c do not inhibit respiration in submitochondrial particles. If the particles are prepared in the presence of the antibody, however, respiration is inhibited. Cytochrome oxidase has been shown to be present on the C-side of the membrane by electron microscopy, but on the M-side by studies with azide. Work with macromolecular probes indicates that both answers are correct, however, as cytochrome a lies on the C-side whilst cytochrome a_3 is on the M-side. Succinic dehydrogenanse is located on the matrix-side of the inner membrane since oxidative phosphorylation may be reconstituted in depleted submitochondrial particles by the additon of succinic dehydrogenase. The cytochromes b presumably lie between succinic dehydrogenase and cytochrome c and this arrangement drawn in Fig. 7.19 is reminiscent of the oxidation–reduction loop proposed by Mitchell.

Other metabolic activities

Apart from the enzymes associated with the TCA cycle and oxidative phosphorylation, a number of other enzymes have been reported associated with the mitochondria. Many of these have been reported

Fig. 7.19   A possible arrangement of the components in the inner mitochondrial membrane. Redrawn from Racker (1972).

only infrequently (e.g. alcohol dehydrogenase, ascorbic acid oxidase, hexokinase) and consequently may be contaminants or only characteristic of the mitochondria of certain tissues. Glutamate oxidation has been reported in a number of cases and both amino transferases (transaminases) and glutamate dehydrogenase appear to occur in the mitochondrial fraction. The formation of glutamate, catalysed by glutamate dehydrogenase:

$$\alpha\text{-Ketoglutarate} + NH_3 + NADH + H^+ = \text{glutamate} + H_2O + NAD^+$$

is an important reaction in terms of amino acid formation, since it is through this reaction that carbon skeletons may be diverted from the TCA cycle. As we have already indicated on p. 293, carbon from the 2- and, particularly 3-position of labelled acetate is not fully recovered as carbon dioxide following oxidation of the acetate. Aspartate and alanine may be produced from glutamate by aminotransferase activity:

$$\text{Oxaloacetate} + \text{L-glutamate} \xrightarrow{\substack{\text{Aspartate} \\ \text{aminotransferase}}} \alpha\text{-ketoglutarate} + \text{aspartate}$$

$$\text{Pyruvate} + \text{L-glutamate} \xrightarrow{\substack{\text{Alanine} \\ \text{aminotransferase}}} \alpha\text{-ketoglutarate} + \alpha\text{-ketoglutarate}$$

Subsequently, aspartate and glutamate, together with glycine (formed by the reaction: glyoxylate + L-glutamate → α-ketoglutarate + glycine) are then able to give rise to a variety of other amino acids, purines, pyrimidines and porphyrins (but see also Ch. 8). Of course, withdrawal of the intermediates of the TCA cycle must ultimately prevent its functioning. However, the pools of oxaloacetate and malate may be supplemented by the activity of phosphoenolpyruvate carboxylase or malic enzyme (see p. 273).

Mitochondria also appear to be able to oxidize glycine to serine in a reaction which requires oxygen:

$$\tfrac{1}{2}O_2 + 2\,\text{glycine} \longrightarrow \text{serine} + CO_2 + NH_3$$

The reaction in mitochondria prepared from green tobacco tissue appears to be mediated by the electron transport system, while the rate of the reaction is increased by the addition of ADP which is phosphorylated to ATP: i.e. the oxidation of glycine to serine is coupled to ATP synthesis (Bird, et al., 1972; Moore et al., 1977). This oxidation of glycine is of importance in producing ATP from the reactions of photorespiration (see Ch. 9).

The β-oxidation of fatty acids by plant mitochondrial fractions has been reported (Stumpf and Barber, 1956), but recently this has been shown to be associated with a separate particle, the glyoxysome (Cooper and Beevers, 1969), the structure and function of which are discussed in Chapter 9.

The other major process which seems to occur in the mitochondria is protein synthesis. As remarked earlier, mitochondria contain RNA, DNA and ribosome-like particles, and although incorporation of amino acids into specific mitochondrial proteins has not so far been demonstrated *in vitro* in higher plants, it is apparent that plant mitochondria can incorporate amino acids into protein. It is possible, therefore, that they are similar to the mitochondria of fungi and animals in that they synthesize some protein in *vivo* (Boulter, 1970; see also p. 219).

Transport processes in mitochondria

Ion transport

The work of Tedeschi and Harris (1955, 1958) showed that animal mitochondria behave as osmometers, i.e. one or both of their membranes has semipermeable properties. Tedeschi and Harris were also able to show that the volume of a mitochondrion is inversely proportional to the osmotic potential (π) of the suspending medium (sucrose) as predicted by the Boyle-van't Hoff osmotic law:

$$V_t = \frac{-K}{\pi} + b$$

Where V_t is the total mitochondrial volume, K is a proportionality constant ($= nRT$ for an ideal solute) and b is the 'osmotic dead space', the volume of the mitochondrion which is not permeated by the solute. Determinations of the sucrose-accessible space have shown that the outer membrane is, in fact, permeable to this solute and that it is the inner membrane which imparts the osmotic properties to the mitochondrion. The osmotic law has been shown to apply not only to animal but also to plant mitochondria. In all such experiments, however, the volume is estimated either by calculation from photomicrographs or by the determination of the water content which is then expressed on the basis of mitochondrial protein. These methods are slow and hence not suitable for determinations of the kinetics of volume changes which may be extremely rapid. Fortunately, rapid changes in volume are readily monitored by following changes in the absorption of the mitochondrial suspension, generally at a wavelength at which absorption changes due to oxidation and reduction are minimal (520–550 nm). It has been shown that the volume is directly proportional to the reciprocal of the absorbance or to the transmittance at high transmittance values (note that absorbance $= -\log_{10}$ transmittance). Thus routine measurements of volume and volume changes are readily made spectrophotometrically: the absorbance decreases as the volume increases. The use of these spectrophotometric techniques has shown that osmotic adjustment of the mitochondrial volume, as the tonicity of the suspending sucrose solution is changed, is extremely rapid with a t of the order of milliseconds. Subsequently the absorbance (and volume) remains relatively constant. Maize mitochondria can be induced to swell in surcose, however, after ageing at room temperature, or by the addition of a chelating agent. Possibly a cation stabilized membrane structure is weakened so that chelation of divalent cations destroys the semi-permeability and allows an influx of sucrose followed by an osmotic equivalent of water (Stoner and Hanson, 1966).

Plant mitochondria are notably different from animal mitochondria because of their spontaneous swelling in buffered KCl and NaCl solutions (Fig. 7.20). This relatively slow change in volume following the rapid osmotic adjustment is apparently not an energy dependent process, since respiratory inhibitors do not prevent swelling while the addition of respiratory substrates does. High concentrations of uncoupling agents promote the rate of swelling, but have little effect on its extent. The passive nature of the swelling in KCl is in direct contrast to the swelling of animal mitochondria which depends on the addition of swelling agents and is apparently mediated by endogenous respiratory substrates. The high permeability of plant mitochondria may be related to high levels of free fatty acids present during the isolation. However, K^+ binding ionophores such as gramicidin D, which promotes energy-linked swelling in animal mitochondria, promotes the rate of passive swelling in plant mitochondria, indicating that it is the cation penetration which is rate limiting.

The spontaneous swelling of plant mitochondria can be reversed by the addition of a respiratory substrate or ATP + Mg^{++} (Fig. 7.20). The ATP-powered contraction is inhibited by oligomycin while the substrate-powered contraction is unaffected by oligomycin, but inhibited by respiratory chain inhibitors. Contraction involves the expulsion of water and the inner membrane does not appear to decrease in area, but collapses around the condensing matrix. The mechanism by which the water is expelled is presumably the same as that of water uptake: the transfer of ions across the membrane followed by an osmotic equivalent of water. This energy dependent movement of ions is generally thought to be linked to an electrochemical potential gradient of protons. We have already seen (p. 310) that the inner membrane is largely impermeable to protons and yet electron transport generates a proton gradient across this membrane. In the presence of ADP and inorganic phosphate the gradient is dissipated in the generation of ATP. However, it is envisaged that apart from driving ATP synthesis, the proton gradient can also be used to drive

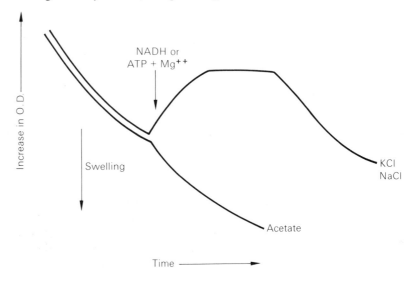

Fig. 7.20 A diagrammatic representation of the volume changes occurring in plant mitochondria in a solution of potassium chloride (or sodium chloride) and of potassium acetate. The different effect of adding of NADH + H^+ or ATP + $MgCl_2$ is shown for chloride and acetate.

ion transport across the membrane. Water movement occurs as an osmotic consequence causing the observed changes in volume.

The transfer of ions across the membrane is thought to occur through specific carriers or *'porters'* (see p. 150) which exchange H^+ or OH^- for other cations or anions. For example, during NADH powered contraction in a KCl medium, electron transport would generate a proton motive force (p. 309) which would enable a H^+/K^+ exchange to occur causing the efflux of potassium ions from the mitochondrion. As the dissociation constant for water is very low, the exchange of hydrogen ions for potassium ions abolishes the gradient of H^+ and turns it into an electrochemical gradient of K^+. Since the exchange is electrically neutral, the membrane potential is unaltered so that if potassium ion efflux is to continue the membrane potential must be collapsed by the efflux of the anion, Cl^-. Thus, in general, energy-dependent salt movement can be envisaged as H^+/cation or OH^-/anion exchange with the accompanying movement of the anion or cation respectively down an electrochemical potential gradient (Fig. 7.21). Where the coupled movement is of two ions of the same charge in opposite directions the carrier is termed an *antiporter*: *symport* is the transport of ions of opposite charge in the same direction.

Fig. 7.21 A schematic representation of the various transport processes occurring across the inner mitochondrial membrane. The respiratory chain is drawn as generating the proton gradient which is then dissipated either through the production of ATP or the transport of anions and cations.

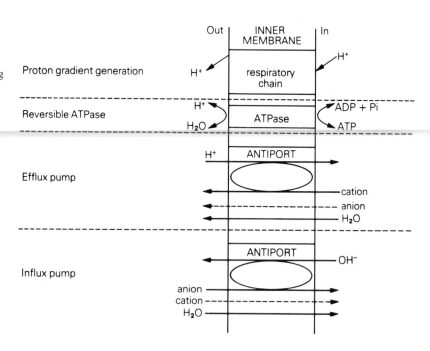

The generation of a proton gradient normally occurs either through the oxidation of substrates or from hydrolysis of ATP by the reversible ATPase. However, it has been found that ammonium salts can also be used to generate a gradient of hydrogen ions. It is envisaged that the neutral NH_3 passes through the membrane and once inside generates NH_4^+ and OH^- through reaction with water, viz.

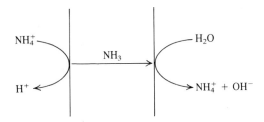

If the anion is permeable as, for example, is acetate, then the mitochondria swell as an osmotic consequence of the transport of the salt to the inner compartment. The distinction between the terms permeable and impermeable is a relative one. The membranes may be rather impermeable to say OH^- but much more permeable to chloride or acetate. However, the permeability may be much increased by the presence of a specific carrier as described above, although it must be established that such a carrier is present. Such a conclusion is normally based on specificity for the substance transported, the fact that transport shows saturation kinetics, and perhaps most usefully, the existence of specific inhibitors. For phosphate transport, which is clearly important for mitochondria carrying out phosphorylation, it has been established that plant mitochondria swell in ammonium phosphate — through the permeation of NH_3 as described above, and of the phosphate anion. However swelling is sensitive to two specific inhibitors, mersalyl and N-ethylmaleimide (NEM), so that it is argued that there is a phosphate/hydroxyl antiport which is sensitive to these inhibitors. Similarly, specific carriers for the dicarboxylic acids, malate and succinate, have been identified by inhibition with butyl- and pentyl-malonate. However, they also require phosphate and may be inhibited by mersalyl and NEM, indicating that OH^- exchanges with phosphate which in turn exchanges with the discarboxylate anion, viz.,

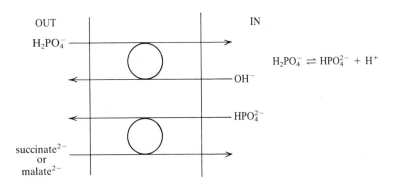

Neither fumarate nor oxalacetate enters on this antiport: oxalacetate apparently penetrates mitochondria with ease. There is some evidence for a tricarboxylate transporter in plants: swelling in ammonium citrate requires phosphate and a transportable dicarboxylic acid and it is argued, largely by analogy with the better studied animal system, that phosphate exchanges for hydroxyl, malate for phosphate, and then citrate for malate. Entry of pyruvate seems to be simply via an antiport with OH^-

Of paramount importance to phosphorylating mitochondria is the import of ADP and the export of ATP. ATP appears to exchange in a 1:1 ratio with ADP in animal mitochondria and the exchange is specifically inhibited by atractyloside and by bongkrekic acid in both plants and animals. Although there is relatively little other information about this antiport from plant systems, work on animal cells indicates that the exchange between the two nucleotides is asymmetric in that it favours the influx of ADP and the efflux of ATP. Since the charges of the two anions differ (ATP^{4-} and ADP^{3-}), the exchange is considered to be electrongenic (see p. 149). Furthermore, since the inside of the mitochondrion is negative with respect to the outside due to the action of the respiratory chain, this would help to explain the asymmetry of transport given the differences in charge between the two nucleotides. Thus the ATP/ADP ratio is higher outside the mitochondrion than within the matrix and this has the consequence that the energy required to phosphorylate ADP is lower inside than outside. However, the maintenance of a difference in ATP/ADP ratio across the inner membrane does itself require energy which is in turn derived from the actions of the respiratory chain.

The various transporters which have been identified in mitochondrial membranes are clearly of fundamental importance in controlling the movement of substances into and out of the mitochondrial matrix. They are therefore capable of controlling the distribution of substances between cytosol and mitochondria and of influencing rates of mitochondrial metabolism. Further discussions of mitochondrial transport systems are available in the reviews of Hanson and Koeppe (1975) and Wiskich (1977).

Ontogeny

The mitotic divisions of meristematic cells results in the passive segregation of the mother cell's complement of mitochondria between the two daughter cells. During the subsequent differentiation of these daughter cells, however, the numbers of mitochondria increase considerably: in *Zea mays* root cap the meristematic initial contains about 200 mitochondria per cell, while during growth this number increases about tenfold (Juniper and Clowes, 1965). This increase in mitochondrial number is not fully understood, although it is apparent that mitochondria can divide. This ability to divide in the absence of nuclear division suggests a degree of autonomy.

If the mitochondria are in any way autonomous then they must be able to control their own replication. If totally autonomous, this means a self contained ability to control their own protein synthesis, although if a degree of semi-autonomy is envisaged, nuclear genetic information and the cytoplasmic machinery for protein synthesis may be involved.

The presence of DNA and ribosomal material within the mitochondrial matrix has already been described (p. 285). Mitochondrial DNA has been demonstrated by electron microscopy in plants, animals and micro-organisms, and in a number of species the use of labelled thymidine has shown that the DNA replicates independently of nuclear DNA. This replication requires the mitochondria to contain DNA polymerase which has been isolated and partially purified for both yeast and rat liver. It is important to note that in most of the species investigated, the mitochondrial DNA is distinct from the nuclear DNA, having a different buoyant density (1.706 to $1.707g$ cm^{-3} for plant mitochondrial DNA) reflecting a difference in base composition (see Rabinowitz and Swift, 1970). Circular DNA molecules have now been isolated from animal, *Neurospora*, *Chlamydomonas* and tobacco mitochondria.

Ribosome-like particles have been seen in the matrices of mitochondria: in *Neurospora* they have been clearly distinguished from cytoplasmic ribosomes. The mitochondrial ribosomes have lower sedimentation coefficients than their cytoplasmic counterparts and distinct RNA components, so should perhaps be considered as a distinct type, differing from cytoplasmic but similar to prokaryotic ribosomes (see Ch. 5). Furthermore, in yeast and certain mammalian mitochondria, *N*-formylmethionyl–tRNA is present in the mitochondria, but not in the cytoplasm. It is *N*-formylmethionyl–tRNA that is involved in the initiation of bacterial protein synthesis (see Ch. 5).

Protein synthesis has now been shown to occur unequivocally in animal and plant mitochondria and in those isolated from *Neurospora* and yeast (Boulter, Ellis and Yarwood, 1972). This protein synthesis is inhibited by D-*threo*-chloramphenicol and lincomycin which also inhibit protein synthesis in bacteria, but not in the eukaryotic cytoplasm; cycloheximide inhibits protein synthesis in the cytoplasm but not in mitochondria or bacteria. However, the amount of information for protein synthesis encoded in the mitochondrial DNA appears to be very limited. From the length of the circular DNA (5μm) found in animal mitochondria, it is possible to calculate the number of base pairs present (bases are 0.33 nm apart) as 15 000 (equivalent to 5 000 amino acids or fifty small proteins). Furthermore, since the DNA present in chick mitochondria has been shown to be homogeneous, it is clear that, at most, the mitochondria may be semi-autonomous. It is interesting to note that plant mitochondria contain considerably more DNA than animal mitochondria although the reasons for this are not known. Genetic evidence confirms that many mitochondrial proteins are coded for in the nucleus. The synthesis of cytochrome *c* has, for example, been shown to be under nuclear control in yeast as are the TCA cycle enzymes. It appears that the majority of polypeptides synthesized by the mitochondrial system are components of the oxidative phosphorylation/electron transport chain, i.e inner membrane proteins (see Schatz, 1976; Whittaker and Danks, 1978; Ch. 5).

The presence of self-replicating DNA and of the ability to carry out at least a limited amount of protein synthesis suggests that during cellular development mitochondria are formed from other mitochondria. Some evidence in favour of this hypothesis has been obtained with *Neurospora* (see Luck, 1965). After labelling mitochondrial lipids by growth on a medium containing ^{3}H-choline and then transferring to unlabelled media, the mitochondria were found to be randomly labelled several generations later (i.e. label appeared in all the mitochondria). This suggested that all of the mitochondria were formed from existing mitochondria. The formation of mitochondria from nuclear evaginations has been proposed and this may be important in the ferns (Bell and Mühlethaler, 1964), although the weight of evidence now suggests that mitochondria are formed by division of existing mitochondria or their sub-units. Since the production of mitochondrial components takes place in both the mitochondrion and the cytoplasm, the turnover of their constituent parts is not uniform. In rat liver the inner membrane appears to turn over with a $t_{\frac{1}{2}}$ of about 5–6 days while the outer membrane has a $t_{\frac{1}{2}}$ of about 4 days.

Further reading

Beevers, H. (1961) *Respiratory Metabolism in Plants*. Harper and Row, New York.

Boulter, D., Ellis, R. J. and Yarwood, A. (1972) Biochemistry of protein synthesis in plants. *Biol. Rev.*, **47**, 113.

Boyer, P. D. *et al.* (1977) Oxidative phosphorylation and photophosphorylation. *Ann. Rev. Biochem.*, **46**, 955.

Clowes, F. A. L. and Juniper, B. D. (1968) *Plant Cells*. Blackwell, Oxford and Edinburgh.

Hanson, J. B. and Hodges, T. K. (1967) Energy-linked reactions in plant mitochondria. *Current Topics in Bioenergetics*, **2**, 65. Ed. D. R. Sandai. Academic Press, New York.

Hanson, J. B. and Koeppe, J. R. (1975) Mitochondria. In *Ion Transport in Plant Cells and Tissues*. Eds. D. A. Baker and J. L. Hall, p. 79. North-Holland.

Ikuma, H. (1972) Electron transport in plant respiration. *Ann. Rev. Plant Physiol.*, **23**, 419.

Palmer, J. M. (1976) The organisation and regulation of electron transport in plant mitochondria. *Ann. Rev. Plant Physiol.*, **27**, 133.

Palmer, J. M. (1979) The 'uniqueness' of plant mitochondria. *Biochem. Soc. Trans.*, **7**, 246.

Pridham, J. B. (Ed) (1968) *Plant Cell Organelles*. Academic Press, London.

Racker, E. (1970) The two faces of the inner mitochondrial membrane. In *Essays in Biochemistry*, Vol. 6, p. 1. Eds. P. N. Campbell and F. Dickens. Biochemical Society.

Roodyn, D. B. (1967) The mitochondrion. In *Enzyme Cytology*, p. 103. Ed. D. B. Roodyn. Academic Press, London.

Storey, B. T. (1980) Electron transport and energy coupling in plant mitochondria. In *The Biochemistry of Plants Vol. 2 Metabolism and Respiration*. Ed. D. D. Davies. Academic Press, New York.

Turner, J. F. and Turner, D. H. (1975) The regulation of carbohydrate metabolism. *Ann. Rev. Plant Physiol.*, **26**, 159.

Whittaker, P. A. and Danks, S. M. (1978) *Mitochondria: Structure, Function and Assembly*. Longman, London and New York

Wiskich, J. T. (1977) Mitochondrial metabolite transport. *Ann. Rev. Plant Physiol.*, **28**, 45.

Literature cited

Baker, J. E., Elfvin, L. G., Biale, J. B. and Honda, S. I. (1968) Studies on ultrastructure and purification of isolated mitochondria. *Plant Physiol.*, **43**, 2001.

Beinhart, H. and Palmer, G. (1965) Contributions of EPR spectroscopy to our knowledge of oxidative enzymes. In *Advances in Enzymology*, Vol. 27, p. 105. Ed. F. F. Nord. Interscience.

Bell, P. R. and Mühlethaler, K. (1964) The degeneration and reappearance of mitochondria in egg cells of a plant. *J. Cell. Biol.*, **20**, 235.

Bhagvat, K. and Hill, R. (1951) Cytochrome oxidase in higher plants. *New Phytol.*, **50**, 112.

Bird, I. F., Cornelius, M. J., Keys, A. J. and Whittingham, C. P. (1972) Oxidation and phosphorylation associated with the conversion of glycine to serine. *Phytochemistry*, **11**, 1587.

Boulter, D. (1970) Protein synthesis in plants. *Ann. Rev. Plant Physiol.* **21**, 91.

Brand, M. D. (1977) The stochiometric relationships between electron transport, proton translocation and adenosine triphosphate synthesis and hydrolysis in mitochondria. *Biochem. Soc. Trans.* **5**, 1615.

De Chadarevjan, S. de Santis, A., Melandri, B. A., Baccarini Melandri, A. (1979) Oxidative phosphorylation and proton translocation in a lipoate-deficient mutant of *Escherichia coli*. FEBS Letts. **97**, 293.

Chance, B. (1954) Spectrophotometry – intracellular respiratory pigments. *Science, N.Y.*, **120**, 767.

Chance, B. and Williams, G. R. (1956) The respiratory chain and oxidative phosphorylation. *Adv. Enzymol.* **17**, 65. Ed. F. F. Nord. Interscience.

Chance, B., Bonner, W. D. and Storey, B. T. (1968) Electron transport in respiration. *Ann. Rev. Plant Physiol.*, **19**, 295.

Clark, L. C. (1956) Monitor and control of blood and tissue oxygen tension. *Trans. Am. Soc. Artificial Internal Organs*, **2**, 41.

Cooper, T. G. and Beevers, H. (1969) Mitochondria and glyoxysomes from castor bean endosperm. Enzyme constituents and catalytic capacity, *J. biol. Chem.* **244**, 3507.

Davies, D. and Ellis, R. J. (1964) Enzymes of the Krebs' cycle, the glyoxalate cycle and related enzymes. In *Modern Methods of Plant Analysis*. Vol. 7, p. 616. Eds. K. Peach and M. V. Tracey.

Day, P. A. and Hanson, J. B. (1977) On methods for the isolation of mitochondria from etiolated corn shoots. *Pl. Sci. Lett.*, **11**, 99.

Douce, R., Christensen, E. L. and Bonner, W. D. (1972) Preparation of intact plant mitochondria. *Biochim. Biophys. Acta.*, **275**, 148.

Dutton, P. L. and Storey, B. T. (1971) The respiratory chain of plant mitochondria, IX. Oxidation-reduction potentials of the cytochromes of mung bean mitochondria. *Plant Physiol.*, **47**, 282.

Fernandez-Moran, H. H. (1962) Cell membrane ultrastructure. Low temperature electron microscopy and X-ray diffraction studies of lipoprotein components in lamellar systems. *Circulation*, **26**, 1039.

Flowers, T. J. (1974) Salt tolerance in the halophyte *Suaeda maritima* L. A comparison of mitochondria isolated from green tissues of *Suaeda* and *Pisum. J. exp. Bot.*, **25**, 101.

Freebairn, H. T. and Remmert, L. F. (1957) The tricarboxylic acid cycle and related reactions catalysed by particulate preparations from cabbage. *Physiol. Plant.*, **10**, 20.

Greville, G. D. (1969) A scrutiny of Mitchell's chemi-osmotic hypothesis of respiratory chain and photosynthetic phosphorylation. *Current Topics in Bioenergetics Vol. 3*, p. 1. Ed. D. R. Sanadi. Academic Press, New York.

Griffiths, D. E. (1976) Studies of energy-linked reactions. *Biochem. J.* **160**, 809.

Hackenbrok, G. R. (1966) Ultrastructural basis for metabolically linked mechanical activity in mitochondria. I. Reversible ultra-structural changes with change in metabolic state in isolated liver mitochondria. *J. Cell Biol.*, **30**, 269.

Henry, M. R. and Nyms, E. J. (1975) Cyanide insensitive respiration. An alternative mitochondrial pathway. *Sub-cell. Biochem.*, **4**, 1.

Jackson, C. and Moore, A. L. (1979) Isolation of intact higher plant mitochondria. In *Plant Organelles*. Ed. E. Reid. Ellis Horwood, Chichester, p. 1.

Juniper, B. E. and Clowes, F.A. L. (1965) Cytoplasmic organelles and cell growth in root caps. *Nature, Lond.*, **208**, 864.

Lambowitz, A. M. and Bonner, W. D. Jr. (1974) The *b*-cytochromes of plant mitochondria. A spectrophotometric and potentiometric study. *J. Biol. Chem*, **249**, 2428.

Lance, C. and Bonner, W. D. (1968) The respiratory chain components of higher plant mitochondria. *Plant Physiol.*, **43**, 756.

Lehninger, A. L. (1970) *Biochemistry*. Worth Pub. Co., New York.

Luck, D. J. L. (1965) Formation of mitochondria in *Neurospora crassa. J. Cell Biol.*, **24**, 461.

Malone, C., Koeppe, P. E. and Miller, R. J. (1974) Corn mitochondrial swelling and contraction – an alternative interpretation. *Plant Physiol.*, **53**, 918.

Meeuse, B. J. P. (1975) Thermogenic respiration in Aroids. *Ann. Rev. Plant Physiol.* **26**, 117.

Millerd, A., Bonner, J., Axelrod, B. and Bandurski, R. S. (1951) Oxidative and phosphorylative activity of plant mitochondria. *Proc. natl. Acad. Sci. U.S.A.*, **37**, 855.

Mitchell, P. (1968) Chemi-osmotic coupling and energy transduction. Glynn Research Ltd., Bodmin, Cornwall, England.

Mitchell, P. (1972) Structural and functional organisation of energy-transducing membranes and their ion-conducting properties. *Febs Symp.*, **28**, 353.

Mitchell, P. (1976) Possible molecular mechanisms of the protonmotive function of cytochrome systems. *J. theoret. Biol.*, **62**, 327.

Moore, A. L. (1978) The electrochemical gradient of protons as an intermediate in energy transduction in plant mitochondria. In *Plant Mitochondria*. Eds. G. Duret and C. Lance, p. 85. Elsevier/North-Holland Biomedical Press.

Moore, A. L., Jackson, C., Halliwell, B., Dench J. E. and Hall, D. O. (1977) Intramitochondrial localisation of glycine decarboxylase in spinach leaves. *Biochem. Biophys. Res. Comm.*, **78**, 483.

Nadakavukaren, M. J. (1964) Fine structure of negatively stained plant mitochondria. *J. Cell. Biol.*, **23**, 193.

Nicholls, P. and Eliot, W. B. (1974) The cytochromes. In *Iron in Biochemistry and Medicine*. Eds. A. Jacobs and M. Worwood, p. 221. Academic Press, London and New York.

Palmer, J. M. (1967) Rapid isolation of active mitochondria from plant tissue. *Nature, London.*, **216**, 1208.

Parsons, D. F. (1965) Recent advances in correlating structure and function in mitochondria. *Int. Rev. exptl. Path.*, **4**, 1.

Parsons, D. F., Bonner, W. D. and Verboon, J. G. (1965) Electron microscopy of isolated plant mitochondria and plastids using both the thin-section and negative-staining techniques. *Can. J. Bot.*, **43**, 647.

Rabinowitz. M. and Swift, H. (1970) Mitochondrial nucleic acids and their relation to the biogenesis of mitochondria. *Physiol. Rev.*, **50**, 376.

Racker, E. (1972) Bioenergetics and the problem of tumor growth. *American Scientist*, **60**, 56.

Rich, P. R., Weigand, N. K., Blum, H., Moore, A. L. and Bonner, W. D. Jr. (1978) Studies on the mechanism of inhibition of redox enzymes by substituted hydroxamic acids. *Biochim. Biophys. Acta,* **525**, 325.

Rich, P. R. and Bonner, W. D. Jr. (1978) EPR studies of higher plant mitochondria. II. Center S-3 of succinate dehydrogenae and its relation to alternative respiratory oxidations. *Biochim. Biophys. Acta.,* **501**, 381.

Rich, P. R. and Moore, A. L. (1976) The involvement of the protonmotive ubiquinone cycle in the respiratory chain of higher plants and its relation to the branchpoint of the alternate pathway. *FEBS Lett.*, **65**, 339.

Sarkissian, I. V. and Srivastava, H. K. (1969) A very simple trick separates plant michondria from starch. *Life Sci.*, **8**, 1201

Schatz, G. (1976) The biogenesis of mitochondria—a review. In *Genetics, Biogenesis and Bioenergetics of Mitochondria*. Eds. W. Bandlow, R. J. Schweyen, D. Y. Thomas, K. Wolf and F. Kaudewitz. de Gruyter, Berlin.

Sjöstrand, F. S., Anderson-Cedergren, E. and Karlsson, U. (1964) Myelin-like figures formed from mitochondrial material. *Nature, Lond.*, **202**, 1075.

Solomos, T. (1977) Cyanide resistant respiration in higher plants. *Ann. Rev. Plant Physiol.*, **28**, 279.

Stoner, C. D. and Hanson, J. B. (1966) Swelling and contraction of corn mitochondria. *Plant Physiol.*, **41**, 255.

Storey, B. T. (1970) The respiratory chain of plant mitochondria VI. Flavoprotein components of the respiratory chain of mung bean mitochondria. *Plant Physiol.*, **46**, 13.

Storey, B. T. (1973) Respiratory chain of plant mitochondria. XV Equilibration of cytochromes c_{549}, b_{553}, b_{557} and ubiquinone in mung bean mitochondria: placement of cytochrome b_{553} and estimation of the mid-point potential of ubiquinone. *Biochim. Biophys. Acta*, **292**, 592.

Storey, B. T. (1976) Respiratory chain of plant mitochondria. XVIII Point of interaction of the alternate oxidase with the respiratory chain. *Plant Physiol.*, **58**, 521.

Stumpf, P. K. and Barber, G. A. (1956) Fat metabolism in higher plants. VII. β-Oxidation of fatty acids by peanut mitochondria. *Plant Physiol.* **31**, 314.

Tedeschi, H. and Harris, D. L. (1955) The osmotic behaviour and permeability to non-electrolytes of mitochondria. *Arch Biochem. Biophys.*, **58**, 52.

Tedeschi, H. and Harris, D. L. (1958) Some observations on the photometric estimation of mitochondrial volume. *Biochim. Biophys. Acta*, **28**, 392.

Walker, D. A. and Beevers, H. (1956) Some requirements of pyruvate oxidation by plant mitochondrial preparations. *Biochem. J.*, **62**, 120.

8 The chloroplast

Chloroplasts are, to the microscopist, perhaps the most obvious of the cytoplasmic organelles present in the cells of higher plants. They are relatively large being about 5–10 μm in diameter and were first described some fifty years prior to the discovery of mitochondria. The discovery of chloroplasts followed the realization, during the latter part of the eighteenth century, that plants were able, in the light, to remove carbon from the carbon dioxide in the air and to incorporate this carbon into their own substance with the concomitant release of oxygen from the cell. During the late nineteenth century the connection between this light dependent carbon assimilation, or photosynthesis, and the chloroplast was appreciated. Engelmann showed in 1894 that oxygen production in the green alga *Spirogyra* occurred only during the illumination of the chloroplasts, while the production of starch within the same organelle was widely realized. Henceforth it was apparent that photosynthesis occurred within chloroplasts under the influence of light, although it had not been established that starch and sucrose were synthesized from hexose.

As we have already seen, the oxidation of carbohydrate is an exergonic process, so that its formation from carbon dioxide requires energy. In photosynthesis, this energy is supplied by solar radiation, which must first be intercepted and then converted to a form which can be utilized by the cell. Consequently one of the primary events of photosynthesis is the absorption of light by suitable pigments (e.g. chlorophyll) and the subsequent production of ATP (photophosphorylation). This ATP is used together with NADPH, which is also generated in the chloroplast under the influence of light, to reduce carbon dioxide to carbohydrate. It should of course be realized that green plants are not unique in their photosynthetic ability which is shared with certain bacteria and with the non-green algae.

Structure

All organisms which carry out the photosynthetic processes contain a system of double membranes or *lamellae* within the cell. In the photosynthetic bacteria, this membrane material is often grouped to form distinct chromatophores although this does not occur in all species (Fig. 8.1). In the blue-green algae the membranes which arise from the outer cell membrane appear to ramify through the cytoplasm,

Fig. 8.1 A photosynthetic bacterium, *Rhodopseudomonas palustris*. The photosynthetic apparatus is visible as a lamellar structure (L) which is not delimited by an envelope. W, cell wall; Ge, genetic material; P polyphosphate deposit. ×123 000. Electron micrograph from Dr G. Cohen–Bazire, Institut Pasteur, Paris: reproduced from Gregory (1971).

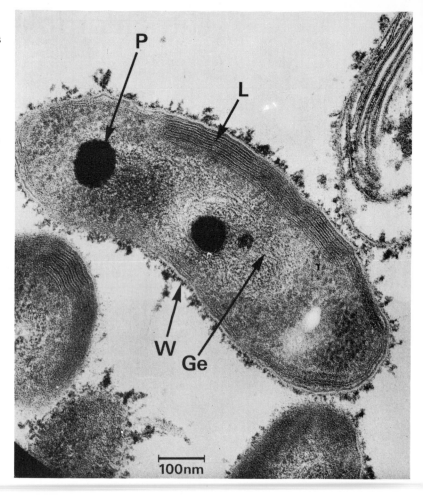

although in the other eukaryotic algae (Fig. 8.2) and in higher plants, the photosynthetic apparatus is separated from the cytoplasm by a bounding envelope.

Higher plant chloroplasts are generally biconvex or plano-convex with a diameter of about 5–10 μm and a thickness of 2–3 μm. In angiosperms there are about 15–20 chloroplasts per photosynthetic cell and they tend to lie with their broad faces parallel to the cell wall. However, neither their position nor their shape is rigidly fixed and they may move in relation to the incident light intensity as well as in the general cytoplasmic streaming. When observed under the electron microscope following osmium fixation, the mature chloroplast can be seen to be bounded by an envelope which appears as two electron dense lines separated by an electron lucent layer (Fig. 8.3). Following permanganate fixation, each dark line can be resolved into two electron dense layers (2 nm thick) separated by an electron transparent layer (1 nm): that is, a unit membrane. Unlike the mitochondria, however, the internal membrane system of the mature chloroplast does not appear to be connected with the envelope. Within the envelope the membranes form lamellae or *thylakoids* which consist of

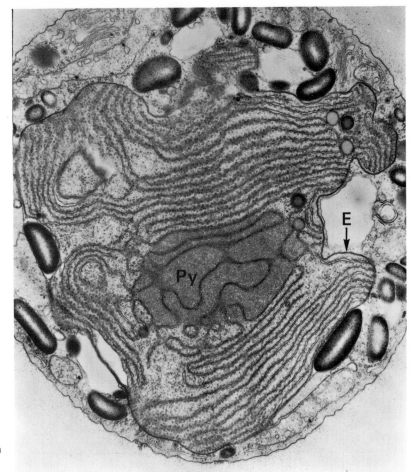

Fig. 8.2   A cell of the red alga, *Phorphyridium cruentum*, in thin section. There is a single chloroplast which is bounded by an envelope (E) and a large pyrenoid (Py). ×22 000. Provided by A. D. Greenwood, Imperial College, University of London; reproduced from Gregory (1971).

Fig. 8.3   A electron micrograph of a section of a broad bean leaf cell. The chloroplast is enclosed by the envelope (E) and the thylakoids (T) run through the stroma (Sr) and in places are arranged to form the grana (G). M, mitochondrion; Mic, microbody; S, starch. ×29 000. Provided by A. D. Greenwood. Imperial College, University of London; reproduced from Gregory (1971).

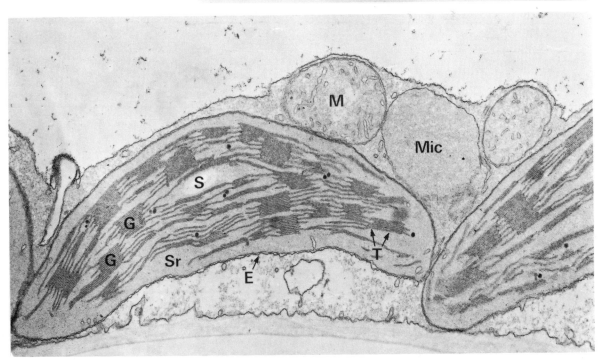

two membranes each about 7·0 nm thick lying adjacent to each other, but separated by an intra-membrane space which may vary from 4 to 70 nm in thickness. These thylakoids run throughout the matrix of the chloroplast which is known as the *stroma* (see Fig. 8.3). In places these lamellae are stacked rather like piles of coins and these layers of membranes produce dark green regions which can be seen under the light microscope and are known as *grana* (Fig. 8.4). These grana are about 0·3–2 μm in diameter and in general there are somewhere between 10 and 100 thylakoids per granum and 40–60 grana per chloroplast in typical photosynthetic cells. The grana thus consist of a series of membranes and spaces although under the electron microscope the exact appearance may be somewhat variable depending on the relative staining of the various components (von Wettstein, 1967).

A three-dimensional picture of the chloroplast lamellae was slow to emerge because of the large focal depth of the electron microscope. It is now realized, however, that the thylakoids run through the stroma (stromal thylakoids) and are perforated and arranged in a three-dimensional network. In places these stromal thylakoids form rounded tongues which are arranged in piles forming the grana (see Fig. 8.5). A single stromal lamella may form a spiral arrangement around a granum and thus be connected with a number of the granal thylakoids in a right-handed helical pattern. The thylakoid system or systems within a chloroplast thus form a membrane bounded space or spaces. Both ends of a granum are in close contact with the stroma and have been

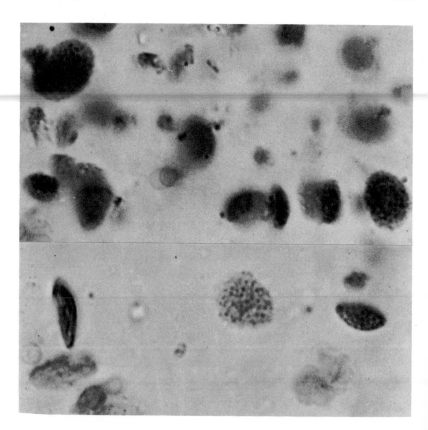

Fig. 8.4 Chloroplasts isolated from spinach leaves and seen under the light microscope. The grana are clearly visible as darker areas within the stroma. ×2250.

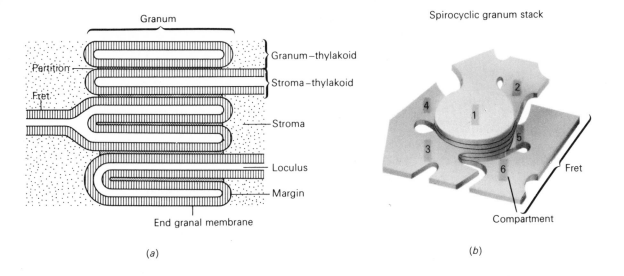

Granum

Partition

Fret

Granum—thylakoid

Stroma—thylakoid

Stroma

Loculus

Margin

End granal membrane

(a)

Spirocyclic granum stack

Fret

Compartment

(b)

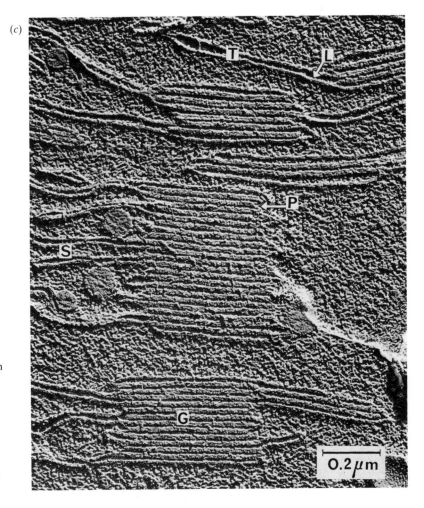

(c)

Fig. 8.5   (a) The terminology and
arrangement of stromal and granal
thylakoids as proposed by Weier,
Stocking, Bracker and Risley (1965).
(b) A three-dimensional representation
of the structure of a granum as
described by Wehrmeyer (1964).
Reproduced from Mühlethaler (1971).
(c) Cross-sectional view of a spinach
chloroplast as revealed by
freeze-etching showing the granal and
stromal thylakoids. T, thylakoid;
L, lumen; P, partition; G, grana;
S, stroma. Reproduced after E. Wehrli
from Mühlethaler (1977).

331

termed *margins:* the end-granal membrane is the name given to a thylakoid at the margin. A partition is the region in which two adjacent thylakoids come together (Fig. 8.5), although the nature of the bonding forces is not known.

The precise nature of the membranes making up the thylakoids is still a matter of controversy. Although it is generally agreed that various sized particles and sub-units are associated with and embedded to various degrees in the membrane, there has been controversy over their size and location. However, there now appears to be some general agreement. Broadly, there appear to be two classes of particles within the membrane which are approximately 17·5 and 11 nm in diameter. The largest occur mainly in the grana while the smallest are found in both granal and fret membranes, clearly suggesting that the two membranes have differences in function. A third loosely bound particle is attached to the membranes and in contact with the stroma. The membranes thus have a bumpy appearance in freeze-etched preparations (see Arntzen and Briantais, 1975). The largest particles are found only in photosynthetic membranes and so are assumed to be associated with the primary processes of photosynthesis. In an attempt to reconcile the various data on the membrane structure, Kirk (1971) proposed the model depicted in Fig. 8.6. This model is not a final solution to the problem, but goes some way to incorporating many of the divergent views. There is a lipid bilayer in which are embedded the larger particles (formed of four sub-units in a square and seen as two of these in cross-section) protruding into the intra-thylakoid space. The smaller particles protrude slightly from the other (external) side of the membrane to which are attached other proteins such as the coupling factor (see p. 356 and analogous to the knobs on the inner mitochondrial membrane).

The position of the chlorophyll molecules in the thylakoid membrane is not depicted in Fig. 8.6 and remains uncertain, although a number of detailed models have been proposed. Chlorophyll, which is located on both stromal and granal thylakoid membranes (the darker colour of the grana under the light microscope is presumably due to the greater concentration of chlorophyll) appears to be bound to protein and/or lipid and a number of chlorophyll-protein complexes can be isolated by detergent extraction of chloroplast membranes. If these extracts are subjected to polyacrylamide gel electrophoresis then three bands, Complexes I, II and III, can be separated with increasing

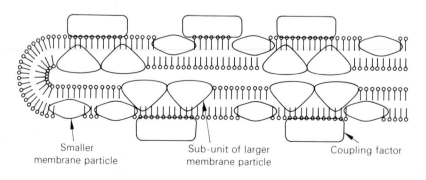

Fig. 8.6   A model of a thylakoid membrane. The figure represents a cross-section through a thylakoid at a margin. Each membrane consists of a lipid bilayer into which are embedded, to various degrees, particulate elements which are mainly protein. Reproduced from Kirk (1971).

Smaller
membrane particle

Sub-unit of larger
membrane particle

Coupling factor

respective electrophoretic mobilities. Complex I is also known as *P 700-chlorophyll a-protein*: both protein and chlorophyll sediment simultaneously in the ultracentrifuge with a sedimentation coefficient of 9 S, indicating that they are in a true complex. Complex II, which is also known as *light harvesting chlorophyll a/b-protein*, has a sedimentation coefficient of some 2–3 S. Complex III appears to lack a protein component. In viewing the chloroplast lamellar membrane from the standpoint of the lipid mosaic model (see Ch. 3), Anderson (1975) has seen these chlorophyll–protein complexes as intrinsic proteins which span the membrane. Each protein is envisaged as having two hydrophilic regions which are exposed at the surfaces of the membrane and a central hydrophobic region which is embedded in the lipid bilayer. The chlorophyll molecule (Fig. 8.12) has a hydrophobic tail which is seen as lying between protein and lipid and a more hydrophilic head which lies partially buried in the hydrophilic region of the protein. Other important intrinsic proteins are those of the photosystems and the photosynthetic electron transport chain and their position in the membrane is summarized in Fig. 8.20. Three important extrinsic proteins are a coupling factor in the chloroplast ATPase, the carboxylation enzyme (ribulose bisphosphate carboxylase), and a reductase involved in the transfer of reducing power to NADP (ferredoxin reductase).

A variety of other structures are also visible within the chloroplast envelope depending on the methods of fixation and staining. These include the plastoglobuli which are osmiophilic granules, 10–500 nm in diameter, and probably lipid in nature, starch grains, phytoferritin (an iron-protein complex), ribosomes and DNA fibrils. The characteristics and function of the chloroplast ribosomes are discussed in Chapter 5.

Other plastids

Although the chloroplast is a very prominent subcellular organelle, it is not the only one of its type present in plant cells. It is one of a group of organelles known as *plastids,* different types being classified in terms of colour and function. They may contain chlorophyll, and other photosynthetic pigments (*chloroplasts*) or non-photosynthetic pigments (*chromoplasts*). Colourless plastids (*leucoplasts*) may store starch (*amyloplasts*), protein (*proteinoplasts*) or fat (*elaioplasts*). Chromoplasts are characterized by a high proportion of carotenoid pigments and their function in some instances seems to be colour production, presumably as an animal attractant. Chromoplasts are generally produced from chlorplasts or amyloplasts, carotenoids accumulating as the lamellar structure breaks down (Fig. 8.7*a*). This change does not appear to be reversible.

Amyloplasts (Fig. 8.7(*b*)) are plastids which synthesize and store starch to the exclusion of other functions. Chloroplasts do contain starch grains, but these never attain the size of those in amyloplasts, where the grain may distend and even burst the envelope. There may be one or more starch grains within an amyloplast and each is enclosed within a membrane which develops from the inner membrane of the

Fig. 8.7 (a) A chromoplast from a ripe red pepper (*Caspicum annuum*) fruit. ×36 000. Reproduced from Clowes and Juniper (1968). (b) Amyloplasts seen in a thin section of a maize root cap cell. ×14 400.

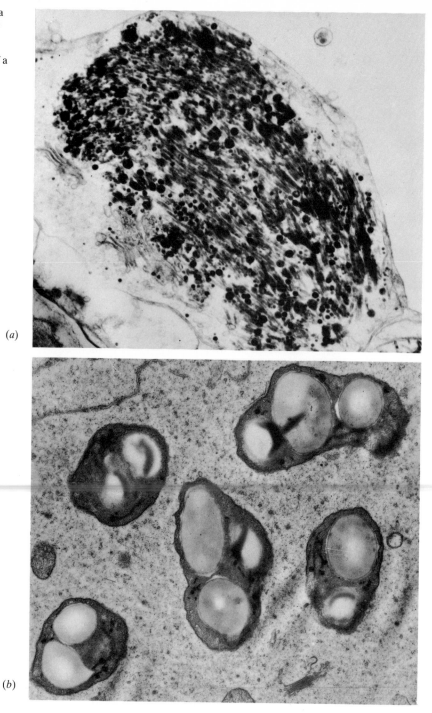

(a)

(b)

plastid envelope: this is then surrounded in turn by the envelope itself (Cronshaw and Wardrop, 1964). In tropical grasses, such as maize or sugar cane, cells of the bundle sheath contain a form of plastid which is intermediate in structure between the chloroplast and the amyloplast:

they store large amounts of starch but also contain rudimentary grana (see p. 368).

Proteoplasts (or proteinoplasts) contain protein which is often in a crystalline form. Little is known about them except that they contain few thylakoids and never grana (Heinrich, 1966). Elaioplasts are plastids containing large amounts of oil and they may be produced from chloroplasts at least in some species. All these forms of plastid appear to bear a developmental relationship to one another or to a common precursor.

Isolation

Although a considerable amount of information about the behaviour of chloroplasts in photosynthesis was obtained with whole plants, much of the detailed knowledge of the process as we understand it today has come from the use of isolated chloroplasts. A number of techniques involving both aqueous and non-aqueous extraction media have been used for this purpose.

Basically the leaves are homogenized by grinding either fresh material in osmotically adjusted aqueous media (for chloroplasts like mitochondria have semipermeable properties) or lyophilized material in a non-aqueous medium. The chloroplasts may be separated from the other cell components by differential centrifugation. In aqueous media, which normally contain sodium chloride (0·35 M) or sugars or sugar alcohols (0·3–0·5 M) buffered with tris-HCl to around pH 8, the cell debris are sedimented by centrifugation for 1 min at 200 g and then the whole chloroplasts precipitated at 1 000 g for something under 10 min. Under non-aqueous conditions (e.g. grinding in hexane and carbon tetrachloride or ether) the speed and time of centrifugation is adjusted somewhat depending on the density of the reagents. Clean separations of chloroplasts from other organelles are not obtained by such methods however, and density gradient centrifugation (see Ch. 1) is always necessary to achieve this.

Tissue that is homogenized in aqueous media using either sodium chloride or sucrose produces two types of chloroplast, which may be separated by density gradient or differential centrifugation. One form of the organelle is devoid of envelope and stroma and consists only of the lamellar systems (Class II), while the other form (Class I) appears to be relatively intact. Normal methods of preparation using differential centrifugation produce both types. Chloroplasts prepared in media based on a non-electrolyte show greater rates of carbon dioxide fixation than do preparations based on sodium chloride.

Biochemical activity

By the early twentieth century it was appreciated that photosynthesis involved chloroplasts in a light dependent conversion of carbon dioxide to carbohydrate with the concurrent evolution of oxygen. The position at that time could be summarized by an equation of the form

$$6CO_2 + 6H_2O \longrightarrow C_6H_{12}O_6 + 6O_2$$

Classic experiments performed by Blackman during the 1920s laid the basis for the conclusion that the reactions of photosynthesis were of two types, those requiring light (the *light reactions*) and those which would proceed in its absence (the *dark reactions*). The effect of light intensity on the rate of photosynthesis was found to depend on the level of carbon dioxide (Fig. 8.8). At very low light intensities the rate of oxygen evolution did not exceed that utilized by respiration: the point of zero net flux was termed the *compensation point*. As the light intensity was increased, the rate of oxygen evolution increased until such time as the level of carbon dioxide present in the atmosphere became limiting. This separation of photosynthesis into light and dark reactions was later amplified by the work of Emerson and Arnold (1932) who showed that the light and dark reactions could be separated in time. Cells of the unicellular green alga *Chlorella* were exposed to brief flashes of light (3 msec) followed by variable periods

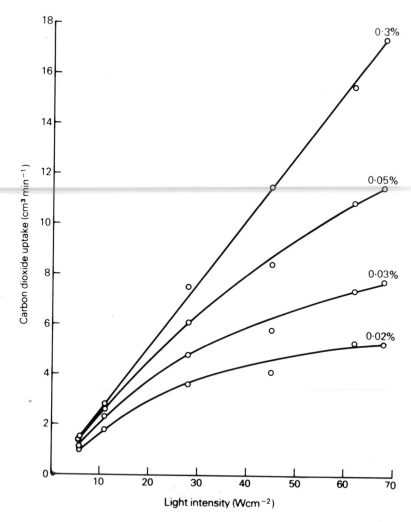

Fig. 8.8   The effect of increasing light intensity on the rate of photosynthesis of wheat leaves at various carbon dioxide concentrations (volume %). Plotted from the data of Hoover, Johnson and Brackett (1934).

of dark. The length of the dark period necessary to sustain continued oxygen evolution (tenths of a second) was found to be much greater than that of the light flash. Furthermore, at low temperatures the yield of oxygen varied considerably with the length of the dark period. It was argued that the light and dark reactions of photosynthesis were consecutive events. During the light, photochemical reactions took place which supplied energy for the subsequent dark reactions in which carbon dioxide was fixed. If the length of the dark period were too short, the photochemical system was not regenerated and the rate of photosynthesis reduced. The observation that the length of the dark period was temperature dependent confirmed the earlier observations of Blackman.

More recently it has been possible not only to separate the light and dark reactions in time, but also in space. Hill (1939) reported that chloroplasts isolated in aqueous media would evolve oxygen in the light provided a suitable hydrogen acceptor (A, sometimes termed a Hill reagent) were present although no carbon assimilation occurred:

$$H_2O + A \xrightarrow[\text{chloroplast}]{h\nu} AH_2 + \tfrac{1}{2}O_2$$

At this time substances such as benzoquinones and indophenol dyes were added to the chloroplast preparations to act as electron acceptors. This clearly demonstrated that oxygen evolution could be separated biochemically from carbon dioxide fixation; the latter could not be demonstrated in chloroplasts at that time.

The origin of the oxygen evolved during photosynthesis has been established with the aid of experiments performed by van Neil (see van Neil, 1941) on photosynthetic bacteria. In these bacteria, carbon dioxide fixation is not accompanied by oxygen evolution but does require the supply of a hydrogen donor from the growth medium. This hydrogen donor may be, for example, an aliphatic acid, hydrogen sulphide or even hydrogen. It is useful to compare carbon dioxide fixation in the purple and green sulphur bacteria which utilize H_2S, with that of a higher plant. The equation describing photosynthesis in these bacteria is:

$$2H_2S + CO_2 = 2S + (CH_2O) + H_2O$$

This compares with that for higher plants:

$$2H_2O + CO_2 = O_2 + (CH_2O) + H_2O$$

Both equations are thus of the general form:

$$2H_2A + CO_2 = 2A + (CH_2O) + H_2O$$

from which it can be appreciated that in higher plant photosynthesis the oxygen arises solely from the water. Corroborative evidence was obtained during the 1940s by following the release of a heavy isotope of oxygen, supplied as water ($H_2^{18}O$), during photosynthesis:

$$2H_2^{18}O + CO_2 = {}^{18}O_2 + (CH_2O) + H_2O$$

That is, the oxygen released during the Hill reaction arises from water which is the electron donor. The search for the natural electron-acceptor continued into the 1950s. In 1958, Arnon and his coworkers (Arnon, Whatley and Allen, 1958) discovered that when $NADP^+$ was provided as a hydrogen acceptor, oxygen evolution could be coupled to the production of ATP:

$$NADP^+ + P_i + ADP + H_2O = NADPH + H^+ + ATP + \tfrac{1}{2}O_2$$

The simplest theory to account for this ATP formation was a process analogous to mitochondrial ATP formation, in which phosphorylation was coupled to free energy changes during electron transport. The process was termed *photosynthetic phosphorylation*. However, the transfer of electrons from water to $NADP^+$ is not a spontaneous process. The standard reduction potential of the H_2O–$\tfrac{1}{2}O_2$ couple is +820 mV while that of the $NADP^+$-NADPH couple is −320 mV at pH 7·0. Electrons would be expected to move spontaneously from the more electronegative to the more electropositive couple. The importance of light must be to raise the electrons to a potential such that this spontaneous flow is possible (Fig. 8.9a).

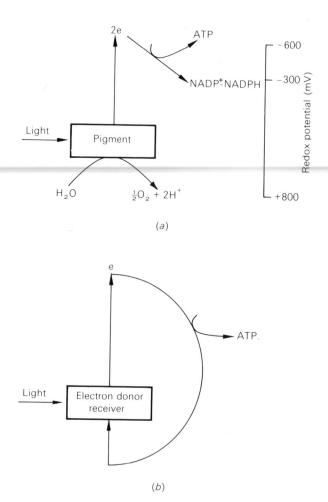

Fig. 8.9 (a) The transfer of electrons from water to $NADP^+$. An electron is raised to a sufficient energy level for it to pass to the $NADP^+$ −NADPH couple. (b) Cyclic photophosphorylation.

Earlier, Arnon, Allen and Whatley (1954) had already shown that ATP may be formed in the light and, provided that certain cofactors were present, oxygen was neither required nor evolved. Since neither electron donor nor acceptor was necessary it was evident that the action of light produced both electron donor and acceptor (Fig. 8.9*b*). Thus two types of photophosphorylation were distinguished; one in which the electron flow was from donor (water) to acceptor and termed *non-cyclic photophosphorylation* and the other in which the electron flow was cyclic (from the electron donor and back to the same compound) and termed *cyclic photophosphorylation*.

It is currently accepted that the primary function of light in photosynthesis is to separate charges; that is, to cause an electron to be removed from a chlorophyll molecule, leaving a 'hole' or positive charge in that molecule. Essentially, although there are many intermediate stages, the electron in non-cyclic photophosphorylation is passed to $NADP^+$ producing a reductant NADPH for carbon dioxide fixation together with the necessary energy as ATP, while the hole left in the chlorophyll is filled with an electron transferred from water. In the next section we shall look in more detail at the light reactions of photosynthesis.

Light reactions

The energy available for photosynthesis in the chloroplast originates from the sun. Of this radiation, 98% has wavelengths between 200 and 4 000 nm. Passage through the atmosphere leads to a total depletion of wavelengths below 300 nm and also a reduction of the infra-red region (particularly at wavelengths greater than 1 000 nm). Consequently the available radiation for photosynthesis lies mostly in the visible and short infra-red wavelengths (300–900 nm, Fig. 8.10).

The energy contents of this radiation depends upon the wavelength of the light such that

Fig. 8.10   The electromagnetic spectrum in relation to light absorption for photosynthesis.

$$E = \frac{hc}{\lambda} = h\nu$$

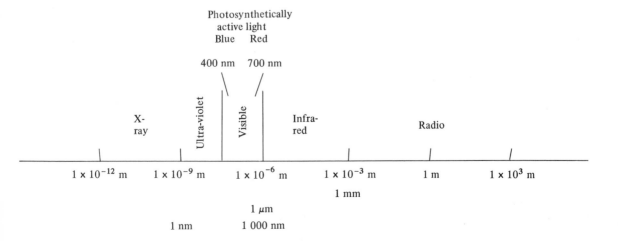

339

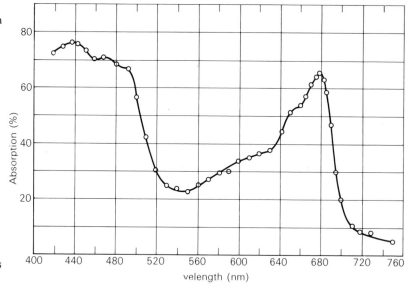

Fig. 8.11 Absorption and action spectra for photosynthesis in the green alga *Ulva taeniata*. Reproduced from Haxo and Blinks (1950).

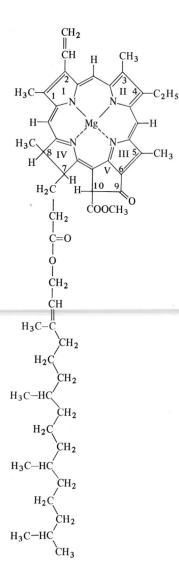

Fig. 8.12 The structure of chlorophyll *a*. The methyl group at position 3 of ring II is replaced by an aldehyde group ($-CHO$) in chlorophyll *b*.

where E is the energy (erg) per quantum of radiation or photon;

h is Planck's constant (6.62×10^{-27} erg sec);

c is the velocity of light (3×10^{10} cm sec^{-1});

λ is the wavelength (cm) and

ν is the frequency (sec^{-1}).

The energy of N quanta, where N is the Avogadro number (6.023×10^{23} molecules mole^{-1}), is termed an einstein. However, in order for the energy of this light to be utilized by the plant, it has to be absorbed and converted to a usable form. This absorption of light is, in fact, brought about by aggregates of pigments which are found in photosynthetic tissues. These pigments fall into three main classes, the *chlorophylls*, the *carotenoids* and the *phycobilins*.

Absorption spectra such as that illustrated in Fig. 8.11 show two main peaks in the red (670–680 nm) and blue (435–438 nm) regions of the spectrum. The absorbing materials may be extracted from the leaves of higher plants (grown in the light) and the chlorophylls *a* and *b* separated by chromatography or solvent partition. These chlorophylls are metallo-compounds each consisting of four substituted pyrrole rings (see p. 119) arranged in a cyclic way such that the nitrogen atoms are complexed with magnesium. A long chain aliphatic alcohol (phytol) is esterified to ring IV (Fig. 8.12). Removal of the phytol side chain results in a substance termed *chlorophyllide* which is an intermediate in chlorophyll biosynthesis. This biosynthesis begins with the formation of δ-aminolevulinic acid from glutamate and α-ketoglutarate and continues via a number of intermediates with the subsequent formation of a macrocyclic compound, uroporphyrinogen III. The tetrapyrrole, protoporphyrin IX, is formed in a further three steps and lies at the branch point for the synthesis of the haems (where iron is inserted into the molecule) and the chlorophylls (with the insertion of magnesium). Chlorophyll *a* lies at the end of this pathway, the previous three intermediates of which are protochlorophyllide,

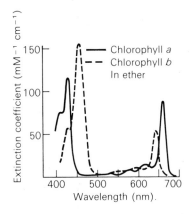

Fig. 8.13 The absorption spectra of chlorophylls *a* and *b* in either. Reproduced from Clayton (1965).

chlorophyllide *a* and chlorophyll *b* (see Granick and Beale, 1978, and Beale, 1978 for further details). Chlorophyll *a* in ether absorbs maximally at wavelengths of 662 410 and 380 nm while chlorophyll *b* (also in ether) absorbs maximally at 644, 455 and 430 nm (Fig. 8.13). French and his co-workers (Smith and French, 1963) have distinguished a variety of forms of chlorophyll *a* with different absorption maxima *in vivo*, although only a single form is found *in vitro*. It is inferred, therefore, that the chlorophyll *in vivo* enters into various molecular associations which cause the change in the absorption maxima. In general, the chlorophyll *a* is present in about three times the amount of chlorophyll *b* and constitutes the main pigment of the chloroplast system. In the algae other chlorophylls substitute for chlorophyll *b*: these are, for example, *c* in the brown algae (Phaeophyta) and *d* in the red algae (Rhodophyta).

The carotenoids are found throughout higher and lower plant phyla and are made up of two classes of compound, the hydrocarbon *carotenes* and their oxygenated derivatives, the *xanthophylls*. The specific names of these carotenes are suffixed with the letters -*ene* while those of the xanthophylls with the letters -*in*. The four commonly occurring carotenoids of higher plants are β-carotene, lutein, violaxanthin and neoxanthin (Fig. 8.14), while a wider range of molecules is found in the lower plant phyla. These carotenoids show broad absorption peaks chiefly in region 450–480 nm. Unlike the carotenoids, the phycobilins have a limited distribution amongst the plant phyla and are to be found only in the Rhodophyta, Cyanophyta (blue-green algae) and the Cryptophyta (Cryptomonads). These phycobilins are variously substituted open chain tetrapyrroles which are conjugated to specific proteins; as in the carotenoids the molecules form conjugated double bond systems.

By comparison of the absorption of spectra of plants with *action spectra* (the measurement of the effectiveness of various wavelengths of light in carrying out a photobiological process) an indication may be obtained concerning which pigments are playing an active rôle in the photosynthesis. A light source is used with some form of monochromator (a prism or grating or some type of interference filter) to illuminate the system under test with known wavelengths and a measurement of energy of the light incident on the experimental material is made with a thermopile or photomultiplier. It is extremely important to determine the incident energy since the energy per quantum of light absorbed is dependent on the wavelength.

Action spectra (Fig. 8.11) for photosynthesis or the Hill reaction are broadly similar to the absorption spectra of the chlorophylls, indicating the primary rôle of the chlorophylls in the absorption of energy for photosynthesis. Careful investigations of such spectra, however, have also revealed that the other photosynthetic pigments play a rôle in light absorption and hence explain the difference between the absorption and action spectra for higher plants (Figs. 8.11, 8.15). The carotenoids absorb chiefly in the region 450–600 nm where chlorophylls absorb relatively little light and this energy is transferred to chlorophyll *a* (see below and Cogdell, 1978). The

Fig. 8.14   The structure of the major carotenoids of the higher plants.

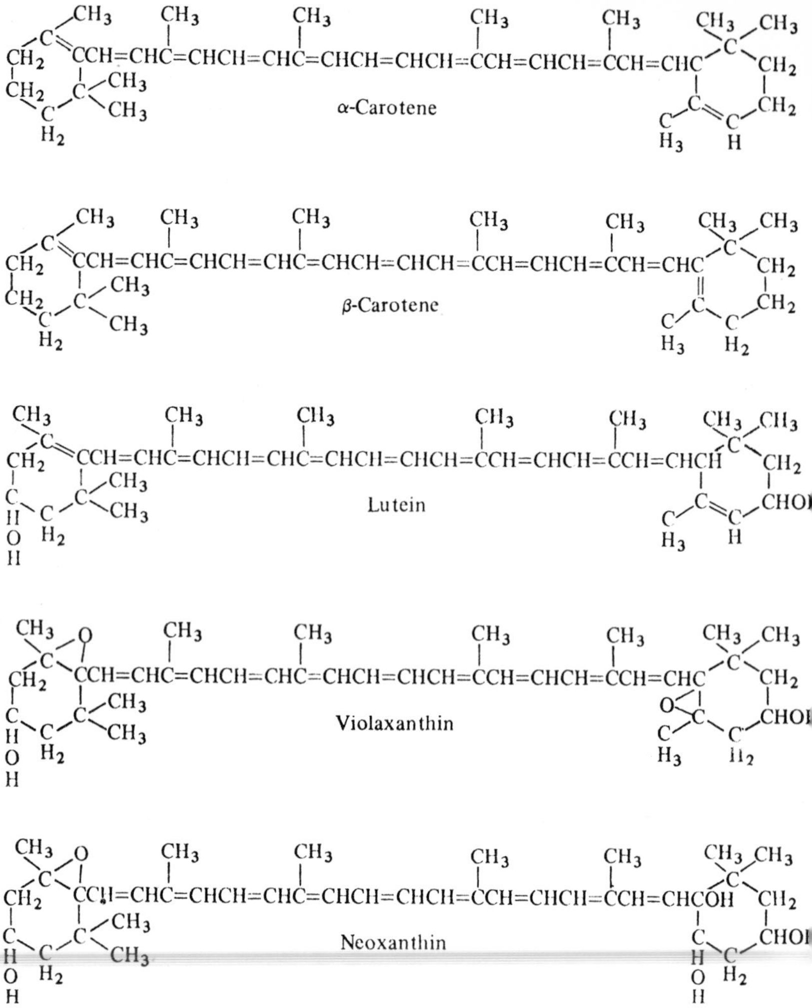

α-Carotene

β-Carotene

Lutein

Violaxanthin

Neoxanthin

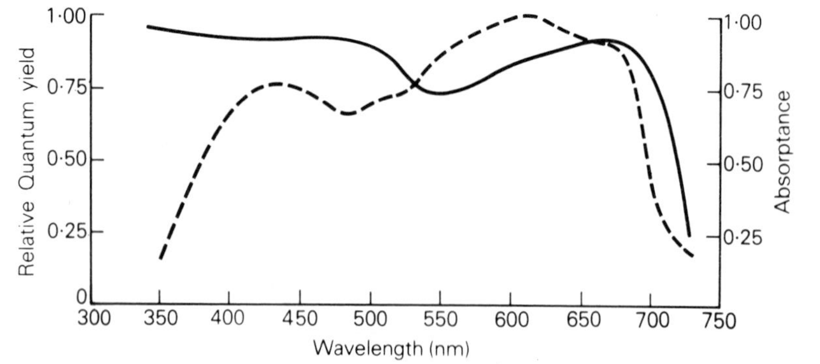

Fig. 8.15   Mean absorptance and relative quantum yield of crop species. Plants of twenty-two different crop species were grown in growth cabinets at 22/20°C (day/night) with an irradiance of 100 w M^{-2} during the 16 h light period. Leaf photosynthesis and light absorptance were measured at various wavelengths. Solid line, absorptance; broken line, relative quantum yield. Plotted from data of McCree (1971, 1972).

carotenoids also appear to play a protective rôle in preventing the photo-oxidation of chlorophyll (see below). Like the carotenoids, the phycobilins also act as *accessory pigments*, transferring energy to chlorophyll *a*, in the red and blue-green algae.

Measurements of the quantum yield of photosynthesis or the Hill reaction are nearly constant with wavelengths less than about 690 nm

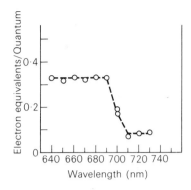

Fig. 8.16 The quantum yield for NADP⁺ reduction by spinach chloroplasts. Redrawn from Kok (1967).

at about 0·3 electron equivalents per quantum. However, as the wavelength is increased to about 710 nm, the yield drops rapidly (Fig. 8.16) to about 0·1 equivalents per quantum (*red drop effect*). The low efficiency of the far red light can be increased by simultaneous illumination with light of shorter wavelengths producing a rate of photosynthesis which is greater than that expected from the sum of the two incident beams alone. This phenomenon is known as the *Emerson enhancement effect.* By illuminating the photosynthetic system with a single wavelength in the far red region and then using a variety of other wavelengths, the action spectrum of enhancement can be determined. Comparison of this action spectrum with the absorption spectra of pigments known to be present in the chloroplast indicates which of these is involved in absorbing the enhancing light. In all higher plants examined this analysis has pointed to chlorophyll *b* and part of the chlorophyll *a* with an absorption maximum at about 673 nm. This evidence is important since it can be inferred that photosynthesis involves the cooperation of two pigment systems or, put another way, two light reactions. Wavelengths in the far red region of the spectrum activate only one of these. Furthermore it can be shown that when far red and shorter wavelengths are provided in alternating flashes, enhancement still occurs even when the two flashes of different wavelengths are several seconds apart. This means that the products of at least one of the light reactions are relatively stable. The two light reactions have been termed *photosystem I (PS I)*, the far red system, for which light is absorbed by that part of the chlorophyll *a* (about 50%) with longer wavelength absorption maxima (680–700 nm) and *photosystem II (PS II)* which, as already indicated, involves, in higher plants, chlorophyll *b* and that part of the chlorophyll *a* with an absorption maximum of about 670 nm.

These two photosystems which have been identified appear to have different functions. It was apparent from earlier work that all oxygen evolving organisms contained both photosystems I and II while the photosynthetic bacteria, which as a group do not evolve oxygen, contained only a long wave photosystem I. *Scenedesmus* mutants have been isolated which lack PS I and cannot photoreduce NADP⁺, i.e. PS I is involved in the photoreduction of NADP⁺ and PS II in the evolution of oxygen. Furthermore, the biochemical separateness of oxygen evolution and NADP⁺ reduction can be shown by the use of DCMU (3-(3,4-dichlorophenyl)-1,1-dimethyl urea) which is able to inhibit oxygen evolution in isolated chloroplasts carrying out the Hill reaction. If an alternative electron donor is provided (e.g. ascorbate) together with a dye to act as a catalyst (e.g. dichlorophenolindophenol, DPIP) reduction of NADP⁺ can still occur.

Primary light reaction

Experiments with light flashes of high intensity have indicated that a single flash cannot produce more than a single oxygen molecule for every 2 000 chlorophyll molecules. Since both the reduction of carbon dioxide to carbohydrate and the evolution of oxygen from water need the participation of four proton equivalents, it can be reasoned that four electrons are removed per 2 000 chlorophyll molecules. In other words, in a single light flash one electron is moved for about 500 chlorophyll molecules. Calculations show that the energy content of four quanta of light is insufficient for both the reduction of carbon dioxide and the photolysis of water and that something between four and eight is the minimum quantum requirement for photosynthesis (Clayton, 1965). Consequently we can calculate (on the basis of two light reactions each requiring four quanta) that one electron is removed per two quanta or at best one quantum is trapped by about 250 chlorophyll molecules. Since single quanta react with single molecules, this is a very low efficiency. However, any single molecule can only collect a few quanta per second and therefore by increasing the numbers of molecules in a complex it is possible to increase the numbers of quanta trapped per unit of time.

Absorption of a light quantum by a molecule raises its energy by an amount equal to that contained within the quantum absorbed and displaces one of its electrons into a higher energy orbital. As already indicated, light is absorbed by a chlorophyll molecule in both the blue and the red regions of the spectrum: the energy contained in the blue quanta ($\lambda = 438$ nm, $E = 4 \cdot 53 \times 10^{-12}$ erg or $2 \cdot 83$ eV $- 1$ eV $= 1 \cdot 602 \times 10^{-12}$ erg) is greater than in the red quanta ($2 \cdot 92 \times 10^{-12}$ erg or $1 \cdot 82$ eV for a wavelength of 680 nm). The absorption of red light causes an electron to be raised into what is termed the *first singlet excited state* (singlet because the electron in the excited orbital has an opposite spin to its paired electron still in the ground state) (Fig. 8.17). Blue quanta contain more energy than the red quanta and raise an electron to a higher energy level (the second singlet state). However, this second singlet state is relatively unstable and decays into the first singlet state in about 10^{-11} sec with a loss of energy in the form of heat. This is an important point since it means that although blue quanta contain more energy than red quanta there is no gain to photosynthesis by the absorption of blue light since the difference in energy is lost as heat: photochemically, blue and red quanta are equivalent.

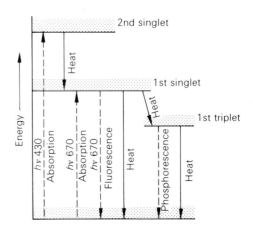

Fig. 8.17   Electron energy levels in the chlorophyll molecule. Reproduced from Kok (1965).

The energy absorbed by the chlorophyll may be lost again in a number of different ways.

1. The molecule may lose the energy by the emission of light as the electron reverts to the ground state: this process is known as *fluorescence*.

2. The molecule can lose the energy as heat by collision or vibration. In some cases the electron may lose only some energy as heat and change to a more stable state known as the *triplet state* in which the electron spin has been reversed. The lifetime of this new state is longer (about 10^{-4} sec) since the chance of spin re-reversal is low. Although this means that chemically the triplet state is more reactive, it appears that its occurrence in chlorophyll is low and so it is unlikely that it plays a direct role in photosynthesis. However, it is in this triplet state (as ^{3}Chl) that chlorophyll can convert oxygen in its ground state to its first singlet stage ($^1O_2$). In this form oxygen is extremely reactive and could, potentially, cause considerable damage to the cell by reacting with unsaturated fatty acids, amino acids and purines. ^{3}Chl is prevented from initiating this chain of events through the protective action of the carotenoids: they are able to convert both the ^{3}Chl and $^1O_2$ to their respective ground states (see Krinsky, 1978).

3. The third way in which the excitation energy can be lost is by transfer of the energy to another molecule, either by electric dipole interactions or electron conductance in overlapping orbitals. Quanta absorbed by one chlorophyll molecule can be transferred to other pigment molecules provided they are in close proximity. This transfer is not random and quanta can only move towards pigments with equal or larger wave absorption maxima (and less energy, since $E = hc/\lambda$). Alternatively, the excited molecule can lose energy in some form of chemical reaction, such as the loss of an electron to an acceptor molecule.

The majority of the chlorophyll present in the chloroplast is engaged in the collection of light quanta, or *light-harvesting* as it has been termed. Photons absorbed by these chlorophylls are able to pass from one molecule to another due to their close packing, until they reach a chlorophyll molecule in a unique position in what is termed the *trapping centre*. The whole complex of molecules is known as a *photosynthetic unit*. It is the 1% of the total chlorophyll (designated P) which is photochemically active and here an electron acceptor is reduced and the chemical consequences of light absorption really begin.

The concept of a trapping centre is central to current view on photosynthesis. At each centre, the P-chlorophyll reacts with an electron donor and an electron acceptor. The light energy causes an electron to be lost from the P-chlorophyll and this is transferred to the acceptor which thus becomes a reductant. The loss of the electron from the chlorophyll leaves it with a net positive charge and it regains an electron from the donor molecule.

Two distinct P-chlorophylls, P-700 and P-680, have been discovered, corresponding to the trapping centres of PS I and PS II respectively. The trapping centres also contain carotenoids, which it is argued by analogy with bacterial systems (see Cogdell, 1978) play a similar protective rôle within the reaction centre to that played by the bulk of the carotenoid pigments. As described earlier, freeze-fracture preparations of chloroplast membranes show repeating structures in the lamellae. The largest particles may be equivalent to the *quantosomes* described earlier in negatively stained or shadowed preparations which were considered to be the morphological expression of the photosynthetic unit (see Park, 1976). However, these particles are not found to any great extent on the fret membranes which nevertheless are photosynthetically active. Thus there is no clear evidence for a structural counterpart of the photosynthetic unit.

Photoreduction of NADP$^+$

The reduction of NADP$^+$ is known to be catalysed by PS I which is located in both the granal and the stromal lamellae and which, on fractionation of the chloroplast, is primarily associated with the smaller particles. It is the light of longer wavelengths which is active in PS I.

During the late 1950s it was shown that the reduction of NADP$^+$ by chloroplasts could be much enhanced by the addition of a soluble protein which could be prepared from spinach chloroplasts. It was subsequently shown that this substance was *ferredoxin*, a low molecular weight (12 000 for plant ferredoxins) non-haem iron protein containing labile sulphur, red in colour, and present at a concentration of about one molecule for every 400 molecules of total chlorophyll. Following its photoreduction, ferredoxin may be oxidized by NADP$^+$ (ferredoxin–NADP reductase), or nitrite (nitrite reductase) or non-enzymically by a number of haem proteins and oxygen. Whatley, Tagawa and Arnon (1963) established the stoichiometry of the NADP$^+$ reductase and showed that one mole of NADP$^+$ oxidized two moles of spinach ferredoxin. During these changes the non-haem iron, of which there are two atoms in spinach ferredoxin, undergoes valency changes such that both atoms are in the ferric state in oxidized ferredoxin. On reduction, half of the iron is in the ferrous state, so that the oxidation–reduction cycle involves one electron and hence the stoichiometry with NADP$^+$. The standard oxidation–reduction potential of ferredoxin is -430 mV at pH $7 \cdot 0$ which is more electronegative than that of NADP$^+$ at -320 mV. However, the use of viologen dyes and of refined spectrophotometric studies has shown that a further substance or substances, the primary electron acceptor, may mediate between P 700 and ferredoxin. The standard reduction potential of P 700 has been shown to be about $+400$ to $+450$ mV in the ground state and therefore it has little tendency to lose electrons. However, during the photochemical reaction, it becomes very electronegative with a potential of around −650 mV. This is considerably more electronegative than ferredoxin which it is consequently able to reduce. The nature of the primary acceptor (X) is still not clear, although recent studies indicated that a bound ferredoxin may mediate between the soluble ferredoxin and P 700. There has also been some speculation, by analogy with bacterial systems, that a quinone may be involved between P 700 and the bound ferredoxin (Bolton and Warden, 1976). The electron donor to PS I is a cytochrome, which is in turn reduced by electrons from PS II (see below). The heart of the PS I reaction centre is isolated as Complex I, the P 700-chlorophyll *a*-protein, mentioned earlier.

Oxygen production

Photosystem II is involved in the evolution of oxygen from water and is activated in higher plants by light absorbed by chlorophyll *b* and forms of chlorophyll *a* which absorb maximally at shorter wavelengths. PS II is found associated with the grana and may be isolated as a heavy fraction from fragmented chloroplasts, (see below). Far less is known, however, about the intermediates in PS II than in PS I or of the nature of the reaction by which oxygen is produced from water. The energy collector, the PS II equivalent of P 700, is P 680. The primary acceptor for PS II has not been identified, although a bound plastosemiquinone has been implicated. It has been speculated that quinones may also be involved in the oxidizing side of PS II (Bolton and Warden, 1976). Absorbed light quanta are presumed to produce a strong oxidant (which is often given the symbol Z) and a weak reductant (symbol, Q), neither of which have been isolated. The strong oxidant oxidizes water and this results in the evolution of oxygen in a reaction which requires the removal of four electrons and whose mechanism remains obscure (see Rodmer and Kok, 1975 for further details); little else is known about this substance. Electrons are transferred, although indirectly, from the weak reductant to the P 700 of PS I (see below). This returns the P 700 to the non-oxidized state and allows PS I to continue to function.

Manganese, of which there are about four to six molecules per 200 PS II chlorophyll molecules in spinach chloroplasts, has been shown to be an integral part of PS II, as has chloride. Following the discovery that chloride was an essential component of PS II, it was confirmed that chloride is, in fact, an essential micronutrient in plant metabolism. The function of manganese and chloride, however, is not clear. The manganese has been assigned a rôle in the evolution of oxygen while chloride is believed to function at a postulated coupling site. Bicarbonate ions seem also to be required for PS II.

Separation of the photosystems

The two photosystems may be physically separated from each other by mechanical disruption of the chloroplast with a French press or treatment with digitonin, triton X-100 or sodium dodecylsulphate followed by differential or density gradient centrifugation. A heavier fraction which is sedimented by centrifugation at about 10 000 g for 30 min is active in the Hill reaction with ferricyanide, but has a low ability to reduce $NADP^+$, while a lighter fraction (sedimented by centrifugation at 144 000 g for 60 min) is inactive in the Hill reaction but catalyses the photoreduction of $NADP^+$ (with ascorbate and dichlorophenolindophenol) with high efficiency, i.e. the heavier particle is enriched in terms of PS II activity, while the lighter one is enriched in PS I. The PS I particle is, as expected, enriched in terms of chlorophyll a and P 700 while being impoverished in chlorophyll b (see Table 8.1). Similar fractions have been examined under the electron microscope by Sane *et al.* (see Park and Sane, 1971) who have claimed that the stromal and end-granal lamellae contain only PS I while the granal lamellae in the partition region contain both PS I and PS II. These two photosystems may be associated with separate particles in the membrane: Arnzen, Dilley and Crane (1969) showed that membranes in the PS II fraction contained mainly 17·5 nm particles while those of the PS I fraction contained 11·0 nm particles. Since many of the membranes in these fractions were about half the normal thickness, it was suggested that the detergent used to prepare the fractions split the membrane and that PS I and PS II were located on opposite sides of the membrane.

The PS I particle is in fact enriched in the Complex I described earlier (p. 347). This chlorophyll-protein appears to be the heart, but not the whole, of PS I and as such contains chlorophyll a (but not chlorophyll b) and carotene (in a ratio of 20–30:1) together with P 700 (1 molecule to 40–50 chlorophyll a), and a quinone which is not plastoquinone. This complex is present in all plants containing P 700 that have been examined. Mutants lacking P 700 lack this complex (see Thornber, 1975). Complex II, which accounts for 40–60% of the total chlorophyll (Complex I contains 10–18% of the total chlorophyll), does not appear to be the PS II equivalent of Complex I. It contains chlorophylls a and b in equal proportions together with representatives of all the carotenoids contained in the chloroplasts, but is not essential to photosynthesis. Since this component contains such a large proportion of the total chlorophyll and yet plants lacking it can still photosynthesize, it is believed that Complex II carries out the major light harvesting process in the chloroplast membranes.

Table 8.1 The distribution of chlorophylls and carotenoids (moles per 100 moles chlorophyll) in chloroplasts, and particles containing photosystems I and II

	Chloroplast	PS I particle	PS II particle
Chlorophyll a	74	84	69
Chlorophyll b	26	16	31
Chlorophyll a/b	2·8	5·3	2·3
Xanthophyll/carotene	2·6	1·7	3·8
P 700	0·23	0·5	0·14

Data from Boardman (1970)

The two photosystems appear to operate in a connected sequence. Light absorbed by PS I causes an electron to be lost from P 700 and transferred to NADP⁺. The electron lost from the P 700 leaves it in the oxidized form with a 'hole' requiring the return of an electron. This is believed to be provided by the weak reductant Q of PS II. The transfer of electrons is postulated not to be direct but to take place down an electron transport chain enabling the formation of ATP coupled to electron transport to take place (Fig. 8.18).

This series formulation (known as the Z scheme) arose from the work of a number of research groups during the early 1960s. It is based on many observations (see Clayton, 1965) which include the evidence listed below.

1. The evidence for the existence of two separate light reactions.
2. The presence of components within the chloroplast whose redox state depends upon which of these light reactions is activated. For example, both cytochrome f and plastoquinone are oxidized by far red light and reduced by shorter wave light. DCMU which

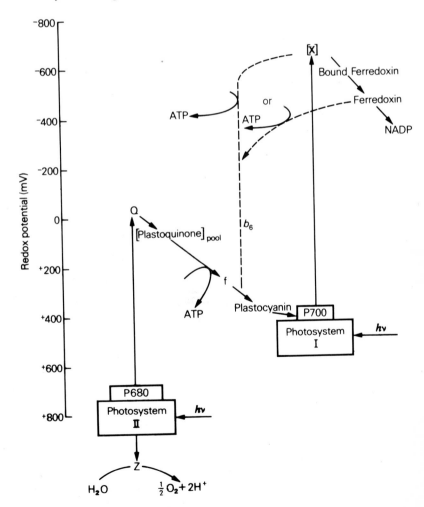

Fig. 8.18   Electron transfer between photosystems I and II (the Z scheme). The position of b_{559} is uncertain and so is not included.

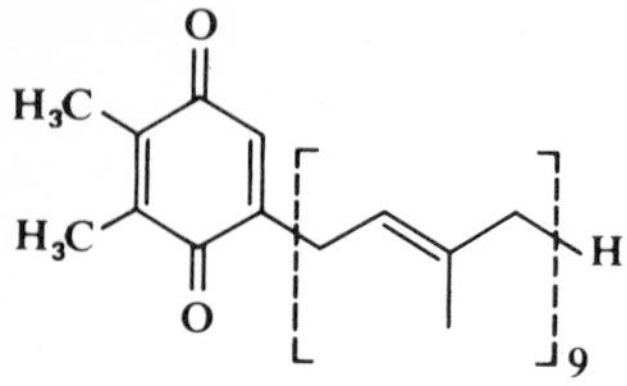

Fig. 8.19   Plastoquinone. There are nine isoprenyl units with 45 C-atoms in the side chain.

inhibits oxygen evolution in PS II also inhibits the reduction of cytochrome f by shorter wavelength light.

3. In the presence of DCMU the photoreduction of $NADP^+$ can be restored by replacing the oxygen evolving system with an artificial electron donor.

It was proposed that electrons are raised from about $+800$ mV to about 0 mV by PS II, to then fall to around $+400$ mV before being promoted to over -500 mV by PS I.

Analysis of the components of chloroplasts has indicated the presence of one class C cytochrome (c.f. Fig. 7.12) known as cytochrome f or c_{555} with two haem groups per molecule and an α band in the reduced form at a wavelength of 555 nm; two class B cytochromes, b_6 (or b_{562} or b_{563}) and b_{559} (or b_{558}); three types of quinone, the napthoquinone vitamin K_1, plastoquinone (analogous to the ubiquinone of mitochondria, see Fig. 8.19) and α-tocopherol-quinone; an acidic protein containing two copper atoms per molecule known as plastocyanin (see Boulter et al., 1977) and probably others as yet unidentified.

The order of the components in the electron transport chain has been investigated by the same means as those applied to the mitochondrial electron transport chain, although with somewhat less success on the whole. This is chiefly because of the problem of measuring difference spectra in the presence of large quantities of light sensitive pigments. In fact, studies during the last decade have in many cases shown the problems to be more complex than previously anticipated (Cramer and Whitmarsh, 1977).

Plastoquinone appears to be closely associated with PS II and chloroplasts lacking this component cannot oxidize water, although they can utilize ascorbate to reduce $NADP^+$. Plastoquinone is thus positioned close to the Q of PS II. One unusual feature of plastoquinone is its high concentration: there is one molecule for about every ten to twenty molecules of chlorophyll while the other components of the photosynthetic chain are present in ratios closer to one molecule for every 400 molecules of chlorophyll. There seems to be a pool of between seven and eight times as many plastoquinone molecules as there are most of the other carriers in the electron transport chain. It is now clear that several photosystem II reaction centres feed electrons, via the reductant Q, into the pool of plastoquinone molecules which in turn channels electrons into several P 700 molecules. This buffering capacity of the plastoquinone allows a number of PS I and PS II reaction centres to interconnect through a number of electron transport chains and thus increases the efficiency of electron transfer between the two photosystems. A similar rôle for ubiquinone was mentioned in the operation of mitochondrial electron transport. We will return later to the central role of plastoquinone in the coupling of electron transport to phosphorylation. As yet no rôle has been clearly defined for vitamin K_1 or for α-tocopherolquinone.

Cytochrome b_{559} is closely associated with PS II particles and and it has been ascribed various roles including (a) an electron acceptor in the water splitting reaction of PS II, (b) a position in the main electron transport chain between photosystems I and II, and (c) in a cyclic chain around PS II. Although it seems to be generally agreed that there is only one spectral form of b_{559} there may be more than one redox form. Potentials ranging between +350 to +400 on the high side and +50 mV on the low side have been reported. Arguments for the low potential form functioning in PS I have largely been rejected (Cramer and Whitmarsh, 1977), and the consensus of evidence at present suggests that b_{559} participates in some way on the high potential side of PS II.

Cytochrome f appears to be closely associated with the PS I particle and kinetic data place it close to P 700: it has a standard potential of about +365 mV.

Plastocyanin, which is essential for the transfer of electrons from ascorbate to $NADP^+$, is also closely associated with PS I particles. Both cytochrome *f* and plastocyanin are present in the ratio of one molecule to about 400 molecules of total chlorophyll: there are two molecules of b_6 and b_{559} present per molecule of cytochrome *f*. Plastocyanin is placed in Fig. 8.18 as following cytochrome *f* with a standard potential of +370 mV. This view is not accepted by all and it has been suggested that cytochrome *f* may accept electrons from plastocyanin or that it is not even on the main pathway from plastoquinone to P 700. There is, however, considerable evidence to support its function on the main pathway between the two photosystems (see Cramer and Whitmarsh, 1977 for details).

Cytochrome b_6, however, does not appear to lie on the electron transport pathway from PS II to PS I. The reduction of b_6 is insensitive to inhibitors of PS II and it appears to be concerned with cyclic electron flow around PS I: cytochrome b_6 is associated with the PS I particle and has a mid-point potential of some 0 mV.

The possibility that there may be two separate photosystems I, one each in the stromal and granal lamellae and one photosystem II in the granal lamellae has been proposed by Park and Sane (1971), (see also Park, 1976; see p. 348). They postulated a variation of the Z scheme, in which the PS I in the stromal lamellae is physically separated from the PS I and PS II of the granal lamellae and possibly primarily involved in cyclic photophosphorylation. The PS I and PS II of the grana may operate in the customary Z scheme with or without contribution from the stromal PS I.

Artificial electron donors and acceptors

So far we have only discussed the naturally occurring components of the photosystems and of the electron transport chain. It is, however, possible to both supply and remove electrons at various points along the electron transport chain by the use of artificial donors and acceptors. These compounds interact with the chain at a point determined by their redox potential, although this may also be influenced by their lipid solubility. For example, hydrophilic compounds are not able, in general, to interact with endogenous electron transport components which are buried in the membrane or located on the inside of a closed vesicle whereas lipophilic compounds are. In freshly prepared thylakoid membranes, ferricyanide (with a mid-point potential of $+360$ mV) can be reduced by both photosystems I and II (with potentials of about $+500$ and $+800$ mV respectively), although the rate of reduction by PS II is about 40% less than that of PS I. Following sonication, when the acceptor site of PS II is uncovered, the rate of reduction by PS II is significantly increased. Other electron donors and acceptors that are commonly used in studies on the photosynthetic electron transport chain include: ascorbate ($E_0' = +58$ mV) which is often provided in conjunction with catalytic amounts of other redox compounds such as dichlorophenolindophenol (DCIP, $E_0' = +217$ mV) or N-tetramethyl-p-phenylene-diamine (TMPD) which it serves to keep reduced; methyl-phenazonium methosulphate (PMS, $E_0' = 80$ mV); and methyl viologen which has a rather negative potential of some -446 mV. In terms of the photosynthetic system, compounds such as benzidine, hydroquinones, hydroxylamine and ascorbate are able to act as donors to PS II while oxidized phenylenediamines act as acceptors to this photosystem. Donors to photosystem I include diaminodurene (DAD), PMS and DCIP + ascorbate (in fact acceptors of PS II may also act as donors to PS I), while only the more electronegative compounds such as benyl- and methylviologen are able to accept electrons from PS I (see Trebst, 1974 for further details).

Photophosphorylation

According to the series formulation (or Z scheme), non-cyclic photophosphorylation, as described by Arnon *et al.* in 1958, occurs during normal photosynthesis involving electron flow from water and the involvement of two photosystems:

$$NADP^+ + H_2O + ADP + P_i \longrightarrow NADPH + H^+ + ATP + \tfrac{1}{2}O_2$$

Cyclic photophosphorylation is believed to involve only PS I, electrons being raised to the level of ferredoxin and then returning to P 700 via cytochromes b_6 and f

$$n(ADP) + n(Pi) \longrightarrow n(ATP)$$

In contrast with oxidative phosphorylation, the study of photophosphorylation suffers from a serious problem. This is the lack of a clear stoichiometry between the amount of ATP synthesized and the number of electrons or electron equivalents transferred down the transport chain. In the case of oxidative phosphorylation, the stoichiometry with various substrates has been fairly well established; as a result of the oxidation of malate, for example, three moles of ATP may be formed for each pair of electrons or their equivalents passing down the respiratory chain, one at each of three coupling sites or three proton translocating loops (p. 307). However, in work with chloroplasts it has not been possible to determine the number of electrons flowing per mole of ADP phosphorylated with cyclic electron flow, since there is no way of measuring the number of electrons flowing around a closed loop.

There is a sufficient energy drop for one site of phosphorylation to lie between plastoquinone and cytochrome f. Application of the crossover theorem (p. 305) to chloroplasts has shown that there is a phosphorylation site prior to cytochrome f and one probably lies between plastoquinone and plastocyanin. Since, during the transfer of electrons from water to ferricyanide the same amount of ATP is produced as during the transfer from water to $NADP^+$, this excludes a further coupling site beyond cytochrome f (ferricyanide is reduced by cytochrome f). There does, however, appear to be a site for the synthesis of ATP between water and the endogenous acceptor for PS II. The discovery of an inhibitor, dibromothymoquinone (DBMIB), which prevents the reoxidation of plastohydroquinone by PS I, showed that in its presence, ATP is synthesized in a PS II-driven Hill reaction in which electrons are transported from water to ferricyanide. In the presence of DBMIB, the stoichiometry of ATP synthesis, i.e. the ATP/2e ratio, is half that found for the complete non-cyclic electron transport from oxygen to PS I (see Trebst, 1978 for details). From these results it has been concluded that there are two coupling sites on the non-cyclic pathway, one lying between plastoquinone and plastocyanin/cytochrome f and the other between oxygen production and the endogeneous acceptor for PS II, Q.

The transfer of electrons from ascorbate also produces ATP, and this is thought to be associated with cyclic electron transport around PS I. There is sufficient energy in the potential drop from ferredoxin to cytochrome b_6 to phosphorylate ADP. Further evidence for the existence of these separate phosphorylation sites comes from the observation that cyclic and non-cyclic phosphorylations are uncoupled by different concentrations of the inhibitor desaspidin. Furthermore, cyclic photophosphorylation is more resistant to heptane extraction of the chloroplast than is non-cyclic photophosphorylation. In summary, there appear to be two sites at which ADP can be phosphorylated during non-cyclic photophosphorylation. One lies between water and PS II and other between plastoquinone and PS I. There is an additional coupling site associated with cyclic electron flow around PS I.

Mechanistically, photophosphorylation is believed to be basically similar to oxidative phosphorylation. The chemical coupling hypothesis is formulated in the same terms for photophosphorylation as for oxidative phosphorylation; only the individual electron carriers are changed. The chemiosmotic hypothesis again explains ATP formation as a consequence of an electrochemical gradient of protons across the chloroplast membrane.

The evidence for the development of an electrical potential gradient across the thylakoid membranes in the light come from studies of electrochromic absorption — that is, the effect of strong electric fields of about 10^7 V m^{-1} (which is equivalent to a potential difference of 100 mV across a membrane of 10 nm) on absorption spectra. Both photosystems contribute to the membrane potential which is positive on the inside, i.e. in the loculus (c.f. Fig. 8.5). This would mean that the primary electron donors to both photosystems were located on the inside of the thylakoid membrane with the primary acceptors on the outside. We will return to the question of the location of the various components of the electron transport chain in more detail below.

Chloroplasts have been shown to catalyse the synthesis of ATP in the *dark* in two circumstances. If ADP, P_i and Mg^{++} are added as the light is turned off, ATP synthesis occurs — i.e. light produces an 'intermediate' from which ATP may be produced in the dark. The half time for the decay of this 'intermediate' is 0·5 sec at pH 6 and at room temperature, although it is more stable at pH 8. The decay is hastened by uncouplers of photophosphorylation. The amount of ATP formed and hence the amount of the 'intermediate' is too large for it to involve one of the various electron carriers and in the Mitchell hypothesis it is equated with a proton gradient. Secondly ATP may be formed in the dark (or if electron flow is prevented by the use of inhibitors) if broken chloroplasts suspended at pH 4 in the presence of a weak organic acid such as succinic are suddenly adjusted to pH 8·5 with the accompanying addition of ADP, P_i and Mg^{++}. The amount of ATP formed is relatively large. Similar experiments have been performed with mitochondria, and ATP is produced with the reverse pH change – from alkaline (pH 9) to acid (pH 4) in the presence of valinomycin – although the yield of ATP is much smaller, possibly because of the very active ATPase and the very much smaller volume of the mitochondrion relative to chloroplast. Such experiments provide strong evidence for the involvement of proton gradients in phosphorylation.

On illumination, broken (Class II) chloroplasts transport protons across the thylakoid membrane. The stoichiometry of this process is still equivocal, but the consensus appears to be that four protons are transferred per electron pair travelling down the electron transport chain (see Junge, 1977). By analogy with the mitochondrial electron transport system, this suggests that, since one ATP is produced for each pair of protons transported across the membrane, there are two coupling sites between water and P 700 (a similar conclusion can be reached on thermodynamic grounds, Nobel, 1975). However, experimental verification of the stoichiometry has not been forthcoming. Values for H^+/ATP of 4, 3 and 2 have been reported, although the idea of an overall fixed stoichiometry is perhaps not very useful. If proton gradients can be utilized for processes other than the synthesis of ATP – for example, in ion transport – the stoichiometry between proton transport and ATP synthesis will depend upon the other processes occurring. It is only in terms of the actual mechanism of ATP synthesis that a precise stoichiometry must be preserved (for an extensive discussion of the relationship between proton transport and photophosphorylation, see Jagendorf, 1975).

Crucial to the operation of the chemi-osmotic hypothesis is the localization of the various components of the photosystems and the electron transport chain. If the hypothesis is to be tenable, the components must be arranged in such a way as to explain net proton transport across the thylakoid membrane. Evidence for the location of the various components of the chain comes from studies with artificial electron donors and acceptors and from antibody labeling studies (see also below). There is substantial evidence to show that the acceptor side of PS I (i.e. the reduction of ferredoxin) is located on the outside of the thylakoid membrane (the side of the stroma/or matrix; Fig. 8.5), whereas the donor of this photosystem (viz. P 700) is located in the inside of the membrane (on the side of the loculus; Fig. 8.5). The acceptor side of PS II (Q) is located on the outside, although the evidence for the location of the donor side of PS II is equivocal (see Trebst, 1974 for a review of the evidence). However, many biochemical and physical properties of the photosynthetic system are more easily explained if the donor side of PS II is taken to be on the inside of the membrane. Furthermore, if PS II is restricted to the grana as has been suggested (p. 348), this may help to explain why evidence is more difficult to obtain for PS II, where the partitions may themselves restrict access to antibodies and artificial electron donors and acceptors. The overall conclusions are summarized in Fig. 8.20, where the components are shown as if functioning in the chemi-osmotic hypothesis. Cyclic electron flow around PS I is omitted from this figure since less is known about the components which are active under these conditions. It does, of course, also make good sense that the final products of the photosystems (ATP and NADPH) are produced on the stromal side of the thylakoid membranes, for it is in the stroma that they are utilized in the reduction of carbon dioxide.

The possibility also remains that plastoquinone functions in a Q-cycle (Mitchell, 1976) as described for ubiquinone in mitochondria, although direct evidence for this in higher plant photosynthesis has not yet been found (Trebst, 1978). The position of the two cytochromes, b_6 and b_{559}, within the membrane is consistent with the operation of such a cycle in which each plastoquinone molecule would carry two protons across the membrane, one electron coming from PS II and the other from cytochrome b_{559} which in turn accepts electrons from b_6. On the inside of the membrane, two protons would be released and one electron transferred to PS I and the second to b_6 and then to the feedback loop to b_{559} (see Fig. 8.20).

These results provide circumstantial evidence for the chemi-osmotic hypothesis although the evidence is equivocal. Supporters of the chemical hypothesis might explain that the respiratory chain is obligately coupled to $X \sim I$. $X \sim I$ may drive ATP synthesis or a proton pump (see Fig. 8.21) which is in turn coupled to ion translocation. The acid bath reverses the proton pump and therefore drives $X \sim I$ to produce ATP. To distinguish between the two explanations, it is necessary to know whether proton gradients are the primary energetic

Fig. 8.20 Topography of the thylakoid membrane. The localization of the various components of the two photosystems and the electron transport chain is that deduced from experimental evidence (see text), although the operation of the Q cycle is speculative.

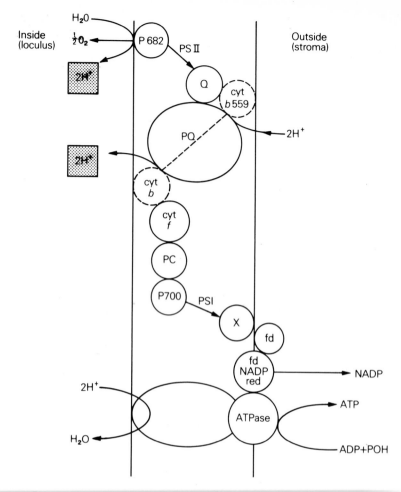

event or whether they are the consequence of hydrogen ion transport. It must be said, however, that in recent years the postulates of the chemi-osmotic hypothesis frequently seem to provide a more tenable explanation of the facts than do either the chemical coupling or conformational coupling hypothesis.

ATPase

In contrast to mitochondria, the ATPase activity of chloroplasts is latent and needs activating in some way. Trypsin treatment produces a calcium-dependent ATPase. EDTA treatment causes the release of a protein from the chloroplasts which can then no longer photophosphorylate, although they do retain the ability for electron transport. This protein has been shown to be identical to the trypsin-activated ATPase. Readdition of the protein to EDTA-treated chloroplasts restores, with the aid of Mg^{++}, photophosphorylation, and hence it has been termed a coupling factor (CF_1). Sulphydryl reagents release a light-triggered ATPase which requires a short period (30 min) of light before it will catalyse the hydrolysis of ATP in the dark. Treatment of the chloroplasts with calcium, on the other hand, produces an ATPase

Fig. 8.21 The relationship between ion transport and ATP synthesis as envisaged by the chemi-osmotic hypothesis and the chemical coupling hypotheses. Redrawn from Greville (1969).

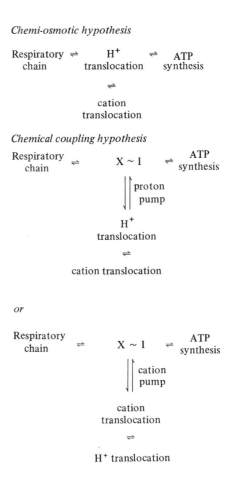

only active in the light. Since removal of the coupling factor inhibits the calcium-dependent, light-triggered and light-dependent ATPases, all three are probably manifestations of the same enzyme and represent part of the phosphorylation system working in reverse (McCarty and Racker, 1966). The coupling factor, CF_1, constitutes some 10% of the thylakoid membrane protein and can be obtained in reasonable amounts with relative ease. It is a colourless, water-soluble protein with a molecular weight of 325 000 and normally exists as a sphere of 9 to 9·25 μm diameter bound to the thylakoid membrane. As might be expected of so large a protein, it is made up of subunits of which there are five designated α, β, γ, δ and ε with molecular weights decreasing from about 60 000 to 15 000, respectively (see McCarty, 1979 for further details). The subunits are not present in equal proportions and although there is controversy over the stoichiometry it seems that their likely ratio is $2:2:1:1:2$.

CF_1 has multiple nucleotide binding sites, but low ATPase activity unless activated by heat or some other treatment. It is likely that it is the enzyme responsible for ATP synthesis and *in vivo* is associated with other proteins which bind it to the membrane and which facilitate proton translocation. Of the individual subunits, α and β appear to contain the active sites for the ATPase, γ may translocate protons to the active site, δ facilitates binding of the CF_1 to the other components of the ATPase complex, while ε appears to be an inhibitor of the ATPase activity (which is lost together with the δ subunit on heating with digitonin, hence explaining the heat activation of the ATPase activity).

The localization of the coupling factor, CF_1, in the membrane has been studied with the use of antibodies. An antibody is only able to react with its antigen if the antigen is freely accessible to the antibody. Where the antigen is part of a membrane system, then if this membrane is vesicular, free access is only available if the antigen is located on the outside of the membrane. If the membrane is fragmented, for example, by sonication, then the inside becomes accessible. The reaction of the antibody with the membrane component which is its antigen can be monitored by the inhibition of activity of that component (if for example it is an enzyme) or by agglutination. Agglutination tests can be performed either directly, when crosslinking of the antibody causes agglutination, or indirectly when either soluble antigen is added after the antibody or antibody against the immunoglobulin is added after the antibody against the membrane component. In the case of CF_1, antibodies against this protein agglutinate thylakoid vesicles: if the vesicles are washed with EDTA to remove the CF_1 there is no agglutination. Antibody prepared against ferredoxin-NADP reductase only agglutinates in the indirect tests if the CF_1 has been removed. It has thus been postulated that both compounds are located on the stroma side of the membrane but with the reductase lying in crevices partially formed by the coupling factor. Electron microscopy has verified the presence of knobs on this side of the membrane, similar to those described for mitochondrial inner membranes (Fig. 7.4). It is noteworthy that the ATPase is not latent in Class I chloroplasts where it is in contact with the stroma. Hence, it has been argued that the inhibitory subunit is bound to the ATPase in Class II chloroplast preparations.

Reduction of carbon dioxide

Light energy is utilized in photosynthesis to produce reduced co-enzyme (NADPH) and ATP. Subsequently, in the so-called dark reactions of photosynthesis, carbon dioxide is reduced to a variety of organic carbon compounds. The pathway of this carbon fixation found in many plants was elucidated particularly during the period 1946 to 1953 by Calvin working with Benson, Bassham, Massini and Wilson on the unicellular algae *Chlorella pyrenoidosa* and *Scenedesmus obliquus*. It is only recently that isolated chloroplasts have been prepared which can fix carbon dioxide at rates comparable with whole cells and it has been found necessary to use aqueous isolation techniques employing sucrose or sorbitol, not sodium chloride, if the ability to fix carbon dioxide is to be preserved. Photosynthesis by chloroplasts isolated from pea and spinach and by leaf discs from these species produce similar results to those described for *Chlorella*.

The sequence of reactions in the unicellular algae was determined by labelling the intermediates with the radioisotope carbon-14 and then separating them by paper chromatography (see p. 39). The compounds were then located on the chromatogram by autoradiography. By using a unicellular alga, label could be added in the solution to all the cells almost instantaneously as ^{14}C-bicarbonate and, furthermore, the reactions could be stopped equally rapidly by running the algal suspension into methanol. This enabled the sequential appearance of label in a variety of intermediates to be determined over short time periods.

After one minute of photosynthesis, sucrose, uridine diphosphate-glucose, 3-phosphoglyceric acid, phosphoenolpyruvic acid, several sugar phosphates and several carboxylic acids and amino acids were labelled (Fig. 8.22b). By reducing the time for photosynthesis to less than 5 sec the number of labelled compounds was very much reduced (Fig. 8.22a) and consisted chiefly of phosphoglyceric acid with small amounts of sugar phosphates, phosphoenolpyruvic acid and malic acid. The importance of phosphoglyceric acid as the first stable product of photosynthesis was confirmed when it was later shown that the rate of incorporation of ^{14}C into phosphoglycerate in very short times was much greater than the rate of labelling of any other compound. Degradation of the 3-phosphoglycerate showed that 90% of the ^{14}C present after only a few seconds photosynthesis was in the carboxyl group (see Table 8.2). The experiments which provided this early evidence suffered from a change in the carbon dioxide concentration during $^{14}CO_2$ addition. However in later work it was possible to maintain the carbon dioxide level constant while adding small amounts of $^{14}CO_2$ in solution as a tracer. This did not alter the effective carbon dioxide concentration. As a result of these steady state experiments the distribution of ^{14}C in phosphoglycerate was confirmed.

Table 8.2   The distribution of radioactivity (in arbitrary units) in Calvin cycle intermediates extracted from *Scenedesmus obliquus* following a 5·4 sec exposure to $^{14}CO_2$

Carbon atom	Phospho-glycerate	Fructose	Sedo-heptulose	Ribulose
1 (—COOH)	82	3	2	11
2 (α)	6	3	2	10
3 (β)	6	43	28	69
4		42	24	5
5		3	27	3
6		3	2	
7			2	

Data of Bassham and Calvin (1957).

Further information on the nature of the formation of phosphoglycerate was obtained from a different type of experiment in which plants were allowed to photosynthesize in $^{14}CO_2$ until such time that all the intermediates were labelled with a specific activity equal to that of the

Fig. 8.22 Radioautographs of two-dimensional chromatograms of the products of (*a*) 2 sec and (*b*) 60 sec of photosynthesis by *Chlorella pyrenoidosa* in $^{14}CO_2$. PEPA, phosphoenolpyruvate; PGA, 3-phosphoglycerate; UDPG, uridine diphosphoglucose. Reproduced from Bassham (1965).

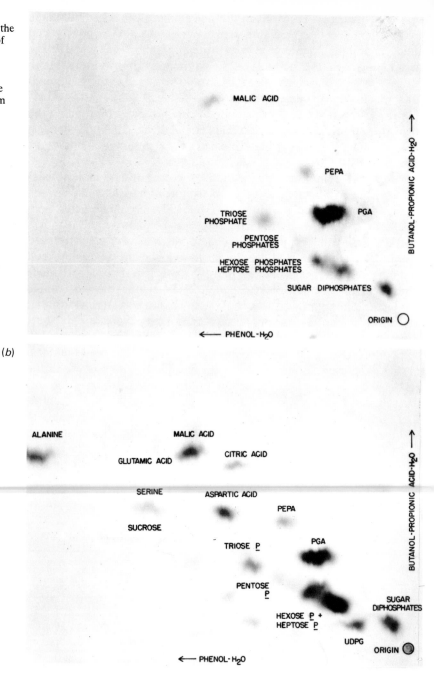

carbon dioxide supplied. The concentration of individual intermediates could then be determined from the amount of radioactivity, providing the specific activity was known. It was also possible in such experiments to distinguish between photosynthetic and non-photosynthetic pools of the intermediates since the former were fully labelled within fifteen minutes while the latter took longer to equilibrate. When cells which had equilibrated with ^{14}C in the light were suddenly placed in the dark, two very important changes were

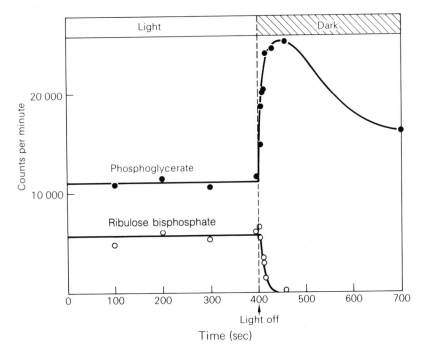

Fig. 8.23 The change in the amounts of label appearing in phosphoglycerate and ribulose bisphosphate in *Scenedesmus obliquus* following a change in conditions from light to dark. Redrawn from Bassham and Calvin (1957).

apparent: the amount of phosphoglyceric acid rose and then fell slowly and the level of ribulose bisphosphate fell rapidly to zero (Fig. 8.23). It was inferred from these results that ribulose bisphosphate was the primary acceptor of carbon dioxide in a reaction that does not require light. This produces phosphoglycerate which builds up in concentration, since its conversion to triose requires reducing power only produced in the light. That is:

$$
\begin{array}{ccc}
\text{*CO}_2 + \begin{array}{l} \text{CH}_2\text{O}\ \textcircled{P} \\ | \\ \text{C}=\text{O} \\ | \\ \text{CHOH} \\ | \\ \text{CHOH} \\ | \\ \text{CH}_2\text{O}\ \textcircled{P} \end{array} & \longrightarrow & \begin{array}{l} \text{CH}_2\text{O}\ \textcircled{P} \\ | \\ \text{HOCH} \\ | \\ \text{*COO}^- \end{array} \quad + \quad \begin{array}{l} \text{COO}^- \\ | \\ \text{HOCH} \\ | \\ \text{CH}_2\text{O}\ \textcircled{P} \end{array} \\
\text{Ribulose bisphosphate} & & \text{Phosphoglycerate}
\end{array}
$$

The asterisk refers to the ^{14}C label added as carbon dioxide and incorporated into the phosphoglycerate (corresponding to Table 8.2). The enzyme catalysing this reaction is known either as ribulose bisphosphate carboxylase, carboxydismutase or Fraction I protein, and as such constitutes nearly half of all soluble leaf protein. This protein is discussed more fully in Chapter 5. The decline in the level of ribulose bisphosphate could be explained by the failure of reduction of phosphoglycerate causing the cessation of a cyclic process which regenerates ribulose bisphosphate. Further evidence for this hypothesis was obtained when cells which had been photosynthesizing in 1% carbon dioxide were suddenly adjusted to 0·003% carbon dioxide. The level of ribulose bisphosphate rose rapidly (Fig, 8.24), as would be

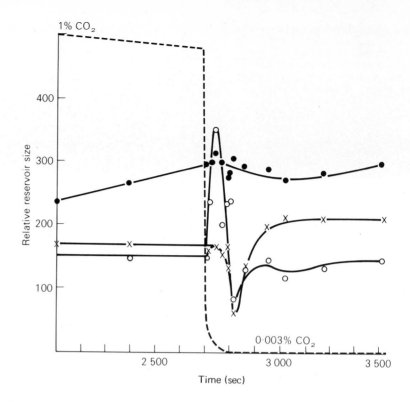

Fig. 8.24 Changes in the amounts of sucrose (●), phosphoglycerate (×) and ribulose bisphosphate (○) in *Scenedesmus* following a decrease in the carbon dioxide concentration (----) at 6°C. Redrawn from Bassham and Calvin (1957).

expected if it were the primary acceptor for carbon dioxide and the level of phosphoglycerate fell as expected if it were formed from ribulose bisphosphate.

The distribution of label in hexose was consistent with its formation from two molecules of triose produced by the reduction of phosphoglycerate, as in the reversal of glycolysis (see p. 245). That is:

Again, the asterisks refer to the level of ^{14}C-label (see Table 8.2). The distribution of label in sedoheptulose (Table 8.2) is not explained by a C-1 + C-6 or C-2 + C-5 addition reaction, but requires a C-3 + C-4 addition. The C_4 compound is generated in a transketolase reaction:

```
CH₂OH          CH₂O-Ⓟ         *CHO              CH₂OH
|              |              |                 |
C=O            HOCH           *CHOH       +     C=O
|              |              |                 |
HO*CH    +     *CHO    ⟶      CHOH              HO*CH
|                             |                 |
*CHOH                         CH₂O-Ⓟ            CHOH
|                                               |
CHOH                          Erythrose         CH₂O-Ⓟ
|                             4-phosphate
CH₂O-Ⓟ                                          Xylulose
                                                5-phosphate
Fructose
6-phosphate
```

And the sedoheptulose formed by an aldolase:

```
*CHO           CH₂O-Ⓟ              CH₂O-Ⓟ
|              |                   |
*CHOH    +     C=O         ⟶       C=O
|              |                   |
CHOH           *CH₂OH              HO*CH
|                                  |
CH₂O-Ⓟ                             *CHOH
                                   |
                                   *CHOH
                                   |
                                   CHOH
                                   |
                                   CH₂O-Ⓟ

                                   Sedoheptulose
                                   1,7-bisphosphate
```

The distribution of label in the pentose sugars is a mean distribution of the label present in three molecules of sugar generated in two separate reactions. One molecule of pentose is produced by the erythrose-forming transketolase. While the other two molecules are formed by a separate transketolase involving sedoheptulose:

```
CH₂OH          CH₂O-Ⓟ              CH₂OH         *CHO
|              |                   |             |
C=O            HOCH                C=O           *CHOH
|              |                   |             |
HO*CH    +     *CHO          ⟶     HO*CH   +     *CHOH
|                                  |             |
*CHOH                              CHOH          CHOH
|                                  |             |
*CHOH                              CH₂O-Ⓟ        CH₂O-Ⓟ
|
CHOH                               Xylulose      Ribose
|                                  5-phosphate   5-phosphate
CH₂O-Ⓟ
```

Following conversion to ribulose 5-phosphate, the sum of the label distribution thus corresponds to that reported in Table 8.2.

```
C       C        *C        *C
|       |        |         |
C       C        *C        *C
|       |        |         |
*C  +  *C   +   *C    =   ***C
|       |        |         |
C       C        C         C
|       |        |         |
C       C        C         C

Xylulose    Ribose       Ribulose
5-phosphate 5-phosphate  5-phosphate
```

The complete cycle (Fig. 8.25) has been described in detail many times (e.g. Bassham, 1965). In essence, however, ribulose 5-phosphate is phosphorylated to ribulose 1,5-bisphosphate by ATP. This ribulose bisphosphate is then carboxylated under the influence of ribulose bisphosphate carboxylase (carboxydismutase, see p. 361) to an unstable compound which breaks down to form two molecules of phosphoglycerate. The second phosphorylation of the pathway converts the 3-phosphoglycerate to phosphoryl 2-phosphoglycerate and then the only reductive step converts the latter to glyceraldehyde 3-phosphate which isomerises to dihydroxyacetone phosphate. In order to complete

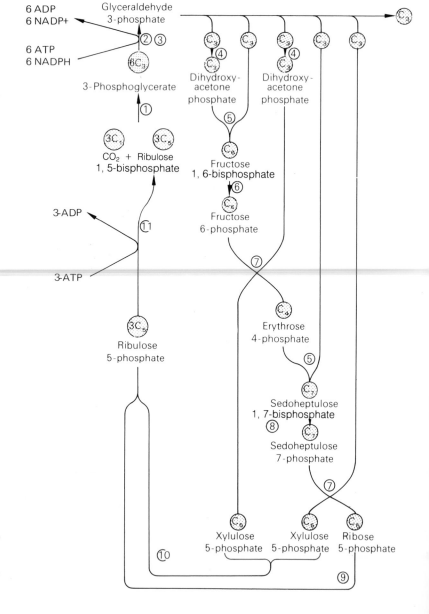

Fig. 8.25   The Calvin cycle. The enzymes involved in the cycle are:
1. Ribulose bisphosphate carboxylase or carboxydismutase or Fraction I protein.
2. Phosphoglycerate kinase.
3. Glyceraldehyde phosphate dehydrogenase. 4. Triose phosphate isomerase. 5. Aldolase. 6. Fructose 1,6-bisphosphatase. 7. Transketolase.
8. Sedoheptulose bisphosphatase.
9. Phosphoribose isomerase.
10. Ribulose 5-phosphate epimerase.
11. Ribulose 5-phosphate kinase.

a cycle, there must be a regeneration of pentose which is achieved by reactions similar to those of the pentose phosphate pathway (see p. 255). Condensation of the dihydroxyacetone phosphate with glyceraldehyde 3-phosphate produces fructose 1,6-bisphosphate which following hydrolysis to the monophosphate is involved in a transketolase reaction with glyceraldehyde 3-phosphate forming erythrose 4-phosphate and xylulose 5-phosphate. Condensation of the erythrose 4-phosphate with dihydroxyacetone phosphate produces sedoheptulose, 1,7-bisphosphate which is subsequently hydrolysed to the 7-monophosphate. The two phosphatases, hydrolysing the fructose 1,6-bisphosphate and the sedoheptulose 1,7-bisphosphate, are not components of the pentose phosphate pathway nor is the aldolase which produces the sedoheptulose 1,7-bisphosphate. One molecule of pentose is regenerated by the fructose 6-phosphate transketolase reaction and a further two molecules by the sedoheptulose 7-phosphate transketolase. The ribose 5-phosphate is isomerized and the two molecules of xylulose 5-phosphate epimerized to ribulose 5-phosphate.

Thus, in summary, the most important reactions in relation to carbon dioxide reduction are the carboxylation of ribulose 1,5-bisphosphate and the only reductive reaction of the cycle, the conversion of 3-phosphoglycerate to glyceraldehyde 3-phosphate. The remainder of the cycle is concerned with the regeneration of pentose. For every step in the cycle to function at least once, three carboxylations must occur and hence six moles of phosphoglycerate (18 carbon atoms) are produced. The reformation of three moles of ribulose 1,5-bisphosphate (15 carbon atoms) requires the carbon from five moles of triose (15 carbons) and thus in a complete cycle three carbon atoms become available for the various end products of photosynthesis. It is the regeneration of the carbon dioxide acceptor which is fundamental to the continued net fixation of CO_2 since any pathway which achieves net fixation must regenerate fresh acceptor in addition to the fixed-carbon product. As far as is known the Calvin cycle is unique in achieving this end. Although other pathways involving carboxylases have been discovered (see below) they do not in themselves carry out the *net* fixation of CO_2. This is always the function of the Calvin cycle. The production of these three 'fixed' carbon atoms involves nine moles of ATP and six moles of NADPH: i.e. for one mole of carbon dioxide to be fixed, the light reactions must provide two moles of NADPH and three moles of ATP.

Ribulose bisphosphate carboxylase consists of eight large and eight small sub-units with an overall molecular weight of about 550 000. The larger sub-unit (molecular weight 51 000 to 58 000) is catalytic and binds ribulose bisphosphate and CO_2 through lysine residues. The chemical intermediate between ribulose bisphosphate and phosphoglycerate which exists on the enzyme is 2-carboxy-3-ketoribitol 1,5-bisphosphate. The smaller sub-units apparently bind Mg^{++} and may play a regulatory rôle. When assayed under conditions for maximum activation (and it is important to note that CO_2 is both substrate and activator), the enzyme is capable of fixing CO_2 at almost 1 000 μmoles h^{-1} mg^{-1} chlorophyll. This is about ten times greater than the maximum for leaf carbon dioxide fixation. However, if the enzyme is assayed immediately post isolation, its activity closely matches that found for CO_2 fixation existing in the leaf before isolation of the enzyme, and Jensen and Bahr (1977) have proposed that the degree of activation of the enzyme *in vivo* will reflect the pH, CO_2 concentration, magnesium ion concentration and temperature in the stroma and so regulate overall CO_2 fixation under light saturating conditions.

In recent years, research into the operation of the Calvin cycle and its component parts has largely dispelled former doubts over its operation (Kelly *et al.*, 1976). Perhaps the most important developments have come in our understanding of the ribulose bisphosphate carboxylase. It had formerly been noted that there was a marked discrepancy between the apparent K_m (CO_2) for ribulose bisphosphate carboxylase of some 70 to 600 μm, the K_m for CO_2 fixation, 10 to 20 μm, and the carbon dioxide concentration in water in equilibrium with air containing 0·03% CO_2, 10 μm. This raised doubts upon the central rôle of the carboxylase. However, it has subsequently been established that once isolated from the chloroplast, the kinetics of carboxylation are rather unstable and only if determined within a short time (3 min) are low K_m (CO_2) values of the order of 10 to 18 μm obtained (Jensen and Bahr, 1977).

The importance of ribulose bisphosphate carboxylase was further enhanced when it was established in the early 1970s that it was also able to function as an oxygenase catalysing the reaction:

$$
\begin{array}{ccc}
\text{H}_2\text{C-O} \; \text{(P)} & & \\
\quad | & & \text{H}_2\text{C-O} \; \text{(P)} \\
\text{C=O} & & \quad | \qquad \text{phosphoglycolate} \\
\quad | & & \text{COO}^- \\
\text{HCOH} & & \\
\quad | \quad + \text{O}_2 \longrightarrow & & + \\
\text{HCOH} & & \\
\quad | & & \text{H}_2\text{C-OH} \\
\text{H}_2\text{CO} \; \text{(P)} & & \quad | \\
& & \text{COO}^- \qquad \text{glycolate}
\end{array}
$$

Oxygen is a competitive inhibitor of the carboxylase activity as is carbon dioxide of the oxygenase and K_m values for CO_2 and O_2 are, within error, the same as the K_i values for O_2 and CO_2 respectively. The importance of the oxygenase reaction to photorespiration is dealt with in the next chapter.

Enzyme activity is closely regulated by a high pH optimum and a requirement for a high magnesium ion concentration for maximum activity. Both of these factors increase in the stroma in the light (see p. 392). They also activate other enzymes of the Calvin cycle.

The function of carbonic anhydrase, which catalyses the reaction $\text{H}^+ + \text{HCO}_2^- \rightarrow \text{H}_2\text{O} + \text{CO}_2$, and which is required for photosynthesis in some species, remains uncertain. The enzyme may operate as a permease, facilitating the transport of carbon dioxide to sites of carboxylation, or it could be physically associated with the ribulose bisphosphate carboxylase and thus increase the local availability of carbon dioxide.

Products of photosynthesis

The formation of starch and sucrose as final storage products of photosynthesis has been known for many years and since the work of Leloir (see Leloir, de Fekete and Cardini, 1961) it has become evident that the synthesis of these molecules involves the sucrose and starch synthetase reactions, the glucose being donated to an acceptor by uridine diphosphate glucose or adenosine diphosphate glucose (see p. 115). The chloroplast starch synthetase appears to be specific for ADP-glucose, although the starch synthetases of storage organs are able to utilize both ADP- and UDP-glucose. Interestingly it has been shown that the 3-phosphoglycerate is an activator of ADP-glucose pyrophosphorylase, the enzyme responsible for ADP-glucose formation. Both of the latter compounds are early labelled products of photosynthesis (cf. Fig. 8.22). For many years it was believed that carbohydrate was the major product of the reactions of photosynthesis. However, following the work of Calvin and his associates it has been shown that a relatively small amount of the fixed carbon may appear as carbohydrate in *Chlorella* depending upon the conditions of culture and there is a rapid appearance of label in amino acids and carboxylic acids.

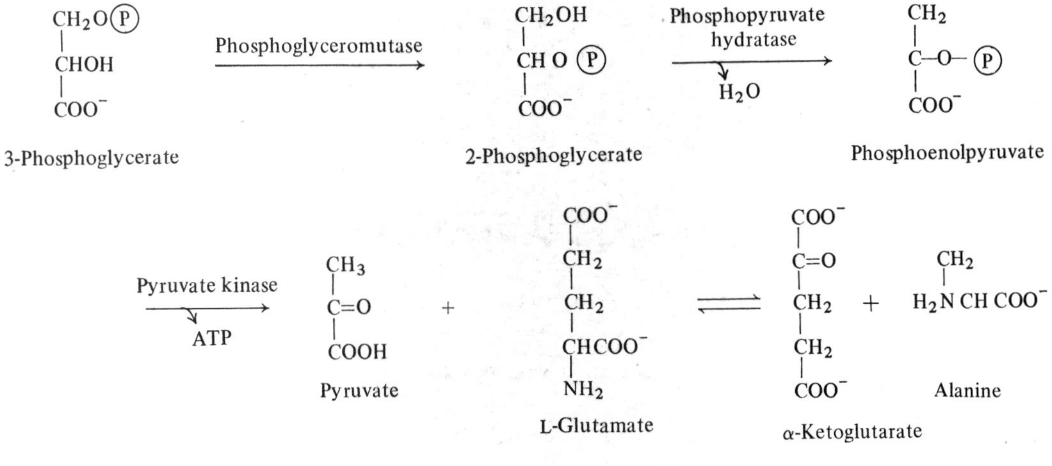

Experiments on the separation of the early products of photosynthesis have shown that there can be a rapid incorporation of label into alanine, aspartic acid, serine, glutamic acid, glycine and glycolate. For example, *Chlorella pyrenoidosa* may incorporate 30% of its photosynthetically fixed carbon into amino acids. The amino acids are formed by transamination and reductive amination reactions with keto acids, derived in turn from the intermediates of the Calvin cycle; e.g. alanine may be formed by the transamination of pyruvic acid derived from phosphoenolpyruvate which in turn is derived from phosphoglycerate. The export of glycolate and formation of glycine and serine seems to involve the reactions of the glycolate pathway and both peroxisomes and mitochondria (Ch. 9). Similar (c.30%) levels of incorporation of label have been reported for lipids during one to two minutes of photosynthesis. This incorporation is light dependent and must involve the conversion of a Calvin cycle intermediate into acetyl coenzyme A, the starting point for fatty acid synthesis, although the precise mechanism of formation of the acetyl coenzyme A is not known. Apart from the amino acids and fatty acids, a rapid appearance of label also occurs in the carboxylic acids, malic, succinic, citric and glycolic. Malic acid may be formed by the reductive carboxylation of phosphoenolpyruvate, whence other di- and tricarboxylic acids may be formed by reactions similar to those of the TCA cycle. Further details of the synthesis of small molecules by chloroplasts can be obtained from the review by Givan and Leech (1971).

Alternative pathways of carbon assimilation

Many, but not all, of the species from a number of families (Graminae, Cyperaceae, Amaranthaceae, Portulaceae, Chenopodiaceae, Euphorbiaceae, Nyctoginaceae, Aizoaceae, Compositae and Zygophyllaceae) show a pattern of carbon dioxide fixation which is not consistent simply with the operation of the Calvin cycle. They are mostly tropical species with high rates of growth and photosynthesis adapted to high temperatures, high irradiance and low water supply. They apparently lack photorespiration (see. Ch. 9) and have a low carbon dioxide compensation point (see p. 336).

Studies on isotopic labelling patterns within such plants have shown that radioactivity does not, at very short times, increase in phosphoglycerate at a much greater rate than in other compounds. In fact, radioactive label increases most rapidly into malate and aspartate. The radioactivity appearing in the C–4 of C_4 acids in very short times accounts for most of the total label incorporated (Fig. 8.26). As the time is extended, the rate of label incorporation into C_4 acids decreases while that appearing in phosphoglycerate increases. This is consistent with the movement of label from C_4 acids to phosphoglycerate and its products. Pulse-chase experiments in which plants were allowed to photosynthesize for 15 sec in $^{14}CO_2$ and then transferred to $^{12}CO_2$ showed that after 90 sec essentially all the label in the C–4 position of C_4 acids had been transferred to phosphoglycerate, triose phosphate and hexose phosphate. The elucidation of a pathway leading to the rapid formation of malate or aspartate has chiefly been the work of Hatch and Slack (see Hatch 1971, 1976) and is often termed either the Hatch-Slack pathway or the C_4 pathway. Plants which show this pathway are often known as C_4 plants while those utilizing the Calvin cycle alone are called C_3 plants.

An important feature of all plants, both monocotyledonous and dicotyledonous, in which the C_4 pathway has been demonstrated is that the vascular bundles are surrounded by two concentric layers of chlorenchyma. The inner layer is a parenchymatous bundle sheath and the outer layer is part of the mesophyll. This arrangement is known as *Kranz-type anatomy*. A further feature of these plants is that the cells of the mesophyll and bundle sheath often contain different types of chloroplast (see Laetsch, 1971). The difference is very pronounced in plants, such as sugar cane and maize, (Fig. 8.27) where the bundle sheath chloroplasts lack grana. Where the bundle sheath chloroplasts

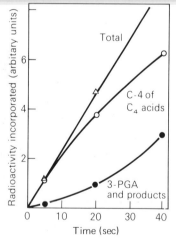

Fig. 8.26   The total amount of ^{14}C uptake in maize leaves (illuminated at 2 000 foot candles) and the amounts of ^{14}C incorporated into C-4 of C_4 acids and into 3-phosphoglycerate. Reproduced from Hatch (1971).

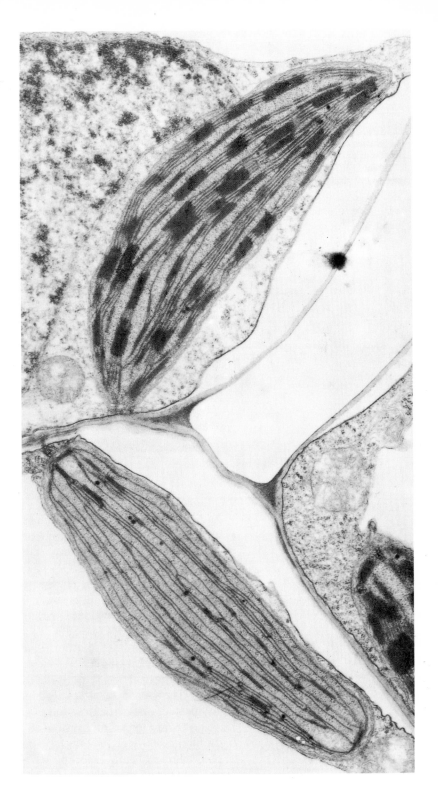

Fig. 8.27 Granal (mesophyll) and agranal (bundle sheath) chloroplasts in cells of a seven day old green maize leaf. ×21 800. Provided by Dr R. M. Leech, University of York.

do develop grana these chloroplasts are generally larger than those of the mesophyll cells. A unique feature of the chloroplasts of C_4 plants is a series of anastomosing tubules associated with the inner membrane of the chloroplast envelope (Fig. 8.28).

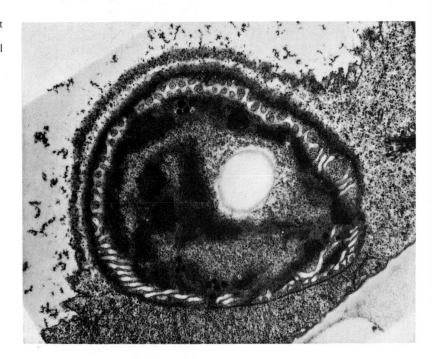

Fig. 8.28   Mesophyll cell chloroplast of *Portulaca oleracae* with anastamosing tubules of the peripheral reticulum ×40 300. Reproduced from Laetsch (1971).

Comparisons of the enzymes present in C_4 plants with those of C_3 plants have indicated a number of striking differences. Plants which utilize the C_4 pathway have characteristically higher levels of phosphoenolpyruvate carboxylase (phosphophenolpyruvate + CO_2 → oxaloacetate) (see p. 273), aspartate amino transferase (oxaloacetate + [NH_2] → aspartate), NADP specific malate dehydrogenase, pyruvate phosphate dikinase (pyruvate + ATP → phosphenolpyruvate + AMP + P_i), adenylate kinase (ATP + AMP ⇌ ADP + ADP) and sometimes malic enzyme (malate + $NADP^+$ → pyruvate + CO_2 + NADPH) (Table 8.3).

Isolation of both mesophyll and bundle sheath chloroplasts by non-aqueous techniques has shown that those enzymes which are not part of the Calvin cycle appear to be located within the mesophyll cells. The chloroplasts of the bundle sheath, although often with an atypical granal structure, have a very similar enzymic complement (with regard to carbon fixation) to those of normal Calvin cycle plants. In fact, as early as 1944, it has been suggested that starch deposited in the bundle sheath chloroplasts was possibly derived from carbon dioxide fixed in the mesophyll. It is currently believed that C_4 acids are formed in the mesophyll cells and then transferred either as malate or aspartate to the cells of the bundle sheath where decarboxylation and refixation by the Calvin cycle occurs (Fig. 8.29). The C–4 of these

Table 8.3 A summary of the activity [mole min^{-1} mg^{-1} chlorophyll] and location of enzymes implicated in the C_4 pathway. The enzyme was designated as either mesophyll or bundle sheath when at least 90% of the activity was located in those particular cells

Enzyme and location	Activity	Average ratio C_4/C_3 species
Mesophyll		
PEP carboxylase	16–21	60
NADP-malate dehydrogenase	2–14	10
Aspartate aminotransferase	4–40	10
Pyruvate P_i dikinase	3–7	∞
Adenylate kinase	17–45	75
Mesophyll and bundle sheath		
Pyrophosphatase	15–60	15
3-Phosphoglycerate kinase	50	—
NADP-triose P dehydrogenase	5–10	—
Bundle sheath		
Fructose bisphosphate aldolase	8	—
Alkaline fructose bisphosphatase	1	—
Ribulose 5-P isomerase	10–25	—
Ribulose bisphosphate carboxylase	2–4	—
Malic enzyme*	10–12	50
	0·1–1	2

Reproduced from Hatch (1971).

    * Plants showing two very different levels of malic enzyme activity have been found.

acids becomes the C–1 of 3-phosphoglycerate. The difference in distribution of ribulose bisphosphate carboxylase in the photosynthetic tissues of C_3 and C_4 plants is strikingly illustrated in Fig. 8.30.

Carbon dioxide is fixed in the mesophyll with the carboxylation of phosphoenolpyruvate by the enzyme phosphoenolpyruvate carboxylase:

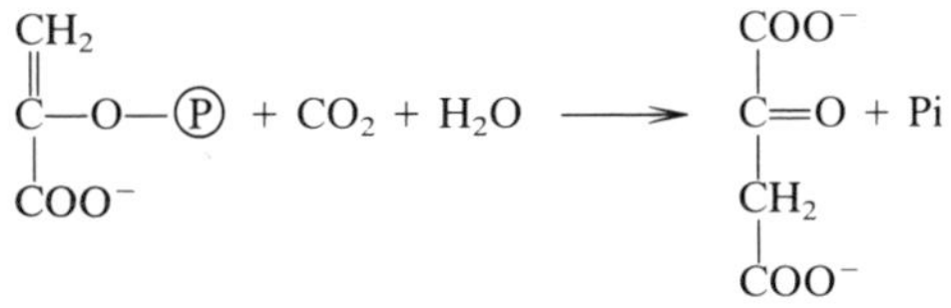

$$CH_2{=}C(-O{-}\textcircled{P})(COO^-) + CO_2 + H_2O \longrightarrow (COO^-)C{=}O(CH_2)(COO^-) + Pi$$

The reaction is claimed to be located in the cytoplasm although there are reports of the enzyme being associated with the chloroplasts. The oxaloacetate so formed is either reduced to malate in the chloroplasts ($NADP^+$ −malate dehydrogenase) or aminated to produce aspartate (transaminase) in the cytoplasm. Normally, the production of one of these compounds predominates in a given species, although both can be formed in the same plant (e.g. *Portulaca oleracea*). Both malate and aspartate are mobile and pass into the bundle sheath chloroplasts. In plants with a high malic enzyme level (recall from Table 8.3 that not all C_4 plants do have high levels of malic enzyme), malate is decarboxylated and the carbon dioxide released is refixed by ribulose bisphosphate carboxylase although it is not known how the CO_2 is prevented from diffusing away prior to refixation. This action of malic enzyme

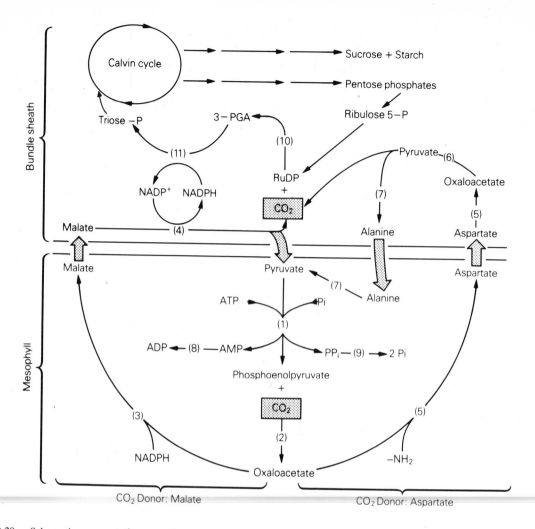

Bundle sheath	Mesophyll

Calvin cycle → → → Sucrose + Starch

→ → → Pentose phosphates

Triose –P   3 – PGA ← Ribulose 5–P

(11)   (10)   Pyruvate (6)

NADP⁺ NADPH   RuDP (7)   Oxaloacetate

+   Alanine   (5)

Malate ——— (4) ——— CO_2 Aspartate

Malate   Pyruvate   Aspartate

ATP → ← Pi   (7) ← Alanine

(1)

ADP ← (8) — AMP ←   → PP$_i$ — (9) → 2 Pi

Phosphoenolpyruvate

+

CO_2

(3)   (2)   (5)

NADPH   –NH₂

Oxaloacetate

CO_2 Donor: Malate CO_2 Donor: Aspartate

Fig. 8.29 Schematic representation of the reactions and intercellular movements (heavy arrows) of metabolites proposed to occur during the operation of the C₄ pathway. The enzymes are (1) pyruvate P$_i$ dikinase; (2) phosphoenolpyruvate carboxylase; (3) NADP⁺ malate dehydrogenase; (4) malic enzyme; (5) aspartate aminotransferase; (6) PEP carboxykinase or NADP⁺-malic dehydrogenase and NADP⁺-malic enzyme; (7) alanine aminotransferase; (8) adenylate kinase; (9) pyrophosphatase; (10) ribulose 1,5-bisphosphate carboxylase; (11) 3-phosphoglycerate kinase and NADP⁺ glyceraldehyde 3-phosphate dehydrogenase. Provided by Dr M. D. Hatch, CSIRO, Canberra.

also regenerates pyruvate

$$
\begin{array}{c}
COO^- \\
| \\
CH_2 \\
| \\
CHOH \\
| \\
COO^-
\end{array}
+ NADP^+ \longrightarrow
\begin{array}{c}
CH_3 \\
| \\
C=O \\
| \\
COO^-
\end{array}
+ CO_2 + NADPH + H^+
$$

which returns to the mesophyll chloroplast for regeneration of phosphoenolpyruvate.

$$
\begin{array}{c}
CH_3 \\
| \\
C=O \\
| \\
COO^-
\end{array}
+ ATP + P_i \longrightarrow
\begin{array}{c}
CH_2 \\
| \\
C-O-\textcircled{P} \\
| \\
COO^-
\end{array}
+ AMP + PP_i
$$

Not all plants have high levels of malic enzyme, as we remarked earlier. Plants without this enzyme generate aspartate from oxalacetate, which then becomes the currency of carbon flow from the mesophyll cells to the bundle sheath. Here the aspartate is reconverted to oxalacetate in the mitochondria and then either reduced to malate

Fig. 8.30 Immunofluorescent labelling (indirect method) of ribulose-1,5-bisphosphate carboxylase (RuP$_2$Case) in sections of leaf blades of a C$_3$ and a C$_4$ plant (adaxial epidermis uppermost). RuP$_2$Case was isolated from leaves and injected into rabbits to make a specific anti-RuP$_2$Case antiserum. These antibodies were, in turn, labelled with sheep antiserum (raised against rabbit antiserum), the sheep antiserum being tagged with the fluorochrome, fluorescein isothiocyanate. Thus the fluorochrome indirectly labels the enzyme *in situ*, and when excited with light of a particular wavelength in a fluorescence microscope, it fluoresces at longer wavelengths in the yellow region. In black and white photography, specific fluoroscence shows as bright white, but so too does some non-specific fluoroscence (e.g. some cell walls) which is not due to the fluorochrome. A. *Digitaria brownii*, a C$_4$ grass, showing specific fluorescence (arrows) associated with the bundle sheath cells, but not with mesophyll cells. × 700. B. *Danthonia bipartita*, a C$_3$ grass, showing specific fluorescence associated with chloroplasts of all mesophyll cells (arrows) with the sparse chloroplasts in bundle sheath cells. × 445. Micrographs provided by Dr P. W. Hattersley, Australian National University. See Hattersley *et al.* (1977) for further details of the technique and results with other species.

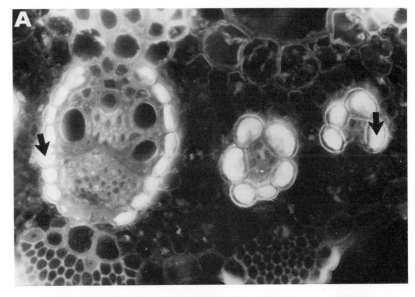

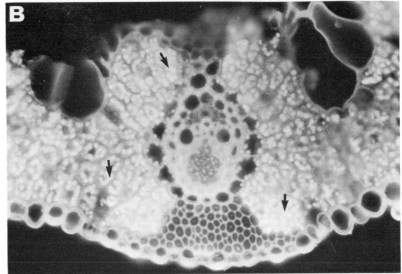

and decarboxylated by malic enzyme or the oxaloacetate is decarboxylated directly by phosphoenolpyruvate carboxykinase (see p. 273). Pyruvate formed by either of these routes is animated and transferred back to the mesophyll cells as alanine where it is recycled through pyruvate and phosphoenolpyruvate. The movement of malate and pyruvate may be correlated with the presence of large numbers of plasmodesmata to be seen between the neighbouring cells. A further feature of the action of malic enzyme is the formation of NADPH in the bundle sheath chloroplasts. It should be remembered that it is these very chloroplasts which have poorly developed grana and therefore probably little PS II activity. NADPH is thus generated by an alternative route. However, only half the necessary NADPH is formed to reduce two moles of phosphoglycerate to triose phosphate. Conversion of phosphoglycerate to triose phosphate does appear to take place in both types of chloroplast while the formation of hexose, sucrose and pentose phosphates appears to be restricted to the bundle sheath.

The importance of the C_4 pathway appears to lie in its ability to raise the carbon dioxide concentration in the bundle sheath cells, a function which is reminiscent of that of Crassulacean acid metabolism (see below). Stomatal resistances are higher for C_4 than for C_3 species, possibly to prevent excessive water loss since these species are commonly found in tropical climates. This means that for a given gradient of carbon dioxide concentration, the diffusion of CO_2 into a C_4 leaf might be expected to be slower than into a C_3 leaf. Since the C_4 leaf is likely to be functioning at a higher temperature, and hence at a faster rate, than the C_3 leaf there would be a tendency for the rapid depletion of carbon dioxide in C_4 species. Since PEP carboxylase has a much lower K_m for carbon dioxide than does ribulose bisphosphate carboxylase, the C_4 pathway effectively raises the carbon dioxide concentration at the site of the ribulose bisphosphate carboxylase and hence not only directly increases the net rate of carbon fixation but also raises the level of carboxylase activity in relation to that of the oxygenase (Hatch and Osmond, 1976). The C_4 acids are also able to act as a buffer in that they provide carbon dioxide for the Calvin cycle under conditions where the stomata may be temporarily closed. Such a situation is more likely to arise in tropical climates and under saline conditions where temporary water deficits can readily occur under the influence of high temperatures and high irradiance. There is also an obvious similarity between the reactions of the C_4 pathway and those of Crassulacean acid metabolism (see Ch. 6). In the latter, PEP carboxylase activity is maximal during the dark while in C_4 metabolism the operation of the two carboxylases is separated spatially. Both pathways appear to be adaptations which have evolved to overcome the problem of obtaining carbon dioxide under conditions when excessive water loss could easily occur.

Measurements of quantum yields of photosynthesis have shown that in a primitive atmosphere with low oxygen concentrations and/or high carbon dioxide levels, C_4 photosynthesis would have been much less efficient than C_3 photosynthesis at non-saturating light intensities (Björkman, 1975). The evidence available suggest that C_4 photosynthesis arose, in evolutionary terms, from C_3 photosynthesis and this has happened on many separate occasions. Species displaying C_3 and C_4 characteristics are found within a genus and the *Atriplex* species are able to hybridize (see Björkman, 1975).

C_4 species generally have higher rates of photosynthesis than C_3 species when measured on maximal values of net CO_2 fixation by an individual leaf. However, although such differences may be large, there is less difference between the two photosynthetic types in seasonal growth means. Figures calculated by Monteith (1978) show that for four C_3 species the highest rates of dry matter production fell in the range 34 to 39 g m^{-2} d^{-1}: for C_4 species, corresponding figures would be 50 to 54 g m^{-2} d^{-1}. Monteith concluded that on sunny, but not cloudless, days the maximum rates of growth of C_4 crops is substantially faster than that of C_3 crops. On cloudless days, however, there is no evidence for faster growth rates perhaps due to the restriction of photosynthesis through a combination of light saturation and water stress. Overall, although there are very large biochemical differences between C_3 and C_4 species, their growth rates are rather similar. However, they remain different with the mean photosynthetic efficiency for the C_4 group being 2·0% of total solar radiation while that of the C_3 species is 1·4%.

In summary, the C_4 pathway appears to involve the complex cooperation of two types of chloroplasts. Carbon dioxide is fixed in the mesophyll by phosphoenolpyruvate carboxylase and then carbon is transferred to the bundle sheath either as malate or aspartate. In both cases carbon dioxide is released in the bundle sheath to be refixed by the Calvin cycle. Carbon dioxide can possibly enter the bundle sheath directly, although to what extent it is not known. The location of the additional carbon dioxide fixation enzyme (phosphoenolpyruvate carboxylase) is somewhat equivocal, however, and it has also been claimed that it is located within the cytoplasm of the mesophyll cells (Coombs and Baldry, 1972). Coombs and Baldry believe that the carbon dioxide transport is not from mesophyll to bundle sheath, but from cytoplasm to mesophyll chloroplasts which *do* contain the enzymes of the Calvin cycle.

Nitrogen metabolism in the chloroplast

Although the chloroplasts are primarily the site of carbon fixation, that is of the Calvin cycle and the two photosystems, they also play an important rôle in the nitrogen metabolism of the plant. Chloroplasts are the site of nitrite reductase, which reduces nitrite to ammonia, and of enzymes which convert the ammonia to amino acids. Thus chloroplasts are involved in the conversion of nitrogen from the inorganic to the organic form, much as they incorporate carbon into organic molecules. The source of inorganic nitrogen available to plants is largely restricted to the nitrate anion and the ammonium cation. For the majority of plants nitrate represents a more favourable and, owing to availability in the soil, a more important form of inorganic nitrogen than does ammonia.

Soil nitrogen, which as we have indicated, primarily exists as NO_3^- or NH_4^+, is derived ultimately from atmospheric nitrogen by dinitrogen fixation. This process is unknown in eukaryotes. It is exclusive to the prokaryotes, where it occurs in free living bacteria (ranging from the photosynthetic through aerobes to strict anaerobes), in bacterial symbionts (found chiefly in leguminous plants but also a number of non-leguminous species) and in the blue-green algae (see Streicher and Valentine, 1973; Burris, 1976). The process of nitrogen fixation, which involves the reduction of molecular nitrogen to ammonia, is carried out by an enzyme, nitrogenase (Fig. 8.31). This enzyme consists of two proteins, both of which contain non-haem iron, while one also contains

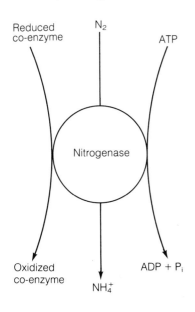

Fig. 8.31   Summary of the reactions of the enzyme nitrogenase.

Reduced co-enzyme

N_2

ATP

Nitrogenase

Oxidized co-enzyme

ADP + P_i

NH_4^+

molybdenum (see Dilworth, 1974). Both are sensitive to oxygen. Nitrogen fixation requires ATP and low-potential electrons which may be supplied as reduced pyridine nucleotide (from a source such as pyruvate) or, in the blue-green algae, as reduced ferredoxin from PS I.

The energy costs of the overall process are relatively high, requiring, in *in vitro* studies of nitrogenase, twelve to fifteen moles of ATP to fix one mole of nitrogen. If the low potential electrons are included and it is assumed, for the sake of the calculation, that two electrons could generate three molecules of ATP, the energy requirement is raised to about 30 ATP/N_2. If the cost of maintaining the basic metabolic rate is also included, this raises the requirement still further to some 35 to 40 ATP/N_2 fixed (Shanmugan *et al.*, 1978).

Nitrogen fixation by symbiotic bacteria occurs essentially by the same route used by the free-living forms, although the organism exists as a bacteroid within the host root cells. The bacteroids consist of one or more bacteria surrounded by a membrane envelope which appears to be developed from the host cell plasmalemma. It is interesting to note that the bacteroids are similar in many respects to mitochondria; they appear to lack a rigid cell wall and are osmotically sensitive. They may contain specific permeases which control the entry of C-compounds supplied by the host (Shanmugan *et al.*, 1978). This similarity may be particularly significant in the light of the endosymbiont hypothesis for the origin of mitochondria and chloroplasts (see p. 397).

The steps involved in the infection of legume roots by *Rhizobium* species and the formation of a root nodule are summarized in Fig. 8.32. The lateral roots release compounds which stimulate the growth of the rhizobia. These compounds include the amino acid, homoserine, which is present in very high amounts in pea seeds. The first symptom of infection is the deformation of root hairs. Changes occur in the wall and an infection thread, consisting of one or more rows of bacteria within a sheath of host cell wall origin, penetrates through the root hair cell and into the root cortex. Cells of the inner cortex begin to divide and become the target for the infection thread. Rhizobia are eventually released from the infection thread and develop into nitrogen-fixing bacteroids. Infected cells lose their ability to divide but the surrounding cells become meristematic and so the nodule continues to grow. The outer layers of the nodule, which are not infected, form a cortex within which vascular elements develop to connect the nodule to the main vascular supply of the root. For a full description of the infection process, see Sprent (1979).

Ammonia formed during nitrogen fixation may be used directly by the plant, although where ammonia is produced by free living organisms, it is likely to be oxidized to nitrate (nitrification) before being absorbed by plant roots. Nitrogen taken up by the plant as nitrate must be reduced firstly to nitrite and subsequently to ammonia before it can be incorporated into an organic molecule. The first step in this reductive process is catalysed by the enzyme nitrate reductase. It is nitrate reductase activity which appears to be responsible for governing the overall flux of nitrogen into protein.

Fig. 8.32 Diagrammatic representation of the stages in the development of a leguminous root nodule. Redrawn from Sprent (1979).

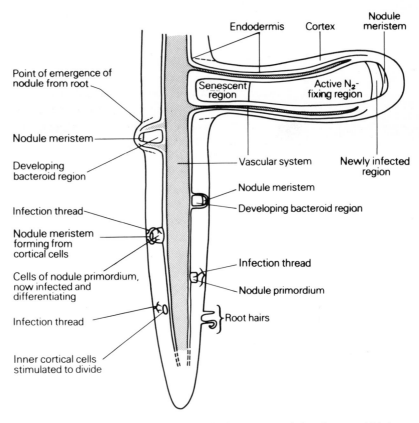

Nitrate reductase is found in both the roots and the shoots of higher plants. The ability to reduce nitrate to nitrite can readily be demonstrated by incubating plant tissue with nitrate in the dark and under anaerobic conditions, when nitrite accumulates. The reductase activity requires the presence of a reduced coenzyme and in most cases the enzyme from higher plants shows a specific requirement for NADH, although NADPH is utilized in some tissues. Reduced FMN or FAD can also act as an electron donor, as can certain synthetic dyes such as methyl- and benzyl-viologen. In fact the native enzyme displays not only NAD(P)H nitrate reductase activity, but also NAD(P)H dehydrogenase activity (with cytochrome c, dichlorophenolindophenol, ferricyanide or tetrazolium acceptors) and methylviologen-, benzyl-viologen-, $FMNH_2$- and $FADH_2$-nitrate reductase activity. Reported molecular weights of the enzyme range from 160 000 in maize to some 600 000 in wheat. Since values of 230 000 and 500 000 have been reported for spinach it is possible that the enzyme may exist as a dimer. The active enzyme from *Aspergilus nidulans* is currently believed to consist of two flavin-containing cytochrome c reductase sub-units, which are linked by a molybdenum-containing protein with a molecular weight of 10 000 to 20 000 (see Hewitt, 1976). Molybdenum has been shown to be an essential micronutrient for plants and to be incorporated into nitrate reductase.

The nitrate reductase enzyme is thought to occur in the cytosol. Although there is a close correlation between the activity of the enzyme and photosynthetic activity in shoots, this is not thought to reflect direct supply of reductant by photosynthesis. Since nitrate reduction generally involves reduced NAD and photosynthesis produces reduced NADP, it is believed that the photosynthetic product glyceraldehyde 3-phosphate, which is effluxed from the chloroplast in exchange for phosphate (and as such constitutes one of the primary currencies of carbon-flow between chloroplast and cytosol, see p. 393), is oxidized to phosphoglycerate in the cytoplasm by the enzyme glyceraldehyde-3-phosphate dehydrogenase, producing NADH. It is this NADH which is utilized in the reduction of nitrate and is unavailable under low-light conditions when the photosynthetic production of glyceraldehyde-3-phosphate is low. In roots, NADH is produced by glycolysis.

Nitrate reductase is notable as an inducible enzyme in plants (see Ch. 2 and Hewitt, 1975). However, correlations between the nitrate contents of leaves and the level of enzyme activity are often poor and it is generally held that it is the flux of nitrate from the roots to the leaves which controls the level of enzyme activity and not the overall leaf content. It has also been inferred from work on cell cultures that nitrate may exist in two pools, 'metabolic' and 'storage', with the metabolic pool possibly acting as inducer and substrate. The enzyme is relatively unstable having a half life of only some 1 to 24h in the dark in the absence of NO_3^-.

Where the bulk of nitrate reductase activity occurs in the roots, little or no nitrate is ever found in the xylem sap. In these plants nitrogen is translocated in an organic form, primarily as the amides glutamine or asparagine, and not as nitrate or ammonia. In such plants not only is nitrate reductase found in the roots, but so is the next enzyme of the reduction pathway, nitrite reductase.

The reduction of nitrite to ammonia is accomplished by a single enzyme, nitrite reductase, with no free intermediates. Since nitrite rarely accumulates in plants it can be deduced that the enzyme is present in excess and this step is not rate limiting in the overall reduction of nitrate to ammonia. Nitrite reduction can be prevented by anaerobiosis (a feature that makes the *in vivo* assay of nitrate reductase possible), although it is not entirely clear how the reduction of nitrite is linked to aerobic metabolism. Uncoupling agents also bring about the accumulation of nitrite, but it does not seem likely that ATP is required in the reduction, since the reaction occurs quite readily in cell free systems.

Nitrite reductase is localized in the plastids and, in particular, in the chloroplasts of green shoots. Reducing power is, in fact, supplied by the light reactions in the form of reduced ferredoxin. The enzyme has a molecular weight of some 60 000 to 70 000, contains iron at a level of two atoms per mole and in the form of a porophyrin (sirohaem) together with acid-labile sulphur. Activity is specific to ferredoxin, benzylviologen or methylviologen radicals. Of the naturally occurring reductants, the specificity to ferredoxin correlates well with the location of the enzyme in the chloroplasts, where it may well occur on the outside of the thylakoid membranes. The enzyme from some microorganisms can also reduce sulphur but this has not been shown to occur in any plant system so far investigated. Purified preparations from plants do show some activity towards hydroxylamine and it remains possible that a derivative of hydroxylamine plays some intermediate rôle in the reaction (Hewitt, 1977).

We have already remarked that in certain plants nitrogen is translocated from roots to shoots in an organic form and in these plants not only nitrate, but also nitrite reductase, must be located in the roots. In these plants the nitrite reductase is probably in plastids, although it is obvious that they cannot generate reduced ferredoxin through the action of light. The root plastids do contain the enzymes of the reductive pentose pathway (see Ch. 6) and could consequently generate NADPH. However, NADPH cannot donate electrons directly for the reduction of nitrite and an intermediate, possibly ferredoxin, is required, although its presence has yet to be demonstrated in the root plastids.

Although a number of plant species translocate nitrogen to the shoots in the organic form, perhaps the majority carry out nitrate reduction in their shoots. In such species there is considerable evidence to support the view that the chloroplasts are the major site of α-amino nitrogen production (see Lea and Miflin, 1974.) Chloroplasts contain two enzymes which are capable of catalysing the incorporation of inorganic nitrogen into an organic molecule. One of these, glutamic dehydrogenase (GDH), catalyses the reaction between α-ketoglutarate (2-oxoglutarate), ammonia and NAD(P)H to form L-glutamate and oxidised coenzyme.

$$
\begin{array}{c}
COO^- \\
| \\
C{=}O \\
| \\
CH_2 \\
| \\
CH_2 \\
| \\
COO^-
\end{array}
\; + NH_4^+ + NAD(P)H + H^+ \longrightarrow
\begin{array}{c}
COO^- \\
| \\
CH \cdot NH_2 \\
| \\
CH_2 \\
| \\
CH_2 \\
| \\
COO^-
\end{array}
\; + NAD(P)
$$

However this enzyme has a low affinity for ammonia (in fact its K_m is higher than the concentration of NH_4^+ which brings about uncoupling in chloroplasts) and is only present in relatively low amounts (it is primarily a mitochondrial enzyme). The second enzyme, which has a much higher activity in the chloroplasts than does GDH and a much lower K_m for ammonia, catalyses the reaction between L-glutamate and ammonia to form L-glutamine, and is known as glutamine synthetase (GS).

$$
\begin{array}{c}
COO^- \\
| \\
CH \cdot NH_2 \\
| \\
CH_2 \\
| \\
CH_2 \\
| \\
COO^-
\end{array}
\; + NH_4^+ + ATP \xrightarrow[Mg^{2+}]{Mn^{2+}}
\begin{array}{c}
COO^- \\
| \\
CH \cdot NH_2 \\
| \\
CH_2 \\
| \\
CH_2 \\
| \\
NH_2{-}C{=}O
\end{array}
\; + ADP + P_i
$$

The enzyme has a molecular weight of some 330 000 to 376 000 and that from root nodules of soybean has been shown to be composed of eight monomers, each with a molecular weight of 47 300. In chlorophyllous tissues, it exists largely in the chloroplasts although there is some evidence for its presence in the soluble phase; in roots it may be in the soluble phase although some appears to be associated with plastids. Although it is not known precisely how GS is regulated the activity is greatly influenced by the concentrations of Mn^{++} and Mg^{++} and by pH and energy charge. Certainly, in chloroplasts, changes in these parameters could increase the activity in the light in parallel with the increase of nitrate reduction which occurs under such conditions.

Although glutamate dehydrogenase activity is able to generate glutamate and then a range of other amino acids by aminotransferase activity (see below), the means by which amide nitrogen from glutamine could be converted to α-amino nitrogen was unknown for many years. Consequently, the major pathway for the synthesis of amino acids was held to be through the action of GDH, despite problems of affinity for ammonia and levels of activity. However, since 1970 and the discovery in a bacterium *Klebsiella* (*Aerobacter*) *aerogenes* of an enzyme which would transfer amide nitrogen to α-ketoglutarate in a reaction yielding two moles of glutamate, an alternative pathway of nitrogen assimilation has been elucidated in plants, particularly through the work of Miflin and Lea.

This enzyme catalyses the reductive transfer of the amide-amino group of glutamine to α-ketoglutarate using either NADPH or reduced ferredoxin as an electron donor depending upon the source of the enzyme. The pyridine nucleotide-dependent enzyme was first discovered in microorganisms while the higher plant enzymes were later, on the whole, found to be specific for reduced ferredoxin.

$$
\begin{array}{ccccc}
\mathrm{COO^-} & \mathrm{COO^-} & & \mathrm{COO^-} & \mathrm{COO^-} \\
| & | & & | & | \\
\mathrm{C{=}O} & \mathrm{CHNH_2} & & \mathrm{CHNH_2} & \mathrm{CHNH_2} \\
| & | & \xrightarrow{\ \mathrm{FdH_2}\ } & | & | \\
\mathrm{CH_2} + & \mathrm{CH_2} & & \mathrm{CH_2} + & \mathrm{CH_2} \\
| & | & & | & | \\
\mathrm{CH_2} & \mathrm{CH_2} & & \mathrm{CH_2} & \mathrm{CH_2} \\
| & | & & | & | \\
\mathrm{COO^-} & \mathrm{NH_2{-}C{=}O} & & \mathrm{COO^-} & \mathrm{COO^-} \\
\end{array}
$$

The enzyme has the trivial name of glutamate synthase; since this is so similar to glutamate synthetase, it is often referred to by an acronymn GOGAT, to avoid confusion. The ferredoxin requiring enzyme is present in the leaves of many plant species and appears, not unexpectedly, to be localized within the chloroplasts. Both NAD (P) and ferredoxin requiring forms have been isolated from plant roots.

We have already indicated that although earlier work suggested the transfer of nitrogen from ammonia to the α-amino nitrogen of amino acids took place through the action of glutamate dehydrogenase, this pathway is not currently believed to be of any significance in higher plants. Miflin and Lea (1976) have presented considerable evidence to show that nitrogen is firstly transferred from ammonia to glutamine by GS and then to glutamate by GOGAT (Fig. 8.33). Glutamate subsequently plays a central rôle in the synthesis of other amino acids.

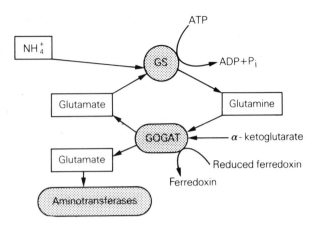

Fig. 8.33   Pathway for the transfer of nitrogen from ammonia to the α-amino nitrogen of amino acids. GS, glutamine synthase; GOGAT; glutamate synthase.

Amino acids are found both in proteins and in soluble form (Ch. 2). There is a far greater variety of free amino acids than those which are found combined in proteins and the latter are restricted to about twenty different types. These include the monoamino monocarboxylic acids, glycine, alanine, valine, leucine and isoleucine; the hydroxy aliphatic acids, serine and threonine; the dicarboxylic amino acids, aspartic acid and glutamic acid; the amides, asparagine and glutamine; the basic amino acids, lysine, arginine and histidine; the sulphur containing amino acids; cysteine, cystine and methionine; the aromatic amino acids, phenylalanine, tyrosine and tryptophan and the imino acid proline. The structure of these compounds is detailed in Fig. 2.27.

Nitrogen, then, is initially incorporated into the glutamine and glutamate molecules and it is particularly from that latter that the other amino acids are derived: the various pathways are dealt with in detail by Beevers (1976), Bryan (1976) and Miflin and Lea (1977).

Amino acids are synthesized from glutamate or from other amino acids by a process of *transamination* which is central to the transformation of these compounds. An amino group is transferred from one amino acid to the keto group of a keto acid in a reaction which is catalysed by an aminotransferase (otherwise known as a transaminase) and involving pyridoxal phosphate as a coenzyme.

$$\underset{\underset{NH_2}{|}}{\overset{\overset{H}{|}}{R_1{-}C{-}COO^-}} + \underset{}{\overset{\overset{O}{\|}}{R_2{-}C{-}COO^-}} \rightleftharpoons \underset{}{\overset{\overset{O}{\|}}{R_1{-}C{-}COO^-}} + \underset{\underset{NH_2}{|}}{\overset{\overset{H}{|}}{R_2{-}C{-}COO^-}}$$

The aminotransferases appear to be relatively unspecific towards their substrate and subject to little regulation by other intermediates. The donor amino acid may be glutamate or another amino acid so that the complete range of twenty protein amino acids may be synthesized providing the relevant keto acid can be provided. A complete range of transaminases have been found in plants and appear to be localized in the soluble phase, chloroplasts and microbodies. The exceptions to their use seem to be that the synthesis of asparagine involves the transfer of the *amide* nitrogen from glutamine and that one of the nitrogen atoms of arginine, tryptophan and histidine are also formed from the amide-amino group of glutamine. In the following part of this

$$\begin{array}{c} \text{CONH}_2 \\ | \\ \text{CH}_2 \\ | \\ \text{H}_2\text{NCH—COO}^- \end{array}$$

Asparagine

$$\begin{array}{c} \text{CH}_3 \\ | \\ \text{H}_2\text{N—CH—COO}^- \end{array}$$

Alanine

Fig. 8.34   The central role of
glutamate in the synthesis of amino
acids and amides.

section, we shall present some details of the synthesis of the more important amino-acids.

The synthesis of asparagine appears to occur though the operation of asparagine synthetase which catalyses the transfer of the amide group from glutamine to aspartate

L-aspartate + ATP + L-glutamine $\rightleftharpoons$

L-glutamate + L-asparagine + AMP + PPi

Asparagine forms the chief means by which organic nitrogen is transported in many legumes (allantoin/allantoic acid may be important in some) and, over short distances, in many tree species. Asparagine is also an important transport compound in many other plants and is catabolized either through asparaginase or amino transferase activity.

As we have already indicated, glutamate plays a central role in the synthesis of amino acids. For example, alanine can be formed directly by transamination with pyruvate

pyruvate + glutamate $\rightleftharpoons$ alanine + α-ketoglutarate

The enzyme from mung bean shoots has virtually no specificity for the amino-group donor, however, failing only to function with serine, glycine and threonine. Glutamate also serves as the starting point for the production of proline, hydroxyproline, ornithine, citrulline and arginine (Fig. 8.34). The synthesis of proline appears to firstly

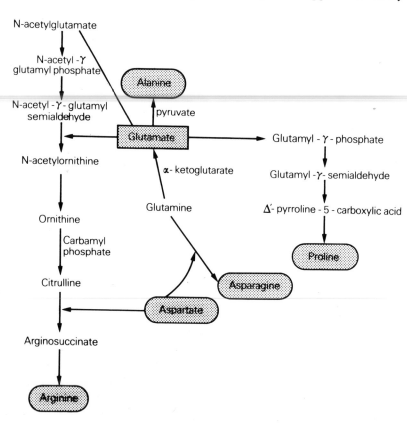

Fig. 8.34   The central role of glutamate in the synthesis of amino acids and amides.

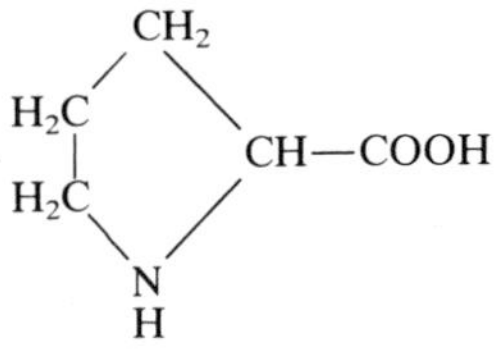

Proline

involve the reduction of glutamate to the semialdehyde via glutamyl phosphate.

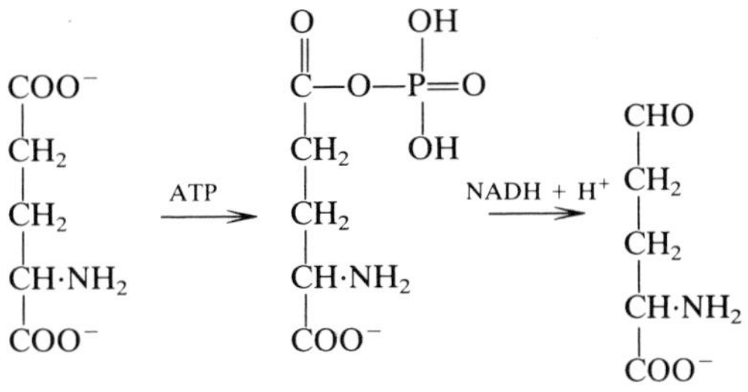

The semialdehyde equilibrates in solution with its cyclic form, Δ'-pyrroline-5-carboxylic acid, which is reduced to proline by an NADH reductase. The synthesis of arginine, on the other hand, first involves the acetylation of glutamate to form N-acetlyglutamate

glutamate + acetyl CoA $\longrightarrow$ N-acetylglutamate + CoA

Plants are also able to utilize acetylornithine as an acetyl donor for this reaction. The acetyl derivative is reduced to the semialdehyde, again through the γ-phosphate, which is transaminated with glutamate to produce N-acetylornithine (N-acetylornithine aminotransferase) and α-ketoglutarate.

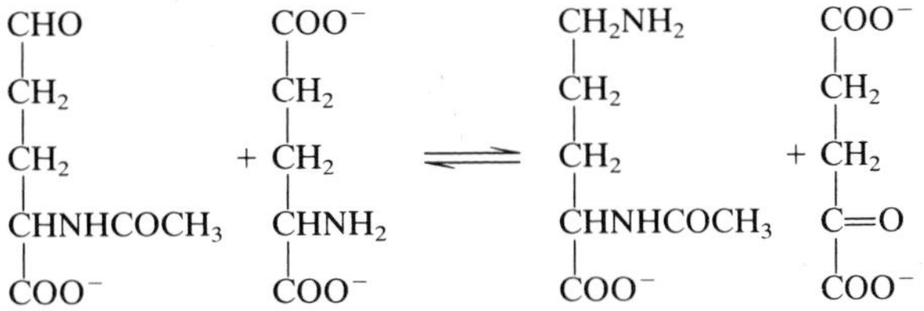

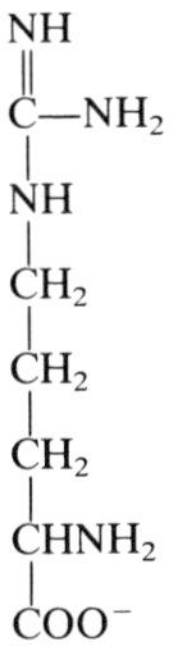

Arginine

The acetyl ornithine may then be cleaved by an acylase to produce the non-protein amino-acid ornithine. The conversion of ornithine to arginine involves firstly its reaction with carbamyl phosphate to produce citrulline

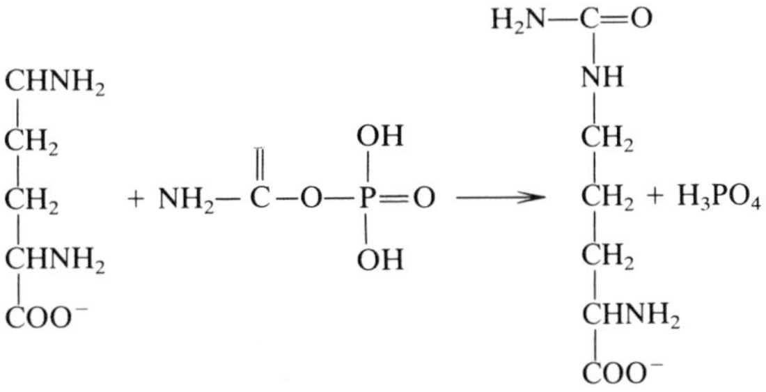

Citrulline is then condensed with aspartate to produce arginosuccinate which is cleaved to produce arginine. This amino acid as well as being a constituent of proteins also plays an important rôle as a storage compound in legumes and in long distance transport of organic nitrogen in tree species (where it also acts as a storage compound for over-wintering). Both asparagine and arginine may be important as

383

COO⁻
|
CH₂
|
H₂N—CH—COO⁻

Aspartate

CH₂OH
|
CH₂
|
H₂N—CH—COO⁻

Homoserine

CH₃
|
HO—CH
|
H₂N—CH—COO⁻

Threonine

CH₃
|
CH₂
|
CH—CH₃
|
H₂N—CH—COO⁻

Isoleucine

NH₂
|
CH₂
|
CH₂
|
CH₂
|
CH₂
|
H₂N—CH—COO⁻

Lysine

CH₂SH
|
CH₂
|
CHNH₂
|
COO⁻

Homocysteine

SCH₃
|
CH₂
|
CH₂
|
H₂N—CH—COO⁻

Methionine

storage and transport compounds due to their high ratio of N/C which means that a minimum of carbon is involved per nitrogen stored or transported.

A further important transamination reaction involving glutamate is that which results in the production of aspartate,

$$\text{glutamate} + \text{oxaloacetate} \rightleftharpoons \text{aspartate} + \alpha\text{-ketoglutarate}$$

Aspartate, like glutamate, is a key intermediate in the formation of yet more amino acids (Fig. 8.35). We have already seen that it is involved in the formation of asparagine; in addition it is the starting point for the synthesis of threonine, isoleucine, lysine, cysteine and methionine. The pathway to all these amino acids begins with the synthesis of γ-aspartyl phosphate from aspartate and ATP under the action of aspartyl kinase. The phosphate is then reduced to the semialdehyde by aspartate semialdehyde dehydrogenase:

$$
\begin{array}{c}
\text{O} \quad\quad \text{OH} \\
\parallel \quad\quad | \\
\text{C—O—P=O} \\
| \quad\quad\quad | \\
\text{CH}_2 \quad\text{OH} \quad + \text{NADPH} \\
| \quad\quad\quad\quad\quad + \text{H}^+ \\
\text{CH.NH}_2 \\
| \\
\text{COO}^-
\end{array}
\longrightarrow
\begin{array}{c}
\text{CHO} \\
| \\
\text{CH}_2 \\
| \\
\text{CHNH}_2 + \text{NADP}^+ + \text{H}_3\text{PO}_4 \\
| \\
\text{COO}^-
\end{array}
$$

The similarity to the initial reactions of glutamate metabolism can readily be seen. The semialdehyde is reduced to homoserine, which is subsequently converted to threonine. Isoleucine biosynthesis is thought to involve the deamination of threonine to α-ketobutyrate and its subsequent conversion to α-keto β-methylvalerate which is trans-aminated to produce isoleucine. The pathway for the synthesis of lysine in higher plants is presumed to be similar to that established in bacterial systems and involves the synthesis of 2,3-dihydropicolinic acid from aspartyl semialdehyde and pyruvate.

Homoserine is also the starting point for the synthesis of methionine. Following phosphorylation of the homoserine, cys-tathione synthetase catalyses the reaction with cysteine to form cystathione

$$
\begin{array}{c}
\text{CH}_2\text{—O—}\textcircled{P} \\
| \\
\text{CH}_2 \\
| \\
\text{CH NH}_2 \\
| \\
\text{COO}^-
\end{array}
+
\begin{array}{c}
\text{CH}_2\text{SH} \\
| \\
\text{CHNH}_2 \\
| \\
\text{COO}^-
\end{array}
\longrightarrow
\begin{array}{c}
\text{CH}_2\text{—S—CH}_2 \\
| \quad\quad\quad | \\
\text{CH}_2 \quad\quad \text{CHNH}_2 \\
| \quad\quad\quad | \\
\text{CHNH}_2 \quad \text{COO}^- \\
| \\
\text{COO}^-
\end{array}
$$

Cystathione is then hydrolysed to produce homocysteine, pyruvate and ammonia and the homocysteine methylated to produce methionine. Cysteine is considered to be formed in plants by the sulphuration of o-acetyl serine. Cystine is not thought to be incorporated directly into protein, but rather formed by oxidation of incorporated cysteine.

The synthesis of the branched chain amino acids, valine and leucine both involve the formation of the relevant keto-acid precursor fol-

Fig. 8.35   The role of asparate in the synthesis of amino acids.

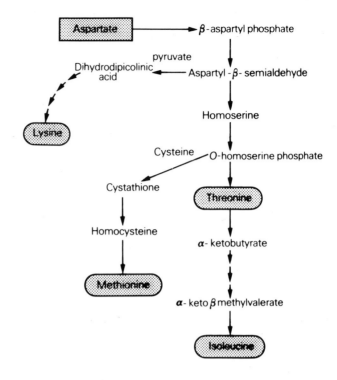

$$CH_2-SH$$
$$H_2N-CH-COO^-$$

Cysteine

lowed by transamination. Thus α-keto-isovalerate is produced from pyruvate and then transaminated to produce valine,

CH₃—CH.
  |
  C=O
  |
  COO⁻

$\longrightarrow$

CH₃—CH
  |
  CHNH₂
  |
  COO⁻

(each with CH₃ above)

while α-keto-isocaproic acid is the precursor of leucine

COO⁻
  |
  C=O
  |
  CH₂
  |
H₃C—CH
  |
  CH₃

$\longrightarrow$

COO⁻
  |
  CHNH₂
  |
  CH₂
  |
H₃C—CH
  |
  CH₃

HC—N
 ‖   ＞CH
C—N′
 | H
 CH₂
  |
 CHNH₂
  |
 COO⁻

Histidine

The pathway of histidine biosynthesis in plants has not yet been established, but it is presumed to be formed from imidazole glycerol phosphate as in microorganisms. The formation and interconversions of glycine and serine will be dealt with in the following chapter since these reactions are an integral part of the process of photorespiration.

Finally, we come to the production of the aromatic amino acids, tyrosine, phenylalanine and tryptophan (Fig. 8.36). There is now reasonable evidence that these amino acids are synthesized from shikimic acid in plants by a pathway originally established in microorganisms. Shikimic acid itself is synthesized from phosphoenolpyruvate, which firstly condenses with erythrose-4- phosphate under the

385

action of a synthase to form 3-deoxy-D-arabino heptulosonic acid-7-phosphate.

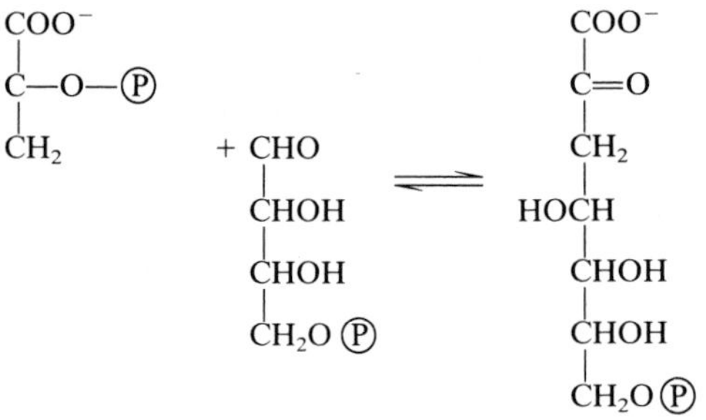

$$
\begin{array}{ccc}
\text{COO}^- & & \text{COO}^- \\
\text{C}-\text{O}-\text{\textcircled{P}} & & \text{C}=\text{O} \\
\text{CH}_2 & + \text{CHO} & \text{CH}_2 \\
& \text{CHOH} & \text{HOCH} \\
& \text{CHOH} & \text{CHOH} \\
& \text{CH}_2\text{O}\,\text{\textcircled{P}} & \text{CHOH} \\
& & \text{CH}_2\text{O}\,\text{\textcircled{P}}
\end{array}
$$

Reduction of this compound produces 5-dehydroquinate which in turn undergoes dehydration and a further reduction to produce shikimate

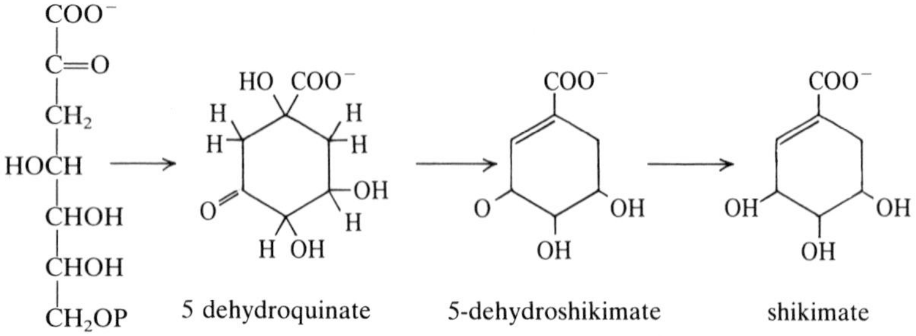

$$
\begin{array}{l}
\text{COO}^- \\
\text{C}=\text{O} \\
\text{CH}_2 \\
\text{HOCH} \\
\text{CHOH} \\
\text{CHOH} \\
\text{CH}_2\text{OP}
\end{array}
$$

5 dehydroquinate   5-dehydroshikimate   shikimate

The synthesis of the amino acids is then believed to follow the pathway established in microorganisms. The shikimate is phosphorylated and then converted in two subsequent steps to chorismate which is either converted to prephenate or to anthranilate.

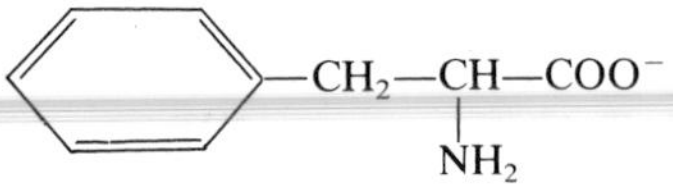

Phenylalanine

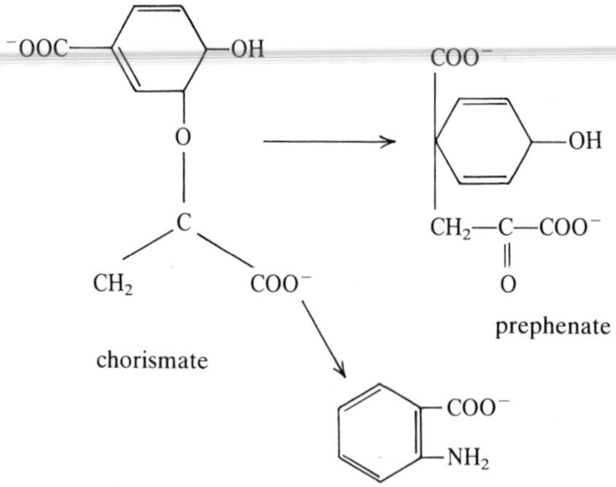

chorismate

prephenate

anthranilate

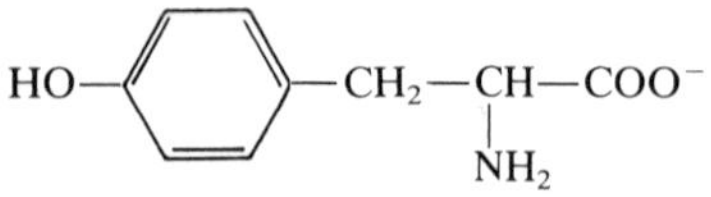

Tyrosine

The precursor for phenylalanine, phenyl pyruvate, is produced by decarboxylative dehydration of prephenate, while the precursor of tyrosine is produced by decarboxylation of prephenate: both amino acids are then produced by transamination. Tryptophan synthesis follows the conversion of anthranilate to N-5 phosphoribosylanthranilate and then to indole-3-glycerol phosphate. The latter, or free indole, reacts with serine to form tryptophan.

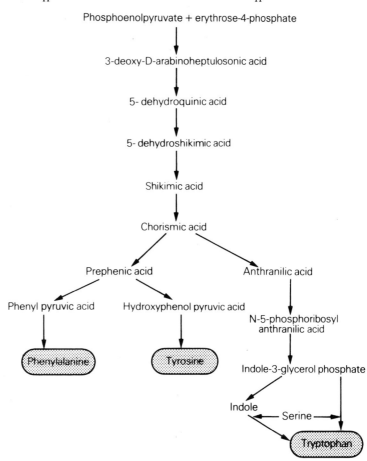

Fig. 8.36   Pathways for the synthesis
of aromatic amino acids.

We have seen how the reduction of nitrite to ammonia and the assimilation of this ammonia into an organic form occurs in the chloroplasts in a manner which is in some ways similar to the incorporation of carbon into an organic form. The chloroplasts however, appear not only to be the site of the enzymes involved in the early reactions of nitrate assimilation, but also the location of many of the reactions of amino acid biosynthesis. To give just two examples, the reactions of methionine synthesis are all to be found in the chloroplasts as are those enzymes necessary to convert chorismate to tryptophan. Thus, although a number of the enzymes involved in the formation of amino acids are found in other organelles, there is increasing evidence for the involvement of the chloroplasts in these reactions. It does not appear, however, that newly fixed carbon can be directly utilized in forming the carbon skeletons, but that carbon must first be exported from the chloroplasts, converted to the relevant precursor and then re-enter the chloroplast for the necessary trans-amination reactions. Amino acids so formed are then available for transport to the cytoplasmic ribosomes for incorporation into protein (Ch. 5). The synthesis of the great variety of non-protein amino acids found in plants is beyond the scope of this book, but is dealt with in the articles and reviews of Fowden (1976), Kjaer and Larsen (1976), and Lea (1977).

The regulation of nitrogen metabolism is undoubtedly very complex and we will only briefly mention some of the processes which may operate in higher plant cells. The enzymes involved are subject to a variety of control processes (see Ch. 2), although work on nitrogen metabolism has been particularly concerned with induction and repression and with control by allosteric feedback inhibition (see Miflin and Lea, 1977; Miflin *et al.,* 1979; Stewart and Rhodes, 1977). For example, many of the enzymes concerned with amino acid biosynthesis exhibit end-product inhibition *in vitro,* although the details may vary between species. Thus aspartate kinase (see Fig. 8.35) from different species may be inhibited by lysine alone, threonine alone, lysine plus threonine, or lysine plus methionine, so that there is no universally applicable control process. In a number of cases, such studies on isolated enzymes have been supported by experiments on the effects of exogenous amino acids, both on the biosynthesis of endogenous amino acids from radioactive precursors and on the overall growth of the plant. For example, it has been shown that the growth of a number of plants is inhibited by lysine and threonine applied together, and that this effect may be reversed by methionine or homocysteine (see Miflin *et al.,* 1979 for further details).

However, any discussion of the regulation of nitrogen metabolism, and indeed of any aspect of metabolism, has to consider much more than the factors producing changes in the level or activity of individual isolated enzymes. Miflin (1977) has pointed out that, in addition to the one-dimensional control described by work on enzymes in the test-tube, control in the intact plant must also be discussed in terms of space and time. For example, if the enzymes are spatially separated in the cell, it is only those metabolities present at the same site as the enzymes that may exercise control. Since a number of the enzymes involved in amino acid biosynthesis are located in the chloroplast, the transport of amino acids and of various co-factors into and out of this organelle will be important in the overall regulation of amino acid biosynthesis and of protein synthesis in the cytoplasm. Finally there is some evidence that the levels and regulatory properties of certain enzymes of nitrogen metabolism change with time as tissues grow, and so could have an important effect on the physiological status of the tissue, e.g. source or sink. For example, some interesting results have been obtained with homoserine dehydrogenase, a regulatory enzyme associated with the biosynthesis of several amino acids from aspartate (Bryan, Lissik and Dicamelli, 1979). The enzyme isolated from young tissues of *Zea mays* is sensitive to inhibition by threonine whereas the enzyme from older tissues is much less so. However, the enzyme from *Vicia faba* is not desensitized during growth, and the factors responsible for the change, or lack of change, are not understood. The reader is referred to the papers quoted above for a more detailed discussion of the possibilities available for the control of nitrogen metabolism in plant cells.

Transport processes in chloroplasts

The chloroplasts of a higher plant cell form the single largest component of the cytoplasm (excluding the vacuole) comprising about 40% of the volume. Consequently, they contain a significant proportion of the cytoplasmic ions and metabolites. Furthermore, since a very high proportion of all fixed carbon crosses the chloroplast membranes at some time, it can be seen that the permeability of the chloroplast envelope plays a significant role in cell metabolism.

The movement of substances into and out of chloroplasts can be studied by techniques similar to those used in the study of mitochondrial ion transport (see Ch. 7). The direct measurement of ion uptake from bathing media into chloroplast pellets, the effects of added metabolites on rates of oxygen uptake (see below), and the determination of volume changes have all been employed. These volume changes, which occur both *in vivo* and *in vitro*, are considered to be both osmotic and non-osmotic or mechanical in origin (Packer, Murakami and Mehard, 1970). The osmotic volume changes or configurational changes are described by similar laws to those descriptive of mitochondrial volume changes, i.e. the volume is proportional to the reciprocal of the absorbance and of the osmotic potential. The non-osmotic changes, however, are believed to result from the direct interaction of ions with the thylakoid membranes and to result in conformational changes (that is, changes in molecular structure). These conformational changes are best monitored by changes in light scattering (measured at 90° to the incident beam), while transmission responses are found to correlate most nearly to the configurational changes described above (see also Murakami *et al.,* 1975). An additional method that has been utilized with chloroplasts relies on their isolation under non-aqueous conditions to retain both ions and metabolites. The tissue is first frozen at the temperature of liquid air or nitrogen then, after subliming off the ice, homogenized and fractionated under non-aqueous conditions using solvents such as carbon tetrachloride and hexane. Subsequently, the solutes present may be analysed by conventional techniques. Finally, since chloroplasts are so large a component of the cell, it is also possible to follow changes in their volume *in vivo*. This may be achieved by a number of techniques, including light microscopy and monitoring changes in the absorbance of leaves. The importance of *in vivo* studies lies in the fact that the organelles are intact. Isolation of chloroplasts tends to strip the outer envelope from the organelle even under the most favourable conditions, so that it is essential during *in vitro* studies to estimate the proportion of intact chloroplasts present in the preparation. This can be done by estimating the oxygen evolution on adding ferricyanide, which does not penetrate the envelope of intact chloroplasts, but interacts directly with the thylakoids of broken chloroplasts (see Walker, 1974, 1976). Only when the proportion of intact chloroplasts is shown to be high can permeability effects be ascribed to the envelope itself. Of course, chloroplasts which have been stripped of their envelopes are important in describing the permeability properties of the thylakoids themselves.

From the study of osmotic volume changes occurring in the dark it has been shown that sucrose, glucose and mannitol will give osmotic support i.e. the membranes, like those of the mitochondria, are relatively impermeable to these substances. We will return to the metabolic significance of these findings and, especially, to the impermeability of sucrose later. However, in contrast to the mitochondria, the chlorides, nitrates and sulphates of univalent cations also provide osmotic support in the dark. Glycerol, on the other hand, is relatively permeable. Evidence obtained with the chloroplasts of *Nitella* shows that the chloroplast membranes are not readily permeated by sodium, potassium, rubidium or bromide. In general, ammonium chloride acts rather as other univalent anions/cations. However, chloroplasts suspended in ammonium acetate swell rapidly as they do in the ammonium salts of a variety of weak acids. The rate of swelling has been shown to be proportional to the concentration of undissociated acid and swelling in ammonium salts, in general, appears to be a function of the presence of undissociated acid and of free ammonia, both of which are freely permeable through the chloroplast membranes.

In the *light*, class II chloroplasts (see p. 335) suspended in sodium chloride or ammonium chloride solutions swell over a period of about 10 min and will recontact if transferred to the dark. Swelling in the two solutes is qualitatively similar, although of lesser magnitude, in the ammonium salt. This light induced swelling is perhaps analogous to the energy dependent swelling of animal mitochondria and is to be contrasted with the energy independent swelling of plant mitochondria. The light induced swelling is envisaged as following inward proton movement, which has been the subject of extensive research and discussion (see, for example, Jagendorf, 1975). We have already mentioned the inward translocation of protons occurring across the thylakoid membranes in the light in relation to photophosphorylation and have seen how an electrochemical gradient of hydrogen ions can be used to drive the synthesis of ATP. The uptake of protons by swollen thylakoid membranes during electron transport in the light is readily demonstrable. If it is to be sustained in the absence of ATP synthesis, an excessive rise in the internal pH must be prevented by internal buffering while an excessive build up in the membrane potential (which seems to lie between 5 and some 100 mV, positive on the inside, during the steady state) must be overcome by counter-ion movement. Experiments with specific ion electrodes and isotopes indicate that K^+ and Mg^{++} ions efflux while Cl^- enters along with the protons in order to dissipate the membrane potential. In the presence of a weak acid anion this behaviour is modified, the inwardly directed proton flow causing the conversion of the anions to the undissociated acid which is freely diffusible in this form. This leaves the internal compartment by diffusion down a concentration gradient and is followed by an osmotic equivalent of water causing a shrinkage of the thylakoids. In an ammonium salt, the uncharged NH_3 enters to combine with a proton to form the NH_4^+ cation. Cl^- entry then follows together with an osmotic equivalent of water.

Much discussion has centred on the primary nature of the proton pump – whether, for example, proton movement is a primary event or, in fact, secondary following cation pumping. If the latter were the case there would have to be potassium and magnesium pumps which were activated in the light. However, such a view cannot account for the continued uptake of protons and foreign organic anions which can be shown to occur when chloroplasts are suspended in a solution of the organic compound in the light. It can be envisaged that the initial uptake of protons might be in exchange for internal cations, but these must soon be exhausted so that the continued uptake of H^+ can only be explained by the presence of a proton pump (the only other and unlikely explanation is a pump for the foreign anion – see Jagendorf, 1975). While results such as these provide good evidence for a proton pump, they do not, of course, allow us to distinguish between the chemi-osmotic and the chemical coupling hypothesis. It is only that many results are more simply explained by the chemi-osmotic hypothesis that adds support to its tenets.

The chloroplast membranes not only control ion movements between the various compartments of the chloroplasts, but also the interaction between the organelle and the cytoplasm. It is not however the thylakoid membranes that are important in this respect, but those of the envelope itself, so that these studies primarily involve the use of intact or Class 1 chloroplasts. The whole organelle does exchange ions with the cytoplasm and evidence has been obtained using isolated organelles as well as non-aqueous extraction techniques and *in vivo* absorbance measurements. Observations on the chloroplasts from a number of different species have shown that chloroplasts flatten in the light *in vivo* – that is, they loose water in such a way that their thickness decreases (Nobel, 1975) – and increase in volume again in the dark. The significance of such a water flux is not yet clear, although it is obvious that any change in volume will alter the concentration of other metabolites present in the stroma. The water movement seems to be activated by photosystem I and is insensitive to the inhibitor DCMU. Since chloroplasts are judged to be in extremely rapid osmotic equilibrium with their surroundings it can be inferred that the water movement follows an ion flux. Estimates of chloroplast ion contents by various techniques including the non-aqueous isolation procedure indicate that they contain some 100 mM K^+ with a slightly higher Cl^- concentration as the chief anion. In addition to the potassium there is approximately one tenth the concentration of Na^+ and similar concentrations of Ca^{++} and Mg^{++}, although much of the divalent cations may be bound and not in free solution. While the evidence for ion fluxes is scant, measurements indicate that light does bring about an efflux of potassium and chloride ions and we can guess that it is this ion flux which leads to the osmotic withdrawal of water. These movements may be a consequence of specific permeases in the envelope or in some way secondary to the alkalization of the stroma which would occur in the light as a result of the pumping of protons across the thylakoid membranes. In addition to the flux of potassium and chloride there is evidence that the movement of magnesium ions may be of particular metabolic significance since they are known to play an important part in activating enzymes such as ribulose bisphosphate carboxylase. Thus, when the light is turned on, Mg^{++} effluxes from the loculus of the thylakoids into the stroma and at the same time enters from the cytoplasm (in contrast to the movement of potassium and chloride). This ion movement together with the loss of water from the chloroplast means that the magnesium concentration could increase from about 5 to 10 mM in the light: at the same time the pH of the stoma is likely to rise from about pH 5 to pH 8. These changes could bring about the well known light-activation of enzymes such as the ribulose bisphosphate carboxylase.

The movement of metabolites across the chloroplast envelope has been followed by many of the techniques used to determine ion movements. In addition it has also been possible to utilize changes in the metabolic activity of the isolated organelles. Isolated chloroplasts do not initially fix carbon dioxide at maximal rates on illumination and there is a lag of some one to three minutes. This lag is believed to result from the intermediates of the Calvin cycle reaching their optimal steady state concentrations and it can be shortened by adding certain intermediates to the reaction medium. In contrast, the lag can be lengthened by the addition of inorganic phosphate to the external medium and in the presence of $10^{-2}M$ P_i to such an extent that it amounts to a suppression of carbon dioxide fixation. Again the suppression can be overcome by the addition of intermediates. In both instances the shortening of the lag time is taken as evidence for the permeation of intermediates across the chloroplast envelope (see Walker, 1974, 1976). The conclusions that have been reached concerning metabolite transport must be prefaced by the important conclusion that the envelope is not freely permeable to sucrose. This means that sucrose is not the means by which fixed carbon is exported to the cytoplasm. Furthermore, the molecules, ribulose-5-phosphate, ribulose-1, 5-bisphosphate, sedoheptulose-7-phosphate, and 1,3-diphosphoglycerate appear not to penetrate the outer chloroplast envelope. The evidence for fructose bisphosphate is somewhat equivocal, but fructose-6-phosphate, ribose-5-phosphate, ribulose-5-phosphate and xylulose-5-phosphate all release the phosphate inhibition of oxygen uptake (see Walker, 1974). However, 3-phosphoglycerate, dihydroxyacetone phosphate and glyceraldehyde-3-phosphate all appear to traverse the envelope with great ease. There is some evidence that glyceraldehyde-3-phosphate does not move *out* of the stroma, but in any event, phosphotriose isomerase (p. 362) brings about the rapid equilibration of the glyceraldehyde-3-phosphate and dihydroxyacetone phosphate in the cytoplasm (note that the equilibrium is strongly in favour of the latter). Thus Heber (1974) has argued that carbon export must occur through the 3-C phosphorylated compounds, which in itself generates a demand for a high capacity phosphate import system. Sucrose synthesis for export from the leaf would occur in the cytoplasm (Fig. 8.37). Phosphate entry into the chloroplast is postulated to take place through a specific translocator which exchanges phosphate for dihydroxyacetone phosphate, 3-phosphoglycerate or glyceraldehyde-3-phosphate. In the presence of excess external inorganic phosphate, the chloroplast can be depleted of these 3-C compounds thus accounting for the phosphate inhibition of photosynthesis. Apart from phosphate, translocators have been postulated for adenylates and dicarboxylic acids. However, the evidence is somewhat controversial, and the envelope appears to be relatively impermeable to the adenine nucleotides: there does not appear to be any rapid exchange of ATP or ADP between the chloroplast and the cytoplasm. Thus, although a specific translocase has been postulated, any major contribution of ATP from the chloroplast to the cytoplasm must be indirect and is thought to occur through the export of dihydroxyacetone phosphate which, through the action of glycolysis and the Krebs cycle, could generate NADH and then ATP. 3-phosphoglycerate would return the carbon to the chloroplast thus forming a DHAP/PGA shuttle. NADP also appears not to permeate the envelope. Thus if the chloropast is to export reducing power this must also be in an indirect manner. This could occur through the operation of the dihydroxyacetone phosphate 3-phosphoglycerate shuttle or through the operation of a malate/oxaloacetate shuttle. Cytoplasmic oxaloacetate would enter

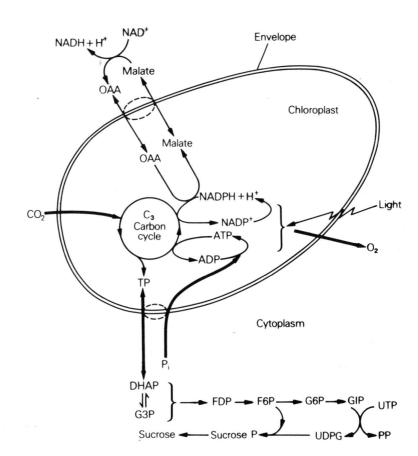

Fig. 8.37   The major metabolite transfer processes that occur across the chloroplast envelope. TP, triose phosphate; DHAP, dihydroxyacetone phosphate; G3P, glyceraldehyde 3-phosphate; FDP, fructose 1,6-bisphosphate; F6P, fructose 6-phosphate; F1P, fructose 1-phosphate, G6P, glucose 6-phosphate; G1P, glucose 1-phosphate; UDPG, uridine diphosphoglucose, OAA, oxaloacetic acid. The heavy arrows indicate metabolites which traverse the envelope with great ease.

the chloroplast and be reduced to malate by an NADP malate dehydrogenase. The malate would then be exported and oxidized in the cytoplasm by the cytoplasmic NAD-dependent malate dehydrogenase, thus providing NADH and oxalacetate which is then available for re-entry into the chloroplast. The exchange of the dicarboxylates is postulated to occur through the operation of the specific translocase mentioned earlier. This translocator is also thought to be responsible for the exchange of glutamate and aspartate between cytoplasm and chloroplasts which we have already seen are a major site of cellular nitrogen metabolism. The efflux of glycolate from the chloroplasts is another process which must take place with ease, since as we shall see in the next chapter, chloroplasts excrete this anion which is subsequently metabolised in the peroxisomes. Finally, the chloroplasts must be permeable to carbon dioxide and, in such a way, that carbon entering the Calvin cycle can match that exported. It is believed that carbon dioxide enters as CO_2, but is hydrated in the stroma

$$CO_2 + H_2O \rightleftharpoons H_2CO_3 \rightleftharpoons H^+ + HCO_3^-$$

As the pH changes from 5 to 8, the proportion of the bicarbonate anion increases from about 2% to about 95%. This reaction is also catalysed by the enzyme carbonic anhydrase, so that in the light there would be a continuing tendency for the entry of CO_2. A problem that remains to be resolved is that the substrate for the ribulose bisphosphate carboxylase appears not to be bicarbonate but CO_2 itself. It has been argued that the production of 3-phosphoglycerate from ribulose bisphosphate, CO_2 and water would cause local acidification at the enzyme surface sufficient to maintain a high local CO_2 concentration (see Walker, 1974).

Ontogeny

Meristematic cells do not contain mature plastids, but these organelles develop during cellular differentiation. The earliest recognizable stage of the development of a plastid is a so-called *initial*, which is in fact indistinguishable from the mitochondrial initial (although it should not be inferred that the two are necessarily identical). The origin of these initials, which are double membrane-bound vesicles less than 500 nm in diameter, is somewhat uncertain, but they may be autonomous in that they divide to keep pace with cell divisions. The mature chloroplasts of algae do divide and there is evidence to suggest that mature higher plant chloroplasts may do the same. They have, in fact, been observed to divide in artificial media after removal from the cell (Ridley and Leech, 1970) (Fig. 8.38). The division appears to take place by constriction to form two daughter chloroplasts and, prior to the completion of separation, DNA containing regions may be found in each half. The first truly characteristic stage of chloroplast development, the *proplastid*, is formed by invagination of the inner membrane of the initial. These invaginations lie parallel to the surface of the

Fig. 8.38   Phase contrast micrograph of dividing spinach chloroplasts. ×2 240. Reproduced from Ridley and Leech (1970)

proplastid and it is only at this stage that they can be clearly distinguished from the developing mitochondrion, where the invaginations lie perpendicular to the surface. Further development seems to be dependent upon the environment in which the plant finds itself. If the cell is in the light, the invaginations of the inner membrane form flattened vesicles which then grow to produce the thylakoid system. Starch grains appear before the completion of this lamellar system. It is not clear, however, why the lamellae fenestrate or what controls the parallel alignment of the small thylakoids to form grana. In the absence of light, the vesicles formed from the inner membrane become regularly spaced tubes which then fuse at junctions of three to form a *prolamellar body*. This is a cubic lattice of 21 nm diameter tubules and the organelle in this form is termed an *etioplast* (Fig. 8.39). Illumination of dark grown plants containing such etioplasts causes a dispersion of the prolamellar body and the development of typical chloroplast structure. The tubes convert to thylakoids at the same time as protochlorophyllide *a*, which is present in about 1/100 to 1/300 of the

Fig. 8.39 Electron micrograph of an etioplast from *Avena sativa* showing the prolamellar body ×42 000. Provided by Drs B. Gunning and M. Steer, The Queen's University, Belfast.

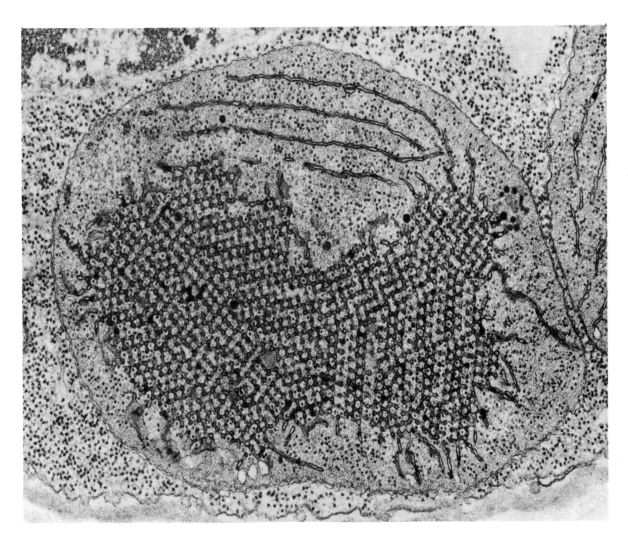

final level of chlorophyll, is converted to chlorophyll *a*. The action spectrum for the conversion of this protochlorophyllide *a* to chlorophyllide *a* (by esterification with phytol) is identical with the absorption spectrum of the protochlorophyllide *a in vivo* – with a maximal absorption at 650 nm. The vesicles firstly become arranged in concentric shells and finally disks are formed and arranged into grana.

The assembly of all the components of the thylakoid membranes necessary to convert the etioplast into the complete chloroplast appears to be a stepwise process (also see Ch. 3 for fuller discussion of membrane synthesis). In the algae, studies with *Chlorella reinhardi* (Hooker, Siekevitz and Palade, 1969) led to the conclusion that there was a stepwise synthesis first of chlorophyll and thylakoid membrane and subsequently of PS II activity. In higher plants the rate of development of chloroplasts may be quite variable. Glyndenholm and Whatley (1968) studying the greening of *Phaseolus vulgaris* leaves showed that PS I activity was first detectable after 10 h and increased rapidly between 15 and 20 h. During this latter period the development of PS I increased about three to four times faster than did the total chlorophyll. Non-cyclic photophosphorylation was not detectable until 15 h. Thorne and Boardman (see Kirk, 1971) who followed chloroplast formation in pea found that once chlorophyllide *a* was formed, the first biochemical change was the appearance of chlorophyll *b* after about 10 min. The photo-oxidation of cytochrome *f* was apparent after 30–40 min and the complete development of PS I by 3 h. Hill reaction activity was apparent after 5 h.

Thus at least during the early stages of greening, chloroplast development is clearly stepwise. In the later stages it is less easy to be certain whether development is stepwise or simultaneous since parts of chloroplasts or particular chloroplasts may be functioning completely while others are developing, thus obscuring the picture.

Amyloplasts also develop from proplastids. Long lamellae form from the inner membrane of the initial, but do not come to lie parallel to the envelope. Starch grains are formed within the lamellae which then become less obvious as the starch grains increase in size. The relationship between amyloplast and chloroplast initials is not simple, however, and it is not merely the presence or absence of light that controls their development: prolamellar bodies are never formed in amyloplasts. The other forms of plastid develop either from amyloplasts or chloroplasts or from an intermediate stage in their development.

As we have already indicated, chloroplasts, like mitochondria, contain DNA, RNA and the machinery for protein synthesis (see also Ch. 5). The chloroplast ribosomes and protein synthesis are discussed in Chapter 5. The DNA of chloroplasts (termed Ct-DNA) has a buoyant density not very different from that of nuclear DNA and lies between 1·694 and 1·698 g/cm^3 (see Kung, 1977). The values are rather constant between different species of higher plants and indicate a CG content of some 37 to 38%, although this conservatism is not now thought to reflect conservation in base sequences. The DNA is circular being about 43 to 55 μm long, which it can be calculated is equivalent to a molecular weight of 0.85 to 1.1×10^8 daltons. Details of the structure of Ct-DNA are reviewed by Bedbrook and Kolodrer (1979). A distinctive feature of the cytosine from both higher plant and algal chloroplasts is that it is never methylated. In contrast, about 25% of the cytosine of nuclear DNA exists as the 5-methyl derivative.

Chloroplast DNA is present in about $2{-}10 \times 10^{15}$ g per chloroplast (equivalent to $1.2{-}6.0 \times 10^9$ daltons) which is about the same as the amount present in a bacterial cell (4×10^9 daltons). However, experimental evidence suggests that the genome size in chloroplasts is equivalent to some 1 to 2×10^8 daltons meaning that the DNA content is ten to thirty times that of the genome content, i.e. there are ten to thirty copies of the genetic material present within chloroplasts. Now the Ct-DNA is confined to distinct areas within the chloroplast (normally three to six, but up to thirty or more in very large chloroplasts), so that it can be deduced that each region contains on average three to five copies of the genetic information. Since there are multiple copies of the Ct-DNA, which does not appear to contain repeated sequences, Kung has argued that chloroplasts are in fact polyploid. However, even $1.1{-}1.8 \times 10^8$ daltons could code for between 180 and 300 proteins with molecular weights of about 40 000 (Kirk, 1971).

In spite of the amount of genetic information present in the chloroplast, the majority of the information required for chloroplast formation appears to reside in the nucleus. There is now good evidence to support the view that genetic information encoded in the Ct-DNA is translated in the chloroplasts while that encoded in the nuclear DNA is translated in the cytoplasm, and that the two systems are separate. Nuclear genes appear to control the synthesis of chlorophyll, carotenoids, chloroplast lipids and photosynthetic and starch synthesizing enzymes. The plastid genes appear to code for chloroplast rRNA and possibly for ribosomal proteins. The Ct-DNA also contains the information for the synthesis of the large sub-unit of ribulose bisphosphate carboxylase (Fraction I protein, which is the single most abundant protein in Nature, being some 50% of the soluble leaf protein). The small sub-unit (see Ch. 5) is encoded in the nuclear DNA so that the overall synthesis of the active enzyme requires the cooperation of nuclear and chloroplast genes. There is also some evidence that the synthesis of the P 700-chlorophyll a-protein is encoded by Ct-DNA, while that of the light harvesting chlorophyll a/b protein is under nuclear control. Three of the five sub-units of the coupling factor, CF_1, also appear to be encoded in the chloroplast as does some membrane protein. In all these cases it is apparent that the synthesis of the active constituents requires the close cooperation of both genomes.

Finally, we can consider the question of why chloroplasts contain genetic information at all? In the early years of this century the view was expressed by a Russian botanist that the chloroplast evolved as a result of a symbiosis between blue-green algae and eukaryotic cells. This hypothesis did not gain any general acceptance until the discovery of chloroplast DNA some sixty years later. Subsequently a considerable number of factors have added to the attractiveness of this *endosymbiont hypothesis*. Apart from the discovery of DNA and RNA in chloroplasts, the presence of 70 S (rather than the eukaryote 80 S) ribosomes on which protein synthesis is initiated by N-formyl methionyl tRNA as in prokaryotes has added weight to the argument that chloroplasts are closely allied to the prokaryotes. Inhibitors of protein synthesis on chloroplast ribosomes are those which inhibit the process in prokaryotes and not those which affect eukaryotes (see Ch. 5). Supporting evidence also comes from the similarities of DNA, RNA and proteins of blue-green algae and chloroplasts (see Kung, 1977). Studies of the amino acid composition of ribulose bisphosphate carboxylase also suggest that the large sub-units from different species are quite homologous while those of the small sub-unit, which are under nuclear control, are rather variable between higher plant species. The simplest interpretation of these observations is that the chloroplasts arose during the evolution of the photosynthetic eukaryotes from a symbiotic association with blue-green algae. Such a view is consistent with many other observations on the behaviour of chloroplasts such as their ability to divide, their ability to swell and contract, and the presence of specific permeases in the outer envelope. The similarity between the symbiotic nitrogen-fixing bacteria present in the roots of some plants, such as legumes, and the mitochondria has already been noted. Mitochondria themselves have been postulated to also have arisen as endosymbionts.

Further reading

Avron, M. (1977) Energy transduction in chloroplasts. *Ann. Rev. Biochem.*, **46,** 143

Barber, J. (ed) (1977) *Primary Processes of Photosynthesis*. Elsevier/North-Holland, Amsterdam.

Bendall, D. S. (1977) Electron and proton transfer in chloroplasts. In *Plant Biochemistry II*. International Review of Biochemistry, Vol. 13. Ed. D. H. Northcote, p. 41. University Park Press, Baltimore.

Bassham, J. A. and Calvin, M. (1957) *The Path of Carbon in Photosynthesis*. Prentice-Hall, New Jersey.

Bauer, W. D. (1981) Infection of legumes by rhizobia. *Ann. Rev. Plant Physiol.*, **32,** 407.

Bishop, N. J. (1971) Photosynthesis: the electron transport system of green plants. *Ann. Rev. Biochem.*, **40,** 197.

Boardman, N. K. (1971) The photochemical systems in C_3 and C_4 plants. In *Photosynthesis and Photorespiration*. Eds. M. D. Hatch, C. B. Osmond and R. O. Slatyer. Wiley-Interscience, New York.

Buchanan, B. B. (1980) Role of light in the regulation of chloroplast enzymes. *Ann. Rev. Plant Physiol.*, **31,** 113.

Clayton, R. K. (1965) *Molecular Physics in Photosynthesis*. Blaisdell Publishing Co.

Devlin, R. M. and Barker, A. V. (1971) *Photosynthesis*. Van Nostrand Reinhold, New York.

Gibbs, M. (ed) (1971) *Structure and Function of Chloroplasts*. Springer-Verlag, Berlin.

Gibbs, M. and Latzko, E. (eds.) (1979) *Photosynthesis II. Photosynthetic Carbon Metabolism and Related Processes*. Encyclo. Plant Physiology, Vol. 6. Springer-Verlag, Berlin.

Govindjee (1975) *Bioenergetics of Photosynthesis*. Academic Press, New York.

Gregory, R. P. F. (1977) *Biochemistry of Photosynthesis*, 2nd edn. Wiley-Interscience, New York.

Guernero, M. G., Vega, J. M. and Losada, M. (1981) The assimilatory nitrate-reducing system and its regulation. *Ann. Rev. Plant Physiol.*, **32,** 169.

Hall, D. O. and Whatley, F. R. (1967) The chloroplast. In *Enzyme Cytology*. Ed. D. B. Roodyn. Academic Press, London.

Hatch, M. D., Osmond, C. B. and Slatyer, R. O. (eds.) (1971) *Photosynthesis and Photorespiration*. Wiley-Interscience, New York.

Hatch, M. D. (1976) Photosynthesis: the path of carbon. In *Plant Biochemistry* 3rd edn. Ed. J. Bonner and J. E Varner, p. 797. Academic Press, New York.

Hewitt, E. J. and Cutting, C. V. (eds.) (1979) *Nitrogen Assimilation in Plants*. Academic Press, London.

Kelly, G. J., Latzko, E. and Gibbs, M. (1976) Regulatory aspects of photosynthetic carbon metabolism. *Ann. Rev. Plant Physiol.*, **7**, 181.

Kirk, J. T. O. and Tilney-Bassett, R. A. E. (1978) *The Plastids*. Elsevier/North-Holland, Amsterdam.

Kok, B. (1976) Photosynthesis: the path of energy. In *Plant Biochemistry* 3rd edn. Ed. J. Bonner and J. E. Varner, p. 845. Academic Press, New York.

Lorimer, G. H. (1981) The carboxylation and oxygenation of ribulose 1,5-bisphosphate: the primary events in photosynthesis and photorespiration. *Ann. Rev. Plant Physiol.* **32**, 349.

McCarty R. E. (1976) Ion transport and energy conservation in chloroplasts. In *Transport in Plants III. Intracellular Interactions and Transport Processes*. Eds. C. R. Stocking and U. Heber, Encycl. Plant Physiol. N. S. Vol. 3, p. 347. Springer-Verlag, Berlin.

McCarty, R. E. (1979) Roles of a coupling factor for photophosphorylation in chloroplasts. *Ann. Rev. Plant Physiol.*, **30**, 79

Packer, L., Murakami, S. and Merhard, C. W. (1970) Ion transport in chloroplasts and plant mitochondria. *Ann. Rev. Plant Physiol.*, **21**, 271.

Possingham, J. V. (1980) Plastid replication and development in the life cycle of higher plants. *Ann. Rev. Plant Physiol.*, **31**, 113.

San Pietro, A., Greer, F. A. and Army, T. J. (eds.) (1967) *Harvesting the Sun*. Academic Press, New York.

Thomson, W. W. and Whatley, J. M. (1980) Development of nongreen plastids. *Ann. Rev. Plant Physiol.*, **31**, 375.

Trebst, A. and Avron, M. (eds.) (1977) *Photosynthesis I. Photosynthetic Electron Transport and Photophosphorylation*. Encycl. Plant Physiol. Vol. 5. Springer-Verlag, Berlin.

Tribe, M. A., Morgan, A. J. and Whittaker, P. A. (1981) *The Evolution of Eukaryotic Cells*. Edward Arnold, London.

Velthuys, B. R. (1980) Mechanisms of electron flow in photosystem II and toward photosystem I. *Ann. Rev. Plant Physiol.*, **31**, 545.

Wildman, S. G. (1979) Aspects of fraction 1 protein evolution. *Arch. Biochem. Biophys.*, **196**, 598.

Wildner, G. F. (1981) Ribulose-1,5-bisphosphate carboxylase-oxygenase: aspects and prospects. *Physiol. Plant.* **52**, 385.

Literature cited

Anderson, J. M. (1975) The molecular organisation of chloroplast thylakoids. *Biochim. Biophys. Acta.*, **416**, 191.

Arnon, D. I., Allen, M. B. and Whatley, F. R. (1954) Photosynthesis by isolated chloroplast. *Nature, Lond.*, **174**, 394.

Arnon, D. L., Whatley, F. R. and Allen, M. B. (1958) Assimilatory power in photosynthesis. *Science, N.Y.*, **127**, 1026.

Arntzen, C. J., Dilley, R. A. and Crane, F. L. (1969) A comparison of chloroplast membrane surfaces visualised by freeze-etch and negative staining techniques; and ultrastructural characterisation of membrane fractions obtained from digitonin-treated spinach chloroplast. *J. Cell. Biol.*, **43**, 16.

Arntzen, C. J. and Briantais, J. M. (1975) Chloroplast structure and function. In *Bioenergetics of Photosynthesis*. Ed. Govinjee, p. 51. Academic Press, New York.

Bassham, J. A. (1965) Photosynthesis: the path of carbon. In *Plant Biochemistry*. 2 edn. Eds. J. Bonner and J. E. Varner. Academic Press, New York, p. 875.

Beale, S. I. (1978) α-aminolevalinic acid in plants: its biosynthesis, regulation, and role in plastid development. *Ann. Rev. Plant. Physiol.*, **29**, 95.

Beevers, L. (1976) *Nitrogen Metabolism in Plants*. Arnold, London.

Bedbrook, J. R., and Kolodner, R. (1979) The structure of chloroplast DNA. *Ann. Rev. Plant. Physiol.*, **30**, 593.

Boardman, N. K. (1970) Physical separation of the photosynthetic photochemical systems. *Ann. Rev. Plant. Physiol.*, **21**, 115.

Bjorkman, O. (1975) Adaptive and genetic aspects of C_4 photosynthesis. In *CO_2 metabolism and Plant Productivity*. Eds. R. H. Burris and C. C. Black, p. 287. University Park Press, Baltimore.

Bolton, J. R., and Warden, J. T. (1976) Paramagnetic intermediates in photosynthesis. *Ann. Rev. Plant Physiol.*, **26**, 127.

Boulter, P., Haslett, B. G., Peacock, D., Ramshaw, J. A. M. and Scowen M.D. (1977) Chemistry, function and evolution of plastocyanin. In *Plant Biochemistry II*. International Review of Biochemistry. Vol. 13. Ed. D. H. Northcote, p. 1. University Park Press, Baltimore.

Bryan, J. K. (1976) Amino acid biosynthesis and its regulation In *Plant Biochemistry*. Eds. J. Bonner and J. E. Varner, p. 525. Academic Press.

Bryan, J. K., Lissik, E. A. and DiCamelli, C. A. (1979) Changes in enzyme regulation in plant growth. In *Nitrogen Assimilation in Plants*. Eds. E. J. Hewitt and C. V. Cutting, p. 423. Academic Press, London.

Burris, R. H. (1976) Nitrogen fixafion. In *Plant Biochemistry*. Eds. J. Bonner and J. E. Varner, p. 887. Academic Press.

Clowes, F. A. C. and Juniper, B. E. (1968) *Plant Cells*. Blackwell, Oxford and Edinburgh.

Cogdell, R. J. (1978) Carotenoids in photosynthesis. *Phil. Trans. Rev. Soc. Ser. B.*, **284**, 569.

Coombs, J. and Baldry, C. W. (1972) C-4 pathway in *Pennisetum purpureum*. *Nature New Biol.*, **238**, 268.

Cramer, W. A., and Whitmarsh, J. (1967) Photosynthetic cytochromes. *Ann. Rev. Plant Physiol.*, **28**, 133.

Cronshaw, J., and Wardrop, A. B. (1964) The organisation of cytoplasm in differentiating xylem. *Aust. J. Biol.*, **12**, 15.

Dilworth, M. J. (1974) Dinitrogen Fixation. *Ann. Rev. Plant Physiol.*, **25**, 81.

Emerson, R. and Arnold, W. (1932) The photochemical reaction in photosynthesis. *J. gen. Physiol.*, **16**, 191.

Fowden, L. (1976) Amino acids occurrence, biosynthesis and analogue behaviour in plants. In *Perspectives in Experimental Biology*, Ed. N. Sunderland. Vol. 2, p. 263. Pergamon, London.

Givan, C. V. and Leech, R. M. (1971) Biochemical autonomy of higher plant chloroplasts and their synthesis of small molecules. *Biol. Rev.*, **46**, 409.

Glydenholm, A. O. and Whatley, F. R. (1968) The onset of photophosphorylation in chloroplasts isolated from developing bean leaves. *New Phytol.*, **67**, 461.

Granick, S. and Beale, S. I. (1978) Hemes, chlorophylls, and related compounds: Biosynthesis and metabolic regulation. In *Advances in Enzymology*. Ed. F. F. Nord, Vol. 46, p. 33. John Wiley and Son, New York.

Gregory, R. P. F. (1971) *Biochemistry of Photosynthesis*. Wiley-Interscience, New York.

Greville, G. D. (1969) A scrutiny of Mitchell's chemiosmotic hypothesis of respiratory chain and photosynthetic phosphorylation. *Current Topics in Bioenergetics*, **3**, 1. Ed. D. R. Sandi. Academic Press, New York.

Hatch, M. D. (1971) Mechanism and function of the C_4 pathway of photosynthesis In *Photosynthesis and Photorespiration*, p. 139. Eds. M. D. Hatch, C. D. Osmond and R. O. Slatyer. Wiley-Interscience, New York.

Hattersley, P. W., Watson, L. and Osmond C. B. (1977) *In situ* immunofluorescence labelling of ribulose-1, 5-diphosphate carboxylase in leaves of C_3 and C_4 plants. *Aust. J. Plant Physiol.*, **4**, 523.

Haxo, F. T. and Blinks, L. R. (1950) Photosynthetic action spectra of marine algae. *J. gen. Physiol.*, **33**, 389.

Heber, U. (1974) Metabolite exchange between chloroplasts and cytoplasm. *Ann. Rev. Plant Physiol.*, **25**, 393.

Heinrich, G. (1966) Die Feinstruktur der Proteinplasten von *Helleborous corsicus*. *Protoplasma*, **61**, 157.

Hewitt, E. J., Hucklesby, D. P. and Nottor, B. A. (1976) Nitrate metabolism. In *Plant Biochemistry*. 3rd edn. Eds. J. Bonner and J. E. Varner, p. 633 Academic Press, New York.

Hewitt, E. J. (1975) Assimilatory nitrate-nitrite reduction. *Ann. Rev. Plant Physiol.*, **26**, 73.

Hill, R. (1939) Oxygen production by isolated chloroplasts. *Proc. Roy. Soc. B.*, **127**, 192.

Hooker, J. K., Siekevitz, P. and Palade, G. E. (1969) Formation of chloroplast membranes in *Chlamydomonas reinhardi* y.1. Effects of inhibitors of protein synthesis. *J. biol. Chem.* **244,** 2621.

Hoover, W. H., Johnson, E. S. and Brackett, F. S. (1934) Carbon dioxide assimilation in a higher plant. *Smithson. misc. Collns.*, **87,** No. 16.

Jagendorf, A. T. (1967) The chemiosmotic hypothesis of photophosphorylation. In *Harvesting the Sun*, p. 69. Eds. A San Pietro., F. A. Greer and J. T. Army. Academic Press, New York.

Jagendorf. A. T. (1975) Mechanism of photophosphorylation. In *Bioenergetics of Photosynthesis*. Ed. Govindjee, p. 413. Academic Press, New York.

Jensen, R. G. and Bahr, J. T. (1977) Ribulose 1,5-biphosphate carboxylase-oxygenase. *Ann. Rev. Plant Physiol.*, **28,** 379.

Junge, W. (1977) Membrane potentials in photosynthesis. *Ann. Rev. Plant Physiol.*, **28,** 503.

Kirk, J. T. O. (1971) Chloroplast structure and biogenesis. *Ann. Rev. Biochem.*, **40,** 161.

Kjaer, A. and Larsen, P. O. (1976) Non-protein amino acids, cyanogenic glycosides and glucosinolates. In *Biosynthesis*, **4,** 179. Chem. Soc., London.

Kok, B. (1965) Photosynthesis: The path of energy. In *Plant Biochemistry*. Eds. J. Booner and J. E. Varner. Academic Press, New York.

Kok, B. (1967) Photosynthesis-physical aspects. In *Harvesting the Sun*. Eds. A. San Pietro, F. A. Greer, and T. J. Army. Academic Press, New York.

Krinsky, N. I. (1978) Non-photosynthetic functions of carotenoids. *Phil. Trans. Roy. Soc. Ser, B.*, **284,** 581

Kung, S. (1977) Expression of chloroplast genomes in higher plants. *Ann. Rev. Plant Physiol,* **28,** 401.

Laetsch, W. M. (1971) Chloroplast structural relationships in leaves of C_4 plants. In *Photosynthesis and Photorespiration.*, p. 323. Eds. M. D. Hatch, C. B. Osmond and R. O. Slatyer. Wiley-Interscience, New York.

Lea, P. J. (1977) Biosynthesis of non-protein amino acids. In *Biosynthesis of Amino Acids and Proteins*. Ed. H. R. V. Arnstein Mech. Techn. Pub.

Lea, P. J. and Miflin, B. J. (1974) Alternative route for nitrogen assimilation in higher plants. *Nature, Lond,* **251,** 614.

Leloir, L. F., de Fekete, M. A. R., and Cardini, C. E. (1961) Starch and oligosaccharide synthesis from uridine diphosphate glucose. *J. biol. Chem.*, **236,** 636.

MaCarty, R. E. and Racker, E. (1966) Effects of a coupling factor and its antiserum on photophosphorylation and hydrogen ion transport. *Brookhaven Symp. Quant. Biol.*, **19,** 202.

McCree, R. J. (1971/72) The action spectrum, absorptance and quantum yield of photosynthesis in crop plants. *Agric. Meterol.*, **9,** 191.

Miflin, B. J. (1977) Modification controls in time and space. In *Regulation of Enzyme Synthesis and Activity in Higher Plants*. Ed. H. Smith, p. 23. Academic Press, London.

Miflin, B. J., Bright, S. W. J., Davies, H. M., Shewry, P. R. and Lea, P. J. (1979) Amino acids derived from aspartate; their biosynthesis and its regulation in plants. In *Nitrogen Assimilation in Plants*. Eds. E. J. Hewitt and C. V. Cutting, p. 335. Academic Press, London.

Miflin, B. J. and Lea, P. J. (1976) The pathway of nitrogen assimilation in plants. *Phytochemistry.*, **15,** 873.

Miflin, B. J. and Lea, P. J. (1977) Amino acid metabolism. *Ann. Rev. Plant Physiol.*, **28,** 299.

Mitchell, P. (1976) Possible molecular mechanisms of the proton motive function of cytochrome systems. *J. theoret. Biol.*, **62,** 327.

Monteith, J. L. (1978) Reassessment of maximum growth rates for C_3 and C_4 crops. *Expl. Agric.*, **14,** 1.

Muhlethaler, K. (1971) The ultrastructure of plastids. In *Structure and Function of Chloroplast*, p. 7. Ed. M. Gibbs. Springer-Verlag, Berlin.

Mühlethaler, K. (1977) Introduction to structure and function of the photosynthesis apparatus. In *Photosynthesis I*. Eds. A. Trebst and M. Avron. Encyl. Plant Physiol., Vol. 5. Springer-Verlag, Berlin.

Murakami, S., Torres-Pereira, J. and Packer, L. (1975) Structure of the chloroplast membrane–Relation to energy coupling and ion transport. In: *Bioenergetics of Photosynthesis*. Ed. Govindjee. Academic Press, New York.

Niel, C. B. van (1941) The bacterial photosyntheses and their importance for the general problem of photosynthesis. *Adv. Enzymol.*, **1,** 263.

Nobel, P. S. (1975) Chloroplasts. In *Ion Transport in Plant Cells and Tissues*. Eds. D. A. Baker and J. L. Hall, p. 101. North-Holland, Amsterdam.

Park, R. B. (1976) The Chloroplast. In *Plant Biochemistry*. 3rd edn. Ed. J. Bonner and J. E. Varner, p. 633. Academic Press, New York.

Park, R. B. and Sane, P. V. (1971) Distribution of function and structure in chloroplast lamellae. *Ann. Rev. Plant Physiol.*, **22**, 395.

Radmer, R. and Kok, B (1975) Energy capture in photosynthesis: photosynthesis II. *Ann. Rev. Biochem.*, **44**, 409.

Ridley, S. M. and Leech, R. M. (1970) Division of chloroplasts in an artificial environment. *Nature, Lond.*, **227**, 463.

Shanmugan, K. T., O'Gara, F . O., Anderssen, K. and Valentine, R. C. (1978) Biological nitrogen fixation. *Ann. Rev. Plant. Physiol.*, **29**, 263.

Smith, J. H. C. and French, C. S. (1963) The major and accessory pigments in photosynthesis. *Ann. Rev. Physiol.*, **14**, 181.

Sprent, J. I. (1979) *The Biology of Nitrogen-Fixing Organisms.* McGraw-Hill, London.

Stewart, C. R. and Rhodes, D. (1977) Control of enzyme levels in the regulation of nitrogen assimilation. In *Regulation of Enzyme Synthesis and Activity in Higher Plants.* Ed. H. Smith, p. 1. Academic Press, London.

Streicher, S. L. and Valentine, R. C. (1973) Comparative biochemistry of nitrogen fixation. *Ann. Rev. Biochem.*, **42**, 279.

Thornber, J. P. (1975) Chlorophyll-proteins: Light harvesting and reaction centre components of plants. *Ann. Rev. Plant. Physiol.*, **26**, 127.

Tolbert, N. E. and Ryan, F. J. (1975) Glycolate biosynthesis and metabolism during photosynthesis. In: CO_2 *Metabolism and Plant Productivity.* Eds. R. H. Burris and C. C. Block, p. 141. University Park Press.

Trebst, A. (1974) Energy conservation in photosynthetic electron transport of chloroplasts. *Ann. Rev. Plant. Physiol.*, **25**, 423.

Trebst, A. (1978) Plastoquinones in photosynthesis. *Phil. Trans. Roy. Soc. Ser. B.*, **284**, 591.

Walker, D. A. (1974) Chloroplast and cell–the movement of certain key substances, etc., across the chloroplast envelope. In *Plant Biochemistry*. Ed. D. H. Northcote. Vol. II MTP International Review of Science, p. 1. Butterworth, London.

Walker, D. A. (1976) Plastids and intracellular transport. In *Transport in Plants III. Intracellular Interactions and Transport Processes.* Eds. C. R. Stocking and U. Heber. Encyclo. Plant Physiol. Vol. 3, p. 85. Springer-Verlag, Berlin.

Wehrmeyer, W. (1964) Zur klarung der strukturellen Variabilitat der Chloroplastengrana des Spinats in Profil and Aufsicht. *Planta*, **62**, 272.

Weier, T. E., Stocking, C. R., Bracker, C. E. and Risley, E. B. (1965) The structural relationships of the internal membrane system of *in situ* and isolated chloroplasts of *Hordeum vulgare. Amer. J. Bot.*, **52**, 339.

Wettstein, von D. W. (1967) Chloroplast structure and genetics. In *Harvesting the Sun.*, p. 153. Eds. A. San Pietro, F.A. Greer and T. J. Army. Academic Press, New York.

Whatley, F. R., Tagawa, K. and Arnon, D. I. (1963) Separation of the light and dark reactions in electron transfer during photosynthesis. *Proc. natl. Acad. Sci. U.S.A.*, **49**, 266.

9   Microbodies

During the early 1950s there were reports of small spherical bodies, about $0{\cdot}3\text{--}1{\cdot}5\ \mu\text{m}$ in diameter, present in electron micrographs of mouse kidney cells. These organelles, whose function at that time was unknown, were termed *microbodies* and have subsequently been observed in liver and kidney cells of other animals (although they are not found generally in vertebrate cells), in protozoa and in a range of fungi and plants. The biochemical function of these organelles has only recently been appreciated to any degree and is very much the subject of current research. All microbodies seem to be characterized by the presence of flavin-linked oxidases which produce hydrogen peroxide with the uptake of molecular oxygen:

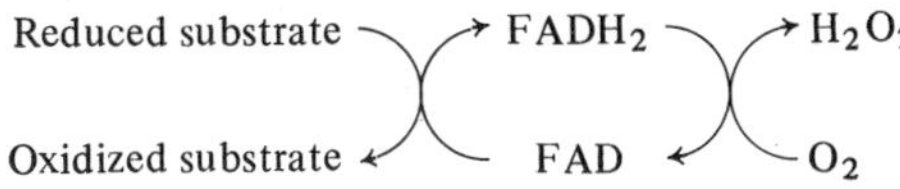

Reduced substrate   FADH$_2$ H$_2$O$_2$

Oxidized substrate   FAD   O$_2$

and of catalase which is subsequently able to break down the peroxide to oxygen and water:

$$H_2O_2 \longrightarrow H_2O + \tfrac{1}{2}O_2$$

As a result of the combined action of the oxidase and the catalase there is a net uptake of one mole of oxygen for the oxidation of two moles of substrate. Thus microbodies, as well as mitochondria, participate in the uptake of oxygen and contribute to the overall respiratory gas exchange of cells. They differ from mitochondria, however, in their lack of cytochromes and in their inability to produce ATP. Energy is apparently released and lost as heat during catalase activity.

As a result of biochemical studies it has been possible to distinguish two types of microbody which have been called *peroxisomes* and *glyoxysomes*. These two organelles differ both in their enzyme complement and in the type of tissue in which they are found. Peroxisomes are found in animal cells and in the leaves of higher plants. Their function is obscure in animals although somewhat clearer in plants. In both they participate in the oxidation of substrates, producing hydrogen peroxide which is subsequently destroyed by catalase activity. In plants they appear to function in close co-operation with chloroplasts and to be the site of many of the reactions of a light stimulated uptake of oxygen known as *photorespiration*. Glyoxysomes have not been described in animal cells but only in plants, where they are particularly abundant in germinating seeds which store fats as a reserve material. They contain the enzymes for the breakdown of fatty acids, together with those for the formation of succinate which may then be used for the synthesis of sugar. A third class of microbody may exist in the roots of higher plants, although these have not yet been clearly characterized. Apart from peroxisomes and glyoxysomes, a number of other terms have been used to describe microbodies, including *cytosome*, *phragmosomes* and *crystal-containing bodies*.

Structure and occurrence

Microbodies are rather variable in size and shape, but usually appear circular in cross-section with diameter of between 0·2 and 1·5 μm. They all have a single limiting membrane which encloses a granular matrix of moderate electron density (Fig. 9.1). In some cases (e.g. in

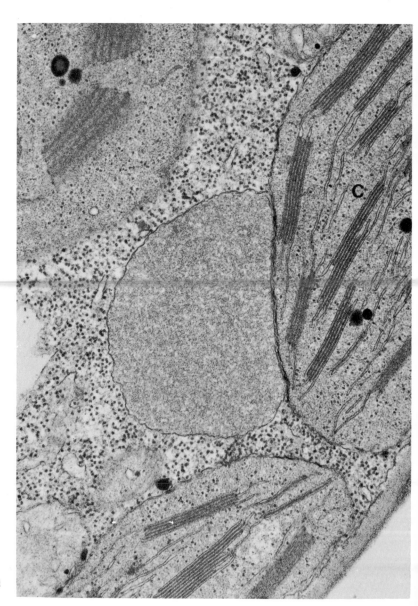

Fig. 9.1   A peroxisome in a mesophyll cell of a tobacco leaf. The peroxisome is closely associated with a chloroplast (C). × 44 000. Reproduced from Newcomb and Frederick (1971).

the festuciod grasses) the matrix contains numerous threads or fibrils while in others they are observed to contain either an amorphous nucleoid or a dense inner core which in many species shows a regular crystalloid structure (Fig. 9.2). In germinating castor bean seeds the crystalloid consists of dense rods 6·0 nm in diameter. Cells may be stained for catalase activity using 3,3′-diaminobenzidine (DAB) as substrate. Under the electron microscope the osmiophilic electron-dense product of the peroxidative activity of catalase on DAB is found largely within the microbodies (Fig. 9.3) and, in particular, when present with the core (Fig. 9.4). Little seems to be known about the function of the core, except that it is the site of the enzyme urate oxidase in rat liver peroxisomes and much of the catalase in some plants. On the whole, the general structure simplicity of the micro-bodies clearly distinguishes them from organelles such as mitochondria and chloroplasts which have internal membrane systems and a double membrane envelope. Lysosomes (Ch. 12) which also have only a single bounding membrane may be less easy to distinguish from the

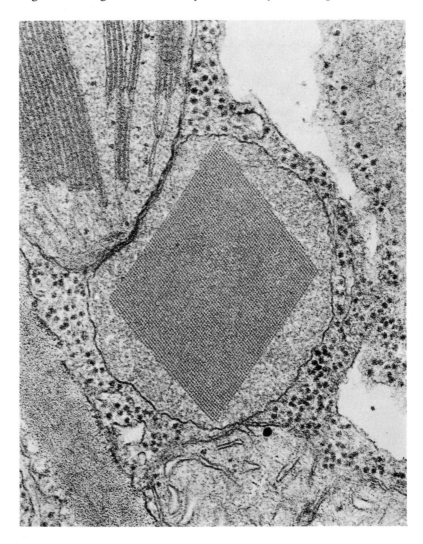

Fig. 9.2   A tobacco leaf cell peroxisome with a large crystalline inclusion. The peroxisome is again closely associated with a chloroplast. × 85 600. Reproduced from Newcomb and Frederick (1971).

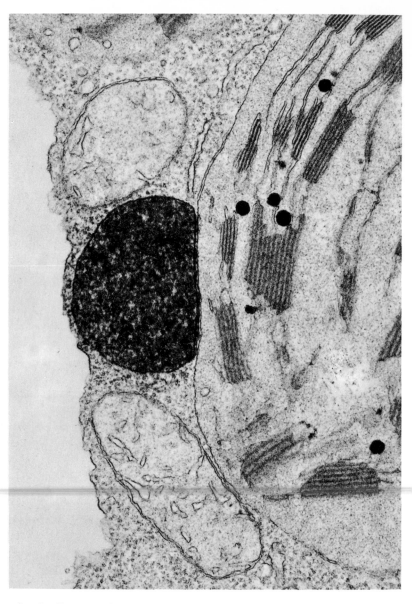

microbodies morphologically, but have a very different complement of enzymes.

Microbodies are, in general, to be found closely associated with the endoplasmic reticulum. In addition, glyoxysomes may be intimately associated with lipid bodies and peroxisomes with chloroplasts.

Peroxisomes have been described in a wide range of plants. They are present in all photosynthetic cells of higher plants which have been examined so far, in etiolated leaf tissue, in coleoptiles and hypocotyls, in tobacco stem and callus, in ripening pear fruits and in the spadix of *Symplocarpus* (skunk cabbage). They have also been described in the Euglenophyta, protozoa, brown algae, fungi, liverworts, mosses and ferns. Glyoxysomes have been extensively studied in germinating castor bean seeds but are also present in the maturing seeds. They

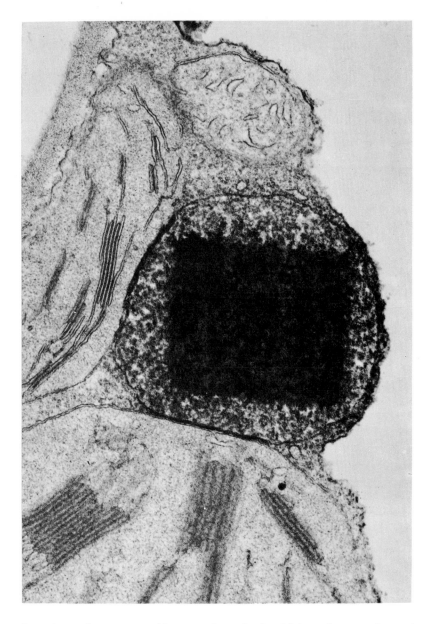

Fig. 9.4   A peroxisome from a tobacco leaf stained with DAB as in Fig. 9.3. Note the catalase activity is particularly associated with the crystalline inclusion. × 54 000. Reproduced from Newcomb and Frederick (1971).

have been demonstrated in a number of other higher plant species and in *Neurospora* and yeast, although in yeast their enzyme complement bears similarities with that of the peroxisomes.

Isolation

Microbodies appear to be very fragile and it is consequently difficult to obtain good yields from plant cells since the methods required to disrupt the cells may damage the microbodies. Generally the tissues are homogenized for a few seconds in a blender or food chopper and then centrifuged at about 10 000–15 000 *g*. The resulting pellet, which corresponds to a crude mitochondrial fraction, is then layered on to a sucrose density gradient. Isopycnic ultracentrifugation (see p. 33) yields a microbody-rich fraction at concentrations of sucrose from 1·5 to 2·2 M. The microbodies which equilibrate in higher concentrations of sucrose than both the mitochondria and the proplastids (Fig. 9.5) have a high specific density of between 1·24 and 1·26 g cm^{-3}. This is attributed to the high permeability of their membranes to sucrose, so that they come to equilibrium at the density of their matrix protein. Yields are generally between 1 and 10% of the total complement of microbodies although values as high as 50% have been reported. As yet, glyoxysomes cannot be separated from peroxisomes and both may be contaminated by proplastids and fragments of mitochondria.

Fig. 9.5 The separation of mitochondria, proplastids and glyoxysomes by centrifugation using (a) a linear and (b) a stepped sucrose density gradient. Reproduced from Cooper and Beevers (1969a).

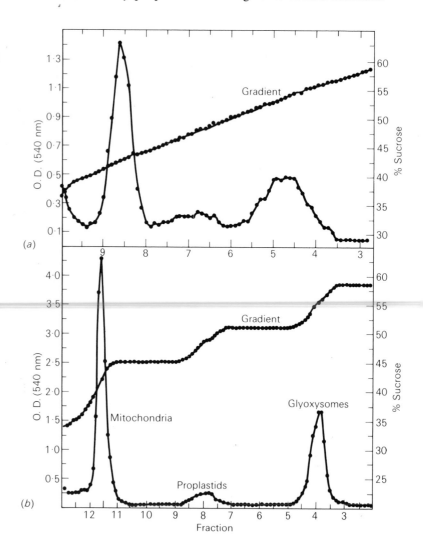

Biochemical activity

As has already been indicated, all microbodies contain flavin-linked oxidases producing hydrogen peroxide which is then broken down by catalase with the net uptake of oxygen by the cell. Glyoxysomes differ from peroxisomes in their complement of enzymes, although both

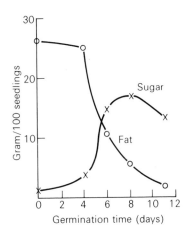

Fig. 9.6 The synthesis of sugar and breakdown of fat during the germination of castor bean seeds. Reproduced from Davies, Giovanelli and Ap Rees (1964); data of Desveaux and Kogane-Charles (1952).

organelles appear to have the potential to transform non-carbohydrate material to carbohydrate.

Glyoxysomes

There are a number of plant species which store lipid materials in their seeds and during germination this lipid is broken down and converted to carbohydrate (Fig. 9.6). Furthermore, it has been shown that there is a correlation between the conversion of fats to carbohydrate during germination and the presence of glyoxysomes within the cells of the tissue. In castor bean seeds, for example, the reserve material is stored in lipid-containing bodies known as *spherosomes* (see p. 508), although it is not metabolized within these organelles. During germination there is a large increase in the number of glyoxysomes and intimate connections with the spherosomes can be seen (Fig. 9.7). As germination proceeds the number of spherosomes decrease.

The breakdown of lipid begins with its conversion to fatty acids. Triglycerides are hydrolysed by the enzyme lipase (glycerol ester hydrolase) to glycerol and fatty acids, while phospholipids are hydrolysed by the enzyme phospholipase. The lipase from castor bean resembles mammalian lipase in that it catalyses the hydrolysis of fatty acids esterified in the 1 and 3 positions of glycerol more rapidly than it does those in position 2. The long chain fatty acids which are released by the hydrolysis are then broken down by the successive removal of 2-C fragments in the process of β-*oxidation*.

β-Oxidation

The process of β-oxidation was elucidated using animal tissues over a period of fifty years starting from the work of Knoop in 1904 and was later confirmed in plants by Stumpf and Barber (1956). The fatty acid

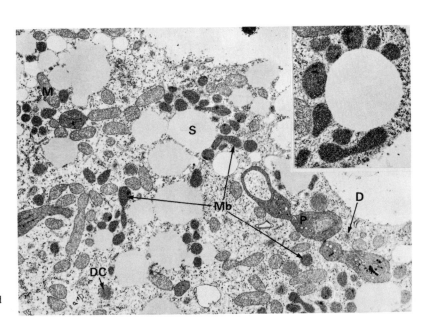

Fig. 9.7 Microbodies (Mb) in a vacuolate cell of castor bean endosperm, showing the close association with spherosomes (S, and inset × 12 100). Other organelles present include a new protein body (DC), mitochondria (M), plastid and dictyosomes (D). × 5 700. Reproduced from Vigil (1970).

released from the glyceride is first activated to a fatty acyl-CoA, then oxidized, hydrated and reoxidized to form a 3-keto acyl-CoA. Finally a thiolytic cleavage yields acetyl-CoA and a fatty acyl-CoA with two less carbon atoms than the original (Fig. 9.8). This new fatty acyl-CoA is then recycled through the same series of reactions until the final two molecules of acetyl-CoA are produced (providing there were an even number of carbon atoms in the original fatty acid).

The reaction of a fatty acid with CoA to form a fatty acyl-CoA occurs under the influence of an enzyme, fatty acid thiokinase.

$$R—CH_2—CH_2—COOH + ATP + ATP + CoASH \rightleftharpoons$$
$$R—CH_2—CH_2—CO—S—CoA + AMP + PP_i \quad (1)$$

The thiokinase is found in the microsomal fraction and hence activation occurs outside the organelle in which subsequent events occur. In animal systems the activated fatty acid penetrates the mitochondrial membrane as a derivative of carnitine although this process does not apparently occur in the glyoxysomes where carnitine derivatives are not metabolized.

$$(CH_3)_3N^+—CH_2—CH—CH_2—COO^-$$
$$\overset{|}{OH}$$
Carnitine

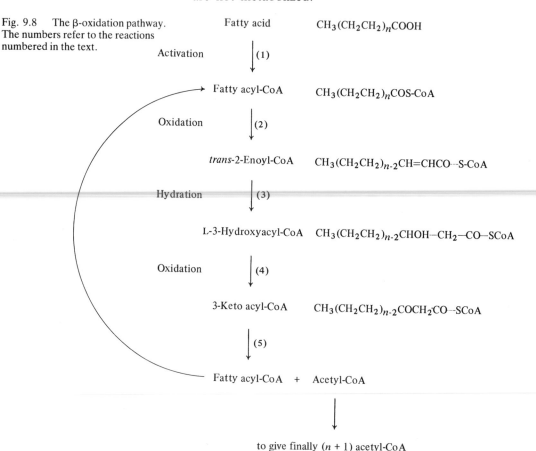

Fig. 9.8   The β-oxidation pathway. The numbers refer to the reactions numbered in the text.

Fatty acid $CH_3(CH_2CH_2)_nCOOH$

Activation   (1)

Fatty acyl-CoA $CH_3(CH_2CH_2)_nCOS\text{-}CoA$

Oxidation   (2)

trans-2-Enoyl-CoA $CH_3(CH_2CH_2)_{n-2}CH=CHCO\text{-}S\text{-}CoA$

Hydration   (3)

L-3-Hydroxyacyl-CoA $CH_3(CH_2CH_2)_{n-2}CHOH—CH_2—CO—SCoA$

Oxidation   (4)

3-Keto acyl-CoA $CH_3(CH_2CH_2)_{n-2}COCH_2CO\text{-}SCoA$

(5)

Fatty acyl-CoA   +   Acetyl-CoA

to give finally $(n + 1)$ acetyl-CoA

The fatty acyl enzyme A is then dehydrogenated at the 2 and 3 (or α and β) carbon atoms, the hydrogen atoms being transferred to FAD under the influence of the enzyme fatty acyl-CoA dehydrogenase.

$$\underset{3}{\overset{\beta}{R-CH_2}}-\underset{2}{\overset{\alpha}{CH_2}}-\underset{\underset{O}{\parallel}}{\overset{}{C}}\underset{1}{\overset{}{-}}S-CoA + FAD \rightleftharpoons R-\overset{}{C}=\overset{\overset{H}{|}}{C}-\underset{\underset{O}{\parallel}}{\overset{\overset{H}{|}}{C}}-S-CoA + FADH_2 \quad (2)$$

Hydrogen peroxide is formed and subsequently broken down by catalase with the resultant net uptake of one mole of oxygen

$$R-CH_2-CH_2-CO-S-CoA \diagdown \nearrow FADH_2 \diagdown O_2$$
$$R-CH=CH-CO-S-CoA \diagup \diagdown FAD \diagup \searrow H_2O_2 \longrightarrow H_2O + \tfrac{1}{2}O_2$$

for every two moles of fatty acyl-CoA dehydrogenated.

The *trans*-2-enoyl-CoA produced as the result of the dehydrogenase activity is then hydrated by an enzyme enoyl hydratase (or crotonase). The hydration is stereospecific, producing only the L-3-hydroxyacyl-CoA.

$$R-\overset{\overset{H}{|}}{C}=\underset{\underset{H}{|}}{C}-\underset{\underset{O}{\parallel}}{C}-S-CoA + H_2O \rightleftharpoons R-\underset{\underset{OH}{|}}{\overset{\overset{H}{|}}{C}}-CH_2-\underset{\underset{O}{\parallel}}{C}-S-CoA \quad (3)$$

The product of this reaction is oxidized by an NAD linked L-3 hydroxyacyl-CoA dehydrogenase.

$$R-\underset{\underset{OH}{|}}{\overset{\overset{H}{|}}{C}}-CH_2-\underset{\underset{O}{\parallel}}{C}-S-CoA + NAD^+ \rightleftharpoons$$
$$R-\underset{\underset{O}{\parallel}}{C}-CH_2-\underset{\underset{O}{\parallel}}{C}-S-CoA + NADH + H^+ \quad (4)$$

It is this 3-keto acyl-CoA, produced in the last reaction, which loses a two carbon fragment under the action of the enzyme thiolase (or β-keto thiolase) to generate a new fatty acyl-CoA and acetyl-CoA.

$$R-CO-CH_2-CO-SCoA + HSCoA \longrightarrow$$
$$R-CO-SCoA + CH_3-CO-SCoA \quad (5)$$

The fatty acyl-CoA is then oxidized by the fatty acyl-CoA dehydrogenase. The sequence of reactions splitting off a two carbon fragment is then repeated until two moles of acetyl-CoA are produced by the final thiolase step in the oxidation of a fatty acid with an even number of carbon atoms.

Where the original fatty acid does not have an even number of carbon atoms, the final step produces one mole of acetyl-CoA plus one mole of propionyl-CoA.

$$CH_3-CH_2-CO-CH_2-CO-SCoA \longrightarrow CH_3-CH_2-CO-SCoA + CH_3-CO-SCoA$$

The propionyl-CoA is dehydrogenated to acrylyl coenzyme A ($CH_2=CH-CO-SCoA$), hydrated to 3-hydroxypropionyl coenzyme A ($CH_2OH-CH_2CO-SCoA$) which is then subjected to the action of an NAD-linked dehydrogenase producing malonic semialdehyde ($HCO-CH_2OCOOH$), coenzyme A and NADH. This sequence of reactions contrasts with that in animals for dealing with

$$CH_3\text{---}\underset{\overset{\displaystyle |}{COOH}}{CH}\text{---}CO\text{---}SCoA$$

Methylmalonyl-CoA

propionyl-CoA, which involves at ATP dependent carboxylation of the propionyl-CoA to methylmalonyl-CoA. This is isomerized to succinyl-CoA ($HOOC\text{---}CH_2\text{---}CH_2\text{---}CO\text{---}SCoA$) in a reaction involving vitamin B_{12} and then metabolized through the TCA cycle.

In cases where the original fatty acid is unsaturated, the β-oxidation sequence can proceed much as described above except that the action of thiolase may in some cases give rise to a *cis*-2-enoyl-CoA. On hydration, this results in the production of D-3-hydroxyacyl-CoA which is not a substrate for the 3-hydroxyacyl-CoA dehydrogenase: this is stereospecific for the L-isomer, and an epimerase must first catalyse the conversion of the D to the L form. Where the thiolase activity gives rise to *cis*- or *trans*-3-enoyl-CoA, this may be converted to the 2-enoyl form enzymically.

The reactions of β-oxidation in animal cells are, as has already been hinted, closely associated with the mitochondria and there is a tight coupling of β-oxidation to the operation of the electron transport chain. The early studies of Stumpf and Barber demonstrated that the reactions of β-oxidation in plants were also associated with mitochondrial fraction, but subsequent work has shown a number of differences between plants and animals. When a crude mitochondrial fraction from a tissue such as germinating castor oil bean is subjected to isopycnic density gradient centrifugation, oxidation activity is not associated with the mitochondria, but with the glyoxysomes (Cooper and Beevers, 1969b, and Table 9.1). This glyoxysomal pathway in plants appears similar to the β-oxidation pathway of mammalian mitochondria, except that penetration of the glyoxysomal membrane by the fatty acyl-CoA does not require carnitine and the glyoxysomes do not themselves reoxidize the NADH produced by the L-3-hydroxyacyl-CoA dehydrogenase. NADH accumulates *in vitro*, while *in vivo* it is presumed to be oxidized by the mitochondria. This may be related to the rapid utilization of exogenous NADH by plant mitochondria, whereas the animal mitochondrial membrane is rel-

Table 9.1   The localization of β-oxidation activity and of various enzymes in mitochondrial, glyoxysomal and proplastid fractions of germinating castor bean seeds

Enzyme	Activity in		
	Mitochondria	Glyoxysomes	Proplastids
	(μ moles of substrate utilized/g fwt/h)		
β-Oxidation of palmitoyl-CoA	0	11·4	2·8
Thiolase	0	22·9	6·3
Catalase	258	3 360	1 039
Isocitritase	8	207	9
Malate synthetase	46	790	27
Citrate synthetase	218	74	2
Malate dehydrogenase	10 700	4 475	311
Succinate dehydrogenase	65	0	0
Fumarase	394	2	2

Data from Cooper and Beevers (1969*a* and *b*)

atively impermeable to this coenzyme (see p. 297). Oxygen is, however, required for glyoxysomal β-oxidation, and is presumably concerned in the oxidation of the reduced flavoprotein produced as a result of the fatty acyl-CoA dehydrogenase activity. Subsequently, the hydrogen peroxide is destroyed by the ever present catalase. This may give direction to what otherwise could be a reversible process.

The stoichiometry for the production of acetyl-CoA, the reduction of NAD^+ and the uptake of oxygen is $1:1:0.5$, respectively. β-Oxidation results in the production of $(n + 1)$ moles of acetyl-CoA from a fatty acid with $(n + 2)$ carbon atoms (where n is an even number: i.e. $CH_3(CH_2CH_2)_nCOOH$). This acetyl-CoA is spatially separated from the enzymes of the TCA cycle and consequently avoids further oxidation. Its participation in gluconeogenic reactions follows a sequence first elucidated by Kornberg and Krebs (1957) which was shown to occur in the endosperm of castor oil bean by Kornberg and Beevers (1957), and is known as the *glyoxylate cycle*. One turn of the cycle yields one mole each of succinate and NADH from two moles of acetyl-CoA.

The glyoxylate cycle

The glyoxylate pathway involves some of the reactions of the TCA cycle in that citrate is formed from oxaloacetate and acetyl-CoA under the action of citrate synthetase (reaction 6, Fig. 9.9), then subsequently converted to isocitrate by aconitase (7). The cycle then involves two enzymic reactions which have not been reported in animal cells. These are the conversion of isocitrate to glyoxylate and succinate by isocitratase:

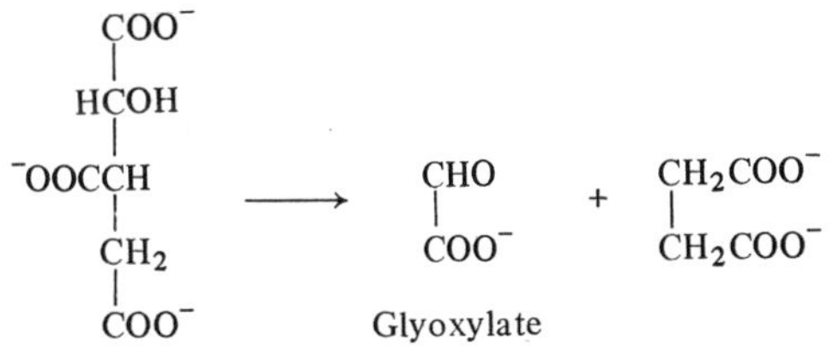

$$
\begin{array}{c}
COO^- \\
|\ \\
HCOH \\
|\ \\
{}^-OOCCH \\
|\ \\
CH_2 \\
|\ \\
COO^-
\end{array}
\quad\longrightarrow\quad
\begin{array}{c}
CHO \\
|\ \\
COO^-
\end{array}
\ +\
\begin{array}{c}
CH_2COO^- \\
|\ \\
CH_2COO^-
\end{array}
\qquad (8)
$$

Glyoxylate

and the formation of malate from glyoxylate and a further mole of acetyl-CoA by malate synthetase:

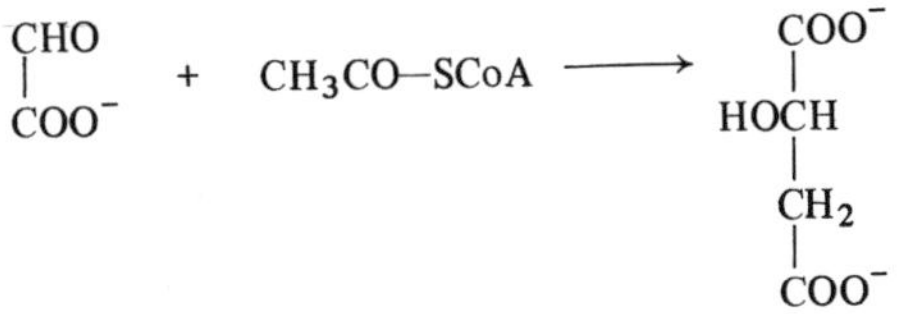

$$
\begin{array}{c}
CHO \\
|\ \\
COO^-
\end{array}
\ +\ CH_3CO-SCoA\ \longrightarrow\
\begin{array}{c}
COO^- \\
|\ \\
HOCH \\
|\ \\
CH_2 \\
|\ \\
COO^-
\end{array}
\qquad (9)
$$

This malate is converted to oxalocetate by malate dehydrogenase (10) for the cycle to the completed (Fig. 9.9). Thus, overall, the pathway involves

2 Acetyl-CoA + NAD^+ $\longrightarrow$ succinate + NADH + H^+

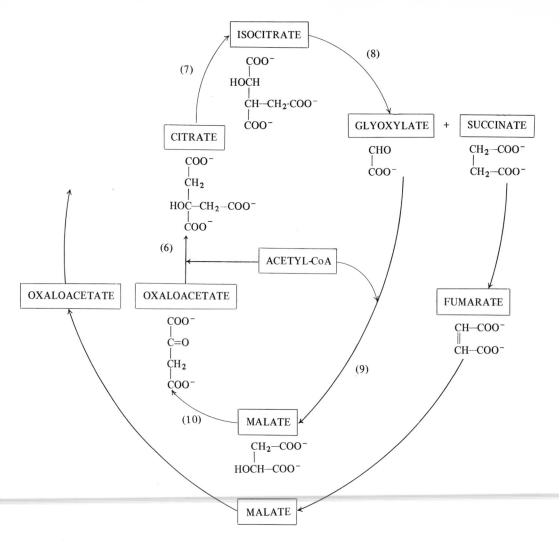

Fig. 9.9   The glyoxylate cycle. The numbers refer to the reactions numbered in the text.

The work of Beevers group (see Cooper and Beevers, 1969a) has shown that the reactions of the glyoxylate cycle occurs in the glyoxysomal fraction (hence the name glyoxysome) (Table 9.1). More than 80% of the particulate activity of the two enzymes unique to the glyoxylate cycle, isocitritase and malate synthetase, were found sedimented in the glyoxysomal fraction, while citrate synthetase and malate dehydrogenase, which occur in both the TCA and glyoxylate cycles, were present in both the mitochondrial and glyoxysomal fractions. Aconitase was relatively unstable and it was not possible to clearly localize its activity. The enzymes of the TCA cycle which are not associated with the glyoxylate cycle (succinic dehydrogenase and fumarase) were present only in the mitochondria. Thus the glyoxysomes of higher plant cells have been shown to be the site of both β-oxidation and of the enzymes of the glyoxylate cycle. In addition, it has been shown that the development of malate synthetase and isocitrate lyase activity during the first five days of germination (see Fig. 9.15) parallels the increase in the number of glyoxysomes seen under the electron microscope. A number of other enzymes are also found in the glyoxysomal fraction and include glutamate: oxaloacetate aminotransferase (60% of the particulate enzyme activity), catalase (66%) and glycolic oxidase (78%) together with urate oxidase and allantoinase (see below).

Succinate is thus the end product of the glyoxysomal metabolism of fatty acid and is not further metabolized within this organelle. The synthesis of hexose involves the conversion of succinate to oxaloacetate, which presumably takes place in the mitochondrion, since the glyoxysomes do not contain the enzymes furmarase and succinic dehydrogenase (see Table 9.1). Two molecules of oxaloacetate are formed from four molecules of acetyl-CoA without carbon loss. This oxaloacetate is converted to phosphoenolpyruvate in the phosphoenolpyruvate carboxykinase reaction with the loss of two molecules of carbon dioxide (also see pp. 252, and 273):

$$2 \text{ Oxaloacetate} + 2\text{ATP} \rightleftharpoons 2 \text{ phosphoenolpyruvate} + 2CO_2 + 2\text{ADP}$$

This is supported by the work of Canvin and Beevers (1961) who showed that by feeding acetate labelled in the 1- or 2-carbon positions ($\overset{2}{C}H_3 \overset{1}{C}OOH$), nearly all carbon dioxide released arose from the C–1 position, while hexose produced by the reversal of glycolysis contained twice as much carbon derived from C–2 as from C–1 (Fig. 9.10). Mammals are unable to synthesize carbohydrate from acetyl-CoA since there can be no net formation of oxaloacetate by the TCA cycle: all the carbon is lost during the two oxidative decarboxylations of the cycle and the phosphoenolpyruvate carboxykinase reaction. Only in the operation of the glyoxylate cycle is a net formation of carbohydrate possible. The important features of the synthesis of hexose by the reversal of glycolysis have already been described (see p. 251).

Other enzymes found in glyoxysomes
Urate oxidase (uricase) which converts uric acid to allantoin

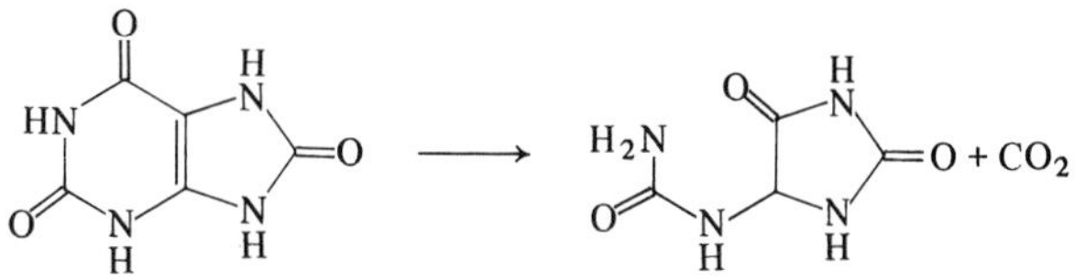

and allantoinase which hydrates allantoin to allantoic acid

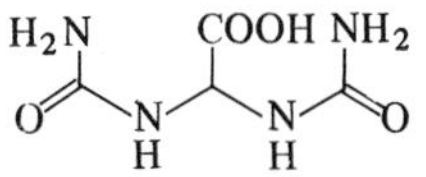

during the degradation of purines have been demonstrated in the glyoxysomes of castor oil bean, although no other enzymes of urate oxidation have been found. de Duve (1969) has theorized that the glyoxysomes are an ancestral form of microbody which has lost many of its enzymic functions in the course of evolution.

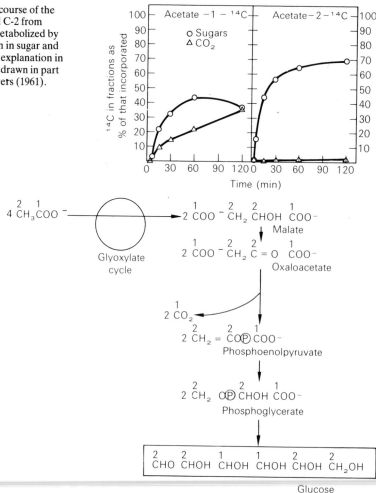

Fig. 9.10 The time course of the appearance of C-1 and C-2 from ^{14}C-labelled acetate metabolized by castor bean endosperm in sugar and carbon dioxide and its explanation in biochemical terms. Redrawn in part from Canvin and Beevers (1961).

Peroxisomes

Microbodies were first described in electron micrographs of animal tissues in the mid 1950s and many of the properties of their biochemical counterparts in animals, the peroxisomes, have subsequently been elucidated (see de Duve, 1969). It has already been indicated that these organelles contain catalase together with one or more enzymes which produce hydrogen peroxide. The primary electron donor for these latter enzymes is usually glycolate, L-lactate or another L-hydroxy acid, and one mole of the hydrogen peroxide is produced with the uptake of one mole of oxygen. Subsequently, the action of catalase causes peroxidations such as that of ethanol to acetaldehyde ($CH_3CH_2OH + H_2O_2 \rightarrow CH_3CHO + 2H_2O$), formic acid to carbon dioxide and water ($HCOOH + H_2O_2 \rightarrow CO_2 + 2H_2O$) or of a further mole of peroxide to oxygen and water ($2H_2O_2 \rightarrow 2H_2O + O_2$). In every case there is a net uptake of oxygen. This contribution to the respiratory oxygen uptake of a cell may be distinguished from that of the mitochondrion by the effect of oxygen concentration on the rate of uptake. The rate of peroxisomal oxygen uptake is almost directly proportional to the oxygen concentration, while mitochondrial oxygen uptake is almost independent of oxygen concentration (above about 5%) due to the extremely high affinity of cytochrome oxidase for oxygen. A further major difference between the respiration of the two particles is, as already stated, that peroxisomal respiration in animals is not coupled to ATP production and in this respect appears to be a wasteful process.

Peroxisomes have also been described in plant leaves, where they contain catalase together with the enzymes of what has been termed the *glycolate pathway* (Fig. 9.11). This pathway is thought to bring about the formation of the amino acids glycine and serine from the non-phosphorylated intermediates of the photosynthetic carbon reduction cycle, i.e. glycerate to serine, or glycolate to glycine and serine in a sequence of reactions which appears to involve chloroplasts, peroxisomes and mitochondria. It was first elucidated during the early 1960s by $^{14}CO_2$ tracer studies.

Fig. 9.11 The glycolate cycle. The numbers refer to the reactions numbered in the text.

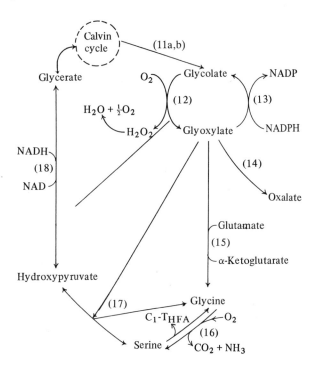

The glycolate cycle

Amongst the products of photosynthesis which are labelled during short periods of exposure to $^{14}CO_2$ are serine, glycine, glyoxylate and glycolate. Glycolate arises as a result of photosynthesis in the chloroplast and its formation is light dependent, although the exact route is not entirely clear. Gibbs (1971) has proposed that the production of glycolate is closely associated with photosystems I and II (see Ch. 8). Under normal light and carbon dioxide levels, hydrogen peroxide is formed by PS II as the result of the interaction of a reduced carrier with molecular oxygen. Subsequently the peroxidation of a TPP-C_2 complex formed by a transketolase reaction (involving fructose 6-phosphate, sedoheptulose 7-phosphate or xylulose 5-phosphate – see p. 122) yields glycolate (the number of the reaction corresponds with those in Fig. 9.11).

$$\begin{bmatrix} -C=O \\ | \\ CH_2OH \end{bmatrix} \xrightarrow{H_2O_2} \begin{array}{c} COO^- \\ | \\ CH_2OH \end{array} \tag{11a}$$

However, evidence obtained using $^{18}O_2$, only one atom of which is incorporated into the carboxyl group of glycolate, has suggested to some workers in the field that this is not a major pathway for the formation of glycolate (Tolbert and Ryan, 1975).

Much recent discussion has centred on the possibility that glycolate is formed from phosphoglycolate which is produced as the result of the oxygenase activity of ribulose bisphosphate carboxylase (see also Ch. 8 and Jensen & Bahr, 1975).

$$\begin{array}{c} CH_2O \,\text{\textcircled{P}} \\ | \\ C=O \\ | \\ CHOH \\ | \\ CHOH \\ | \\ CH_2O \,\text{\textcircled{P}} \end{array} \;+\; O_2 \;\longrightarrow\; \begin{array}{c} COO^- \\ | \\ CH_2O \,\text{\textcircled{P}} \end{array} \;+\; \begin{array}{c} COO^- \\ | \\ CHOH \\ | \\ CH_2O \,\text{\textcircled{P}} \end{array}$$

The action of phosphoglycolate phosphatase, which is an abundant enzyme in the chloroplast, would produce glycolate.

$$\begin{array}{c} COO^- \\ | \\ CH_2O\,P \end{array} \;\longrightarrow\; \begin{array}{c} COO^- \\ | \\ CH_2OH \end{array} \;+\; P_i \tag{11b}$$

It is interesting to try to evaluate the effect of the oxygenase activity on the net fixation of carbon dioxide. In Chapter 8, we showed that if every step of the Calvin cycle were to function at least once, then three carboxylations must occur and six moles of phosphoglycerate are formed. Of these, five are used to regenerate the ribulose 1,5-bisphosphate resulting in the net fixation of three carbon atoms. For ten complete cycles, the net reaction would be:

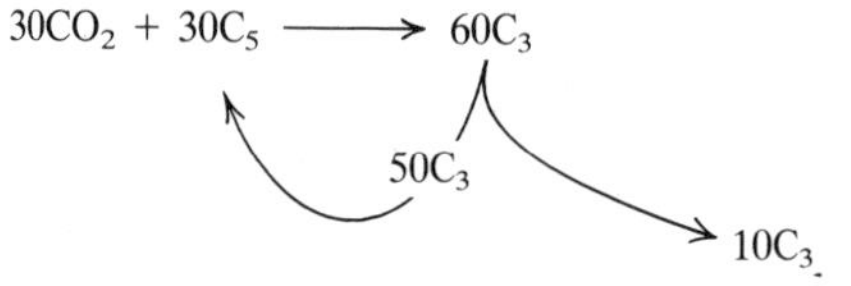

$$30CO_2 + 30C_5 \longrightarrow 60C_3$$
$$50C_3$$
$$10C_3$$

In air, the ratio of carboxylase to oxygenase activity has been found to be about 4 (Schrader, 1975). If we assume the ratio to be exactly 4, then under such conditions, 30 moles of ribulose 1,5-bisphosphate would react with 24 moles of CO_2 and 6 of O_2. This would produce 48 moles of phosphoglycerate as a result of the carboxylase activity and a further six from the oxygenase, together with six moles of phosphoglycolate. Of the net 54 moles of phosphoglycerate, 50 would again be required to regenerate the ribulose 1,5-bisphosphate, leaving a net of only 12 fixed C-atoms from the Calvin cycle:

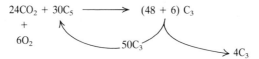

Put another way, only 50% of the carbon that is initially fixed appears as a direct product of Calvin cycle activity. We shall describe below how some of the carbon which is initially fixed into phosphoglycolate is salvaged and how about 14% may be lost as carbon dioxide.

Under conditions where the oxygen concentration is reduced to near zero and yet the carbon dioxide concentration remains close to that in air, all the carbon fixed would be assimilated through the Calvin cycle. However, at the CO_2 compensation point (see p. 336) when the carboxylase to oxygenase ratio is theoretically 0·5 (and assuming the dark respiration is zero), there would be no net fixation of carbon.

Although the balance between carboxylase and oxygenase activities appears to explain the observed amounts of carbon fixation, there is evidence for the operation of other mechanisms and some workers (Zelitch, 1975a) believe that there is a significant proportion of glycolate which is not accounted for by the operation of ribulose bisphosphate oxygenase activity and is synthesized from CO_2 or organic acids.

However, whatever may be the mechanism of glycolate biosynthesis, experiments involving the infiltration of excised leaves with glycolate have shown that its oxidation to glyoxylate and the subsequent formation of glycine and serine could both occur in the dark and in the presence of DCMU (which inhibits oxygen evolution in photosynthesis). Separation of chloroplasts from other organelles has indicated that the metabolism of glycolate first takes place in the peroxisome.

The oxidation of glycolate to glyoxylate is catalysed by the enzyme glycolate oxidase, which like other peroxisomal and glyoxysomal oxidases has a flavoprotein as its prosthetic group:

$$\begin{array}{c}CH_2OH \\ | \\ COO^-\end{array} \quad + \quad O_2 \quad \longrightarrow \quad \begin{array}{c}CHO \\ | \\ COO^-\end{array} \quad + H_2O_2 \tag{12}$$

Glycolate   Glyoxylate

α-Hydroxysulphonates are specific inhibitors of the utilization of glycolate and it builds up in their presence.

A number of fates may befall the glyoxylate. For example, it may return to the chloroplast to be reduced to glycolate by glyoxylate reductase (13) so that a glycolate/glyoxylate shuttle may be set up between the two organelles. This would have the effect of oxidizing any excess NADPH produced during photosynthesis although it has yet to be established that such a shuttle functions *in vivo* while the glyoxylate reductase in the chloroplast is apparently only present at a low activity (Tolbert and Ryan, 1975). Alternatively, the glyoxylate may be further oxidized by glycolate oxidase to oxalate, which does not undergo any subsequent metabolism in the peroxisome:

$$
\begin{array}{c}
\underset{|}{\text{CHO}} \\
\text{COO}^-
\end{array}
\longrightarrow
\begin{array}{c}
\underset{|}{\text{COO}^-} \\
\text{COO}^-
\end{array}
\tag{14}
$$

Oxalate

A third possibility is that the glyoxylate may be transaminated with L-glutamate to yield glycine (glutamate-glyoxylate aminotransferase):

$$
\begin{array}{c}
\underset{|}{\text{CHO}} \\
\text{COO}^-
\end{array}
+
\begin{array}{c}
\underset{|}{\text{COO}^-} \\
\underset{|}{\text{CH}_2} \\
\underset{|}{\text{CH}_2} \\
\text{H}_2\text{N}\cdot\text{CH—COO}^-
\end{array}
\longrightarrow
\begin{array}{c}
\underset{|}{\text{CH}_2\text{NH}_2} \\
\text{COO}^-
\end{array}
+
\begin{array}{c}
\underset{|}{\text{COO}^-} \\
\underset{|}{\text{CH}_2} \\
\underset{|}{\text{CH}_2} \\
\text{CO—COO}^-
\end{array}
\tag{15}
$$

L-Glutamate   glycine   α-Ketoglutarate

Glycine subsequently appears to be converted to serine.

The formation of serine from glycine seems then to occur within the mitochondrion, not the peroxisome, and is coupled to ATP synthesis (Bird *et al.*, 1972; Moore *et al.*, 1977).

$$\tfrac{1}{2}O_2 + 2 \text{ Glycine} \longrightarrow \text{ serine} + CO_2 + NH_3 \tag{16}$$

One molecule of glycine may be split, transferring a one-carbon fragment to tetrahydrofolic acid (THFA) in a reaction involving pyridoxal phosphate (see p. 123)

$$\text{Glycine} + \text{FH}_4 \rightleftharpoons N^5,N^{10}\text{-methylene FH}_4 + CO_2 + NH_3$$

The second molecule of glycine reacts with the N^5N^{10}-methylene tetrahydrofolic acid to form serine under the action of serine transhydroxy-methylase.

$$N^5,N^{10}\text{-methylene FH}_4 + \text{glycine} \rightleftharpoons \text{serine} + \text{FH}_4$$

The serine may, in fact, be reconverted to glycine and the C–1 tetrahydrofolate complex, thus forming a system by which the sustained release of carbon dioxide from glycolate may be effected. It is important to note that of the reactions so far described, it is only this glycine–serine interconversion that is, in practice, reversible: the ribulose bisphosphate oxygenase, phosphoglycolate phosphatase, glycolate oxidase and glutamate-glyoxylate aminotransferase reactions are all physiologically irreversible.

Transmination of serine with a further mole of glyoxylate yields hydroxypyruvate (serine-glyoxylate aminotransferase).

$$
\begin{array}{c}
\underset{|}{\text{CHO}} \\
\text{COO}^-
\end{array}
+
\begin{array}{c}
\underset{|}{\text{CH}_2\text{OH}} \\
\underset{|}{\text{CH COO}^-} \\
\text{NH}_2
\end{array}
\rightleftharpoons
\begin{array}{c}
\underset{|}{\text{COO}^-} \\
\underset{|}{\text{CO}} \\
\text{CH}_2\text{OH}
\end{array}
+
\begin{array}{c}
\underset{|}{\text{CH}_2\text{NH}_2} \\
\text{COO}^-
\end{array}
\tag{17}
$$

Glyoxylate   Serine   Hydroxypyruvate   Glycine

This reaction is reversible and so does allow the synthesis of serine from hydroxypyruvate and glycine.

The final step of the glycolate cycle which is again reversible is postulated to regenerate glycerate by the reduction of hydroxypyruvate (by hydroxypyruvate reductase).

$$\begin{array}{ccc}
\text{COO}^- & & \text{COO}^- \\
| & & | \\
\text{CO} & + \text{NADH} + \text{H}^+ \rightleftharpoons & \text{CHOH} + \text{NAD}^+ \\
| & & | \\
\text{CH}_2\text{OH} & & \text{CH}_2\text{OH}
\end{array} \tag{18}$$

The glycerate apparently leaves the peroxisome to return to the chloroplast and complete the cycle (Fig. 9.12). NADH is regenerated by an NAD specific malate dehydrogenase, which is found in the peroxisome and converts malate to oxaloacetate. This allows the formation of aspartate through aspartate-α-ketoglutarate aminotransferase.

The glycolate pathway is complex in that it is not confined to a single organelle. Parts of the pathway occur in the chloroplasts, peroxisomes and mitochondria (Fig. 9.13). The genesis of glycolate appears to be confined to the chloroplast. Some of the enzymes involved in glycolate

Fig. 9.12   The distribution of the elements of the glycolate cycle between peroxisomes, mitochondria and chloroplasts.

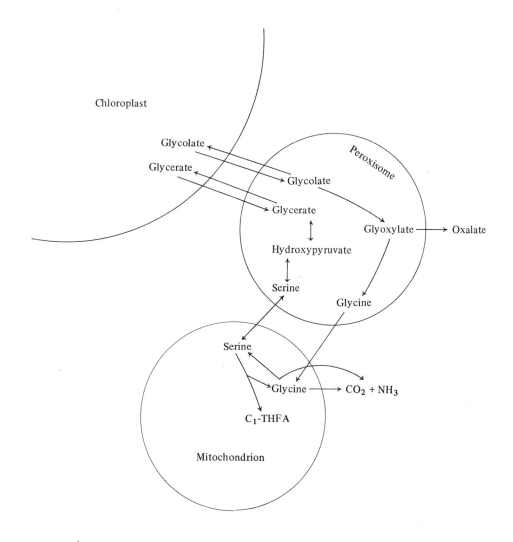

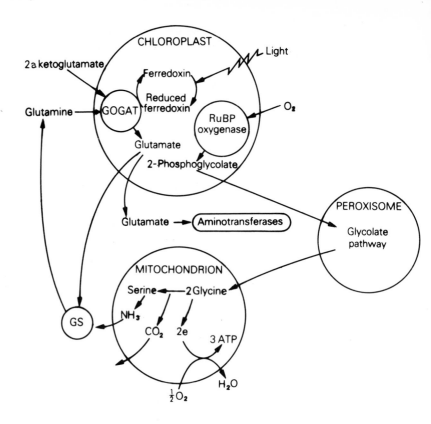

Fig. 9.13   The photorespiratory nitrogen cycle. Redrawn after Keys *et al.* (1978).

metabolism have been found in the peroxisome (glycolate oxidase, catalase, glutamate-glyoxylate, serine-glyoxylate and aspartate-α-ketoglutarate aminotransferases, hydroxypyruvate reductase and malic dehydrogenase) while others may occur in the mitochondrion (serine hydroxymethyl transferase). The first and last reactions of the cycle occur in the chloroplasts (glycolate biosynthesis and glycerate kinase, respectively). The reactions which we have described as returning carbon to the chloroplast and which are largely reversible are sometimes known as the glycerate pathway (Tolbert and Ryan, 1975). This allows the synthesis of serine from 3-phosphoglycerate which is formed in the chloroplasts and then hydrolysed by a unique phosphatase to form glycerate. In all cases the metabolism of glyoxylate is confined to the peroxisomes, perhaps because of its reactivity.

The emphasis of our discussion so far has centred on the flow of carbon, but it must be pointed out that the glycine–serine interconversion not only releases carbon dioxide, but also ammonia, which the plant can ill-afford to lose. Recent work has shown that this nitrogen is recycled and it has been proposed that the ammonia released is refixed by glutamine synthetase(GS) in the cytoplasm (Keys *et al.*, 1978). If glutamine and α-ketoglutarate then entered the chloroplast this would allow the synthesis of two molecules of glutamate under the influence of glutamate synthase (GOGAT, see p. 381) in the presence of reduced ferredoxin (Fig. 9.13). One molecule of glutamate would then be available for the further synthesis of glutamine and the second for any of the variety of amino-transferase reactions into which glutamate is known to enter (Ch. 8): this in itself may regenerate α-ketoglutarate.

Photorespiration

There is a net uptake of oxygen and evolution of carbon dioxide during the operation of the glycolate cycle; this is known physiologically as *photorespiration* since it is stimulated in the light. This respiratory oxygen uptake is believed to occur in the peroxisome during the glycolate oxidase step of the glycolate pathway, during the formation of phosphoglycolate in the chloroplast and in the production of serine which occurs in the mitochondrion. Its response to ambient oxygen concentration is similar to that of the oxygen uptake already described for animal peroxisomes i.e. it is stimulated by increasing the oxygen concentration up to 100% in contrast to dark (mitochondrial) respiration which saturates at about 2% oxygen (Fig. 9.14). It is also stimulated by relatively high temperatures and inhibited by high concentrations of carbon dioxide. The mechanism by which carbon dioxide is released is not entirely clear and could occur by a number of separate reactions. Studies with 1-^{14}C labelled glycine and glycolate have indicated that both substrates may produce equal amounts of ^{14}C-labelled carbon dioxide (Kisaki and Tolbert, 1970). Since the pathway from glycolate to glycine is irreversible, the evolution of carbon dioxide is thought to be associated with the later stages of the glycolate pathway. The most likely reaction is during the formation of serine from two moles of glycine. Although the N^5,N^{10}-methylene tetrahydrofolic acid is available as a C–1 donor it may, however, also be converted to N^{10}-formyl tetrahydrofolic acid and hence to formic acid and carbon dioxide. There has been no evidence of the peroxidation of glycolate to formaldehyde and formate in the peroxisome *in vivo*, although *in vitro* studies indicate that such reactions are certainly a possible source of photorespiratory carbon dioxide. However, the action of catalase should rapidly destroy any excess peroxide. The release of carbon dioxide has not been demonstrated in animal peroxisomes, nor need it occur during glyoxysomal respiration.

The release of carbon dioxide into carbon dioxide free air in the light has been used as a measure of photorespiration (see Jackson and Volk, 1970). This process is demonstrable in many species which, since they are generally unable to deplete external carbon dioxide levels much below 50 ppm, are known as high compensation point species. The carbon dioxide compensation point is defined as that concentration of carbon dioxide at which there is no net exchange of carbon dioxide between the atmosphere and the plant (see p. 336). The carbon dioxide fluxes involved are photosynthetic towards the plant and respiratory, both light and dark, towards the atmosphere. In a number of other species, however, photorespiration is difficult to detect and these plants deplete carbon dioxide levels in the atmosphere to about 5–10 ppm, i.e. they have a low carbon dioxide compensation point of 5–10 ppm. Since the compensation point can be measured both quickly and easily, it has often been used as a

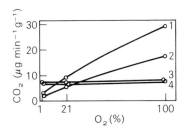

Fig. 9.14   The effect of oxygen concentration on the rates of photorespiration at different light intensities (curve 1, 120 × 10³ erg cm⁻¹ sec⁻¹ curve 2, 30 × 10³ erg cm¹ sec⁻¹ and on dark respiration curves 3 and 4). Reproduced from Poskuta (1968).

diagnostic character for the presence of photorespiration. More recent work, however, has shown that compensation points are not as constant as had at one time been thought, making this a less reliable indicator of photorespiration than other measures of gas flux (see Zelitch, 1975b). Nonetheless, it is now recognized that, on the whole, these two groups of high and low compensation point species correspond with those utilizing the Calvin cycle (C_3 plants, e.g. spinach, *Chlorella* spp.), and the C_4 dicarboxylic acid pathway (C_4 plants, e.g. maize, sugar cane, some *Atriplex* spp.) respectively. Thus the low compensation point species tend to be those well adapted to conditions of high irradiance, high temperature and low water supply. They have high maximum rates of photosynthesis and growth. It is interesting in this respect that Björkman *et al.* (1968) grew both C_3 and C_4 plants for 17 days in low oxygen concentrations (from 2 to 5%) and found that although this procedure had little effect on the overall growth of the C_4 plants, there was a doubling in the rate of dry matter production of the C_3 plants in comparison with their growth in air (20% oxygen). The low compensation point species are not entirely without photorespiration, however. The leaves of many of these plants contain microbodies in both the bundle sheath and mesophyll cells and do have significant levels of glycolate oxidase activity. In species such as maize however, the synthesis of glycolate occurs at about one tenth of the rate at which it is synthesized by C_3 species. This suggests that the lower rates of photorespiration shown by C_4 species reflects reduced synthesis of the substrate, glycolate. It is also possible that the PEP carboxylase in the mesophyll cells effects refixation of CO_2 released in photorespiration, and it is interesting in this respect that it is the bundle sheath cells which contain the preponderance of microbodies present in such species.

The magnitude of photorespiration in C_3 species often appears quite remarkable, although it is still technically extremely difficult to obtain reliable estimates of photorespiration during photosynthesis. In sunflower, photosynthesizing in normal air at a light intensity of 9 000 ft-c ($1,700 \ \mu E \ m^{-2} \ sec^{-1}$), photorespiration was measured at 20% of the true rate of photosynthesis and 2·7 times the rate of dark respiration. If the carbon dioxide all arose from the glycine to serine interconversion without the completion of the glycolate cycle, then carbon flow through glycine would have to be close to 80% of the rate of carbon fixation (Canvin *et al.*, 1975). Operation of the complete cycle, returning carbon to the chloroplasts, would reduce the carbon flow through the pathway if CO_2 were also released by other reactions (see also Zelitch, 1975a).

The measurement of the gas exchange of leaves by a variety of techniques (see Jackson and Volk, 1970) has indicated that mitochondrial respiration may be reduced in the light. The reaction of glyoxylate and oxaloacetate to produce α-hydroxy-α-carboxy-glutarate could be the means by which this inhibition is brought about since this compound is a powerful inhibitor of aconitase, isocitric dehydrogenase and α-ketoglutarate dehydrogenase activities. Alternatively, the peroxisome may drain NADH from the mitochondrion in the light, and thus inhibit dark respiration. Such speculations have not been confirmed experimentally however, and since glycine is converted to serine in the light, the mitochondrial electron transport chain must presumably be functioning under such conditions.

Functions of peroxisomes

There are probably more questions than answers in any discussion of the functions of plant and animal peroxisomes. Clearly, in plants, they are involved in the synthesis of glycine and serine and possibly of C-1 donor from the primary products of photosynthesis. However, since glycine and serine and C-1 tetrahydrofolic acid can be formed directly from 3-phosphoglycerate, the glycolate route to these compounds is somewhat enigmatic, unless the importance of the reaction is the production of ATP during the mitochondrial oxidation of glycine to serine (Bird *et al.*, 1972; Moore *et al.*, 1977). If the formation of glycolate in the chloroplast is an end product of a protective mechanism against hydrogen peroxide or the unavoidable consequence of photosynthetic activity in normal air, then the conversion of glycolate back to glycerate salvages 75% of the carbon. This notion of a protective mechanism against excess peroxide is in accord with a postulated function for the animal peroxisome by de Duve (see below). Tolbert and Ryan (1975) have also suggested that photorespiration is a complex terminal oxidase system for consuming excess ATP and NADPH produced in photosynthesis. Unlike respiration, where the oxidation of substrates can be regulated through the product (ATP) in the process termed respiratory control, photosynthesis cannot be regulated at source. Plants have no means of shutting off the light or removing water from the chloroplasts. The operation of the glyoxylate/glycolate shuttle between the chloroplast and the peroxisome may destroy excess reducing power in the chloroplast. Glyoxylate returning to the chloroplast is reduced back to glycolate by glyoxylate reductase with the concomitant oxidation of NADPH. However, such a mechanism is speculative at the present time.

Ontogeny

Less is known of the development of microbodies than of mitochondria and chloroplasts, although work carried out during the mid-1970s has provided an insight at least into the development of glyoxysomes. Neither peroxisomes nor glyoxysomes contain their own genetic machinery as do the chloroplasts and mitochondria and, in consequence, they rely entirely on the cytoplasmic system for the synthesis of their enzymes. Experiments conducted on the incorporation of ^{14}C-choline into the phospholipids of the castor bean endosperm have shown that label firstly appeared in the endoplasmic reticulum and later in the glyoxysomal membranes. Furthermore, analysis of the individual phospholipid components of both the membranes of the endoplasmic reticulum and of the glyoxysomes has shown that the relative amounts of these phospholipids are very similar, as are their constituent fatty acids, suggesting a similarity of origin. Other evidence also supports the concept that the glyoxysomes arise from the ER. It has been shown that enzymes, such as malate synthetase and citrate synthetase, are associated with the glyoxysomal membranes rather than with their matrix. Malate synthetase is associated with the ER early in germination, but as germination proceeds, the level of malate synthetase present in the ER falls to a very low level while that in the glyoxysome rises. Furthermore, the enzyme in the two organelles has been shown by antiseral studies to be identical. Evidence is more difficult to obtain with citrate synthetase since this enzyme is also present in the mitochondria, while the soluble enzymes of the glyoxysomal matrix, such as catalase, are easily solubilized and lost to the supernatant during fractionation.

The evidence currently available has lead Beevers and his co-workers (see Beevers, 1979) to propose that the glyoxysomes are derived directly from the ER (see also Chs. 3 and 11). Proteins are synthesized on the ribosomes attached to the ER and permeate into its lumen. The membranes then vesiculate to form the developing glyoxysomes. Little is known however of the development of the leaf peroxisomes, although it might be presumed that their origin is basically similar.

Another question that has been asked is whether there are in fact two separate organelles, the peroxisome and the glyoxysome, or whether the two particles represent a change in the enzyme complement of a single organelle. In some species (e.g. members of the Cucurbitaceae) the cotyledons are not just the site of fat storage in the seed, since they later expand and turn green. Glyoxysomal activity builds up to a maximum and corresponds with maximal rates of β-oxidation during the early stages of germination. During greening, however, glyoxysomal activity declines as peroxisomal activity increases (Beevers, 1971). Catalase activity does not decrease to zero along with the other glyoxysomal marker enzymes, since it is also a constituent of the peroxisomes (Fig. 9.15). If these plants are grown in the dark then the glyoxysomal enzymes decline as they normally would in the light, but the peroxisomal enzymes do not increase. A subsequent light period causes the expected increase in peroxisomal activity. It has been reported, however, that during such a transfer from dark to light there are no corresponding changes in the numbers of microbodies visible under the electron microscope (Trelease, Becker, Graber and Newcomb, 1971). It was suggested that either the enzyme complement of a single organelle changed (the one-population model) or that the microbodies that were observed during germination were always of two types, only one of which was functional at any one time (the two-population model). Of the two possibilities Trelease et al. favoured the former, since there was no morphological evidence of two populations of microbody. However, Kagawa and Beevers (1970) found a decline in the amount of microbody protein present in sucrose gradients and interpreted this result to mean that glyoxysomes are destroyed following fat utilization. This view has some support from

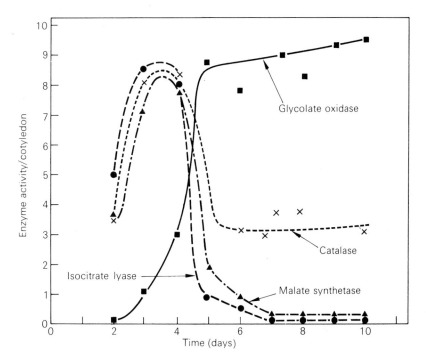

Fig. 9.15 Changes in glyoxysomal enzymes in homogenates of cucumber cotyledons. Reproduced from Trelease et al. (1971).

electron micrographs of castor oil bean endosperm (Vigil, 1970) in which microbodies appeared to be visible within autophagic vacuoles. However, Trelease et al. could find no electron micrographic evidence for this in cucumber. They suggested that the microbodies become more fragile during germination and therefore fewer survive extraction to be analysed by the methods used by Kagawa and Beevers. The considerable evidence now available is summarized by Beevers (1979), but the controversy still awaits resolution.

It has been speculated (see de Duve, 1969) that microbodies arose as respiratory particles during the increase in oxygen levels occuring in the earth's atmosphere. This oxygen was and still is able to react with certain metabolic intermediates to produce toxic amounts of hydrogen peroxide. The peroxisome could therefore have served a dual rôle in both destroying the peroxide and also acting as a terminal oxidase for the products of fermentation, thus providing a sink for electrons which would otherwise produce a pool of reduced metabolites. Mitochondria were envisaged as having originated from parasitic bacteria which had since evolved to a state of symbiosis with the host cell. Once both mitochondria and microbodies were established in the cell, competition between the two must have taken place. The fact that the microbodies have survived such competition indicates that they may well have functions in addition to those already postulated, particularly in animal cells.

Further reading

Beevers, H. (1979) Microbodies in higher plants. Ann. Rev. Plant Physiol., 30, 159.
Burris, R. H. and Black, C. C. (eds.) (1975) CO₂ Metabolism and Plant Productivity. University Park Press, Baltimore.

Hatch, M. D. Osmond C. B. and Slatyer, R. O., (eds.) (1971). *Photosynthesis and Photorespiration* . Wiley-Interscience, New York.

Jackson, W. A. and Volk, R. J. (1970) Photorespiration. *Ann. Rev. Plant Physiol.,* **21,** 385.

Schnarrenberger, C. and Fock, H. (1976) Interactions among organelles involved in photorespiration. In *Transport in Plants.* III. Eds. C. R. Stocking and U. Heber. Encyl. Plant Physiol. p. 185. Springer-Verlag, Berlin.

Tolbert, N. E. (1971) Microbodies–peroxisomes and glyoxysomes. *Ann. Rev. Plant Physiol.,* **22,** 45.

Tolbert, N. E. (1981) Metabolic pathways in peroxisomes and glyoxysomes. *Ann. Rev. Biochem.,* **50,** 133.

Zelitch, I. (1975b) Pathways of carbon fixation in green plants. *Ann. Rev. Biochem.,* **44,** 123.

Literature cited

Beevers, H. (1971) Comparative biochemistry of microbodies (glyoxysomes and peroxisomes). In *Photosynthesis and Photorespiration.* Eds. M. D. Hatch, C. B. Osmond and R. O. Slatyer, Wiley-Interscience, New York.

Bird, I. F., Cornelius, M. J., Keys, A. J. and Whittingham, C. P. (1972) Oxidation and phosphorylation associated with the conversion of glycine to serine. *Phytochemistry,* **11,** 1587.

Björkman, O., Gaul, E., Hiesey, W. M., Nicholson, F. and Nobbs, M. A. (1968) Growth of *Mimulus, Marchantia* and *Zea* under different oxygen and carbon dioxide levels. *Carnegie Inst. Washington Yearb.,* **67,** 447.

Canvin, D. T. and Beevers, H. (1961) Sucrose synthesis from acetate in the germinating castor oil bean. *J. biol. Chem.,* **236,** 988.

Canvin, D. T., Lloyd, N. D. H., Fock, H. and Przybylla, K. (1975) Glycine and serine metabolism and photorespiration. *In CO_2 Metabolism and Plant Productivity.* Eds. R. H. Burris and C. C. Black, p. 161. University Park Press, Baltimore.

Cooper, T. G. and Beevers, H. (1969a) Mitochondria and glyoxysomes from castor bean endosperm. *J. biol. Chem.,* **244,** 3507.

Cooper, T. G. and Beevers, H. (1969b) β-oxidation in glyoxysomes from castor bean endosperm. *J. biol. Chem.,* **244,** 3514.

Davies, D. D., Giovanelli, J. and Ap. Rees, T. (1964) *Plant Biochemistry,* Blackwell, Oxford.

du Duve, C. (1969) The peroxisome: a new cytoplasmic organelle. *Proc. Roy. Soc. B.,* **173,** 71.

Desveaux, R. and Kogane-Charles, M. (1952) Etudé sur la germination de quelques graines oleogineuses. *Ann. Inst. Natl. Rech. Agron.,* **3,** 385.

Gibbs, M. (1971) Biosynthesis of glycolytic acid. In *Photosynthesis and Photorespiration.* Eds. M. D. Hatch, C. B. Osmond and R. O. Slatyer. Wiley-Interscience, New York.

Jensen. P. G. and Bahr, J. T. (1975) Regulation of CO_2 incorporation via the pentose phosphate pathway. In *CO_2 Metabolism and Plant Productivity.* Eds. R. H. Burris and C. C. Black, p. 3. University Park Press, Baltimore.

Kagawa, T. and Beevers, H. (1970) Glyoxysomes and peroxisomes in water-melon seedlings. *Plant Physiol.,* **46s,** 38.

Keys, A. J., Bird, I. F., Cornelius, M. J., Lea, P. J., Wallsgrove, R. M. and Miflin, B. J. (1978) Photorespiratory nitrogen cycle. *Nature, Lond.,* **275,** 741.

Kisaki, T. and Tolbert, N. E. (1970) Glycine as a substrate for photorespiration. *Plant Cell Physiol.,* **11,** 247.

Kornberg, H. L. and Beevers, H. (1975) The glyoxylate cycle as a stage in the conversion of fat to carbohydrate. *Biochim. Biophys. Acta.,* **26,** 531.

Kornberg, H. L. and Krebs, H. A. (1975) Synthesis of cell constituents from C_2-units by a modified tricarboxylic acid cycle. *Nature, Lond.,* **179,** 988.

Moore, A. L., Jackson, C., Halliwell, B., Dench, J. E. and Hall, D. O. (1977). Intramitochondrial localisation of glycine decarboxylase in spinach leaves. *Biochem. Biophys. Res. Comm.,* **78,** 483.

Newcomb, E. H. and Frederick, S. E. (1971) Distribution and structure of plant microbodies (peroxisomes). In *Photosynthesis and Photorespiration.* Eds. M. D. Hatch, C. B. Osmond and R. O. Slatyer. Wiley-Interscience, New York.

Poskuta, J. (1968) Photosynthesis, photorespiration and respiration of detached spruce twigs as influenced by oxygen concentration and light intensity. *Physiol. Plant.,* **21,** 1129.

Schrader, C. E. (1975) CO_2 metabolism and productivity in C_3 plants: an assessment. In *CO_2 Metabolism and Plant Productivity.* Eds. R. H. Burris and C. C. Black, p. 385. University Park Press, Baltimore.

Stumpf, P. K. and Barber, G. A. (1956) Fat metabolism in higher plants. VII β-Oxidation of fatty acids by peanut mitochondria. *Plant Physiol.,* **31,** 304.

Tolbert, N. E., and Ryan, F. J. (1975) Glycolate biosynthesis and metabolism during photosynthesis. In *CO₂ Metabolism and Plant Productivity.* Eds. R. H. Burris and C. C. Black, p. 141. University Park Press, Baltimore.

Trelease, R. N., Becker, W. N., Gruber, P. J. and Newcomb, E. H. (1971) Microbodies (glyoxysomes and peroxisomes) in cucumber cotyledons. *Plant. Physiol.,* **48,** 461.

Vigil, E. L. (1970) Cytochemical and developmental changes in microbodies (glyoxysomes) and related organelles of castor bean endosperm. *J. Cell Biol.,* **46,** 435.

Zelitch, I. (1975a) Biochemical and genetic control of photorespiration. In *CO₂ Metabolism and Plant Physiology.* Eds. R. H. Burris and C. C. Black, p. 343. University Park Press, Baltimore.

10   Cell walls

A carbohydrate-rich cell wall has always been regarded as one of the features of plant cells which distinguished them from animals. It should be realized, however, that most animal cells are now known to possess a cell coat which contains a large proportion of carbohydrate (Martinez-Palomo, 1970), though in many cases this is an extension of the outer membrane, and is often visible on electron micrographs as a 'fuzz' around the cell surface. In plants, the wall is secreted as an organized extraprotoplasmic layer which is often quite rigid and strong. Where extensive thickening occurs, the wall may occupy most of the lumen of the cell and it always contributes significantly to the total dry weight of the organism. A considerable portion of the economy of the cell must be directed, therefore, towards a synthesis of its wall. Furthermore this structure, though isolated from the cytoplasm by the plasmalemma, grows to accommodate increases in cell size, contains enzymes, and may even have a complement of ribosomes (see p. 215), though they may have been occluded there by chance at cell division and have no function. The wall, therefore, can be reasonably regarded as an organelle and, like the other organelles which we discuss, it is not autonomous so that we must always consider its interaction and relationship with the rest of the cell. Most of its component parts, for example, are synthesized intracellularly. Its growth and architecture are also directed from within the cell. In its turn, however, it limits and presumably controls cell growth, it forms specific interactions with neighbouring cells to bind them together in the tissue complex and provides a protective barrier against injury and infection. Finally, of course, it lends skeletal support to the whole organism, and provides a source of food, fibre and fuel for man.

Cell plate formation and the origin of the cell wall

The formation of the new cell wall begins at cell division after the two sets of chromosomes have receded to their respective poles at late anaphase. An organized zone of cytoplasm known as the *phragmoplast* appears in the equatorial plane of the dividing mother cell. This zone is barrel-shaped and populated by filamentous structures recognizable under the electron microscope as microtubules (see p. 231 and p. 485) which are arranged in parallel to each other and perpendicular to the axis of the cell division (Fig. 10.1). The cell plate, which forms the

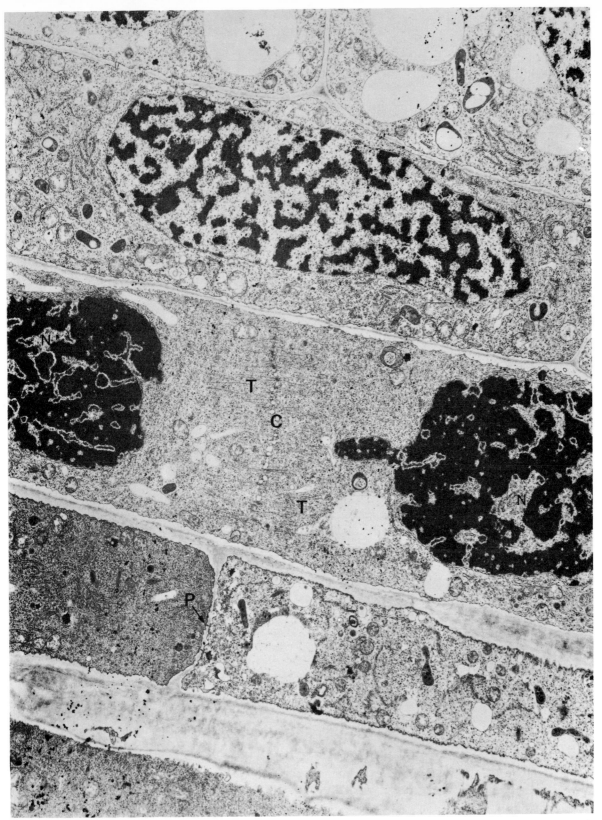

Fig. 10.1   Electron micrograph of wheat coleoptile (×11 000) showing telophase of a cambial cell. Note the large numbers of microtubules (T) between daughter nuclei (N). These probably guide various vesicular components into the developing cell plate (C). Microtubules are most numerous at the margins of the cell plate, while in the central region, where consolidation is almost complete, they are infrequent. The cells in the lower part of the micrograph have plasmodesmata (P) passing through the cross walls. Provided by Dr J. D. Pickett-Heaps, University of Colorado.

base on which the new cell walls are built, is assembled within the phragmoplast. It is formed from small cytoplasmic vesicles about 100 nm in diameter, which collect in a narrow zone approximately mid-way between the two daughter nuclei (Fig. 11.10). The vesicles, which contain polysaccharide, seem to originate from the Golgi bodies which are found around the margins of the phragmoplast and from elements of the endoplasmic reticulum which frequently invade this interzone region. They fuse to form a semi-solid layer between the two new protoplasts. Their membranes give rise to the plasmalemma, while their contents form the cell plate sandwiched between two layers of membrane. The process of plate assembly begins in the mid-zone of the dividing cell and develops outwards. Microtubules are always most numerous in the zone marginal to the extending plate and it is usually assumed that they direct the movement of the vesicles to the plate region either by forming channels or by providing contractile forces which draw the vesicles into position. Individual protoplasmic threads about 400 nm in diameter usually persist in certain regions between the two daughter cells where the vesicular layer has not fused completely (Fig. 10.1), and form tubular connections known as plasmodesmata which we have discussed in an earlier chapter (see p. 9).

Primary and secondary cell walls

Before the cell plate has reached the side walls of the mother cell, the process of wall consolidation usually begins. New material consisting largely of non-cellulosic polysaccharide is deposited from Golgi vesicles as two layers on each side of the plate, which itself persists as the middle lamella between the two young cells. Even at this early stage of development, some cellulose is incorporated into the walls as they usually show a weak but positive birefringence when viewed microscopically under polarizing light. The young walls, however, though they undoubtedly possess great mechanical strength, are not rigid structures, but will continue to grow throughout their full length in order to accommodate any further increases in cell size. A wall is usually recognized as *primary* during all the stages of cell development that accompany increases in cell surface area. After growth stops, the cell wall often becomes thickened by apposition (layering) of material onto the primary wall. During this process, the so-called *secondary cell wall* is built up, which differs both microscopically and chemically from the original primary wall. It is usually composed of a number of distinctive layers, each characterized by the particular orientation of its cellulose microfibrils. This thickened wall gives the cell its final shape and forms the main basis of the plant's mechanical support. It is also of technological importance, forming the main structural component of wood, numerous plant fibres, paper and cork.

It should be noted that a few cells such as collenchyma do continue to grow after the deposition of the apposition layers which usually are characteristic of secondary walls. They cannot, therefore, be regarded as true primary structures.

Isolation

The isolation of cell walls sufficiently free of contaminating cytoplasm to allow a reasonable study of their composition has proved to be a difficult and controversial task. It is not easy, for example, to ensure that all of the cells in a tissue have been broken open, even when the most stringent methods of breakage are employed. Any unopened cells and wall inclusions such as plasmodesmata will, of course, contain cytoplasmic materials trapped within them which will contaminate the final preparation, and even broken walls can probably adsorb protein in a non-specific manner. A number of techniques have been employed to rupture the plant cell, including tissue homogenization, freezing and thawing, shaking with glass beads, crushing in a pressure press and grinding with abrasive materials such as sand. Sonication is often a method of choice for unicellular organisms such as the alga *Chlorella*, but is not usually successful on large tissue pieces.

Once the cells are open, the cytoplasmic contents can be washed out into the homogenizing medium, which is usually aqueous solution of salts. Sometimes this is hypertonic so that the cells are quickly plasmolysed when immersed in the solution, drawing the cytoplasm and its limiting membrane away from the wall. The broken cell walls are then separated from other cell constituents by centrifuging at low speed (usually $500\ g$ or less). They are then usually washed several times to remove contaminating debris, treated with organic solvents to remove lipids, and allowed to dry.

Living walls are hydrated and likely to contain between 60 and 70% of their weight as water. The dry matter consists largely of polysaccharide material, which is itself divided up into a number of subfractions, plus smaller amounts of lignin, protein, lipid, and other compounds such as tannins. Mineral salts are also deposited to varying extents in walls and in certain diatoms the structure is so heavily impregnated with crystalline silica that it remains intact even after the polysaccharide is removed. In the next sections, we shall discuss in some detail, firstly the nature of the polysaccharide and protein portions of the cell wall, and then how the composition of walls can change during growth and development. Finally, we shall deal with the biosynthesis of the individual components.

Cell-wall polysaccharides

The generally accepted model of the primary cell wall is of cellulose fibres embedded in an amorphous mixture of polysaccharides and proteinaceous material. Although this concept may be accurate, it clearly lacks detail, and much of the present work on cell walls is aimed at refining the model and analysing the component polymers and their interactions in greater detail.

The polysaccharides of cell walls are an extremely complicated group of compounds. Historically, they have always been classified in an empirical way on the basis of their solubility characteristics in a number of extracting, inorganic solvents. The actual isolation procedures are often laborious, however, and only rarely do they yield polymers which could be considered homogeneous. A typical procedure is outlined in Fig. 10.2. More recently, specific endoglycosidases (enzymes which cleave internal glycosidic bonds within a polysaccharide) have been used to release fragments from the cell wall in soluble form and to break up particular polymeric associations (Talmadge *et al.*, 1973; Bauer *et al.*, 1973). Such an approach has led to more rapid progress in our understanding of wall composition and to the realization that several different types of polymer may be linked together through covalent bonds.

Fig. 10.2 Flow diagram showing one method for separating a number of cell wall polysaccharide fractions.

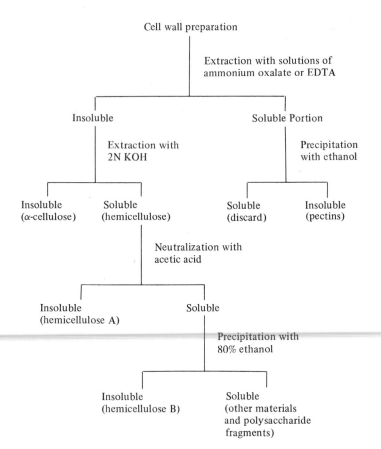

Pectins

The pectins are usually extracted first because they are considered to be the most soluble polysaccharides. A variety of methods have been used including treatment with hot water, with dilute acid or with dilute solutions of ammonium oxalate, EDTA, or other chelating agents which will readily bind magnesium or calcium ions in the cell wall. Those ions are believed to form cross-links between carboxyl groups on adjacent polysaccharide chains, so binding the molecules together into a massive complex. The pectins, which are rich in galacturonic acid, are loosened when these binding cations are removed and rendered soluble. Albersheim and his colleagues (Talmadge *et al.*, 1973) have employed a purified endo-α-1, 4-galacturonidase to hydrolyse a few internal linkages of unesterified galacturonic acid residues within the core of the polysaccharides, a procedure which extracted about 20% of the wall substances. The soluble polymers were then purified further by ion-exchange and gel chromatography and shown to consist of a variety of polymeric types, including acidic molecules rich in galacturonic acid plus more neutral molecules containing large amounts of arabinose and galactose.

The pectins are generally characterized by the presence of galacturonic acid and by their gel-like behaviour. They are typically found in relatively high amounts (up to about 30%) in primary cell walls. Nevertheless, they still remain a loosely defined group of polysaccharides and usually contain L-arabinose, D-galactose and L-rhamnose, plus a variety of minor components, as well as D-galacturonic acid itself (Aspinall, 1970). Older textbooks and reviews have usually suggested that three distinct homopolysaccharides occurred together as a *pectic triad,* consisting of a galactan (polygalactose), an arabinan and a polygalacturonan. The latter was often termed *pectic acid,* a polymer comprised entirely of $\alpha(1 \rightarrow 4)$ linked galacturonic acid residues (Fig. 10.3). This substance was regarded as the most characteristic component of, and frequently synonymous with pectin. Some or all of its uronic acid carboxyls were methyl esterified, depending upon the source of the material. However, more recent attempts to purify such homopolymers using milder extraction procedures and careful separative techniques have generally failed to yield homogeneous material. L-Rhamnose, for example, is frequently found linked to galacturonic acid at intervals along the main chain of the polymer. The polysaccharides based on galacturonic acid are also somewhat unstable, particularly under alkaline conditions and often break down during extraction. Some of the side chains are also readily hydrolyzed at pH 4 and below and can be lost from the molecule as small molecular fragments which are overlooked during analysis. In this way, large heteropolysaccharides can probably be cleaved into smaller components of a more uniform composition. In the intact wall, however, side chains of neutral sugars consisting mainly of arabinose and galactose are attached as branches to the polyuronide backbone, and the pectic substances probably comprise several types of polysaccharide chain varying considerably in composition and size. On the one extreme, there are the molecules that are highly acidic, rich in unesterified galacturonic acid and relatively unbranched. On the other hand, other

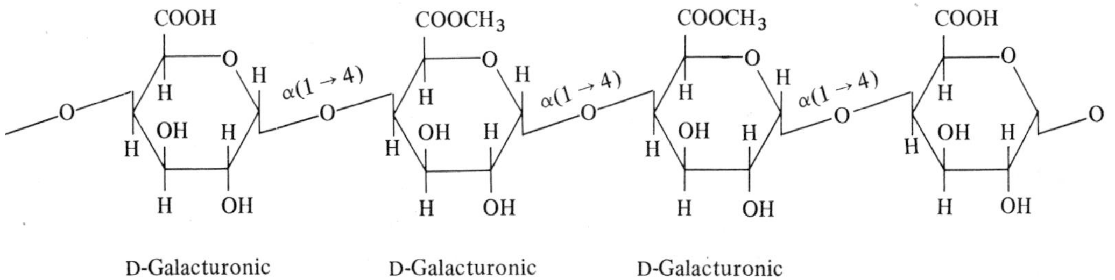

Fig. 10.3 α-D-(1 → 4)-Linked galacturonosyl residues as found in pectin. Some of the carboxyl groups shown are methyl esterified. Rhamnose residues, singly or in groups, are often found interposed between long chains of uninterrupted galacturonic acid residues. Side chains of galactose and arabinose are also usually encountered.

D-Galacturonic acid   D-Galacturonic acid   D-Galacturonic acid types are only weakly acidic with many of their carboxyl groups methyl esterified, and these molecules probably bear large neutral side chains.

The arabans that have been extracted from primary walls and seeds also appear to be highly branched with degrees of polymerization of up to 90 residues. The arabinose units are invariably furanose in ring form and linked to each other through α-glycosidic bonds. In addition, polymers consisting largely of linear chains of $\beta 1 \rightarrow 4$ and $\beta 1 \rightarrow 6$ linked galactosyl units have been isolated from citrus pectin and a variety of secondary walls, and probably also exist in primary walls. However, most of the galactose extracted with the pectins is thought to exist in the form of arabinogalactans, molecules which are highly variable in the ratios of their component sugars and whose structures, for the most part, are poorly understood. Many of these different types of polysaccharide within the pectin group are probably associated together in giant macromolecular complexes by both covalent bonds, hydrogen bonds and calcium cross-bridges between carboxyl groups on adjacent galacturonan chains (Grant *et al.*, 1973).

Hemicellulose

After the pectin has been extracted, the hemicellulose can be removed from the cell wall using solutions of potassium or sodium hydroxide. Polysaccharide may then be fractionally precipitated from this alkaline extract. Neutralization, for example, yields a water-insoluble hemicellulose A, while addition of ethanol or acetone to this neutral solution (usually to 80% by volume) precipitates further amounts of polysaccharide which is referred to as hemicellulose B. The alkali treatment undoubtedly modifies the polysaccharides. The more labile linkages are probably broken and the exposed reducing ends of the chains slowly degraded. Furthermore, the fractions obtained in this way are not pure compounds, but usually consist of a series of polysaccharides of great complexity in terms of the numbers of individual sugar residues present, the nature of the linkages and the extent of the branching. They by no means represent a biochemically homogeneous group. However, we shall here consider three of the more important structural types, namely those based on a central chain of xylose, commonly known as xylans, the glucomannans which, as their name suggests, are made up predominantly of glucose and mannose, and the xyloglucans which are composed largely of glucose and xylose.

The xylans consist of a central chain of $\beta(1 \rightarrow 4)$ linked xylopyranose units which form the core of a molecule (Fig. 10.4). Other sugar residues, particularly L-arabinose, D-glucuronic acid, 4-O-methyl-D-glucuronic acid and D-galactose, may be attached to the main chain usually as single units or occasionally as complex branches (Aspinall, 1970). The total number of monosaccharides present in a molecule (i.e. the degree of polymerization) is usually about 150–200 and in some plants many of the exposed hydroxyl groups of the xylose units are acetylated. The typical xylans of hardwoods have single unit side chains of 4-O-methyl-glucuronic acid joined to about one in every ten

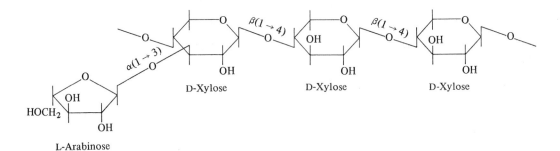

$\beta(1 \rightarrow 4)$ $\beta(1 \rightarrow 4)$

OH   OH   OH

OH   OH   OH

D-Xylose   D-Xylose   D-Xylose

$\alpha(1 \rightarrow 3)$

OH

HOCH$_2$

OH

L-Arabinose

Fig. 10.4   Partial structure of an arabinoxylan of angiosperms.

xylose units. The linkage to the main chain is by $\alpha(1 \rightarrow 2)$ bonds which are difficult to hydrolyse in dilute acid, so that disaccharides, known as aldobiouronic acids containing both a uronic acid and a xylose unit are usually obtained.

Other xylans, typically those of grasses, have L-arabinofuranosyl groups attached by $\alpha(1 \rightarrow 3)$ linkages to the xylose units of the main chain. $\alpha(1 \rightarrow 2)$ 4-O-Methyl-D-glucuronic acid and D-glucuronic acid itself are also usually present on the molecules as single unit side chains.

Another group of hemicelluloses are the mannan–glucomannans. These occur extensively in coniferous woods and are made up of molecules containing variable amounts of both D-glucose and D-mannose, linked by $\beta(1 \rightarrow 4)$ glycosidic bonds. There is probably a random arrangement of these units along the chain, although single unit chains of β-D-galactopyranose may also be encountered.

The most thoroughly studied hemicellulose polymer for primary walls is the xyloglucan from *Acer pseudoplatanus* suspension cultures (Bauer *et al.*, 1973). This polysaccharide resembles the so-called amyloid component of seeds. The backbone of the molecule consists of linear chains of $\beta 1 \rightarrow 4$ glucosyl residues while the side chains are composed of single $\alpha 1 \rightarrow 6$ linked xylose units interspersed with more complicated branches containing xylose, galactose and fucose (Fig. 10.5). Similar xyloglucans are found in primary walls of most dicotyledonous plants, although they do not seem to be present in monocots where an arabinoxylan predominates in the hemicellulose fraction. Both types of molecule can probably form hydrogen bonds with cellulose.

A number of polymers made up entirely of glucose residues have also been recognized as components of the hemicellulose fraction of plant cell walls. One of these, callose, is a $\beta 1 \rightarrow 3$ polyglucan. It is commonly associated with the end plates of phloem sieve tubes and may also be deposited as a wounding response. Other glucans have been identified which contain both $\beta 1 \rightarrow 3$ and $\beta 1 \rightarrow 4$ bonds.

Cellulose

Once the hemicellulose has been removed from the delignified cell walls by alkali, the cellulose is left as an insoluble residue which is largely resistant to hydrolysis by dilute acids but can be dissolved in phosphoric acid or in high concentrations of sulphuric acid. This

Fig. 10.5 Tentative structure for a portion of the xyloglucan of *Acer* cell walls (see Albersheim, 1978). Glc, Xyl, Ara, Gal and Fuc are abbreviations for glucose, xylose, arabinose, galactose and fucose respectively.

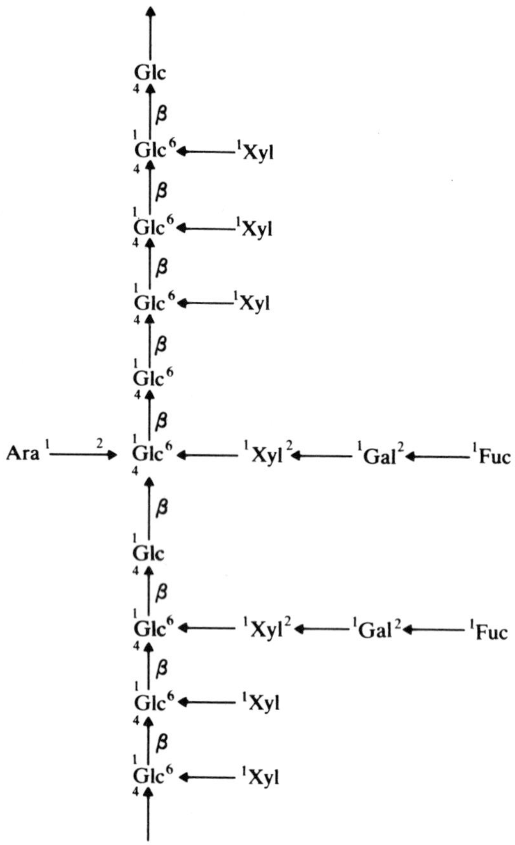

treatment leads to partial depolymerization of the polysaccharide chains, so that after standing and subsequent dilution of the acid, the cellulose can be hydrolysed and the constituent sugars then analysed. Usually, more than 90% of cellulose is made up of glucose, though other sugar units are often present. Whether these are covalently bonded to and are, therefore, essential constituents of the cellulose molecule, or whether they have arisen from contaminating polysaccharide, trapped within the fibrillar matrix, is still not certain.

The glucan portion of extracted cellulose is a linear molecule made up of $\beta(1 \rightarrow 4)$ linked glucopyranose units (Fig. 10.6). The degree of

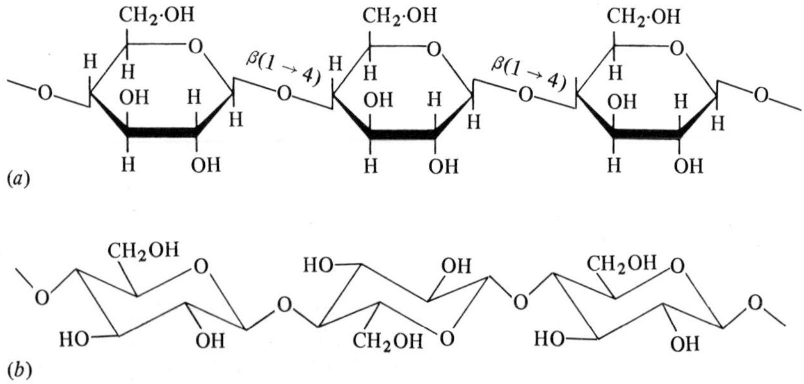

Fig. 10.6 The structure of $\beta(1 \rightarrow 4)$. glucan chains as found in cellulose. In (a) the glucose molecules are shown by the planar convention; in (b) they are shown in the chair configuration.

438

polymerization probably varies from source to source and according to the method used for isolating it. The cellulose from primary walls is probably of lower molecular weight than that from secondary structures, but most authorities estimate a value within the range 2 000–14 000 glucose units for chains of natural cellulose. This corresponds to a chain length of 1–7 μm and a molecular weight approaching 1 million or more. $\beta(1 \rightarrow 4)$ Links give rise to relatively rigid linear molecules which are able to align closely with their neighbours, and the free hydroxyl groups of carbons 2, 3 and 6 probably form hydrogen bonds with neighbouring chains which bind the molecules together. In this way, the glucan chains are thought to be packed in orientated aggregates with a definite structure, known as *microfibrils*.

Where the chains are highly ordered within the microfibril, crystalline regions, usually called micelles, are found which are responsible for the distinctive X-ray pattern given by cellulose and the positive birefringence of cell walls containing the polymer. Microfibrils are particularly resistant to degradation and account for the stability of cellulose towards both microbial and direct chemical attack.

Possibly up to 70 glucan chains are packed together in these microfibrils giving rise to a flattened, thread-like structure approximately 5–8 nm wide and several μm in length (Preston, 1974). They can be viewed by electron microscopy on shadowed sections of the cell wall once the encrusting substances have been removed (Fig. 10.7). They can also be visualized by freeze-etch techniques. Several theories have been proposed to account for the structure of microfibrils. There was, until recently, considerable controversy as to whether adjacent chains were arranged in a antiparallel (i.e. head to tail) or parallel arrangement. However, X-ray diffraction studies on the highly crystalline cellulose from the alga *Valonia ventricosa* have shown that the glucan chains are packed in parallel arrays (Gardner and Blackwell, 1974; Sarko and Muggli, 1974). It seems likely, therefore, that the

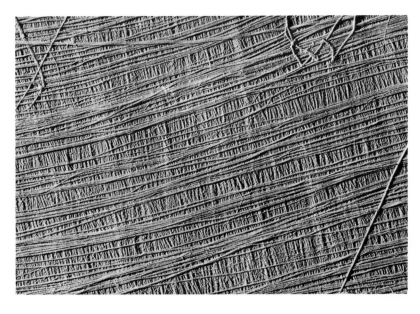

Fig. 10.7 Cellulose lamellae from the side walls of the green alga *Cladophora rupestris* (shadowed Pt/Au) ×19 500. The cellulose microfibrils in two crossed arrays are each 20 nm wide. Provided by Drs Eva Frei and R. D. Preston, University of Leeds.

cellulose molecules within the microfibrils of higher plants have a similar, parallel alignment.

A microfibrillar form as depicted on dehydrated sections shadowed by heavy metals may be misleading. Frequently, they appear to form a closely woven mat, whereas it can be calculated that even in secondary walls they occupy no more than 15% of the total wall volume. The microfibrils are probably separated from their nearest neighbours by 50–100 nm (i.e. at least 2–4 times their own diameter) and the space between them is filled by hydrated, encrusting materials of various kinds.

The mechanical strength of plant cell walls is believed to be due to the presence of the cellulose microfibrils which form a skeletal framework embedded in an amorphous, encrusting matrix. This fibrous material is of great technological value. Cellulose, for example, is the major component of paper, cotton and a number of other fibres, and wood is derived from cellulosic secondary cell walls that have been further strengthened by lignification. Cellulose is in fact present in almost all plant cell walls, and in those species in which it is absent, other polysaccharides often assume the strengthening rôle and may form microfibrils themselves. For example, the cell walls of a number of algae are low in cellulose and the microfibrils are composed of either xylose or mannose. In fungi, chitin is the characteristic strengthening element in the cell wall and this too will form microfibrils. Even in higher plants, the hemicellulosic xylans are believed to form long strands parallel to the direction of the microfibrils (Preston, 1964).

Uniqueness of the cell-wall polysaccharides

Of the polysaccharide fractions isolated from the cell wall, cellulose is the best defined. Because it is fibrillar, it can be recognized on sections as a separate phase of the cell wall. The encrusting polysaccharides, however, are separable only by a series of rather arbitrary procedures. Occasionally, relatively pure polysaccharide types can be isolated from the fractions by gel-filtration, ion-exchange chromatography, fractional precipitation, electrophoresis or by a combination of these techniques. Nevertheless, the distinction between what is pectin and what is hemicellulose is blurred and has prompted many workers to classify the pectin and hemicelluloses together as non-cellulosic polysaccharides. This issue has encouraged much recent debate because of the studies of Albersheim and his colleagues who, at one time, suggested that the entire matrix material of the growing primary wall could be regarded as one giant, covalently linked molecule such that some lengths of polysaccharide acted as bridges between other polymers (Keegstra et al., 1973). In this chapter, however, we have preferred to retain the terms pectin and hemicellulose because we believe that the two fractions are separated and show distinctive changes within the wall according to how the cell develops. Moreover, there is good reason to believe that much of the cohesion between polysaccharides is due to hydrogen bonding. This view is taken into consideration in a later section when we discuss cell wall growth.

What has become increasingly clear in recent years, however, is that, even within polysaccharide fractions of the same general type (the polygalacturonans, for example), there may be considerable variation in the sizes of the molecule, the extent and frequency of the branching, and the type of substituents involved. Furthermore, a wide spectrum of molecular types may be isolated from even a single tissue. Indeed, it is not clear how precisely each polysaccharide has to be assembled in order to be functional, for unlike nucleic acids and proteins, they are molecules of relatively low informational content (see p. 45) and presumably do not have to be built up so precisely. At present, we assume that polysaccharides are not transcribed directly from a 'code' as are proteins, but depend for their assembly on the specificity of the various glycosyl transferase enzymes for the end group of the growing chain. It is also impossible to decide whether observed differences in size, substituents or linkage are due to 'errors' in biosynthesis or to genetically predicted differences between related molecules. Such genetic variation in wall composition may be one explanation for the differences that are encountered in disease resistance between varieties of a plant species. Fungal and bacterial pathogens, for example, have to gain entry to the tissues by penetrating the cell wall of the host. For this purpose they produce a broad spectrum of hydrolytic enzymes which is presumably capable of breaking some but not necessarily all of the glycosidic bonds of the cell-wall polysaccharides. Resistant plants might well be those in which some crucial change in polysaccharide structure has occurred.

Polysaccharide breakdown

As pointed out in the previous section, cell-wall polysaccharides can be broken down by appropriate hydrolytic enzymes. The organisms which produce such enzymes include not only those which are responsible for the destruction of the plant after its death, but also those which have the capacity to infect it during its lifetime. Plants themselves also seem capable of breaking down their own polysaccharide products. End walls, for example, are dissolved during the formation of xylem vessels and phloem sieve tubes. Leaf abscission, fruit ripening, and, in some instances, seed germination are accompanied by whole or partial dissolution of cell walls. Pollen tubes also produce polysaccharide hydrolases and will often descend stylar tissues by dissolving the wall or middle lamellar region between obstructing cells in their path. The enzymes responsible for polysaccharide breakdown, whether of fungal, bacterial or plant origin, are highly specific for particular glycosidic bonds; and while some (exoglycosidases) will attack only terminal non-reducing ends of the polysaccharide, others (endoglycosidases) are more catholic and will hydrolyse bonds within the molecule. Cellulases are highly specific $\beta(1\rightarrow4)$ endoglucosidases, though they have rarely been purified. By contrast, the 'pectinases' and 'hemicellulases' of commercial origin are invariably as complex a mixture as the polysaccharides they attack and contain a myriad of enzymes.

However, by careful fractionation, it is often possible to isolate pure enzymes from such a mixture.

Lignin

Lignin is always found in cell walls closely associated with cellulose or some other carbohydrate. Although this is generally believed to be a physical association, it is difficult to remove all the lignin by mild extraction procedures and some covalent linkage to sugars may exist. Lignin is often estimated by weight after extracting the polysaccharide from the cell wall in concentrated sulphuric acid, in which the lignin is itself insoluble. But it can be removed from wood, a process very necessary in paper making, by extraction with solutions of hot sodium bisulphite or with alkaline sulphate-sulphide. The so-called protolignin, an immature, presumably low-polymeric form, can be solubilized by ethanol, or other organic solvents such as dioxane. A third method that has been used with some success in analytical work is to digest away the polysaccharide, preferentially by means of a brown-rot fungus, such as *Lentinus lepideus*. By such means, it has been possible to purify some forms of lignin and identify the primary building units. If non-oxidative techniques are employed to degrade the lignin, *p*-hydroxyphenylpropane units of various kinds are recovered (Fig. 10.8). These are usually methoxylated on the aromatic ring.

Fig. 10.8   Building blocks of the lignin molecule. Lignin is a highly cross-linked polymer based on the phenylpropane units illustrated. Type (c) are relatively rare.

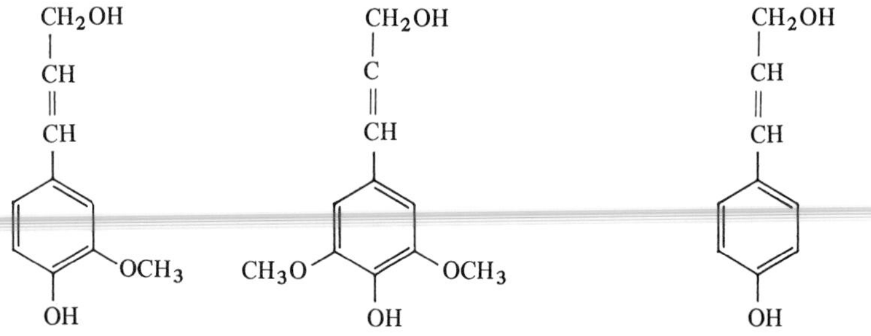

(a) Coniferyl alcohol (Guaiacyl)   (b) Sinapyl alcohol (Syringyl)   (c) *p*-Hydroxycinnamyl alcohol

Guaiacyl units are found in both soft and hard woods, while syringyl units are confined to hard woods. If lignin is oxidized with nitrobenzene, however, a mixture of aromatic aldehydes is produced (Fig. 10.9).

Fig. 10.9   Typical aromatic aldehydes released from lignin after treatment with nitrobenzene.

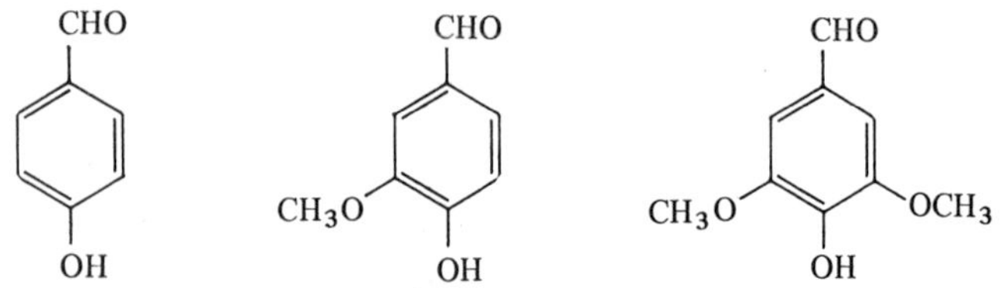

p-Hydroxybenz-aldehyde   Vanillin   Syring-aldehyde

The whole lignin molecule is made up of many phenylpropanoid units associated in a complex cross-linked molecule of uncertain structure. The polycondensations probably occur randomly in the presence of a peroxidase in the cell wall. An example showing the association of a number of units into a complex polymer is shown in Fig. 10.10. The double bonds of the propane side chains are easily oxidized, and both inter-chain and inter-ring condensation can occur. The linkages are variable and as many as 100 units may be accommodated in a single molecule. It is this lack of a consistent type of linkage between units which has made lignin so difficult to study.

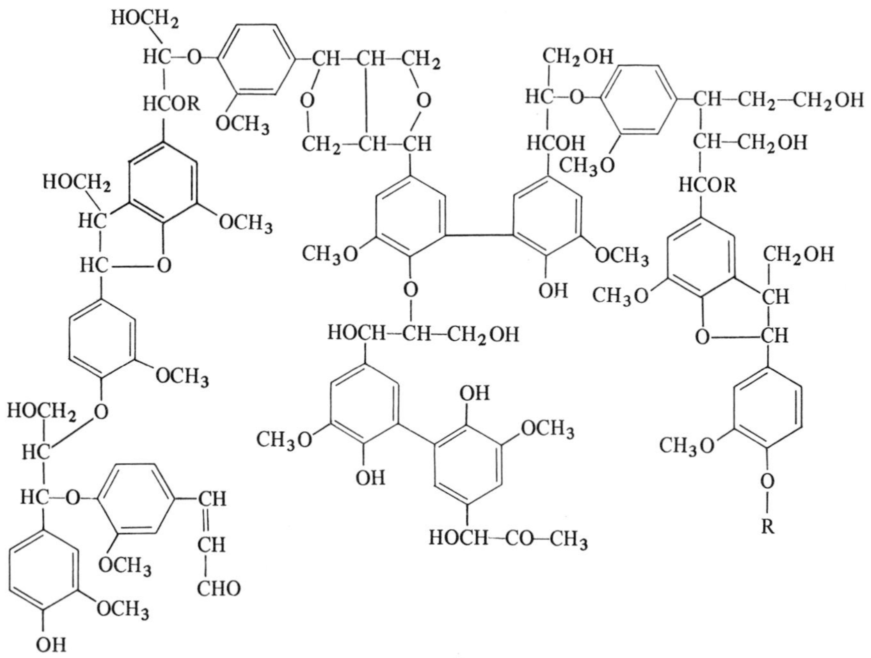

Fig. 10.10   Structural elements of part of a molecule of guaiacyl lignin. R represents the rest of lignin molecule. This is not a structural formula in the true sense, but rather a representation of linkages believed to exist in spruce lignin, and should not be interpreted quantitatively. Reproduced from Neish, 1965.

Lignification appears to fulfil two main functions. It cements and anchors the cellulose fibrils together and, because of its hardness, it stiffens the wall and prevents chemical and physical damage to the cell. Although gross lignification can be demonstrated histochemically (using phloroglucinol, for example), the extent of minor lignification in primary or immature secondary walls is probably underestimated, and the process may be more widespread than hitherto suspected.

Cell-wall protein

Young cell walls invariably contain a small but significant amount of residual protein material, even after meticulous attempts to remove any contaminating cytoplasmic material from the preparations. In dicotyledons, this protein is unusually rich in the uncommon imino-acid, hydroxyproline (Fig. 10.11) and it is now generally assumed that a significant portion of the total hydroxyproline of the cell is confined to the wall (Lamport, 1970). Hydroxyproline, however, is not a

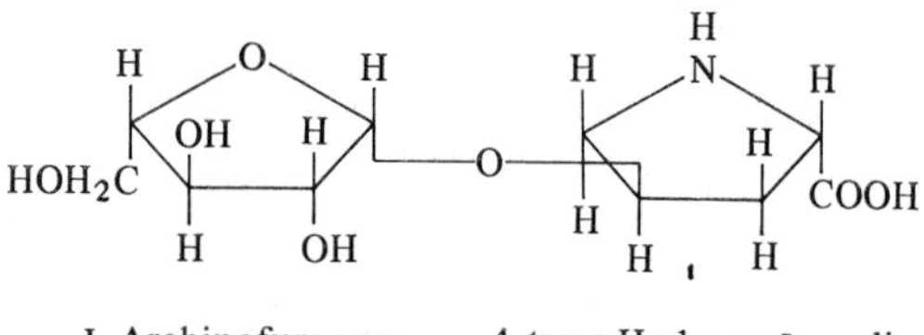

Fig. 10.11   Arabinosylhydroxyproline.
This structure shows the likely linkage
between the arabinose of the
carbohydrate portion of the cell-wall
glycoprotein and the hydroxyproline on
the peptide portion. The
stereochemistry of the linkage is
believed to be in the β-configuration.

L-Arabinofuranose   4-*trans*-Hydroxy-L-proline

widespread component of proteins in general, and in animals it is found only in collagen, the protein of the connective tissues. Consequently, there has been considerable speculation as to whether the protein of the wall resembles collagen in its physical properties. There is, however, little evidence that it forms the fibrous complexes typical of collagen and, except for its unique content of hydroxyproline, it shows little chemical similarity to the animal protein.

The hydroxyproline-rich protein has not been purified in intact form from cell walls because of its insolubility. However, fragments of the molecule have been isolated from walls of suspension cultured sycamore (*Acer pseudoplatanus*) and tomato (*Lycopersicon esculentum*) cells using either alkaline hydrolysis or digestion with a crude mixture of proteases and polysaccharide-degrading enzymes (Lamport, 1970; Talmadge *et al.*, 1973). It was found that all of the hydroxyproline residues in these fragments were linked covalently to arabinose residues. In general, the arabinose side chains were 4 residues in length and, attached via a β-O-glycosidic bond to the 4-position of hydroxyproline (Akiyama and Kato, 1977) (Fig. 10.11). In tomato and carrot cells, it has also been established that galactosyl residues are attached to the peptide through O-glycosidic linkages to the amino acid serine (Lamport, Katona and Roerig, 1973). A recent model for a portion of the hydroxyproline-rich, structural glycoprotein of primary cell walls is shown in Fig. 10.12. It should be emphasized here that the linkages between the carbohydrate chains and the peptide are unique to the plant kingdom. They do not resemble those found in collagen, where a glucose–galactose disaccharide is linked to hydroxylysine, or those in proteoglycans where the bond involves xylose and serine. Further, although Lamport (1970) has named this molecule 'extensin', there is, as yet, no evidence that it is involved in controlling elongation of the plant cell wall. Moreover, all available evidence to date indicates that the glycoprotein is not *covalently* attached to any of the other cell wall polymers (McNeil, Darvill and Albersheim, 1978).

A hydroxyproline-rich glycoprotein, distinct from that of the cell wall is secreted into the culture medium of sycamore suspension cells (Pope, 1977). Presumably, in the intact plant this would also be deposited in the polysaccharide matrix between the cells, rather than being dispersed into the medium in soluble form. This molecule has large carbohydrate side chains, resembling an arabinogalactan, which are attached to hydroxyproline residues through an arabinose residue. It seems likely that the arabinose-hydroxyproline linkage group might be of wide occurrence in the plant kingdom. A parallel can be drawn with the N-acetylglucosamine-asparagine linkage of mammalian

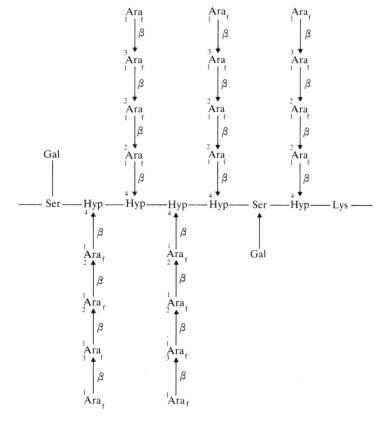

Fig. 10.12 Tentative structure of a portion of the hydroxyproline-rich structural glycoprotein of primary cell walls of *Acer pseudoplatanus* cultures. Ara_f, Hyp, Gal, Ser, Lys are abbreviations for L-arabinofuranose, hydroxyproline, galactose, serine and lysine respectively.

glycoproteins which is encountered on many hundreds of distinct molecules.

Although the enzymes involved in polysaccharide synthesis do not appear to be associated with the wall, a number of other enzyme activities can often be found in carefully washed and sedimented cell-wall preparations. These include pectin methylesterase, ascorbic acid oxidase, acid phosphatase, invertase and peroxidase. It should be pointed out that these activities are not confined to the wall, but they do occur there in significant amounts. They may be structural proteins and form an integral part of the wall and, in this case, their enzyme activities may be incidental to their true rôle. Alternatively, they could be useful in releasing nutrients from the soil or in preventing infection. Finally, of course, they could be there by chance, merely occluded in the wall during its deposition.

Chemical and physical changes in the primary cell walls during growth and development

The cell wall varies considerably in structural complexity from cell type to cell type. Furthermore, the primary wall has also to fill apparently paradoxical requirements. It must be sufficiently extensible for rapid growth and yet strong enough to provide support. It is the purpose of the following sections to investigate briefly some of the more general changes that occur in cell walls from the moment they form after cell division to the stage when they start to develop a thickened secondary wall, after growth has stopped.

Gross composition of primary walls

Wall composition can vary considerably from cell to cell and from species to species. It is difficult to draw meaningful conclusions about the relationship between composition and function from gross analyses of whole tissues or organs. Table 10.1 lists analytical data on walls from three well investigated, though not necessarily typical, primary tissues. In two of these, the content of pectin is rather low, while hemicellulose which is mainly in the form of complex xylans, comprises over 30% of the cell wall. Cellulose, which is generally assumed to be the main strengthening element of the wall, comprises about one-half of the total polysaccharide.

Table 10.1 Carbohydrate and lignin composition of the primary cell walls of growing seedling tissues.

	Percent of total dry cell wall material		
Component	*Oat coleoptile*	*Wheat coleoptile*	*Sunflower hypocotyl*
Cellulose	42	36	38
Pectic substances	8	13	46
Hemicellulose	38	30	8
Lignin	—	—	8

Adapted from Bonner (1950)

More recent studies have emphasized the use of suspension cultured cells in order to provide a more uniform preparation of cell wall material in, what is often claimed to be, a totally undifferentiated state. It remains questionable as to whether these can necessarily be employed to draw meaningful conclusions about the walls of cells in intact tissues whose growth properties and tissue interactions are clearly different. Nevertheless, the wall of suspension-cultured sycamore cells contains polysaccharides similar to those extracted from a variety of whole plant organs, and may serve as a useful model for young primary walls. The polymer composition of such cell wall is listed in Table 10.2. Note, in particular, the very high content of pectin and glycoprotein and the relatively low amount of cellulose. Furthermore, the primary walls of several, unrelated dicotyledonous plant cells in culture have similar compositions to that of sycamore. On the other hand, the walls of cultured cells from a number of grass species contain little hydroxyproline (Burke *et al.*, 1974). This suggests that there may be fundamental differences in the composition and possible organization of primary walls of monocotyledonous and dicotyledonous plant species.

Table 10.2   Polymer composition of the walls of suspension-cultured sycamore (*Acer pseudoplatanus*) cells*

	Wall component	Weight % of wall
A.	*Pectic polysaccharides*	*34*
	Rhamnogalacturonans	10
	Homogalacturonan	6
	Arabinan	9
	Galactan and arabinogalactan	9
B.	*Hemicelluloses*	*24*
	Xyloglucan	19
	Glucuronarabinoxylan	5
C.	*Cellulose*	*23*
D.	*Hydroxyproline-rich glycoprotein*	*19*

* From McNeil *et al.*, 1978.

The lignin content of primary walls is usually low, often so low in fact that it is easily overlooked during analysis. Gross lignification is confined to particular tissues, such as elements of xylem and phloem, and seems to occur as a widespread process only after growth has stopped during secondary cell-wall formation. Indeed, lignification may be a means whereby continued growth of the cell wall is prevented.

The protein content of primary walls is a highly controversial subject. It now seems clear that significant though relatively small amounts (usually less than 10% of the total dry weight) are recovered from most primary walls, and that much of this protein may be linked to carbohydrate. Cell walls of tissue cultures are relatively rich in protein (15% or more of the dry weight), but these walls are anomalous in many respects and should not be considered typically representative of primary walls.

Changes in cell-wall composition associated with growth

The cell wall has received special attention in relationship to growth and development for a number of reasons. It is, for example, a conspicuous feature of the cell, and its polysaccharides are major cellular products. Furthermore, the wall appears to be the structure which immediately restricts and, hence, controls cell extension.

Many of the observations pertaining to the development of primary cell walls have been made on the growing tips of stems and roots. These organs are particularly useful in this respect, because the meristem is confined to a relatively narrow zone in the apex, so that the distance of any cell from this apical region relates directly to its age and hence gives an approximate measure of its stage of development. The growth of these cells is also distributed evenly along their lengths and not confined to the tips, as in pollen tubes and root hairs.

In the case of the shoot and roots, growth seems to occur in three stages (Clowes and Juniper, 1969). There is an early phase of radial enlargement, followed by a transition period in which both radial enlargement and elongation occur (Fig. 1.1). Finally, there is a stage of rapid cell elongation. During this sequence of events, the small isodiametric cells at the very apex, which may measure only 8 μm by 8 μm or less, can reach a size of 30 μm $\times$ 200 μm or more in the cortex and pith of corn roots by the time they have completed their growth 7 mm from the tip. Growth in other tissues shows essentially similar features, although the final size of the cells may be less. The cell may divide either longitudinally or transversely at any of these stages, although meristematic activity is generally confined to the tip region during which the first two growth stages described above occur.

It has become clear by using histological procedures at the level of both the electron and light microscope, by analysing thin segments of tissue cut at increasing distances along the axis of such organs as roots (Jensen and Ashton, 1960) and by using specific autoradiographic techniques for following the incorporation of particular sugar units (Roberts and Butt, 1967, 1968), that wall composition changes as cells grow and mature. It is also clear that different tissues even within the same organ can differ markedly in the composition as well as the architecture of their walls.

However, some consistent trends are evident as cells mature. Generally, very young cell walls are richer in pectin and lower in cellulose than cells in the later stages of growth. The cell plate or middle lamella region, which is laid down first between daughter cells after cell division, contains little or no cellulose yet stains readily for polyuronic acid materials. The young primary walls deposited on the surface of this cell plate have also only a sparse number of microfibrils and the bulk of the wall is made up of amorphous, non-cellulosic materials. As growth proceeds, more material is incorporated into the wall to maintain its structure, although complete synchrony is not necessarily maintained. Indeed there is good evidence that during rapid growth the wall becomes thinner as elongation outstrips biosynthesis (Roberts and Butt, 1968). During this stage, the content of cellulose increases while the synthesis of pectins markedly declines.

The development of a new cell wall has also been investigated as it reforms around isolated protoplasts prepared from intact cells. The protoplasts are prepared by dissolving existing walls by means of polysaccharidases in an isotonic incubation medium (see Ch. 13). The protoplasts, after washing, are then placed in an appropriate medium and allowed to form new walls. Hanke and Northcote (1974) have shown that during the initial phase of wall regeneration soybean protoplasts excrete soluble pectins into the growth medium. A mesh-work of cellulose microfibrils then develops which seems to trap the outwardly migrating matrix material. Again these results suggest that a cell's earliest response in the generation of a new wall is the production of matrix polysaccharides, followed by the generation of a woven mat of cellulose.

The orientation of cellulose microfibrils within growing longitudinal walls has been observed to change even after they have been deposited. Those nearest the plasmalemma are deposited as a loose mat and have a more or less transverse orientation (i.e. a flatter patch) than those embedded deeper in the wall (Figs. 10.13 and 10.14). They seem

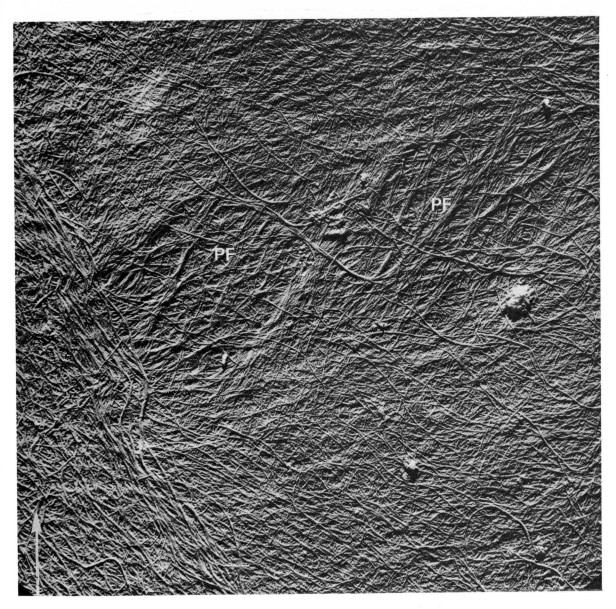

Fig. 10.13   Electron micrograph of the primary wall of a parenchyma cell from wheat (*Triticum vulgare*) coleoptile, 10 mm from the tip (×18 800). The microfibrils are in the form of a loose mat but with a predominantly transverse orientation, except in the lower left which represents an older part of the wall. Here the microfibrils are arranged in parallel and have more longitudinal orientation. Two pit fields (PF) are visible within which are localized numbers of plasmodesmata. The arrow indicates the long axis of the cell. Provided by Dr R. D. Preston, University of Leeds.

to act like the hoops of a barrel preventing the cell from increasing in girth, but allowing longitudinal stretching. As the cell grows, more material is deposited on the surface of this network and the orientation of the older microfibrils becomes more longitudinal in response to the stretching of the wall (Fig. 10.15).

The qualitative changes in the hemicellulose fraction appear to be very complicated and too difficult to interpret at present. Many different monosaccharide units and several polysaccharides are present in this fraction, and although it is clear that hemicellulose contributes significantly both to very young and to fully grown primary walls, the precise developmental patterns in which the individual components are laid down are far from understood.

Fig. 10.14 Wall structure of *Valonia ventricosa*. A–C (all ×12 000) show the wall of the spherical aplanospore showing the increasing degree of order as it grows. Note that the microfibrils become more parallel and change their orientation. D (×18 000) shows the secondary wall of a mature vesicle in surface view showing three lamellae. Reproduced from Steward (1968).

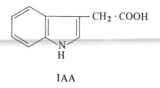

IAA

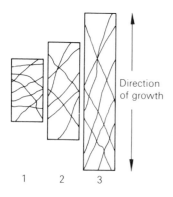

Direction of growth

1   2   3

Fig. 10.15 Diagram illustrating the change in orientation of the microfibrils during three stages of elongation of primary wall. After Roelofsen (1959).

Control of cell wall growth. Cleland (1977) has pointed out that if the stem cells of trees behaved like a liver cell and enlarged to only 15–20 μm in length, redwoods would reach a maximum height of less than 2 m. Clearly the process of cell enlargement is one vital to the plant. It is a process which is highly complex and controlled at least in part by hormones such as auxins, gibberellins and cytokinins as well as by several internal factors including water conductivity, the difference in osmotic potential between the aqueous surroundings and the cell, the yield turgor (the value of turgor pressure that must be exceeded before any wall extension can take place) and wall extensibility. It is now believed that auxins, such as indole acetic acid (IAA), have their major effect on cell growth by influencing wall extensibility by a series of as yet unidentified, wall loosening events. This process will now be considered in further detail, even though it must be emphasized that the generation of an osmoticum and the water status of the cell may, in the long term, be the most important factors controlling cell extension.

Auxin and wall loosening. Auxin does not appear to act on the cell wall directly. Rather it is thought to act on the cell membrane or the cytoplasm, one result of which is that some 'factor' is produced, to which the wall then responds. The best candidate for the wall loosening factor so far proposed is the proton, H^+. Thus, acidic conditions promote wall loosening and growth in a variety of stem tissues (Fig. 10.16). Moreover, auxin-sensitive tissues excrete protons in response to auxin, and the rate and onset of proton excretion correlate with the rate and onset of wall extension (Fig. 10.17).

Let us assume therefore that chemical bonds between polymers in the wall become broken in response to the local excretion of protons. Unfortuntely, we still do not know which linkages or, for that matter which macromolecules, are involved.

There appears to be, as we discussed earlier, four major types of polymer in the primary wall: cellulose, hemicellulose, pectic polysaccharides and glycoprotein. The pattern of organization which is

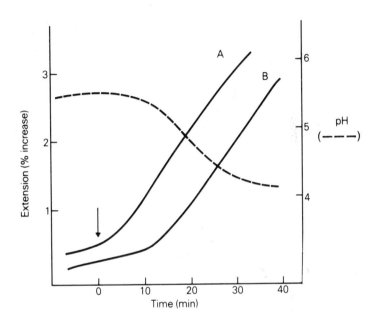

Fig. 10.16 Kinetics of elongation growth in response to H$^+$ (A) or auxin (B). The growth curve A was obtained in 10 mM citrate buffer (pH 3·0) without hormones. The auxin concentration in B was 10 μM. The broken line shows the kinetics of H$^+$ excretion in presence of 10 μM auxin. Data is compiled from Rayle and Cleland (1972) and Cleland (1976). Note the delay in the growth response to auxin and the correlation with H$^+$ excretion. Arrow indicates the point of addition.

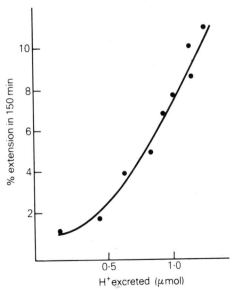

Fig. 10.17 Correlation between growth rate and H$^+$ excretion for *Avena* coleoptiles given varying amounts of auxin (from Cleland, 1975).

beginning to emerge is that the more rigid, linear hemicellulose molecules (the glucoxylans of dicots and the arabinoxylans of monocots) form tight, hydrogen-bonded associations with the cellulose microfibrils and with each other (McNeil et al., 1978). These coated fibrils are then thought to be interconnected by the pectic polysaccharides. The latter substances are probably linked together in a complex network within the hydrated matrix of the wall by both covalent (glycosidic) and non-covalent bonds. The extent of covalent attachment between the hemicelluloses and the pectins is unknown. There is, in addition, no evidence which demonstrates a covalent linkage between the glycoprotein and the polysaccharides. Nevertheless, because of its very firm association with the wall, it is tempting to speculate that strong, non-covalent associations hold the glycoprotein to the polysaccharides. A number of workers (McNeil et al., 1978; Kauss, 1977) have speculated that these glycoproteins have carbohydrate binding properties (ie. they are *lectins*), with high affinities for specific glycosyl groups. Provided such molecules are multivalent (i.e. possessing more than one binding site) they could promote efficient cross-linking between adjacent polysaccharides chains. Interestingly, a number of lectins extracted from plants have been shown to be rich in hydroxyproline and to possess arabinose and galactose in their carbohydrate groups (Allen and Neuberger, 1973; Anderson et al., 1977), thus suggesting they are of a structural type similar to the cell wall glycoprotein.

Loosening of the cell wall structure by H^+ sufficient to allow osmotic extension of the cell and the accompanying reorientation of the wall microfibrils could possibly result from a weakening of any of the above type of interactions. It is possible that a hydrolytic enzyme with a sharp pH profile, such that it is inactive at pH 6·0 but fully active at pH 4·0, cleaves some essential link in a wall polysaccharide. As yet, no such enzyme obtained from any source has been shown to promote wall loosening when added back to wall preparations under acidic conditions. Lowering of pH is also expected to weaken certain ionic interactions, such as those involving carboxyl or other groups with pKa's between 3 and 6, to influence hydrogen bonding, and to disrupt certain types of hydrophobic association. Because of this, the conformation of both proteins and polysaccharides might be anticipated to alter significantly as the pH falls. Such physical changes possibly form the basis of auxin-induced wall loosening, events which in turn allow the cell to expand. It should be emphasized that this ability of stem tissues to grow in length in the presence of acid, IAA, or synthetic growth substances such as 2,4 dichlorophenoxy acetic acid (2,4-D), particularly in the absence of the carbohydrate source, is relatively short-lived. In part, this is because the tissues are dying and presumably losing their selective permeability properties. However, it would also seem that the introduction of new wall material is also required for continued stretching of the wall.

However it should be noted that, although there is much evidence in support of the acid-growth theory, not all reported observations are consistent with the concept. Particular difficulties are encountered with respect to the timing of the pH change and short-term growth response, and the means by which the pH is lowered. The latter could result from a number of processes including proton pumping, OH^- influx, and CO_2 from respiration. The evidence, both for and against, has been carefully reviewed by Penny and Penny (1978) and Zeroni and Hall (1980).

The reader should also be aware that not all tissues enlarge by means of auxin-induced H^+ excretion. The upper nodes of mature grass stems are insensitive to auxin but grow in response to gibberellins (Adams, Kaufman and Ikuma, 1973), while the alga *Nitella* flourishes and grows at alkaline pH.

Secondary Cells

Large numbers of analyses have been performed on different types of woods and mature grasses and these have provided much of the information on the composition of plant cell walls after extensive secondary thickening has occurred and after the wall has developed to its maximum strength (Table 10.3). It can be seen that the percentage contents of the cellulose and lignin are much higher in secondary than in primary walls and that grasses have a higher percentage of hemicellulose than do woods. The content of pectin is very low in both tissues. The hemicelluloses are mainly of the acidic xylan type in the grasses, whereas considerable amounts of glucomannans are also found in the trees, particularly in the gymnosperms.

Table 10.3 The cell-wall composition of mature tissues from two grasses and two species of tree

	% of dry tissue			
	Wheat straw	*Oat straw*	*Douglas fir*	*Beech*
Cellulose	39	44	57	44
Hemicellulose	32	32	14	24
Pectin	1	1	–	–
Lignin	17	19	28	22

Adapted from Browning (1963) and Bonner (1950)

Changes in composition during development

The crudest way in which cell-wall composition has been measured during the transition from primary to secondary cell wall growth has been to analyse whole organs of increasing developmental age. More refined observations have been made on the maturation of woods. In these experiments, plugs of tissues are removed from living trees and dissected into a number of portions, e.g. the youngest cambial elements, the recently matured sap wood, found just inside the cambial ring, and the heartwood at the centre of the trunk. As the walls thicken, the average weight of each cell increases due to the deposition of α-cellulose, lignin and hemicellulose (Fig. 10.18). The pectins remain in about constant amounts during sapwood formation, but decline as a percentage of the total wall content. The most striking increases are in lignin. The amount found in sapwood cells of *Acer*, for example, is as much as ninety-fold greater than in the cambial cells from which they originate. Lignification seems to spread into the layers of the secondary wall from the primary region, where it is always most concentrated, and cements the whole structure together as a rigid complex.

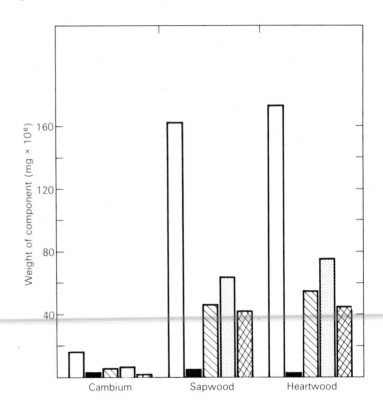

Fig. 10.18   Histogram showing changes in weight of each cell component at three stages of cambial development. Weight of total ethanolbenzene extracted cell (□), of pectin (■), of hemicellulose (▨), of α-cellulose (▧) and of lignin (▦) are shown during cambial to heartwood development. Adapted from Thornber and Northcote 1961(a).

Lignin probably replaces much of the space originally occupied by water, converting the matrix phase from a viscous gel to relatively hard, inelastic cement in which the cellulose microfibrils, with their high tensile strength, are embedded.

Organization of the secondary wall

The secondary wall is usually built up as a series of layers on top of the primary wall. Occasionally this wall may be so thickened that it occupies most of the lumen of the cell. The protoplasmic contents often disappear as the wall thickens and as the cell becomes more impermeable to water and other nutrients.

The arrangement of secondary walls is often complex but in woods it usually exists as three layers, referred to as the S_1, S_2 and S_3 layers

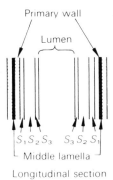

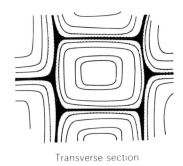

Primary wall

Lumen

$S_1 S_2 S_3$ $S_3 S_2 S_1$

Middle lamella

Longitudinal section

Transverse section

Cell lumen

S_3

S_2

S_1

Primary wall

Fig. 10.20   The orientation of the microfibrils in the secondary wall of cells from flax fibres.

(Fig. 10.19), a nomenclature which was introduced by Kerr and Bailey from microscopic observations of woods and fibres. The layers can usually be distinguished by the particular orientation of the cellulose microfibrils. This is illustrated in Fig. 10.20, which shows the approximate arrangement of microfibrils in the S_1, S_2 and S_3 layers of phloem fibres from flax. In some cells, such as primary xylem, the secondary thickening is laid down in a reticulate or spiral manner, and not evenly over the entire wall. In these instances, the cellulose microfibrils follow the course of the thickened band. Other secondary walls develop elaborate bordered pits which are thin-walled regions through which liquid can flow. Each pit develops from a pit field, a region of primary cell wall originally occupied by clusters of plasmodesmata (see Fig. 10.13), while the pit borders are overgrowths often characteristic of the tissue or species.

The biosynthesis of the macro-molecules of the cell wall

When radioactive glucose is supplied to growing plant organs, polysaccharides soon become labelled, while protein, lipid or other macromolecules acquire radioactivity much more slowly. If we make the assumption that glucose is the natural starting point for most biochemical interconversions in the cell, then it would appear that the pathway to polysaccharides is very direct. The glucose is also incorporated into cell-wall hexose and uronic acid with only a negligible amount of rearrangement of its carbon skeleton. The introduction of radioactivity into cellulose and pectin probably occurs, therefore, by a fairly direct route without label passing through intermediates other than hexose derivatives.

The formation of pentosyl units, however, involves the loss of carbon atom 6 of glucose as carbon dioxide, though carbons 1 to 5 are retained as an intact unit. This contrasts with the direct oxidation pathway of glucose (the pentose phosphate pathway) in which there is a loss of carbon atom 1 from glucose in the formation of pentose phosphate (see Ch. 6). These labelling patterns can now be explained in terms of well defined biochemical pathways (see Nikaido and Hassid, 1971; Hassid, 1966).

The formation of glycosidic bonds

Most polysaccharides are made up entirely of sugars linked *O*-glycosidically (see p. 56). The standard negative free energy of hydrolysis of these bonds is relatively high, varying from around 8-21 kJ. Thus polysaccharides or other glycosides cannot be synthesized by spontaneous condensation of free monosaccharides and are invariably formed by transglycosylation reactions from 'activated' donors, in which the energy of the glycosidic linkage of the donor compound is partially conserved in the product. The donor compounds involved in cell-wall polysaccharide biosynthesis are nucleoside diphosphate sugar derivatives, the best known of which is uridine diphosphate glucose (see p. 115), a compound first described in a yeast by LeLoir in 1951 (see Hassid, 1966).

A transglycosylation proceeds by a general reaction of the following kind:

Donor—O—sugar + H—O—acceptor $\longrightarrow$ sugar—O—acceptor + donor

The net change in free energy of the overall reaction and, of course, the concentration of the reactants will determine the position of the equilibrium, and only those reactions in which a considerable amount of free energy is liberated can be expected to proceed towards completion. For this reason, UDP-glucose and other nucleoside diphosphate sugars are excellent glycosyl donors for polysaccharide biosynthesis because the standard negative free energy of hydrolysis of their glycosidic bond is high ($-31 \cdot 8$ kJ) compared with that of the glycosidic bond they give rise to in the polysaccharide product.

Occurrence and biosynthesis of nucleoside diphosphate sugars

Following the discovery of UDP-glucose, a large number of other uridine diphosphate nucleosides bearing different glycosyl groups were isolated from plants, animals and microorganisms. In addition, nucleoside diphosphate sugars with bases other than uracil were detected. A list of the more common members of this group of compounds is given in Table 10.4. Although widespread, they are

Table 10.4 Nucleoside diphosphate sugars commonly encountered in plants

Uridine series	*With bases other than uracil*
UDP-D-Glucose	GDP-D-Glucose
UDP-D-Galactose	ADP-D-Glucose
UDP-L-Rhamnose	GDP-D-Mannose
UDP-D-Glucuronic acid	GDP-L-Fucose
UDP-D-Galacturonic acid	GDP-D-Mannuronic acid
UDP-*N*-Acetyl-D-glucosamine	
UDP-D-Xylose	
UDP-L-Arabinose	
UDP-D-Apiose	

often found in only trace amounts. However, because they are believed to be involved in polysaccharide biosynthesis and are therefore important precursor compounds, their rate of turnover is probably high. In plants, derivatives of uridine are the most widespread and most of the later discussion on interconversion of nucleoside diphosphate sugars will be limited to this group.

There are three main ways in which nucleoside diphosphate sugars are formed in plants. The first is by *nucleotidyl transfer* from a nucleoside triphosphate to an aldose 1-phosphate by the following type of general reaction. The enzymes are commonly known as *pyrophosphorylases*.

Nucleoside triphosphate $\rightleftharpoons$ nucleoside diphosphate sugar
+ +
sugar 1-phosphate pyrophosphate

The inorganic pyrophosphate arises from the two terminal phosphates of the nucleoside triphosphate; the glycosidic linkage is therefore not broken, although it assumes a higher negative free energy of hydrolysis. Quantitatively, the most important of these reactions and the one studied in most detail is that by which UDP-glucose is formed:

$$\text{UTP} + \alpha\text{-D-glucose 1-phosphate} \underset{}{\overset{Mg^{++}}{\rightleftharpoons}} \text{UDP-}\alpha\text{-D-glucose} + \text{pyrophosphate}$$

The enzyme responsible is UDPG pyrophosphorylase, which is known properly as α-D-glucose 1-phosphate uridylyl transferase. This reaction is freely reversible, with an equilibrium constant of about 0·35. In practice, it might be rendered essentially irreversible by the removal of pyrophosphate as soon as it is formed by the action of inorganic pyrophosphatases, which are widespread in plants. In fact, UDP-glucose, the product, inhibits the reaction as it accumulates and so controls its own formation (Hopper and Dickinson, 1972).

Many nucleoside diphosphate sugars containing different bases and different sugar moieties can be synthesized from their aldose 1-phosphates by similar reactions in plants. Some of the more important are listed in Table 10.5.

Table 10.5 Some pyrophosphorylase activities demonstrated in plants

Name of enzyme	Reaction
UDP-Glucose pyrophosphorylase	D-Glucose 1-P + UTP → UDP-glucose + PP_i
GDP-Glucose pyrophosphorylase	D-Glucose 1-P + GTP → GDP-glucose + PP_i
ADP-Glucose pyrophosphorylase	D-Glucose 1-P + ATP → ADP-glucose + PP_i
UDP-Galactose pyrophosphorylase	D-Galactose 1-P + UTP → UDP-galactose + PP_i
UDP-Glucuronic acid pyrophosphorylase	D-Glucuronic acid 1-P + UTP → UDP-glucuronic acid + PP_i
UDP-Galacturonic acid pyrophosphorylase	D-Galacturonic acid 1-P + UDP → UDP-galacturonic acid + PP_i
UDP-Arabinose pyrophosphorylase	L-Arabinose 1-P + UTP → UDP-arabinose + PP_i
GDP-Mannose pyrophosphorylase	D-Mannose 1-P + GTP → GDP-mannose + PP_i
UDP-N-Acetylglucosamine pyrophosphorylase	N-Acetyl-D-glucosamine 1-P + UTP → UDP-N-acetylglucosamine + PP_i

UDP-Glucose can be formed from sucrose, the main transport sugar in plants, by a reaction catalysed by the enzyme, sucrose synthetase.

Sucrose + UDP $\rightleftharpoons$ fructose + UDP-glucose

Although it is difficult to assess the importance of this reaction as a mechanism for the breakdown rather than the synthesis of sucrose within tissues, the free energy of the glycosidic bond in sucrose is conserved and UDP-glucose is immediately available for further metabolism. ADP and GDP can substitute for UDP, yielding ADP-glucose and GDP-glucose, respectively, though the reactions in these instances proceed at considerably slower rates and the enzyme shows less affinity for the substrates.

A third way in which nucleoside diphosphate sugars may be formed is by interconversion from existing compounds (Fig. 10.21). For example, UDP-glucose is the parent compound of a whole family of related uridine diphosphate sugars. It may be converted to UDP-galactose by an epimerase which catalyses the inversion of the hydroxyl group on carbon atom 4 of glucose. This enzyme bears a single molecule of NAD^+ as a cofactor which undergoes a reversible reduction–oxidation upon addition of substrate, and the interconversion probably involves the intermediate formation of an enzyme-bound 4-keto derivative.

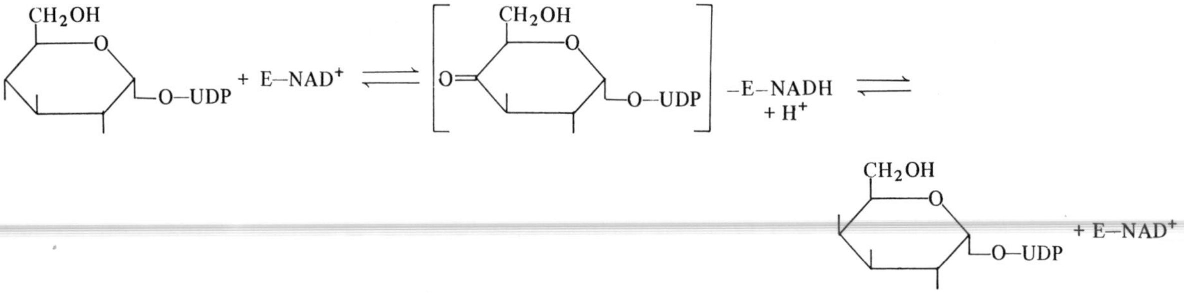

Separate but similar enzymes catalyse the interconversions of UDP-D-glucuronic acid with UDP-D-galacturonic acid and UDP-D-xylose with UDP-L-arabinose.

UDP-Glucose may also be oxidized to UDP-glucuronic acid. This is a two-step oxidation which involves two molecules of NAD^+ as oxidant and is essentially irreversible. A further enzyme then catalyses the C-6 decarboxylation of UDP-glucuronic acid, giving UDP-xylose. The oxidation of UDP-D-galactose and decarboxylation of UDP-galacturonic acid have not been observed in plants to date.

By such a series of reactions, UDP-glucose can be converted to several related UDP-sugars. Two of the steps described, namely the formation of UDP-glucuronic and UDP-xylose, are not reversible, so that carbon flow is expected to be mainly unidirectional from UDP-glucose. However, monosaccharides such as glucuronic acid, galacturonic acid, arabinose and galactose can enter the scheme at various points by the successive action of specific kinases, which phosphorylate these compounds on the glycosidic C–1 positions, and pyrophosphory-

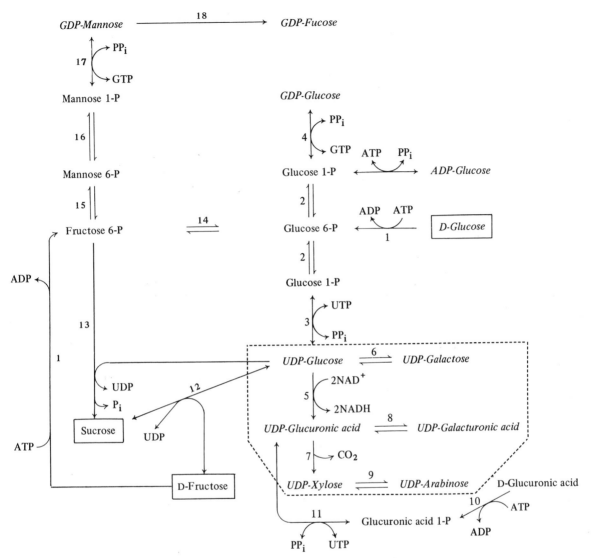

Fig. 10.21   Diagram to illustrate the biosynthesis and interconversion of certain important nucleoside diphosphate sugars in plants. The enzymes involved are: 1, hexokinase; 2, phosphoglucomutase; 3, UDP-glucose pyrophosphorylase; 4, GDP-glucose pyrophosphorylase; 5, UDP-glucose dehydrogenase; 6, UDP-glucose 4-epimerase; 7, UDP-glucuronic acid decarboxylase; 8, UDP-glucuronic acid 4-epimerase; 9, UDP-xylose 4-epimerase; 10, D-glucuronic acid kinase; 11, UDP-glucuronic acid pyrophosphorylase; 12, sucrose synthetase; 13, sucrose phosphate synthetase and sucrose phosphatase; 14, glucose 6-P isomerase; 15, mannose 6-P-isomerase; 16, phosphomannomutase; 17, GDP-mannose pyrophosphorylase; 18, multi-step conversion.

lases which convert the 1-phosphates to the nucleoside diphosphate sugar. All of these monosaccharides, however, with the possible exception of D-glucuronic acid, are only minor metabolites, and the reactions are unlikely to be of a significance except as salvage pathways to retrieve the small amounts of sugars that are released from time to time when more complex carbohydrates, such as seed reserves, are being broken down. D-Glucuronic acid, however, as we shall see in a later section, is formed from *myo*-inositol, which is an ubiquitous metabolite in plants, and it may represent an important intermediate on a pathway leading to the biosynthesis of UDP-D-glucuronic acid, alternative to the reaction catalysed by UDP-glucose dehydrogenase.

Rather more complicated interconversions than those described here have also been recognized in plants. UDP-Glucose, for example, is converted to UDP-rhamnose by a process that must involve epimerizations at C–3, C–4 and C–5, plus reduction at C–6. The proposed enzymic mechanism for this is clearly complicated, is incompletely understood and will not be discussed here. Similarly, GDP-mannose is thought to be the parent compound of GDP-L-fucose, GDP-D-mannuronic acid and GDP-L-galactose. The amino sugar N-acetyl D-glucosamine (Fig. 10.22) has been identified as a common component of plant glycoproteins. It is also, of course, the main structural unit of fungal chitin (a $\beta1 \to 4$ polymer) and bacterial peptidoglycan and an important constituent of mammalian glycoproteins and proteoglycans (such as hyaluronic acid). Its epimer, N-acetyl D-galactosamine, seems much less common in plants. Since hydrolysed cell walls have been shown to contain small amounts of D-glucosamine, it seems likely that the amino sugar is a component of a cell wall glycoprotein (Roberts et al., 1972). The activated precursor of polymeric N-acetyl D-glucosamine is believed to be UDP-N-acetyl D-glucosamine whose formation is illustrated in Fig. 10.22.

Fig. 10.22   Formation of UDP-N-acetyl D-glucosamine. The natural precursors of D-glucosamine are probably D-fructose and L-glutamine. The latter transfers its amido group to the fructose 6-phosphate in what is thought to be the rate limiting step in the pathway, since it is strongly inhibited by the end product UDP-N-acetyl D-glucosamine. Provision of D-glucosamine by-passes this reaction and causes the rapid build-up of the nucleoside diphosphate sugar. Because of this, D-glucosamine acts as a highly specific precursor of the amino sugar residues of complex saccharides. Enzyme types involved in the pathway are (1) amidotransferase; (2) acetyltransferase; (3) mutase; (4) pyrophosphorylase.

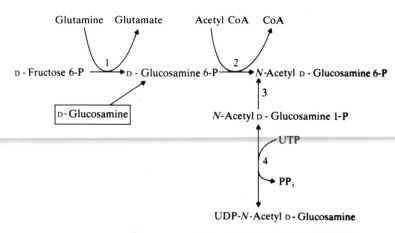

Biosynthesis of polysaccharide chains

Cellulose

Glaser first demonstrated a cell-free synthesis of cellulose in 1957 using an enzyme system from the bacterium *Acetobacter xylinum*, which was specific for UDP-glucose. The early attempts to repeat these experiments with an enzyme from plants failed. In 1964, however, Elbein, Barber and Hassid isolated a preparation from macerates of mung bean seedlings by high speed centrifugation; this was capable of utilizing GDP-glucose rather than UDP-glucose for the formation of a polysaccharide which was alkali insoluble and had all the chemical characteristics of cellulose (for detailed review see Shafizadeh and McGinnis, 1971). The enzymically active particles could be collected in the ultracentrifuge between about 20 000 and 40 000 g. By contrast, the cell wall, which can be precipitated below 1 000 g, was synthetically inactive. The site of polysaccharide synthesis was, therefore, inferred to be a function of a membrane-bound enzyme within the protoplasm.

Cellulose is believed to be formed by repetitive D-glucose transfer to the non-reducing end of an endogenous acceptor within the particles by the following type of reaction:

$$\text{GDP-Glucose} + [\beta(1 \to 4)\ \text{glucose}]_n \longrightarrow [\beta(1 \to 4)\ \text{glucose}]_{n+1} + \text{GDP}$$

When GDP-glucose and GDP-mannose are presented simultaneously to the particles from mung beans, a different product is formed which resembles a glucomannan. After partial hydrolysis of the radioactive polysaccharide with β-mannanase, oligosaccharides have been purified which contain random arrangements of both glucose and mannose, suggesting that GDP-glucose may not be the true precursor of cellulose at all. This controversy has deepened further, because there have been a large number of reports claiming that UDP-glucose could serve as a substrate for cellulose synthesis by particulate enzyme preparations of higher plants. Many of these claims have been disputed, however. Moreover, in many instances, the reaction product was poorly characterized or was shown to contain $\beta 1 \to 3$ as well as $\beta 1 \to 4$ bonds. It should also be recalled that the glucoxylans and glucomannans both contain extensive amounts of $\beta 1 \to 4$ linkage, yet are assuredly non-cellulosic. In addition, the amount of GDP-glucose and the level of the enzyme responsible for its formation, GDP-glucose pyrophosphorylase, are very low in most plants, suggesting they are poor candidates for a role in cellulose synthesis. Indeed, some workers have questioned whether the synthesis of cellulose has ever been attained *in vitro* (Delmer, 1977). Possibly the extreme lability of the enzyme, some missing essential co-factor, or the use of a wrong substrate provide reasons why this key biochemical pathway of plants has not been elucidated.

Biosynthesis of non-cellulosic polysaccharides

The same crude particulate preparation which catalyzes the synthesis of glucans will also incorporate radioactivity from UDP-D-galacturonic acid-^{14}C into a polysaccharide resembling pectic acid. The methyl esterification of the carboxyl groups seems to occur subsequent to polymerization, with S-adenosylmethionine-[14]CH$_3$ acting as methyl donor.

Particulate enzymes from corn and a number of other tissues will also catalyse the incorporation of labelled D-glucuronic acid, D-xylose, L-arabinose and D-galactose from their corresponding uridine diphosphate derivatives into polymeric materials which are similar to polysaccharides of the cell wall. For example, an enzyme from corn cobs will use UDP-xylose to form polysaccharide which resembles the plant xylan in almost all of its chemical and physical properties. Mixed polymers containing a number of different glycosyl units have also

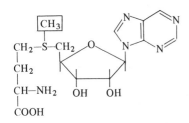

S-Adenosylmethionine with 'active' methyl group shown

461

been formed by incubating the particles simultaneously with more than one precursor compound.

The 4-*O*-methylation of D-glucuronic acid seems to occur after the unsubstituted uronic acid has been incorporated into the polymer. Again, the best methyl donor tested was *S*-adenosylmethionine. It seems to be a fairly general rule that the addition of methyl, acetyl, or sulphate groups which are found *O*-substituted onto polysaccharides occurs after polymerization.

Lipid intermediates in plant polysaccharide biosynthesis

Although the biosynthesis of several bacterial wall and membrane polysaccharides occurs by sequential transfer of monosaccharide residues from nucleoside diphosphate sugar derivatives to the growing polysaccharide chain, this transfer is probably not direct but involves lipid-linked intermediates acting as cofactors. The lipid involved in peptidoglycan synthesis in *Staphylococcus aureus*, in *o*-antigen synthesis in *Salmonella*, and in mannan synthesis in *Micrococcus lysodeikticus* is a phosphomonoester of a C_{55} polyisoprenoid alcohol:

$$HO-\overset{\overset{\text{O}}{\|}}{\underset{\underset{\text{OH}}{|}}{P}}-O-(CH_2-CH=\overset{\overset{\text{CH}_3}{|}}{C}-CH_2)_{11}H$$

The nucleoside diphosphate derivative transfers its glycosyl moiety to the lipid through either a pyrophosphate or phosphodiester bridge by a reversible reaction. The reversibility indicates that the high free energy of hydrolysis of the glycosidic linkage in the sugar nucleotide is conserved, and that the glycolipids are likely in turn to be good glycosyl donors themselves. An oligosaccharide is built up on the lipid carrier and then transferred to the growing wall polymer in one piece. As lipids, these compounds might be anticipated to penetrate or cross membranes, or participate easily in reactions occurring at or across membrane interphases where polysaccharide synthesis is believed to occur. Similar polyisoprenoid lipids, but with longer chain lengths, have been shown to transfer oligosaccharide chains to *N*-asparagine residues of mammalian glycoproteins and to the mannan-rich proteins of yeast cell walls.

Pyrophosphate   Phosphodiester

$$\text{Lipid}-O-\overset{\overset{\text{O}}{\|}}{\underset{\underset{\text{OH}}{|}}{P}}-O-\overset{\overset{\text{O}}{\|}}{\underset{\underset{\text{OH}}{|}}{P}}-O-\text{sugar} \quad or \quad \text{Lipid}-O-\overset{\overset{\text{O}}{\|}}{\underset{\underset{\text{OH}}{|}}{P}}-O-\text{sugar}$$

There is now some evidence to implicate lipids as intermediates in polysaccharide biosynthesis in plants, for when the synthetase preparations from mung bean were fed GDP-mannose, labelled with ^{14}C in the glycosyl moiety, radioactivity passed rapidly into lipid-like material, but only relatively slowly into polysaccharide during the initial exposure to substrate. The former, however, soon reached a constant

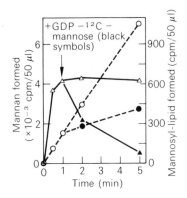

Fig. 10.23 Chase experiment indicating the likely turnover of ^{14}C-mannosyl residues of the mannosyl lipid. GDP-Mannose-^{14}C was provided to subcellular particles from mung bean shoots and radioactivity in lipids (△) and in mannan (○) followed. Note the rapid rise in radioactivity in lipid which levels off within 1 min. Note also that label in polysaccharide increases at a steady rate. When particles were 'pulsed' for 1 min and then chased with non-radioactive GDP-mannose, ^{14}C in lipid (▲) rapidly declined while polysaccharide (●) continued to accumulate more labels. Data from Kauss (1969).

specific activity while the polysaccharide accumulated label at a linear rate (Fig. 10.23). Furthermore, when particles were 'pulsed' with radioactive precursor for a brief period and then incubated with unlabelled GDP-mannose, label passed rapidly from the lipid to the polysaccharide, indicating a likely precursor-product relationship between the two.

Polyprenols with between 6 and 13 isoprene units have been purified from higher plants and have been postulated to be intermediate, oligosaccharide carriers in both polysaccharide and glycoprotein biosynthesis (see Delmer, 1977). There is also some evidence that glycosyl-phosphoryl-polyprenols participate in the formation of cellulose (Forsee and Elbein, 1973), although this has not been clearly established.

The site of polysaccharide biosynthesis

The isolation of the particles which catalyse polysaccharide biosynthesis involves breakage of cells under conditions in which many organelles burst and the membrane systems become so disrupted that the particles cannot be identified or related to any recognizable part of the cell. As we shall discuss in the next chapter, however, Ray, Shininger and Ray (1969) purified an active enzyme preparation by isopycnic gradient centrifugation in sucrose from carefully chopped tissues of pea under conditions in which organelle disruption was minimized. They showed that at least some of the enzymes involved in polysaccharide biosynthesis were associated with the Golgi bodies and there is now general agreement that certain of the non-cellulosic polysaccharides are synthesized within the Golgi bodies, or the associated membranes of the endoplasmic reticulum, and that vesicles containing the completed product are budded from the Golgi and moved across the cytoplasm to the plasmalemma where they release their contents by reverse pinocytosis (see Ch. 11). The organization of the polysaccharide into the defined architecture of the wall presumably occurs at the plasmalemma interface or within the wall itself. There is some controversy, however, about the site of cellulose biosynthesis. With one exception, significant levels of cellulose-like polymers have not been detected during analyses of vesicles and other membrane-associated material derived from the inside of plant cells. The exception is the unusual alga *Pleurochrysis*, discussed in Ch. 11, which synthesizes whole cellulosic scales within its Golgi apparatus. However, most enzymic evidence is consistent with the view that the biosynthesis of cellulose in higher plant cells occurs on the plasmalemma at the cell surface, while the matrix polysaccharides are formed internally and released from the cell as secretions. There is, in addition, little doubt that the synthesis of chitin, also a $\beta 1 \rightarrow 4$ fibrillar polymer, is organized on the plasmalemma surface of fungal cells (Ruiz-Herrea *et al*, 1975). There appear to be organized zones of spherical particles on the outer face of the plasmalemma from which microfibrils radiate (Fig. 10.24). It is still not clear whether these particles are the site of cellulose synthesis, the location at which

Fig. 10.24   The innermost lamella of a side wall of the green alga *Chaetomorpha melagonium* (shadowed Pt/Au; ×16 000). The arrays of close packed granules, each about 30 nm diameter, are derived from the cytoplasm and most probably from the outer surface of the plasmalemma. These granules may be enzyme complexes involved in cellulose microfibril synthesis. Provided by Drs Eva Frei and R. D. Preston, University of Leeds.

completed cellulose molecules are organized into microfibrils, or artefacts of some kind. However, bearing in mind that any over-all scheme for polysaccharide metabolism can only be tentative, Fig. 10.25 attempts to represent and summarize the possible sites of polysaccharide synthesis, transport and deposition within a growing plant cell.

Fig. 10.25   Diagram to show possible sites of polysaccharide synthesis and transport within a growing plant cell. Redrawn after Northcote (1968).

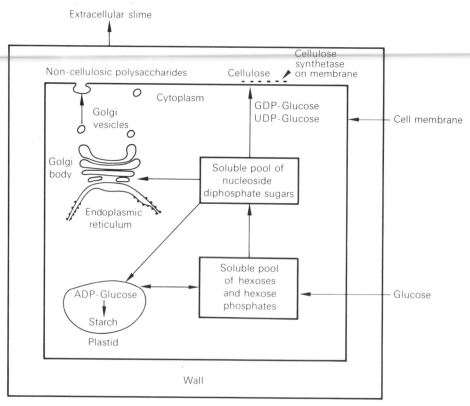

The rôle of microtubules in cellulose deposition

We have pointed out earlier how microtubules seem to play a rôle in directing the movement of vesicular material to the developing cell plate after nuclear division. They also seem to be involved in the control of polysaccharide deposition in older walls. For example, in cells that are growing predominantly in a longitudinal direction, the cellulose microfibrils are laid down with a more or less transverse orientation, and the microtubules adjacent to the wall are also orientated in such a transverse manner and run continuously around the cell as a series of hoops (Fig. 10.26).

There are numerous studies which show that orientated microtubules are not confined to cells with primary walls, but occur adjacent to regions of secondary thickening (Newcomb, 1969). In parenchyma cells of wounded *Coleus*, for example, which were differentiating into tracheary elements, a system of microtubules was associated with the thickening bands of the wall. A similar morphological association of microtubules with developing wall thickenings has been noted in maturing xylem of such organs as coleoptiles or roots, and in vessels and tracheary elements differentiating from cambium. In section, the

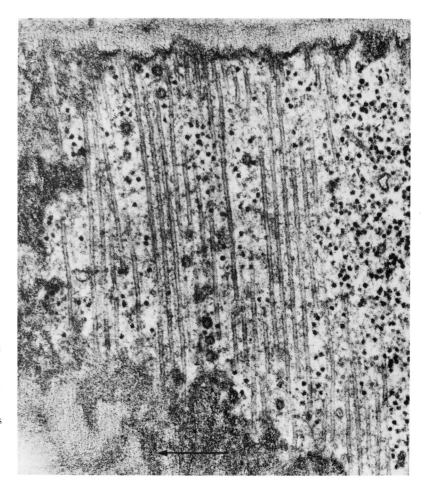

Fig. 10.26 Microtubules beneath the primary wall of cells in the root tip of bean (*Phaseolus vulgaris*) shown in grazing section. ×52 000. Note that the microtubules run parallel to one another and are orientated parallel to the microfibrils in the cell wall. The axis of cell elongation (indicated by arrow) is at right angles to the orientation of both microtubules and microfibrils. Reproduced from Newcomb (1969).

Fig. 10.27   Microtubules adjacent to a secondary wall thickening (both in transverse section) in a tracheary element in the leaf of *Petunia hybrida*. ×77 000. Reproduced from Newcomb (1969).

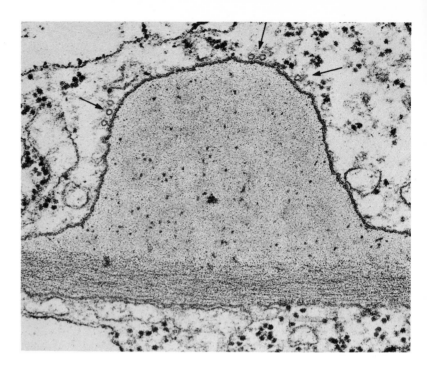

microtubules can be seen to run parallel to the orientation of the microfibrils and to mirror the pattern of the wall thickening within the cell (Fig. 10.27).

There is no evidence to implicate microtubules in the enzymic synthesis of cellulose, however, as they are usually found at least 10 nm away from the plasmalemma and well away from the site of polysaccharide deposition. In addition, the drug colchicine which inhibits microtubule assembly does not prevent cell wall synthesis when it is applied to plant tissues. It does, however, have profound effects on the organization of the wall. In secondary walls, a regular pattern of thickening will become disrupted. It also inhibits the formation of the cell plate and causes a disorganized deposition of cellulose in primary walls.

Microtubules, therefore, seem to play some part in controlling the precise way in which cell walls are laid down. Nevertheless, we have pointed out earlier that their appearance is not restricted to the development of the wall and they have been implicated in a number of other morphogenic processes in the cell. They are protein fibrils which are used to maintain local regions of stress or rigidity or to guide material such as organelles or vesicles through the cytoplasm in a precise manner. The exact way in which they control microfibril orientation is not known.

myo-Inositol and its conversion to cell-wall polysaccharides

myo-Inositol seems to be ubiquitous in plant cells, sometimes occurring in relatively large amounts as the free cyclitol, though it is also commonly encountered as a component of phospholipids and as its hexaphosphoric acid ester, phytic acid. The latter is frequently found as a storage product in seeds in the form of an insoluble calcium or magnesium salt (see p. 509). Loewus and his associates (Loewus, 1971) have shown that *myo*-inositol-^{14}C is rapidly metabolized to cell-wall polysaccharide in a wide number of plant tissues. The radioactivity is recovered mainly in the pentose units D-xylose and L-arabinose and in the uronic acids D-glucuronic acid, 4-*O*-methyl D-glucuronic acid, and D-galacturonic acid. As the first step in its metabolism, the ring of *myo*-inositol is oxidatively cleaved between carbon atoms 1 and 6, yielding D-glucuronic acid, and it is the further metabolism of this compound which accounts for the labelling in the cell wall, for in plants, D-glucuronic acid is converted successively to its 1-phosphate and UDP-glucuronic acid by successive action of kinase and specific pyrophosphorylase (Fig. 10.28) (Roberts, 1971).

Fig. 10.28   Alternative pathways for the formation and metabolism of UDP-D-glucuronic acid in plants. The enzymes (trivial names) involved are: D-glucose 6-P-cycloaldolase (1); *myo*-inositol l-P-phosphatase (2); *myo*-inositol oxygenase (3); D-glucuronic acid kinase (4); UDP-D-glucuronic acid pyrophosphorylase (5); phosphoglucomutase (6); UDP-D-glucose pyrophosphorylase (7); UDP-D-glucose dehydrogenase (8); *myo*-Inositol is a cyclitol or hexahydrocyclohexane. Cyclization of the glucose 6-phosphate between C_1 and C_6 leads directly to the formation of *myo*-inositol 1-P.

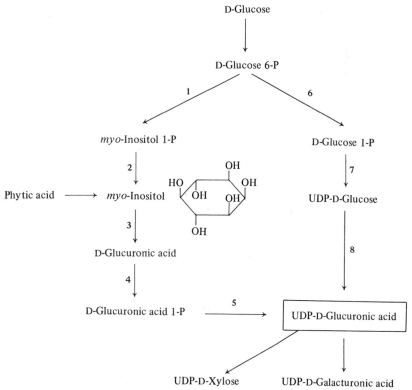

The evidence gathered so far has indicated that *myo*-inositol can act as a major intermediate in the biosynthesis of cell-wall polysaccharides. However, UDP-glucuronic acid can also be formed from UDP-glucose by action of UDP-glucose dehydrogenase (Figs. 10.21, 10.28). As yet, the relative contribution of the two pathways to UDP-glucuronic acid biosynthesis has not been determined. Clearly, however, the formation of UDP-glucuronic acid from *myo*-inositol can circumvent UDP-glucose as a necessary precursor for all uridine diphosphate sugar derivatives.

The formation of hydroxyproline rich protein

Protein-bound hydroxyproline is not derived from free hydroxyproline but is formed by hydroxylation of proline after the latter has become incorporated into polypeptide. This reaction involves atmospheric oxygen, and requires Fe^{++}, ascorbate, α-ketoglutarate and peptide-bound proline as substrate. The rôle of the cofactors is not understood. The hydroxylation step probably occurs soon after the peptide has been assembled. The hydroxyproline residues are probably glycosylated within the endoplasmic reticulum as the polypeptide emerges into the intraluminal space, and transported to the Golgi for secretion. Nucleoside diphosphate sugars provide the initial activated sugar precursors, but lipid intermediates may again be involved in the ultimate glycosyl transferase reaction in much the same way that they participate in mammaliam glycoprotein synthesis (see Cooke, 1977).

The biosynthesis of lignin

Early experiments with ^{14}C tracers established that shikimic acid and other precursors of aromatic amino acids were readily converted into lignin (Fig. 10.29) (Neish, 1965). Furthermore, the carbon skeletons of phenyl pyruvic acid, phenolic cinnamic acids, phenylalanine and tyrosine (the latter in grasses only) were incorporated as intact units, indicating that C_6-C_3 units were the most likely precursor compounds of lignin. The amino acid phenylalanine is deaminated and probably serves as the main source of the cinnamic acid derivatives which are the immediate precursors of lignin (Fig. 10.29). These derivatives, the alcohols of ferulic acid (coniferyl alcohol) and sinapic acid (sinapyl alcohol), arise by stepwise hydroxylation and methylation of cinnamic acid. The nature of the final condensation reaction is not understood, but the apparent lack of optical activity of natural lignins prompted Freudenberg (1968) to suggest that lignin is formed by a relatively unspecific polymerization in the presence of peroxidase, which promotes oxidation of the double bonds in the side chain and allows complex interchain and chain-ring condensation to occur.

The enzyme responsible for the deamination of phenylalanine, phenylalannie ammonia-lyase, is only found in high concentrations in tissues that are synthesizing lignin and is almost completely absent in tissues where there is no lignin formation (Rubery and Northcote, 1968). The lyase has also attracted much attention because it shows remarkable changes in activity in leaves, stems and other tissues in response to light, wounding, disease, radiation damage and growth factors including ethylene. This emphasizes that the production of lignin precursors and possibly lignin itself may in part be controlled by a number of previously unsuspected environmental and physiological factors.

Outlook. The last six years have seen a great increase in our knowledge of the composition of primary cell walls, particularly in the detailed structural analyses of their component polysaccharides and glycoproteins. Studies on biosynthesis have languished somewhat. Key

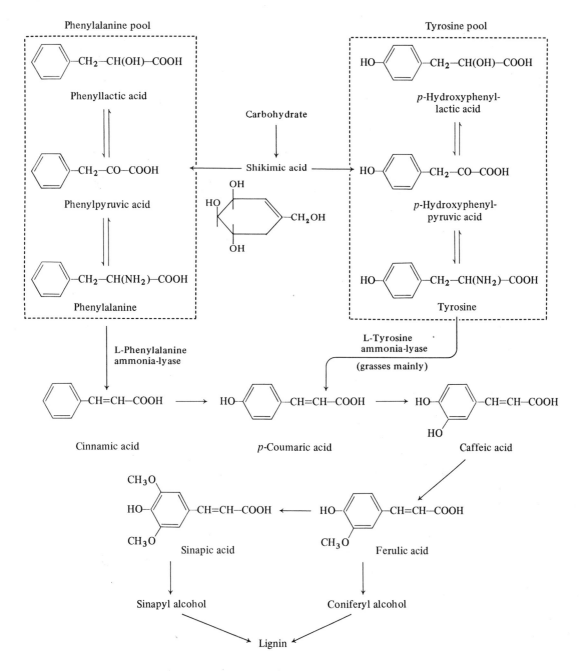

Fig. 10.29   Likely pathway for lignin biosynthesis in vascular plants. The details of the final polymerization and precise nature of the intermediates are still not understood. Adapted from Neish (1965).

questions to be asked include, How are lipid intermediates involved in polymer assembly? How can we achieve respectable rates of cellulose synthesis *in vitro*? What processes control the rates at which different polysaccharides and other wall polymers are synthesized? Where are the different biosynthetic reactions localized in the cell? In addition we know very little about the macromolecular interactions involved in cell wall assembly and growth, how the wall changes as cells develop, or the role of the wall in resisting infection. All these areas will undoubtedly receive great attention within the next decade.

Further reading

Albersheim, P. (1976) The primary cell wall. In *Plant Biochemistry*. Eds. J. Bonner and J. E. Varner. Academic Press, New York, p. 225.

Albersheim, P. (1978) Concerning the structure and biosynthesis of the primary cell walls of plants. In *Int. Rev. Biochem.* **16**. Ed. D. J. Manners. University Park Press, Baltimore, p. 127.

Chrispeels, M. J. (1976) Biosynthesis, intracellular transport and secretion of extracellular macromolecules. *Ann. Rev. Plant Physiol.,* **27**, 19.

Cleland, R. (1971) Cell wall extension. *Ann. Rev. Plant Physiol.,* **22**, 197.

Cleland, R. (1977) The control of cell enlargement. In *Integration of Activity in the Higher Plant*. Symp. Soc. Exp. Biol., Vol. 13, Cambridge University Press, Cambridge, p. 101.

Colvin, J. R. (1980) The biosynthesis of cellulose. In *The Biochemistry of Plants. Vol. 3. Carbohydrates: Structure and Function*. Ed. J. Preiss. Academic Press, New York.

Frey-Wyssling, A. and Mühlethaler, K. (1964) *Ultrastructural Plant Cytology*. Elsevier, Amsterdam.

Labavitch, J. M. (1981) Cell wall turnover in plant development. *Ann. Rev. Plant Physiol.,* **32**, 385.

McNeil, M., Darvill, A. G. and Albersheim, P. (1979) The structural polymers of the primary cell walls of dicots. In *Progress in the Chemistry of Organic Natural Products*. Ed. H. Grisebach.

Northcote, D. H. (1972) Chemistry of the plant cell wall. *Ann. Rev. Plant Physiol.,* **23**, 113.

Preston, R. D. (1974) *The Physical Biology of Plant Cell Walls*. Chapman and Hall, London.

Roelofsen, P. A. (1965) Ultrastructure of the wall in growing cells and its relation to the direction of growth. In *Adv. Bot. Res*. Ed. R. D. Preston. Academic Press, New York.

Roland, J-C. and Vian, B. (1979) The wall of the growing plant cell: its three-dimensional organization. *Int. Rev. Cytol.,* **61**, 129.

Literature cited

Adams, P. A., Kaufman, P. B. and Ikuma, H. (1973) Effects of GA and sucrose on the growth of oat stem segments. *Plant Physiol.,* **51**, 1102.

Allen, A. K. and Neuberger, A. (1973) The purification and properties of the lectin from potato tubers, a hydroxyproline—containing glycoprotein. *Biochem. J.,* **135**, 307.

Akiyama, Y. and Kato, K. (1977) Structure of hydroxyproline-arabinoside from tobacco cells. *Agric. Biol. Chem.,* **41**, 79.

Anderson, R. L., Clark, A. E., Jermyn, M. A., Knox, R. B. and Stone, B. A. (1977). A carbohydrate-binding arabinogalactan protein from liquid suspension cultures of endosperm from *Lolium multiflorum*. *Aust. J. Plant Physiol.,* **4**, 143.

Aspinall, G. O. (1970) *Polysaccharides*. Pergamon Press, Oxford.

Bauer, W. D., Talmadge, K. W., Keegstra, K. and Albersheim, P. (1973) The structure of plant cell walls. II. The hemicellulose of the walls of suspension-cultured sycamore cells. *Plant Physiol.,* **51**, 174.

Bonner, J. (1950) *Plant Biochemistry*. Academic Press, New York, p. 134.

Browning, B. L. (1963) In *The Chemistry of Wood*. Ed. B. L. Browning. Interscience Publishers, New York. p. 57.

Burke, D., Kaufman, P., McNeil, M. and Albersheim, P. (1974) The structure of plant cell walls. VI. A survey of the walls of suspension cultured monocots. *Plant Physiol.,* **54**, 109.

Cleland, R. E. (1975) Auxin induced hydrogen ion excretion: Correlation with growth, and control by external pH and water stress. *Planta,* **127**, 233.

Cleland, R. E. (1976) Kinetics of hormone-induced H^+ excretion. *Plant Physiol.,* **58**, 210.

Clowes, F. A. L. and Juniper, B. E. (1969) *Plant Cells*. Blackwell Scientific Publications, Oxford.

Cocking, E. C. (1972) Plant cell protoplasts–isolation and development. *Ann. Rev. Plant Physiol.,* **23**, 29.

Cooke, G. M. W. (1977) Biosynthesis of plasma membrane glycoproteins. In *The Synthesis, Assembly and Turnover of Cell Surface Components*. Eds. G. Poste and G. L. Nicolson. North-Holland, Amsterdam, p. 85.

Delmer, D. P. (1977) The biosynthesis of cellulose and other plant cell wall polysaccharides. In *Recent Advances in Phytochemistry, II.* Eds. F. A. Loewus and V. C. Runeckles. Plenum, N.Y., p. 45.

Forsee, W. T. and Elbein, A. D. (1973) Biosynthesis of mannosyl- and glucosyl-phosphoryl polyprenols in cotton fibres. *J. Biol. Chem.,* **248,** 2858.

Freudenberg, K. (1968) The constitution and biosynthesis of lignin. In *Molecular Biology, Biochemistry and Biophysics, Vol. 2.* Ed. A. Kleinzeller. Springer-Verlag, Berlin, p. 45.

Gardner, K. H. and Blackwell, J. (1974) The structure of native cellulose. *Biopolymers,* **13,** 1975.

Grant, G. T., Morris, E. R., Rees, D. A., Smith, P. and Thom, D. (1973) Biological interactions between polysaccharides and divalent cations: The egg-box model. *FEBS Letters,* **32,** 195.

Hanke, D. E. and Northcote, D. H. (1974) Cell wall formation by soybean callus protoplast. *J. Cell Sci.,* 14, 29.

Hassid, W. Z. (1966) Some aspects of sugar nucleotide metabolism. In *Current Aspects of Biochemical Energetics.* Eds. N. O. Kaplan, and E. P. Kennedy. Academic Press, New York, p. 35.

Hopper, J. E. and Dickinson, D. B. (1972) Partial purification and sugar nucleotide inhibition of UDP-glucose phosphorylase from *Lilium longifolium* pollen. *Arch. Biochem. Biophys.,* **148,** 523.

Jensen, W. A. and Ashton, M. (1960) The composition of the developing primary wall in onion root-tip cells. *Plant Physiol.,* **35,** 313.

Kauss, H. (1969) A plant mannosyl lipid acting in reversible transfer of mannose. *FEBS Lett.,* **5,** 81.

Kauss, H. (1977) The possible physiological role of lectins. In *Cell wall biochemistry related to specificity in host-plant pathogen interactions.* Eds. B. Solheim and J. Rag. Scandinavian University Books, Oslo, p. 347.

Keegstra, K., Talmadge, K. W., Bauer, W. D. and Albersheim, P. (1973) The structure of plant cell walls. III. A Model of the walls of suspension-cultured sycamore cells based on the interconnections of the macromolecular components. *Plant Physiol.,* **51,** 188.

Lamport, D. T. O. (1970) Cell wall metabolism. *Ann. Rev. Plant Physiol.,* **21,** 235.

Lamport, D. T. O., Katona, L., and Roerig, S. (1973). Galactosylserine residues in extensin. *Biochem. J.,* **133,** 125.

Loewus, F. (1971) Carbohydrate interconversions. *Ann. Rev. Plant Physiol.,* **22,** 337.

Neish, A. C. (1965) Coumarins, phenylpropanes and lignin. In *Plant Biochemistry.* Eds. J. Bonner and J. E. Varner. Academic Press, New York.

Newcomb, E. H. (1969) Plant microtubules. *Ann. Rev. Plant Physiol.,* **20,** 253.

Nikaido, H. and Hassid, W. Z. (1971). Biosynthesis of saccharides from glycopyranosyl esters of nucleoside pyrophosphates (sugar nucleotides). *Adv. Carbohyd. Chem. and Biochem.,* **26,** 351.

Northcote, D. H. (1968) The organization of the endoplasmic reticulum, the Golgi bodies and microtubules during cell division and subsequent growth. In *Plant Cell Organelles.* Ed. J. B. Pridham. Academic Press, London and New York.

Northcote, D. H. (1969) The synthesis and metabolic control of polysaccharides and lignin during differentiation of plant cells. In *Essays in Biochemistry,* Vol. 5. Eds. P. N. Campbell and G. D. Greville. Academic Press, London.

Northcote, D. H. (1974) Sites of synthesis of the polysaccharide of the cell wall. In *Plant Carbohydrate Chemistry,* Ed. J. B. Pridham. Academic Press, London and New York, p. 165

Penny, P. and Penny, D. (1978) Rapid responses to phytohormones. In *Phytohormones and Related Compounds—A Comprehensive Treatise.* Vol. 2. Eds. D. S. Letham, P. B. Goodwin and D. V. Goodwin. Elsevier/North-Holland, Amsterdam. p. 1.

Pope, D. G. (1977) Relationships between hydroxyproline-containing proteins secreted into the cell wall and medium by suspension-cultured *Acer pseudoplatanus* cells. *Plant Physiol.,* **59,** 894.

Preston, R. D. (1964) Structural and mechanical aspects of plant cell walls with particular reference to synthesis and growth. In *The Formation of Woods in Forest Trees.* Ed. M. Zimmerman. Academic Press, New York.

Ray, P. M., Shininger, T. L. and Ray, M. M. (1969) Isolation of β-glucan synthetase particles from plant cells and identification with Golgi membranes. *Proc. Natl. Acad. Sci. U.S.A.,* **64,** 605.

Rayle, D. L. and Cleland, R. E. (1972) The *in vitro* growth response: Relation to *in vito* growth response and auxin action. *Planta,* **104,** 282.

Roberts, R. M. (1971) The formation of uridine diphosphate-glucuronic acid in plants. *J. biol. Chem.,* **246,** 4995.

Roberts, R. M. and Butt, V. S. (1967) Patterns of cellulose synthesis in maize root-tips. *Exptl. Cell Res., 46,* 495.

Roberts, R. M. and Butt, V. S. (1968) Patterns of incorporated of pentose and uronic acid into the cell walls of maize roots. *Exptl. Cell Res., 51,* 519.

Roberts, R. M., Cetorelli, J. J., Kirby, E. G. and Ericson, M. (1972) Location of glycoproteins that contain glucosamine in plant tissues. *Plant Physiol., 50,* 531.

Roberts, R. M. and Loewus, F. (1966) Inositol metabolism in plants. Conversion of myo-inositol-2-^{3}H to cell-wall polysaccharide in sycamore (*Acer pseudoplatanus L.*) cell culture. *Plant Physiol., 41,* 1489.

Rubery, P. H. and Northcote, D. H. (1968) Site of phenylalanine ammonia-lyase activity and synthesis of lignin during xylem differentiation. *Nature, Lond., 219,* 1230.

Ruiz-Herrera, J., Sing, V. O., Van der Woude, W. J. and Bartnicki-Garcia, S. (1975) Microfibril assembly of granules of chitin synthetase. *Proc. Natl. Acad. Sci. U.S.A., 72,* 2706.

Sarko, A. and Muggli, R. (1974) Packing analysis of carbohydrates and polysaccharides. III. *Valonia* cellulose and cellulose II. *Macromolecules, 7,* 486.

Shafizadeh, F. and McGinnis, G. D. (1971) Morphology and biogenesis of cellulose and plant cell walls. *Adv. Carbohyd. Chem. and Biochem., 26,* 297.

Steward, F. C. (1968) *Growth and Organization in Plants.* Addison-Wesley, New York.

Talmadge, K. W., Keegstra, K., Bauer, W. D. and Albersheim, P. (1973) The structure of plant cell walls. 1. The macromolecular components of the walls of suspension-cultured sycamore cells with a detailed analysis of the pectic polysaccharides. *Plant Physiol., 51,* 158.

Thornber, J. P. and Northcote, D. H. (1961a) Changes in the chemical composition of a cambial cell during its differentiation into xylem and phloem tissue in trees. 1. Main components. *Biochem. J., 81,* 449.

Thornber, J. P. and Northcote, D. H. (1961b) Changes in the chemical composition of a cambial cell during its differentiation into xylem and phloem tissue in trees. 2. Carbohydrate constituents of each main component. *Biochem. J., 81,* 455.

Zeroni, M. and Hall, M. A. (1980) Molecular effects of hormone treatment on tissue. In *Hormonal Regulation of Plant Development. I.* Ed. J. MacMillan. Encyclo. Plant Physiology, Vol. 9. Springer-Verlag, Berlin.

The Golgi body

Golgi bodies or dictyosomes, as they are frequently called, are recognized on electron micrographs of plant cells as stacks of smooth-membraned, disc-shaped sacs known as *cisternae* (Fig. 11.1). The name Golgi body is derived from animal nomenclature because they resemble similar organelles found in animal cells. They are also thought to function similarly in plant cells, acting as controlling agents for the movement of cellular material both within and to the outside of the

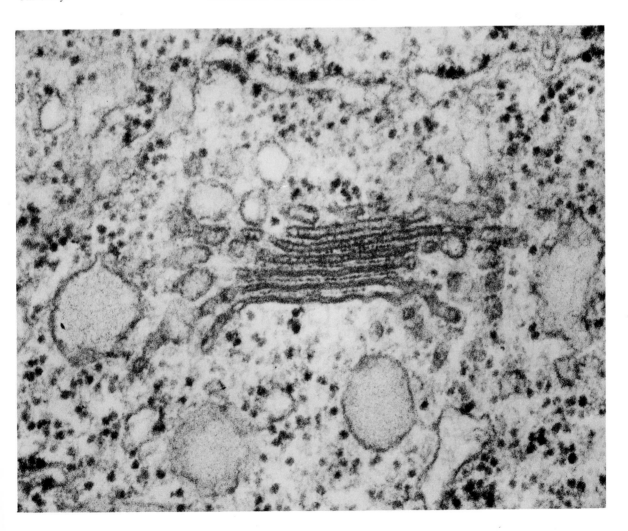

Fig. 11.1 Electron micrograph of a Golgi body from epidermal cells of the root of *Zea mays*. ×126 000. Provided by Dr H. Mollenhauer, Texas A & M University.

cell. As we shall see, they can also have a rôle in the synthesis of this material and are sites of polysaccharide formation within the cell. Their intermediate functional position between the membrane systems of the endoplasmic reticulum and the cell surface also means that they play a rôle in membrane transformation and turnover.

Golgi bodies were first described by Camillo Golgi in 1891 long before there was any evidence for the endoplasmic reticulum. He observed a meshed system of associated threads localized close to the nucleus in neurons of owls and cats. Although their function was unknown, they did have the unique ability to reduce silver nitrate stain after fixation and so could be recognized histochemically on fixed sections. As techniques in microscopy improved, Golgi bodies were observed in many kinds of animal cells, and, because of basic similarities in their structure, these organelles were regarded, not as fixation artefacts but as real cellular entities. A rôle in secretion of mucus from intestinal goblet cells was first suggested as early as 1914 by Cajal.

They are not easily seen, however, by light microscopic observations of most fixed or living plant cells, even though their dimensions $(1-3\mu m)$ are well within the resolving power of the light microscope. Their existence was, therefore, doubted for many years and it was only with the development of thin sections for the electron microscope that they were recognized as a general feature of the eukaryotic plant cell. Our knowledge of Golgi bodies in plants, therefore, derives largely from the use of the electron microscope, with all the limitations that this entails. Fortunately, however, there are certain algae (for example the diatoms *Pinnularia* and *Microsterias*) which have particularly large Golgi bodies with many cisternae that can be easily observed both in the living state by phase microscopy and, after fixation, by electron microscopy. The Golgi bodies of these algae undergo little change in appearance or size during processing for the electron microscope. It is assumed, therefore, that what is observed in thin section by the electron microscope does give a reasonable impression of the organelle as it occurs *in vivo*. Its structure has also been confirmed by freeze-etch techniques (Fig. 11.2) on both fixed and non-fixed materials.

Distribution

In cells of higher plants, the individual Golgi bodies are usually found scattered throughout the cytoplasm and their distribution does not seem to be ordered or localized in any particular manner. Their number per cell can vary from several hundred, as in tissues of corn root, to a single organelle in some algae. They have been reported absent in mature sieve tubes, sperm cells of bryophytes and pteridiophytes, and in blue-green algae which, of course, resemble bacteria rather than plants. Morré, Mollenhauer and Chambers (1965) noted that in their preparations of Golgi, isolated by differential centrifugation from cauliflowers, membrane continuities could be traced between individual organelles. By careful serial sectioning, some Golgi bodies have also been shown to be continuous with the membrane

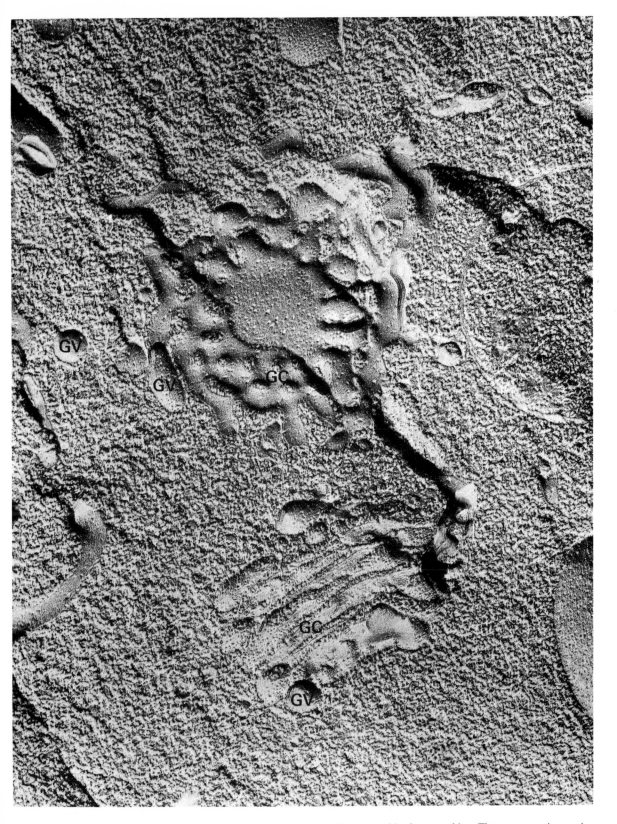

Fig. 11.2   Electron micrograph of two Golgi bodies from onion root tip prepared by freeze-etching. The upper one is seen in face view and the lower one in cross-section. GV, Golgi vesicles, GC, Golgi cisternae. ×107 000. Provided by Dr D. Branton, University of California, Berkeley.

sheets of the endoplasmic reticulum. The Golgi bodies of the intact cell are often envisaged, therefore, as being part of a more complex, interconnected smooth-membraned system. Electron microscopy of whole cells, however, indicates that the system is exceedingly dispersed, and not strictly comparable to the situation in those animal cells where a number of individual Golgi bodies become closely assembled in the form of a localized Golgi complex. However, as we shall see, some investigators prefer to regard the Golgi bodies as being functionally related to, or even a differentiated part of the endoplasmic reticulum. In this case, one must suppose some degree of association between these organelles in different parts of the cell (see Mollenhauer, Morré and Van der Woude, 1975).

Structure

Golgi bodies in sections of plant cells are usually 1–3 μm in length and about 0·5 μm high. Each organelle is comprised of a number of plate-like cisternae which in sections appear like flattened sacs (Fig. 11.1). These are enclosed by the usual three-layered unit membranes, about 6 nm in width, which appear similar to those of smooth endoplasmic reticulum. There are usually four to seven, but sometimes up to twenty hollow cisternae in each Golgi body. The association is relatively firm, as many of the organelles retain their characteristic morphology when isolated by sucrose density gradient centrifugation from extracts of broken cells. Bundles of fibres can often be seen to extend across the intercisternal spaces and may serve as strengthening elements maintaining the regular spacing between the individual lamellae.

The cisternae are usually dilated towards their periphery, and there appear to be vesicles either attached to or situated close to these marginal regions as if they had been budded off from the rim. It must be emphasized, however, that is not always easy, and sometimes exceedingly difficult, to interpret single sections of cells without considering the structure in three dimensions. By making serial sections, it has now been recognized that the outer edges of the cisternae consists of a reticulate network of tubules which in single section can appear like a system of closely associated vesicles (Fig. 11.3). The innermost core region, usually about 1 μm in diameter, consists of a reasonably compact disc of associated cisternae and it is this structure which on electron micrographs is usually recognized as the typical Golgi body. The peripheral region extending out for a further 1–2 μm consists of a flat network of interconnecting tubes which Clowes and Juniper (1969) compared to a disc of lace. This basic structure is clearly demonstrated for plant Golgi bodies in the negatively stained cisterna shown in Fig. 11.4. The central plate, fenestrated outer region and a number of budding vesicles can all be seen on this micrograph.

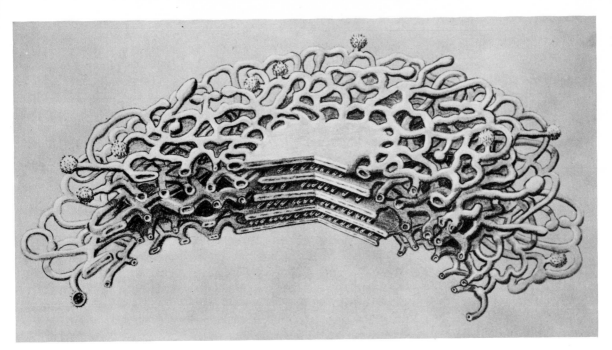

Fig. 11.3   A diagrammatic interpretation of a portion of a Golgi body composed of six cisternae. The diagram shows the fenestrated outer region bearing vesicles at the end of tubules. Reproduced from Mollenhauer and Morré (1966a).

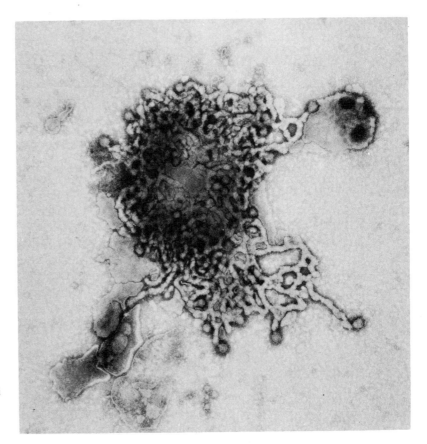

Fig. 11.4   A negatively stained Golgi body isolated from bean roots. Provided by Drs H. Mollenhauer and J. Morré.

Production of vesicles

The smooth vesicles which are believed to contain secretory material are budded from the ends of cisternal tubules within the net. They vary in diameter from 20 to 80 nm. Often more than one tubule connects to, and presumably fills, a single forming vesicle. They do not usually take up stain easily, but phosphotungstic acid or uranyl acetate may be deposited within the sac depending upon their content. A second type of vesicle, the coated vesicle, is also found associated with the Golgi bodies. These appear as spherical protuberances, about 50 nm in diameter, with a rough surface which stains heavily with phosphotungstic acid. They are found at the periphery of the organelle, usually at the ends of single tubules, and are morphologically quite distinct from the secretory vesicles. Their function is unknown.

Vesicles leaving the cisternae have, in some instances, been observed to fragment into smaller members; in other cases, several have been seen to coalesce into larger bodies filled with secretory material. This process is particularly conspicuous in pancreatic exocrine cells of mammals where proteins to be discharged are temporarily accumulated in large zymogen granules, which are formed from Golgi vesicles. In some flagellate algae, which produce a number of distinct kinds of surface scales, the individual scales are first formed in separate vesicles, which coalesce soon after or even prior to release from the Golgi body, then migrate as a composite body to the plasmalemma.

Mollenhauer and Morré (1966a) have suggested that the cisternae of an individual Golgi body are not all equivalent however, and that there is a maturing face and a forming face to each organelle. New cisternae are added continuously to the forming face, possible by input of membrane material from the endoplasmic reticulum, and as these mature they move progressively across the stack. At the maturation face, the Golgi cisternae are more swollen, and it is from here that the majority of the larger secretion vesicles are shed. This presumably leads to fragmentation and loss of mature cisternae. It has been estimated that in the mucus-producing goblet cells of mammalian intestine, it takes only about 40 min for a stack of cisternae to be completely turned over and transformed into secretion vesicles. This also implies that Golgi membrane is very rapidly transformed into plasma membrane. The process of turnover is likely to occur much more slowly in cells which are synthetically less active, and so the speed of the process in goblet cells cannot be assumed typical of all secretory cells. It has been calculated that each Golgi body in gland cells of leaves of the insectivorous plant *Drosophyllum* produces about three vesicles per minute, but this gives no indication of the rate of membrane turnover (see Mollenhauer and Morré, 1966).

Differentiation of Golgi bodies

The structural polarity of Golgi bodies might explain some controversial observations that have been made on Golgi body structure and histochemistry. Asymmetric stain reactions for acid phosphatase activities have been noted in Golgi from both plants and animals, and different cisternae in the same organelle have also been seen producing quite distinct kinds of secretory product. For example, Pickett-Heaps (1967) noted, that in stomatal cells from wheat leaves, some cisternae were apparently giving rise to vacuolar fluid, while others were simultaneously involved in secretion of smaller darker vesicles destined for the cell wall. The simultaneous formation of more than one type of product by a single Golgi body indicates that there must be some degree of compartmentalization within the organelle. A dual function has also been postulated in a number of animal systems. Conceivably, therefore, a biochemcial as well as a morphological gradation of function occurs across the cisternal stack. Changes in the dimensions of the membranes have also been noted in some instances (Fig. 11.5). Whereas the lamellae on the forming face resemble the membrane sheets of the endoplasmic reticulum and have a thickness of about 4 nm, those on the maturing face and on the vesicles have similar dimensions to the outer membrane of the cell (thickness about 7–8 nm). This is not perhaps unexpected because as we shall see later, Golgi membranes presumably originate from the endoplasmic reticulum or other internal membrane systems of the cell while their vesicles, which are shed from the more mature parts of the organelle, ultimately fuse with plasmalemma. Golgi membranes are also intermediate in lipid, protein and enzymic composition between the endoplasmic reticulum on the one hand and the outer membrane on the other.

As cells differentiate, so the morphology of the Golgi bodies can change, though the basic stacked form is usually preserved. In the root-cap, for example, the Golgi in the older cells on the cap periphery become much larger, with swollen cisternae that produce large vesicles. (Fig. 11.6). This change is relatively abrupt and seems to occur when these outer cells start to secrete slime (Mollenhauer, Whaley and Leech, 1961.) It also correlates with the breakdown of starch in the amyloplasts, and it is possibly significant that, when the carbohydrate is used up, usually after the cells detach themselves from the cap, the Golgi return to a non-swollen condition (Juniper and Roberts, 1966) (see Fig. 11.7).

Isolation

In order to understand the function of the Golgi bodies within the cell, uncontaminated preparations must be isolated and their enzymic and chemical content investigated. Unfortunately, the isolation of Golgi bodies from both plant and animal cells has not proved easy. For example, when cells are homogenized under conditions in which other organelles, including mitochondria, plastids and nuclei, remain intact, the Golgi bodies often fragment. They seem to be highly susceptible to injury, particularly to osmotic shock, which leads to swelling and rupture of the cisternae, and to the shearing forces which occur during homogenization and resuspension of the pellets. Furthermore, the

Fig. 11.5   Differentiation of membranes across the cisternal stack of Golgi bodies from the fungus *Pythium ultimum*. Those at the pole adjacent to the nuclear envelope or endoplasmic reticulum are thinner than those at the opposite pole which resemble the plasma membrane. A. A single Golgi body close to the nucleus. ×128 000. B. Nuclear envelope. C. Endoplasmic reticulum. D. Membrane of secretory vesicle adjacent to Golgi body. E. Membrane of secretory vesicle free in cytoplasm. F. Plasma membrane. G. Golgi body showing progressive increase in membrane thickness across the stack. (N = nucleus; NE = nuclear envelope; D_p = forming face of Golgi body; D_d = maturing face; SV = secretory vesicle; CW = cell wall). B-G are at the same magnification, ×256 000. Fixed and stained with glutaraldehyde-OsO$_4$ with a Ba(MnO$_4$)$_2$ post stain. Reproduced from Grove, Bracker and Morré (1968). Copyright © 1968 by the American Association for the Advancement of Science.

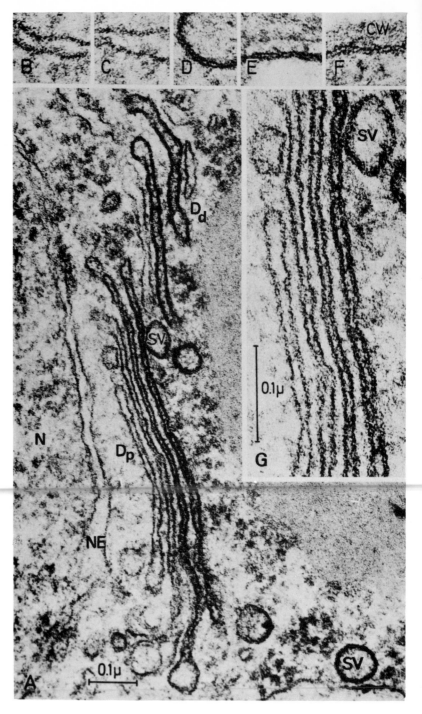

membranes of the Golgi bodies show a great tendency to coalesce, so that the diagnostic stacked structure becomes unrecognizable. For these reasons, even in the most careful work, membrane fragments, vesicles (often of unknown origin) and occasionally other organelles such as light mitochondria may contaminate the fractions.

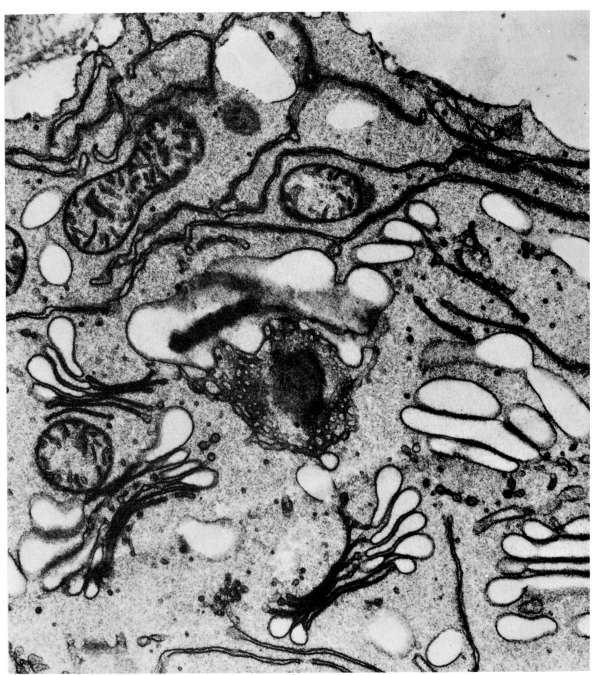

Fig. 11.6   Electron micrograph of a cell in the outer region of the root cap of *Zea mays*. Note the large vesicles and swollen cisternae. One Golgi body has been cut transversely across the cisternal stack and clearly resembles the isolated organelle shown in Fig. 11.8, except for the presence of a large secretory vesicle on the edge of the fenestrated plate. Reproduced from Mollenhauer and Morré (1966b).

Recognizable Golgi were first isolated from macerated plant cells by Morré and Mollenhauer (1964), using finely chopped onion bulbs. They also obtained reasonably pure preparations from corn roots, radish roots and cauliflower inflorescences. The tissues were chopped in a medium which included 0·5 M sucrose and 1% dextran plus various salts. Organelles were collected by differential centrifugation and identified by electron microscopy. Many of the Golgi bodies

Fig. 11.7 Electron micrograph of parts of three peripheral cells and one detached cell from the root cap of *Zea mays*. The Golgi bodies (G) are apparently pinching-off hypertrophoid vesicles which are migrating to the cell boundary and coalescing (X) between the plasmalemma and the wall. The remains of a dissolved cell wall (W) indicate where the free cell (bottom left) with a modified cell wall (MCW) was once attached. This free cell now lies embedded in the mucus which surrounds the whole root tip. m, mitochondria. ×3 800. Reproduced from Juniper and Roberts (1966).

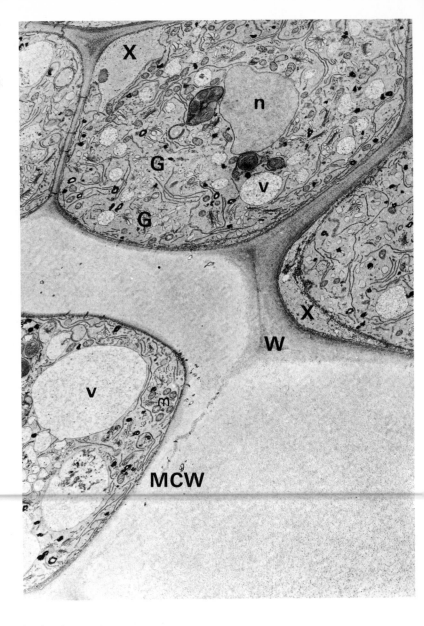

obtained were broken, and the pellet was contaminated by vesicles and other unrecognizable debris. By including the tissue fixative, glutaraldehyde, in the homogenization medium, however, whole Golgi bodies were isolated in an improved state of preservation (Morré, Mollenhauer and Chambers, 1965) (Figs. 11.4 and 11.8). The integrity of the membranes and the characteristic fenestrated structure of the peripheral layers were preserved during the isolation. However, although the preparations are suitable for chemical analysis, most biological activities are probably destroyed by such a fixation procedure. Glutaraldehyde presumably stabilizes the rather loose structure of the organelle by cross-linking protein molecules. Later, Ray, Shininger and Ray (1969) isolated a discrete, subcellular fraction from

Fig. 11.8   Isolated Golgi bodies from
onion stems. ×24 000. Reproduced
from Morré, Mollenhauer and
Chambers (1965).

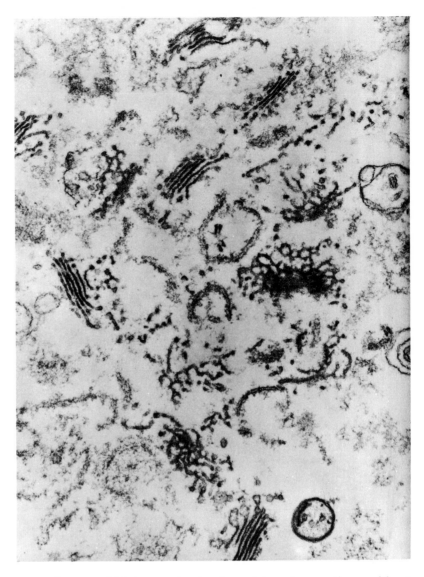

mung beans, which was active in glycosyl transfer reactions, without
using such destructive additives. The fraction contained fragmented,
and occasionally whole, Golgi bodies. It was clearly distinct both
enzymically and morphologically from the mitochondria, the nucleus
and other recognizable cell organelles. Methods and markers for the
isolation of Golgi bodies from plants are described by Morré and
Buckhout (1979).

Function

Electron microscopy has shown that in many cells which are involved
in secretion, the Golgi bodies and their associated vesicles are particularly well developed. The pattern of secretion seems to be similar in
both plants and animal cells. The product is accumulated within the

membrane bound vesicles, which are attached to Golgi bodies by one or more cisternal tubules, until the vesicles ultimately become detached, possibly after they have reached a critical size. They move across the cytoplasm to the cell membrane, where the secretory material is discharged by reverse pinocytosis. During the latter process, the membrane of the transported vesicle fuses and becomes continuous with the plasmalemma. Simultaneously, the contents of the vesicle are released. In some instances, numbers of vesicles coalesce during migration, and it is a relatively large body which eventually discharges its contents from the cell. Of course, this pattern of events as visualized by electron microscopy is static, and essentially ambiguous. The direction of movement and the interrelationship of the vesicles to the Golgi bodies can only be inferred indirectly. However, a reverse movement in which vesicles traverse the cell from the plasmalemma to the Golgi bodies, is clearly a remote possibility.

The most clear cut results which implicate the Golgi bodies in secretion have been obtained in cells which are exporting relatively large amounts of extracellular product. In such hypersecretory cells, the Golgi bodies have large swollen cisternae and produce numerous vesicles which ultimately empty their contents outside the plasmalemma. In addition, there is frequently a temporary accumulation of material between the membrane and the cell wall (Fig. 11.7). Good examples of this type of Golgi body behaviour are found in the leaf glands for the insectivorous plants *Drosera* and *Pinguicula*, which produce a sticky slime on which insects become trapped prior to digestion. This slime seems to be conveyed from the Golgi bodies via vesicles. Large swollen Golgi bodies are also observed in the outer root cap cells of many terrestrial plants (Figs. 11.6 and 11.7). Such cells typically secrete a hydrated slime, rich in polysaccharide, which passes through the cell wall, coating and presumably lubricating the tip of the root as it penetrates the soil (Fig. 11.7). In corn (*Zea mays*), the existing walls of these root cap cells simultaneously start to break down, and it is possible that the Golgi bodies are secreting enzymes which dissolve the cell wall, as well as polysaccharides which contribute to the slime (Fig. 11.7).

Golgi bodies in cell-wall deposition

Microscopical evidence

The cell wall of plant cells can itself be regarded as resulting from the extracellular deposition of polysaccharide and protein. There is now good evidence that the synthesis of the component materials takes place within the cell away from the final site of cell-wall organization, possibly within the Golgi bodies or its associated membranes. In many of these cells that are rapidly producing new cell wall material, the Golgi bodies are visibly swollen and the vesicles appear to be contributing materials to the wall by reverse pinocytosis. There are numerous examples. Conspicuously active Golgi bodies have been noted in growing pollen tubes, in epidermal walls of marine algae which secrete large amounts of polysaccharide slime, in xylem cells depositing thickened secondary walls and in expanding root hairs. Directed movement of vesicles is most obvious where wall synthesis is localized, as in the pollen tube (Fig. 11.9) and root hair, which elongate by tip growth, or in the xylem cells where the bands of secondary thickening are regularly spaced along the wall.

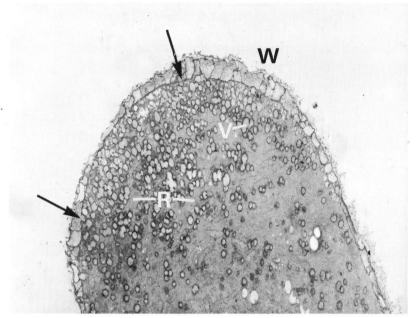

Fig. 11.9 Electron micrograph of a pollen tube tip from *Lilium*. A compartmented wall (W) is formed from Golgi-derived vesicles (V) which coalesce with each other and then fuse to make the wall. R (Endoplasmic reticulum). Fixation by glutaraldehyde. Stained by phosphotungstic acid and lead hydroxide. ×7 400. Provided by Drs W. Dashek and W. G. Rosen, Virginia Commonwealth University and University of Massachusetts, Boston.

Golgi bodies have also been implicated in the formation of the new cross walls laid down between daughter cells during late telophase of mitosis. Vesicles, which apparently originate from nearby Golgi bodies, can be seen to first aggregate and then coalesce in an orderly manner across the region of the cell plate (Fig. 11.10). This layer of material is usually considered to be the rudimentary middle lamella, on which the young primary walls are built (see Ch. 10). It should be pointed out, however, that there is now little doubt that the endoplasmic reticulum can also contribute material, enclosed in membranes, to the developing cell plate, and that such a secretory rôle is not an exclusive property of the Golgi bodies.

Cytochemical evidence

Clearly, the visual evidence related so far which implicates the Golgi bodies in the secretion of polysaccharides and in the synthesis of the cell wall is circumstantial. What concrete evidence is there to link these organelles with the secretion of particular extracellular products? Occasionally, the material within the Golgi vesicles does stain characteristically and can be recognized intracellularly, prior to extrusion, as well as in the secretory product. Often, however, the histochemical bases of these reactions, such as staining with phosphotungstic acid or

Fig. 11.10 (*a*) Electron micrograph (*a*)
of a forming cell plate in a cell of *Acer
pseudoplatanus* grown in liquid culture.
Golgi vesicles with a dense central spot
can be seen migrating towards the cell
plate region where they fuse to form
larger vesicles and finally coalesce to
create the cell plate. A number of
microtubules can be seen on the
micrograph and may be guiding the
Golgi vesicles into position. ×26 000.
Provided by E. G. Kirby and
R. M. Roberts. (*b*) Meristematic cell
of corn (*Zea mays*) root at cell division
showing the partially completed cell
plate almost joining the side-walls. A
Golgi body (GB) and fragments of ER
can be seen close to the developing
wall. Microtubules (MT) are present in
the cytoplasm close to the side walls
where vesicular material is being
incorporated into the plate. ×15 000.
(Glut/OsO$_4$/Pb/U). Provided by Dr
B. E. Juniper, University of Oxford.

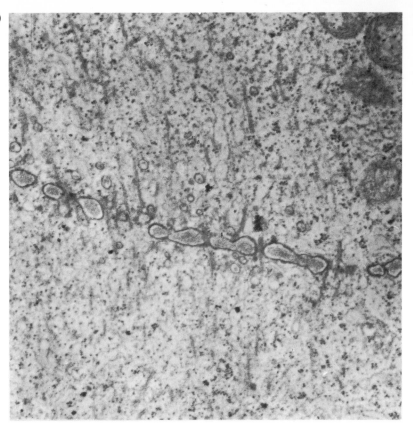

(*b*)

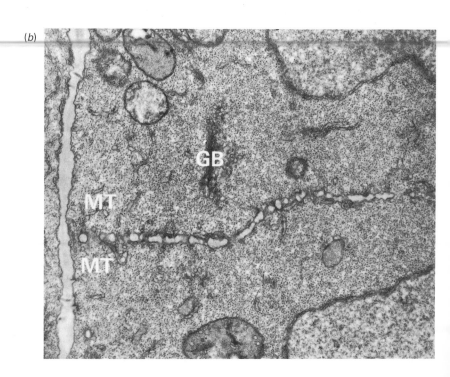

uranyl acetate, are unknown. It is unfortunate, too, that the vesicles derived from the Golgi bodies are not easily recovered for analysis when isolated from cell homogenates; they tend to rupture, be variable in size and not usually distinguishable from other membrane bound vesicles or organelles, which may or may not have their origin in the Golgi bodies (see p. 480). There are a few examples in animal tissues where the Golgi vesicles have been isolated. For example, the zymogen granules of pancreatic endocrine cells originate from fusion of several vesicles and are particularly large. They can be collected by differential centrifugation, and shown to contain the enzymes typically found in pancreatic secretory product. However, for the most part, this has not been achieved in plants. Neither have isolated Golgi bodies themselves been analysed in any detail

However, improvements with specific histochemical staining reactions at the subcellular level have, on some occasions, allowed the contents of the vesicles to be partially characterized. Pickett-Heaps (1968) used a staining reagent containing alkaline silver hexamine, which deposited metallic silver on aldehydic groups on tissue sections. Such aldehydic groups could be exposed on polysaccharides by permanganate fixation or by periodate oxidation. Post-treatment with the silver stain then deposited a fine coating of silver over areas of the cell which contain the partially oxidized polysaccharides. The Golgi bodies, their vesicles and the cell wall were all clearly stained by this procedure in several tissues of root and shoot of wheat seedling, indicating that each contained polysaccharide (Fig. 11.11). The cisternae on the maturing face could also frequently be distinguished from those on the forming face because they stained more heavily. These experiments support the argument that these organelles possess a functional polarity. Dashek and Rosen (1966) have used a hydroxylamine-ferric chloride procedure to detect polyuronic acid material of pectin in germinating pollen. The initial reaction with hydroxylamine yielded the pectic hydroxamic acids which form insoluble complexes with ferric ions. The complexes appeared electron dense on micrographs, and were localized exclusively in Golgi vesicles and in the cell wall. If the sections were treated with pectinase prior to staining, the reactive materials were removed, again indicating that the Golgi bodies were concerned with the movement of cell-wall polysaccharides.

There are one or two remarkable instances where the secreted materials can be recognized within the cell prior to extrusion, not by a specific staining reaction but by their distinct morphology. For example, in dividing cells of *Acer pseudoplatanus* cultures, the Golgi vesicles possess a dark inclusion and can be recognized up to the point they fuse to form the cell plate (Roberts and Northcote, 1970) (see Fig. 11.10a), strongly suggesting that they contribute to the young wall. A second, more striking example is provided by the polysaccharide scales produced by Chrysophycean flagellate algae which can be identified in section, both outside the plasmalemma as an organized layer around the organelle, and inside the cell within Golgi vesicles. These scales are assembled in a stepwise manner within the cisternae of the single Golgi apparatus (see Dodge, 1973; Brown *et al.*, 1970; Manton and Leedale,

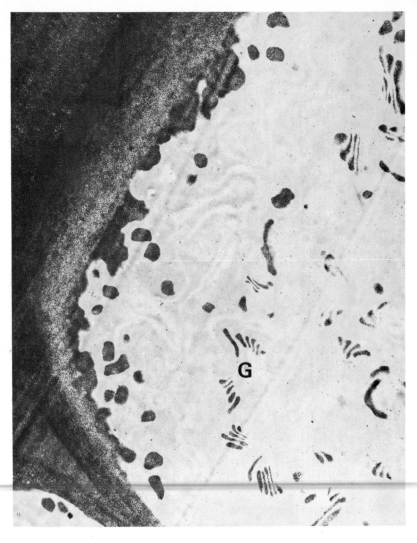

Fig. 11.11 Electron micrograph of outer root cap cells of wheat (*Triticum vulgare*). The sections were fixed in permanganate and then post-treated with periodate before being embedded. The sections were finally stained with 1% buffered hexamine containing silver nitrate. Silver is deposited over areas oxidized by periodate. Note the staining of Golgi bodies (G) and the heavy staining of the slime layer within the cell wall. Some of the Golgi bodies show a gradation of staining across the cisternal stack. ×6 400. Reproduced from Pickett-Heaps (1968).

1961 (Fig. 11.12). Because of their large size, Brown has been able to record the formation of such scales by cinematography (Brown, 1969), so that the magnitude of the process and the direction of secretion can, in this instance, hardly be disputed. The scales themselves consist of an organized assembly of non-cellulosic fibrillar material, microfibrillar cellulose linked covalently to protein, and an amorphous polysaccharide matrix (Brown *et al.*, 1973) (Fig. 11.13); at a certain stage of growth, calcium carbonate crystals are deposited on the margins of the scale. The entire process takes place within the confines of a single cisterna, and provides what is probably the best existing evidence implicating the Golgi apparatus in cell wall deposition.

Autoradiographic evidence

A dynamic picture of Golgi body secretion has also been obtained in experiments using autoradiography. A radioactive compound which is

Fig. 11.12. The Golgi apparatus of *Pleurochrysis* shown in median section. Note the differentiating stack of cisternae beginning with compact cisternal development at the upper right hand corner and progressing toward the cell surface with inflation. Two types of central dilations are seen in the proximal region of the Golgi apparatus: arrow 1 depicts a large dilation with visible product and arrow 2 shows a smaller electron transparent dilation that is surrounded by heavy electron-dense staining of material next to the membrane. The first visible scale elements (radial microfibrils) appears as a double layer in the cisterna at arrow 3. Mature scales are shown at arrow 4 where amorphous material coats the microfibrillar scale network. Scale secretion occurs at arrow 5 where the distal-most cisterna fuses with the plasma membrane and releases the scale to the cell surface where is stratifies to form the cell wall. Arrow 6 depicts a cluster of polysomes in between the cisternal stacks probably involved in a specific stage of scale assembly. Whirls of membrane (arrow 7) appear in the forming region of the Golgi apparatus. Polysomes and vesicles appear along the cisternae of varying stages of differentiation along the cisternal stack. ×18 000. Stained with lead citrate and uranyl acetate. Reproduced from Brown *et al* (1973).

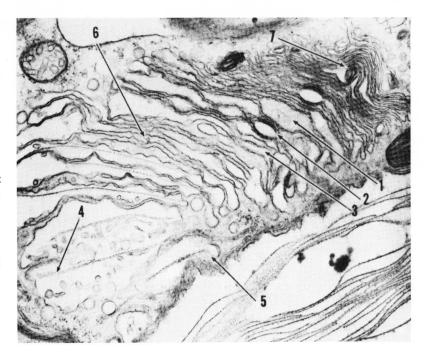

a known precursor of the macromolecular material under investigation is administered to the tissues for a brief period of time (the 'pulse'). This is followed by a 'chase' with nonradioactive material to dilute out the label (i.e. reduce its radioactive concentration) in precursor pools. At various times after the pulse, samples are fixed and prepared for microscopy and autoradiography (see p. 25). By this method, the site within the cell at which isotope is initially incorporated into macro-molecular material can be identified, and its subsequent movement through the cell and out into the secretory product followed. North-cote and Pickett-Heaps (1966) supplied D-glucose-^{3}H to roots of wheat seedlings. Non-cellulosic polysaccharide containing labelled galactose and a number of other sugars were rapidly formed, but protein and nucleic acids did not immediately become radioactive in these cells after such a pulse. The autoradiographs therefore gave a reasonable picture of the synthesis of certain types of polysaccharide but not of other macromolecules within the cell. Radioactive product was de-tected first in the Golgi bodies (Fig. 11.14), subsequently in the vesicles derived from them, and finally in the cell wall and mucous material surrounding the outer root cap. These experiments indicated that insoluble polysaccharide was formed initially in the Golgi bodies and that the movement of vesicles represented an outward rather than inward migration of material. Within 10 min of the pulse, polysaccha-ride had already started to accumulate outside the cell, and by 30 mins the Golgi bodies had lost most of their radioactivity. Synthesis and transport of product was therefore very rapid.

More recently, Northcote and coworkers (see Northcote, 1974) performed a parallel type of study in which roots were given a 'pulse' of radioactive glucose and then homogenized, rather than sectioned,

Fig. 11.13. Negatively stained scale from a vegetative cell of *Pleurochrysis* in surface view with amorphous material in between the radial "concentric" cellulose microfibrils. Note the two axes of symmetry for the radial network. The "concentric" microfibrils appear to spiral and some of them end on the periphery at arrows. ×84 000. From Brown *et al* (1973).

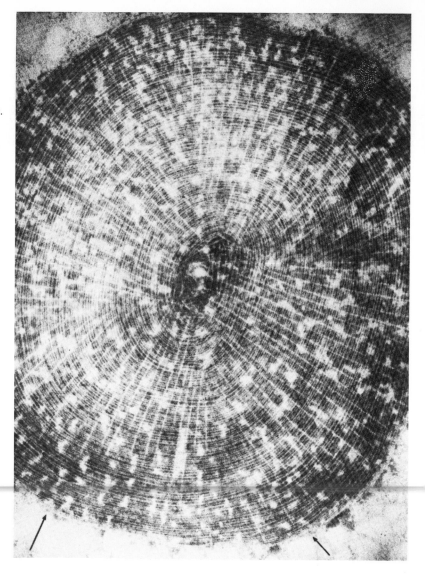

and discrete sub-cellular membrane fractions analysed for their content of complex carbohydrates. These studies were also consistent with the view that the root-cap slime was elaborated intracellularly within the Golgi bodies, and then moved in membrane-bound vesicles to the cell surface.

Dashek and Rosen (1966) have also used autoradiography to show that label from *myo*-inositol-2-^{3}H, a specific precursor of pectin in pollen and other tissues (see p. 466), and from methionine-methyl-^{3}H, an established methyl donor to polygalacturonic acid, accompanies the Golgi vesicles to the growing tip of pollen tubes, and is subsequently deposited in the cell wall. Treatment of the sections with pectinase prior to autoradiography removed most of the label from the sections. A crude preparation of Golgi-derived vesicles from such germinating pollen tubes have been analysed and shown to contain the

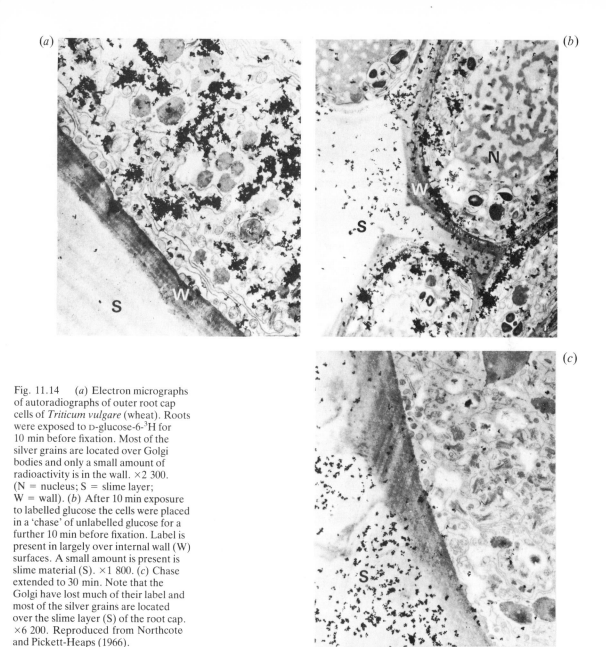

(a)

(b)

(c)

Fig. 11.14   (a) Electron micrographs of autoradiographs of outer root cap cells of *Triticum vulgare* (wheat). Roots were exposed to D-glucose-6-^{3}H for 10 min before fixation. Most of the silver grains are located over Golgi bodies and only a small amount of radioactivity is in the wall. ×2 300. (N = nucleus; S = slime layer; W = wall). (b) After 10 min exposure to labelled glucose the cells were placed in a 'chase' of unlabelled glucose for a further 10 min before fixation. Label is present in largely over internal wall (W) surfaces. A small amount is present is slime material (S). ×1 800. (c) Chase extended to 30 min. Note that the Golgi have lost much of their label and most of the silver grains are located over the slime layer (S) of the root cap. ×6 200. Reproduced from Northcote and Pickett-Heaps (1966).

non-cellulosic polysaccharide components typical of the elongating wall (Van der Woude, Morré and Bracker, 1971).

In other experiments, Pickett-Heaps (1967) showed that label from glucose was massively incorporated into the cell plate region between wheat epidermal cells that had recently divided and not just into the more mature wall. His micrographs suggest that the nearby Golgi bodies and their vesicles were simultaneously labelled and were contributing to the wall. However, in this case, the organelles were considerably smaller than those found in the outer root-cap, and the autoradiographic resolution obtained was not as precise, so that the β-particles could not be unequivocally said to have originated from the Golgi bodies. Nevertheless, the results do seem to confirm the earlier, purely microscopical, evidence that the cell plate in epidermal cells was at least partly derived from the Golgi bodies. It is also significant that, in the meristematic cells of the deeper lying cortex and stele, where there is some controversy as to the origins of the vesicles which fuse to form the cell plate, there was no obvious localization of radioactivity over the Golgi bodies during the build up to the young wall. In these cells, radioactive polysaccharide seemed to originate in the endoplasmic reticulum. Pickett-Heaps (1968) has also shown that Golgi bodies of these cells do not stain histochemically for polysaccharide. He points out, however, that partially polymerized material and the more soluble types of polysaccharide might well have been leached from the sections prior to both staining and to autoradiography.

Secretion of cellulose

There is, therefore, reasonably definitive evidence, in certain instances, to implicate the Golgi bodies in the packaging and possible synthesis of non-cellulosic polysaccharide material destined for secretion into the wall or into extracellular slime. There have been some doubts expressed, however, as to whether cellulose is transported via the Golgi vesicles (see p. 463). Its characteristic microfibrils are not observed intracellularly and so the molecules are presumably organized, even if not synthesized, at the cell surface. The scales of the flagellate algae mentioned previously, however, do often contain a $\beta(1 \rightarrow 4)$ glucan as well as polysaccharides which contain fucose and galactose, and these are undoubtedly secreted from the Golgi bodies. Nevertheless the nature of this cellulose is peculiar since it is covalently linked to protein. In general, it seems that the Golgi and possibly other endomembrane systems are responsible for the elaboration of pectins and hemicelluloses, and not cellulose. The $\beta(1 \rightarrow 4)$ glucan synthetase activity, which has been attributed by some workers to be present in the Golgi bodies of higher plants, is probably due to an enzyme system involved in the synthesis of a non-cellulosic glucan or to contamination by plasmalemma.

Other secretory activities

Cell walls usually contain a composite mixture of different kinds of polysaccharide and possibly glycoproteins, as well as several other components such as lignin (see Ch. 10). These compounds may also originate in the cytoplasm, and are moved out of the cell in some way. In pollen tubes, Dashek and Rosen (1966) have presented autoradiographic evidence that, in addition to polysaccharide, macromolecules containing hydroxyproline can move to the growing cell wall in Golgi vesicles. Because this imino acid is a major component of a recently characterized cell-well glycoprotein, their results imply that the Golgi bodies may be responsible for the outward movement of glycoproteins as well as of the more conventionally recognized type of cell-wall polysaccharide. There is evidence that lignin may also be secreted by the Golgi bodies, because material that stained deeply with silver-hexamine (prior to periodate or permanganate oxidation) could be detected in Golgi vesicles, as well as in the lignified cell wall of xylem cells from wheat seedlings. Golgi secretions are not restricted to organic materials, however. The diatom *Amphipleura pellucida,* for example, deposits silica around its exterior surface, and, because this material is electron dense, its movement can be traced from the Golgi bodies, where it is probably concentrated, by way of vesicles to the wall. Water can also be transported. The algae *Glaucocystis* and *Vacuolaria* produce vacuoles derived from the Golgi which maintain the organism at a constant size. Similarly, in wheat stomatal cells, the Golgi vesicles contribute fluid to the main vacuole. Presumably, therefore, the vacuolar membranes are also contributed by the Golgi bodies (see Morré and Mollenhauer, 1976). However, despite many suggestions, based largely on evidence derived from studies of electron micrographs, a definitive account of how vacuoles form is still lacking. This is discussed more fully in Chapter 12.

The extent of the involvement of the Golgi apparatus in internal movement of materials is not known. Autoradiography has indicated that the canals of the endoplasmic reticulum in mammals do serve as intracellular channels for the movement of protein. The protein is usually assembled into globules, within vesicles or tubules, and then moved to the Golgi bodies prior to extrusion. However, there is no evidence to rule out the possibility that protein and other material cannot be distributed internally by the Golgi apparatus. Furthermore, little is known about the movement and secretion of small molecules. They probably do not require packaging prior to crossing membranes, but they may well diffuse along and even be concentrated within the canals and cisternae of the endoplasmic reticulum and Golgi bodies.

Lomasomes

Various vesicular and membranous structures have been seen between the cell wall and the plasmalemma of both higher and lower plants, particularly in fungi, which may have some sort of rôle in cell-wall elaboration (Fig. 11.15). Such formations do not seem to be artefacts as they can be observed using several different methods of fixation. Although they are extremely diverse in appearance and their ontogeny and precise function are unknown, they are often classified together as *lomasomes*; they may be formed either as an elaboration of the plasmalemma or directly from cytoplasmic vesicles which pass out into the cell wall through the plasmalemma. In either instance, they probably originate from the Golgi bodies and are biochemically related to them. No definite rôle in wall formation has been established, however, but recent experiments using autoradiography have established that high concentrations of newly formed non-cellulosic polysaccharides are found in cell walls of celery collenchyma close to the lomasomes (Fig. 11.15).

Fig. 11.15   (a) Lomasome (or paramural body) in celery collenchyma (×44 000). This complex body consisting of membranes in the form of vesicles and complex myelin type figures is found close to the cell wall and is believed to have some rôle in wall elaboration. Such a rôle is suggested in (b) which is an autoradiograph of an adjacent section. The celery petiole had, in this experiment, been supplied with *myo*-inositol-^{3}H, a precursor of cell-wall pentose and uronic acid. The presence of silver grains over the cell wall close to the lomasome and over the lomasome itself, suggests that this structure is involved in polysaccharide formation. However, an alternative hypothesis is that the *myo*-inositol is incorporated into the complex lipids of the membranes. (×22 000). Provided by D. G. Cox, University of Oxford.

(a)

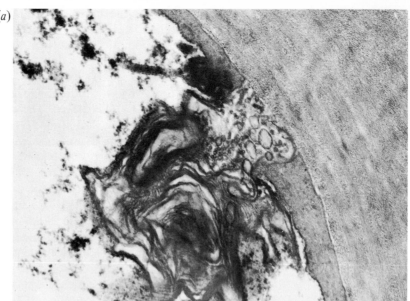

(b)

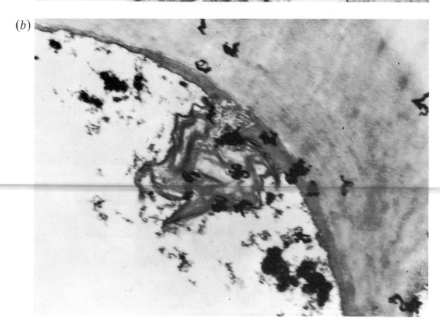

Enzymic activity and chemical content of the Golgi bodies

The Golgi bodies that have been so far isolated have a relatively low enzyme content compared to the large number of enzymes found in such organelles as the mitochondrion and chloroplast. The membranes contain much phospholipid and lipoprotein, but only very low amounts of nucleic acid. When the particles are resuspended, this latter material can usually be washed away, and there is no reason to believe that the Golgi bodies are semi-autonomous organelles, possessing their own ribosomes, DNA and capacity for DNA-dependents RNA synthesis.

Polysaccharide synthetase activity

Because Golgi bodies are involved in the secretion of polysaccharide material into the cell wall, it is clearly relevant at this stage to question whether they are involved in the enzymic synthesis of polysaccharide. Alternatively, are they merely packaging the material and concentrating it in such a way that it can be exported from the cell? Several plant polysaccharides, albeit incomplete, have been synthesized enzymically by glycosyl transfer from nucleoside diphosphate sugars, using the enzymes present in particulate preparations isolated from a variety of plant tissues, which have included corn, beans, oats and lupin. The enzymes which are membrane bound have usually been sedimented in the ultracentrifuge somewhere between 20 000 and 40 000 g (see p. 461). The particles have not been well defined cytologically, and there has been considerable speculation about their nature. Are they part of the Golgi bodies or fragmented parts of the endoplasmic reticulum? When such particulate preparations from mung beans were analysed, they contained all the representative types of polysaccharides which are normally recovered from the cell walls, including a well-defined fraction soluble in ammonium oxalate, which had a typically pectic composition.

Ray and co-workers (1969) later showed that a number of these polysaccharide synthetases (i.e. enzymes which transferred the sugar moiety of nucleoside diphosphate sugars to polysaccharide—see p. 456) can be isolated from etiolated pea stem segments in a discrete fraction by isopycnic density gradient centrifugation in sucrose solutions. The particles had an effective density of about 1·15 and showed polysaccharide synthetase activity using ^{14}C-labelled UDP-glucose, GDP-glucose and UDP-galactose as precursors. Electron micrographs of particles isolated on such sucrose density gradients revealed a population of segments of smooth membranes bearing vesicles at their edges. When the synthetase particles were isolated from chopped rather than ground tissues, a procedure which led to less fragmentation of the particles, occasional whole Golgi bodies were isolated in the synthetase peaks, and the membrane fragments were, on the whole, larger than those obtained from homogenized tissues. They concluded that the synthetase particles were Golgi membranes, and that the Golgi bodies are not only the sites of polysaccharide secretion within the cell, but also the sites of polysaccharide biosynthesis. Further evidence for such a synthetic rôle has been given by Harris and Northcote (1971) who purified Golgi bodies from tissues that had been fed D-glucose-^{14}C for a short period of time. The isolated fraction contained a number of newly synthesized, non-cellulosic polysaccharides, but not cellulose itself.

Enzymic markers

In cell fractionation studies, the premise has frequently been made that each morphologically identifiable cell component is earmarked by some defined biochemical characteristics or set of characteristics which distinguish it from other cell components (see Ch. 1). For example, the enzyme succinate dehydrogenase seems to be uniquely associated with the inner mitochondrial membrane, and is, therefore, a very useful marker for mitochondria, both in its purification and in assessing the degree to which mitochondria contaminate other cell fractions. Unfortunately, the Golgi body is less well defined in terms of specific enzymes. In animal cells, a glycosyl transferase which catalyses the transfer of D-galactose from UDP-galactose to either N-acety-glucosamine (giving rise to N-acetyllactosamine, an analogue of the milk sugar lactose) or to a glycoprotein acceptor, is probably the most widely used enzymic marker. Nucleoside diphosphatases, which cleave off orthophosphate from IDP, UDP, GDP, etc., are also employed, but these enzymes have an unknown function. Until now, the latter have probably provided the best markers for plant Golgi bodies. IDPases, for example, have been shown by histochemical procedures to be constituents of the cisternae. Moreover, plant fractions enriched in identifiable Golgi membranes are consistently enriched in an IDPase which exhibits 'latency' (Morre, Lembi and Van der Woude, 1977). That is, the enzyme increases markedly in activity upon cold storage of homogenized preparations.

Because Golgi bodies are highly susceptible to breakage, such enzymic markers will continue to assume importance in diagnosing isolated membrane fractions for Golgi content. However, it must be stressed that no marker seems to be confined exclusively to this organelle. This is not altogether surprising since the Golgi bodies seem to form a central part of a functional continuum of cytoplasmic membranes (see p. 497). Such a model predicts that absolute markers will be the exception rather than the rule.

Glycosylation of protein

As pointed out in Chapter 10, the polysaccharide material of animal cells is usually linked to protein. The protein moiety of the molecule is synthesized on the ribosomes, and the carbohydrate becomes linked only after polypeptide formation. Several workers have isolated Golgi-rich fractions from liver and other tissue by differential centrifugation and by discontinuous sucrose density gradient centrifugation. These fractions contained enzymes which can catalyse a variety of glycosyl transfer reactions. Some investigators prefer to believe that most of the cellular potential for glycosyl transfer is confined to the Golgi bodies. Others consider that the enzymes have a more general distribution within the membranes of the cell. Recently, it has been shown that the enzymes involved in terminating the oligosaccharide chains of plasma glycoproteins are localized specifically in this Golgi-rich fraction, although core sugars, linked nearer to the peptide backbone, are added as soon as the nascent peptide begins to emerge through the membrane into the lumen of the rough endoplasmic reticulum. Thus, although polysaccharide and glycoprotein synthesis may be initiated on the endoplasmic reticulum, it is probably completed in the Golgi bodies, prior to transport of the macromolecules from the cell.

The same general situation may well be true in plant systems. Polysaccharides may not be synthesized as a whole within the Golgi bodies, but they may be terminated there, and accumulated temporarily in semi-liquid form. Clearly, an understanding of integrated polysaccharide biosynthesis within the plant cell will depend upon knowing the interrelationship between the endoplasmic reticulum and the Golgi bodies, particularly as both are believed to contribute materials to the cell wall. Furthermore, there must be some link, even in plant cells, between the respective sites for protein and polysaccharide biosynthesis, because glycoproteins have now been shown to occur in cell wall, and many extracellular (secreted) enzymes bear covalently bound carbohydrate.

Origin of Golgi bodies and relationship to endoplasmic reticulum and other membrane systems

Because the number of Golgi bodies per cell is either maintained at a constant level or even increased as cells grow and divide, there is presumably some way in which the organelles are reproduced. In a number of tissues, the total number of Golgi bodies per cell has been shown to increase several-fold as the physiological state of the tissues changes. There are numerous explanations as to how the Golgi bodies replicate, but no definite experimental evidence to support these theories, and so none will be discussed at length. Because they are often found localized close to the cell plate after mitosis, it has been suggested that the Golgi arise *de novo* from the phragmoplast, but this theory has not been substantiated in any way. Alternative explanations have included division by longitudinal splitting, by fragmentation, or by constriction of an existing organelle. Most of the evidence for the latter processes is based on observing organelles, apparently paired or separating, on electron micrographs. These pictures could well represent chance artefacts, obtained in sectioning, or in over rigorous selection of data. However, in rapidly dividing tissues, many of the Golgi seem rudimentary, consisting of only a few cisternae, and appearing as though they had only recently been formed by replication from existing, larger bodies or from an existing membrane system such as the endoplasmic reticulum. The most likely possibility is that Golgi bodies arise as a differentiated outgrowth of the endoplasmic reticulum (see Fig. 11.16). The latter appears to produce vesicles which fuse to form Golgi cisternae on the forming face. Some tubular connections may also exist since it is difficult to distinguish the two types of element on electron micrographs. Thus, the endoplasmic reticulum not only supplies secretory product to the Golgi bodies, but membrane as well.

At the maturing face vesicles are released which, in secretory cells, fuse with the plasmalemma. Thus, the Golgi bodies exist as a transition zone in a functional continuum between the internal membrane system of the cell, which includes the endoplasmic reticulum and probably the nuclear envelope as well, and the cell surface. It is known that a good deal of the phospholipid and much of the membrane protein synthesis for this *endomembrane* system is localized in the endoplasmic reticulum from where it is distributed to other parts of the cell, yet the

Fig. 11.16. Schematic representation of portions of the internal membrane system of a secretory cell in an attempt to show the possible, but still hypothetical relationship between the nuclear envelope (NE), the rough and smooth endoplasmic reticulum (RER and SER) and a dictyosome or Golgi body (D). A, Plane view as seen in thin section; B, space filling view shown to illustrate the three dimensional complexity. Cisternae of the endoplasmic reticulum adjacent to the forming face (FF) may be characterized by a part rough, part smooth surface. Small vesicles (transition vesicles, TV) seem to bud from the endoplasmic reticulum and fuse with the forming face. Tubular connections may also join the SER with Golgi cisternae. The maturing face (MF) gives rise to secretory vesicles (SV) which migrate to the cell surface and fuse with the plasma membrane (PM). Other abbreviations: CV, coated vesicle, ECS, extra-cellular space. This model is based largely on observations made in animal cells, but may be equally valid for plants. (Reproduced from Morré and Ovtracht, 1977).

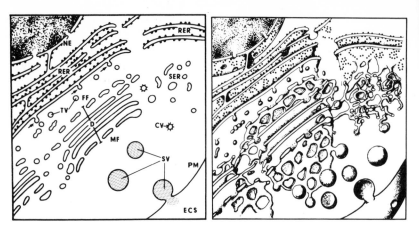

appearance, composition and functional properties of the membrane changes in the progression from the ER to the cell surface. Presumably, movement of membrane material destined for permanent incorporation into the plasmalemma is selective or else extensive turnover and modification of constituents must occur during the transition. For a detailed discussion of the concept that the Golgi bodies are organelles involved in membrane differentiation, the reader is referred to a number of recent reviews by Morré and his associates, which are listed below in the section on *Further Reading*.

The extensive fusion of Golgi-derived vesicles with the surfaces of secretory cells has other important implications with regard to membrane dynamics. In many instances, such cells are neither dividing nor elongating. They have, therefore, a relatively constant surface area, yet are continually receiving large quantities of membrane with fuses with the plasmalemma. How, under these circumstances, can a constant membrane function be maintained and excess membrane be disposed of? There is certainly no evidence in mammalian or plant cells that surface membrane proteins or lipids have a high rate of degradation. This suggests that a considerable portion of the membrane added to the plasmalemma through fusion with secretory vesicles is rapidly and selectively returned inside for reuse. Although strong evidence for such a process is lacking, it is tempting to speculate that some sort of cycling of membrane between the Golgi and the plasmalemma is always occurring in secretory cells. Certainly the idea of a one-way system of membrane flow cannot accommodate the high rates of vectorial transport that occurs in the non-growing outer root cap cells of grasses or other secreting cells. Presumably, there is also a cyclic movement of membrane between the Golgi and ER. Thus, the different membrane systems are able to maintain relative constancy of structure and function in the face of massive movement of secretory material through the cell.

A reverse, but very similar, conceptual difficulty arises when one considers the process of endocytosis (see p. 154). In animal, and probably plant, cells also, considerable amounts of material can be taken up in small vesicles budded from the surface membrane. If this process is extensive, recycling of membranes back to the surface must occur. Whether, in this case, the Golgi bodies are involved directly is unclear. Nevertheless, the reader should bear in mind that the relationship between the different membrane systems and organelles in eukaryotic cells is extremely complex and far from understood, although we anticipate that considerable progress will be made on this topic during the next decade.

Further reading

Mollenhauer, H. H. and Morré, D. J. (1966a) Golgi apparatus and plant secretions. *Ann. Rev. Plant Physiol.*, **18**, 27.

Morré, D. J. (1977) Membrane differentiation and the control of secretion: A comparison of plant and animal Golgi apparatus. In *International Cell Biology*. Eds. B. R. Brinkley, and K. R. Porter. Rockefeller Press, New York, p. 293.

Morré, D. J. (1977) The Golgi apparatus and membrane biogenesis. In *Cell Surface Reviews*, Vol. 4. Eds. G. Poste, and G. L. Nicolson, North-Holland, Amsterdam–New York–Oxford, p. 1.

Morré, D. J. and Mollenhauer, H. H. (1976) Interactions among cytoplasm, endomembranes and the cell surface. In *Transport in Plants*. Encyclopedia of Plant Physiology, New Series, Vol. 3. Eds. C. R. Stock, and U. Heber, Springer-Verlag, Berlin, Heidelberg and New York, p. 288.

Morré, D. J. and Ovtracht, L. (1977) Dynamics of the Golgi apparatus: Membrane differentiation and membrane flow. *Int. Rev. Cytol.*, **5**, 61.

Northcote, D. H. (1971) The Golgi apparatus. Endeavour, **30**, 26.

O'Brien, T. P. (1972) The cytology of cell wall formation in some eukaryotic cells. *Botan. Rev.*, **38**, 87.

Rothman, J. E. (1981) The Golgi apparatus: two organelles in tandem. *Science*, **213**, 1212.

Whaley, W. G. (1975) *The Golgi Apparatus*. Cell Biology Monographs, Vol. 2. Springer-Verlag, Vienna.

Whaley, W. G. and Dauwalder, M (1979) The Golgi apparatus, the plasma membrane and functional integration. *Int. Rev. Cytol.*, **58**, 199.

Whaley, W. G., Dauwalder, M. and Kephart, J. E. (1972) Golgi apparatus: Influence on cell surfaces. *Science N. Y.*, **175**, 596.

Literature cited

Brown, R. M. (1969) Observations on the relationship of the Golgi apparatus to wall formation in the marine Chrysophycean alga *Pleurochrysis scherffelii* Pringsheim. *J. Cell Biol.*, **41**, 109.

Brown, R. M., Franke, W. W., Kleinig, H., Falk, H. and Sitte, P. (1970) Scale formation in Chrysophycean algae. *J. Cell Biol.*, **45**, 246.

Brown, R. M., Herth, W. W., Franke, W. W. and Ramanovicz, D. (1973) The role of the Golgi apparatus in the biogenesis and secretion of a cellulosic glycoprotein in *Pleurochrysis*. A model system for the synthesis of structural polysaccharides. In *Biogenesis of Plant Cell Wall Polysaccharides*. Ed. F. Loewus, Academic Press, New York. p. 207.

Clowes, F. A. A. and Juniper, B. E. (1969) *Plant Cells*. Blackwell Scientific Publications, Oxford.

Dashek, W. V. and Rosen, W. G. (1966) Electron microscopical localization of chemical components in the growth zone of lily pollen tubes. *Protoplasma*, **61**, 192.

Dodge, J. D. (1973) *The Fine Structure of Algal Cells*. Academic Press, New York.

Grove, S. N., Bracker, G. E. and Morré, D. J. (1968) Cytomembrane differentiation in the endoplasmic reticulum-Golgi apparatus-vesicle complex. *Science, N. Y.*, **161**, 171.

Harris, P. J. and Northcote, D. H. (1971) Polysaccharide formation in plant Golgi bodies. *Biochim Biophys. Acta*, **237**, 56.

Juniper, B. E. and Roberts, R. M. (1966) Polysaccharide synthesis and the fine structure of root cells. *J. Roy. micr. Soc.*, **85**, 63.

Manton, I. and Leedale, G. F. (1961) Observations on the fine structure of *Paraphysomonas vestita* with special reference to the Golgi apparatus and origin of scales. *Phycologia*, **1**, 37.

Mollenhauer, H. H. and Morré, D. J. (1966b) Tubular connections between dicyto-somes and forming secretory vesicles in plant Golgi apparatus. *J. Cell Biol.*, **29**, 373.

Mollenhauer, H. H., Morré, D. J. and Van der Woude, W. J. (1975). *Microscopic*, **31**, 257.

Mollenhauer, H. H., Whaley, W. G. and Leech, J. H. (1961) A function of the Golgi apparatus in outer root cap cells. *J. Ultrastruct. Res.*, **5**, 193.

Morré, D. J. and Buckhout, T. J. (1979) Isolation of Golgi apparatus. In *Plant Organelles*. Ed. E. Reid. Ellis Horwood, Chichester. p. 117.

Morré, D. J., Lembi, C. A. and Van der Woude, W. J. (1977) A latent inosine-5′-diphosphatase associated with Golgi apparatus-rich fractions from onion stem. *Cytobiologie*, **16**, 72.

Morré, D. J. and Mollenhauer, H. H. (1964) Isolation of the Golgi apparatus from plant cells. *J. Cell Biol.*, **23**, 295.

Morré, D. J., Mollenhauer, H. H. and Chambers, J. E. (1965) Glutaraldehyde stabilization as an aid to Golgi apparatus isolation. *Expl. Cell Res.*, **38**, 672.

Northcote, D. H. (1974) Membrane systems of plant cells. *Phil. Trans. R. Soc. London. B.*, **268**, 119.

Northcote, D. H. and Pickett-Heaps, J. D. (1966) A function of the Golgi apparatus in polysaccharide synthesis and transport in the root-cap cells of wheat. *Biochem. J.*, **98**, 159.

Pickett-Heaps, J. D. (1967) Further observations on the Golgi apparatus and its functions in cells of the wheat seedling. *J. Ultrastruct. Res.*, **18**, 287.

Pickett-Heaps, J. D. (1968) Further ultrastructural observations on polysaccharide localization in plant cells. *J Cell Sci.*, **3**, 55.

Ray, P. M., Shininger, T. L. and Ray, M. M. (1969) Isolation of β-glucan synthetase particles from plant cells and identification with golgi membranes. *Proc. natl. Acad. Sci. U.S.A.*, **64**, 605.

Roberts, K. and Northcote, D. H. (1970) The structure of sycamore callus cells during division in a partially synchronized suspension culture. *J. Cell Sci.*, **6**, 299.

Van der Woude, W. J., Morré, D. J., and Bracker, C. E. (1971) Isolation and characterization of secretory vesicles in germinated pollen of *Lilium longiflorum*. *J. Cell Sci.*, **8**, 331.

12 Lysosomes and vacuoles

The history of the discovery and characterization of lysosomes is a short but exciting one. The particles were first identified in mammalian cells by de Duve and co-workers in 1955. Unlike many other cellular organelles, they were not discovered by microscopic methods but by the technique of cell fractionation. It was noticed that certain hydrolytic enzymes with acid pH optima, such as acid phosphatase, appeared in a mitochondrial fraction isolated by centrifugation from homogenates prepared from rat livers. These hydrolytic enzymes were not characteristic of the oxidative and respiratory processes normally associated with the mitochondria. Further centrifugation of the mitochondrial fraction separated two sub-fractions which were originally called light and heavy mitochondria, the light fraction being found to contain most of the hydrolytic enzyme activity. It is now known that this activity is contained in a special type of cellular organelle, similar in size to mitochondria but bound by a single membrane, which was given the name of *lysosome*, meaning lytic or digestive body. Since these initial experiments, lysosomes have been found in many different types of animal cell and are known to play an important rôle in a variety of physiological and pathological processes. Similarly, there are a number of stages in the growth and development of plant cells and tissues in which lysosome-like organelles could have an important function; there is now considerable biochemical and cytological evidence for their existence.

The lysosome concept in animal cells

Lysosomes are now known to contain a wide range of enzymes which catalyse the breakdown of all the major classes of biological molecules (Table 12.1). This general digestive action raises the question of why these enzymes do not normally break down these compounds within the cell. In their original studies de Duve and co-workers noticed that the hydrolytic enzymes of isolated lysosomes were only fully active if the organelles were damaged by alternate freezing and thawing, osmotic shock or detergents. These methods were all known to disrupt biological membranes. They therefore concluded that the lysosomes are normally surrounded by a lipoprotein membrane, making the internal enzymes latent or inaccessible to external substrates and so giving protection to the cell itself. Thus lysosomes were defined as

Table 12.1 Enzymes reported to be concentrated in animal lysosomes

Enzyme	Substrate	Products of reaction
Acid phosphatase	o-Phosphoric monoesters	Alcohols and inorganic phosphate
Acid ribonuclease	RNA	Large oligonucleotides
Acid deoxyribonuclease	DNA	Large oligonucleotides
Aryl sulphatases A and B	Phenol sulphate esters	Phenols and inorganic sulphate
Glycosidases	Polysaccharides and glycosides	Alcohols and sugars
Cathepsins	Proteins and peptides	Smaller peptides
Collagenase	Peptides containing proline, e.g. collagen, gelatin	Smaller peptides
Phosphatidic acid phosphatase	α-Phosphatidic acids	Diglycerides and inorganic phosphate
Lipase	Triglycerides	Diglycerides and fatty acids
Phospholipase	Phospholipids, e.g. lecithin	Lysolecithin and fatty acids
Esterase	Carboxylic esters	Alcohols and carboxylic acids

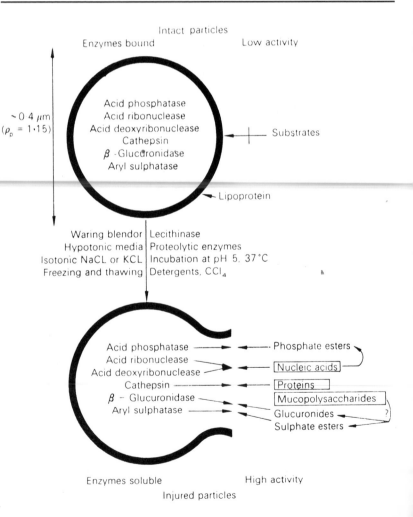

Fig. 12.1   Early model of the lysosome concept in animal cells. The membrane is normally impermeable to the passage of both substrates and enzymes. Isolated and undamaged lysosomes may thus show enzyme latency. When the membrane permeability is increased by natural or artificial causes, substrates can penetrate and enzymes may be released. The membrane need not be completely ruptured as indicated here. Redrawn from de Duve (1959).

cytoplasmic particles containing high concentrations of hydrolytic enzymes with acid pH optima and showing latency of these enzymes. This concept is illustrated in Fig. 12.1. Electron microscopy of isolated lysosome preparations showed them to be spherical particles surrounded by a single membrane, with no detailed internal structure (Fig. 12.3). The use of a cytochemical stain for the marker lysosomal enzyme, acid phosphatase (see p. 27), has allowed these particles to be easily recognized in tissue sections and has shown them to be of widespread occurrence in animal cells (Fig. 12.3).

Lysosome polymorphism

Electron microscopy has demonstrated that animal cells contain other particles with hydrolytic activity in addition to the relatively simple structure described above, and various sub-types of lysosomes have now been defined. In fact, more than twenty-five names have been used to describe the various types of lysosome-like bodies found in animal cells. De Duve and Wattiaux (1966) have attempted to classify these various forms and while still retaining the word lysosome, they recognize the following sub-types in animal cells.

Fig. 12.2   Electron micrograph of isolated rat-liver lysosomes. ×47 000. Provided by Dr P. Baudhuin, Catholic University of Louvain.

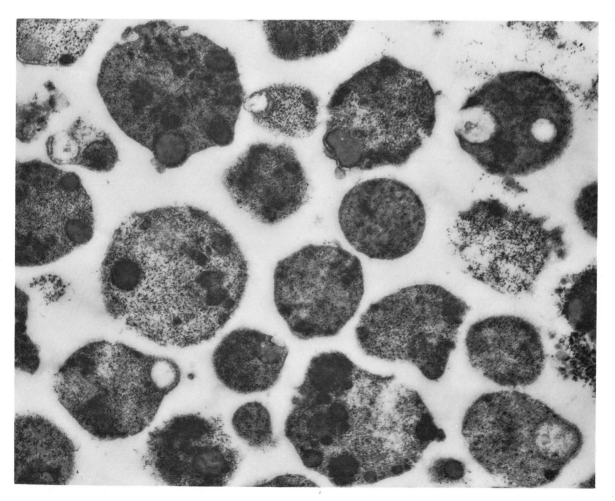

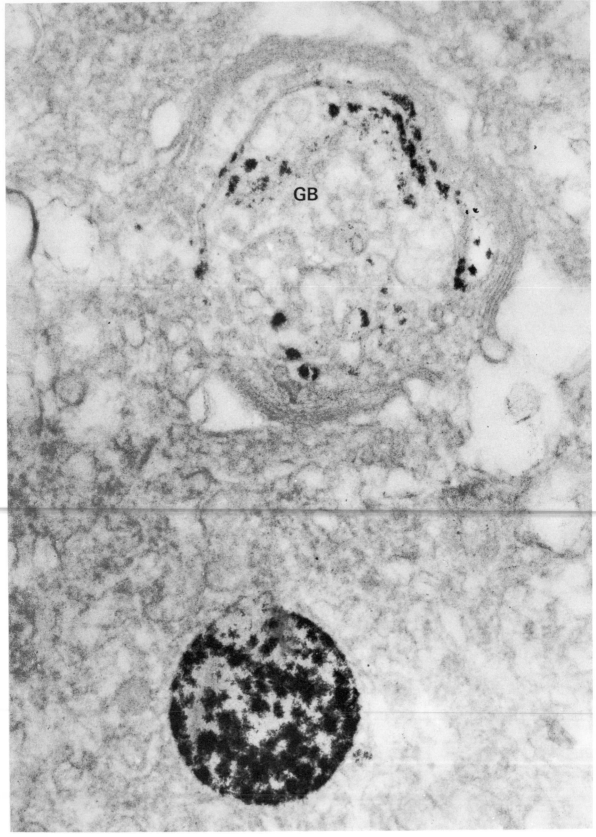

Fig. 12.3   Electron micrograph of part of a grasshopper neuron stained for acid phosphatase showing reaction product in a lysosome and in several Golgi saccules. GB, Golgi body. ×88 000. Reproduced from Lane (1968).

1. *Primary lysosomes* are particles whose enzymes have never been involved in a digestive event. The origin of these bodies is not fully understood, although evidence points to the Golgi system as the site of formation in certain cells, while parts of the ER close to the Golgi apparatus appear to be the origin in other cases (see below).

2. *Secondary lysosomes* are sites of present or past digestive activity and may be concerned with the digestion of material of both exogenous and endogenous origin.

3. *Pre-lysosomes* are enzyme-less members of the system. The most widely known are *phagosomes* which are phagocytic or pinocytic vacuoles which later fuse with lysosomes to produce digestive vacuoles.

4. *Post-lysosomes* or *residual bodies* are degenerate bodies which have lost their enzymes but may contain undigested residues.

The possible relationships between these various forms are illustrated in Fig. 12.4 and excellent discussions of the origin and interrelationship of these various types are to be found in the Further Reading list at the end of the chapter.

Fig. 12.4 Suggested origins, interrelationships and roles of lysosomes. Shaded areas indicate hydrolase activity. In animal cells, primary lysosomes arise from the Golgi apparatus via route I, and as vesicles derived from the ER (route II(1)) or from the GERL (route II(11)). See p. 514 for explanation. Autophagic bodies (secondary lysosomes) are thought to originate through route III. Route IV is also a possible primary lysosome source. In plants, route I and II are thought to predominate. Evidence for the existence of GERL is discussed in the text. In fungi, a special pathway (route V) may be important in extracellular secretion. Redrawn from Pitt (1975) after Novikoff and Holtzman (1970).

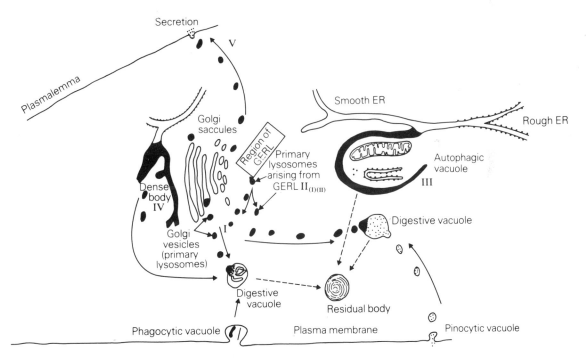

Functions of lysosomes

There is now a considerable volume of biochemical and cytochemical evidence which implicates lysosomes in a wide range of physiological processes and pathological disorders in animal cells. As these functions may be compared later with the possible functions of lysosomes in plant cells, some of them are discussed briefly below.

Lysosomes are involved in the digestion of a wide range of extracellular materials taken in by animal cells. These materials are initially stored in phagosomes which are devoid of acid hydrolase activity but later these bodies fuse with lysosomes to form secondary lysosomes and so allow digestion to begin. This mechanism may take the form of heterotrophic nutrition or may be an important defence against invading microorganisms or toxic macromolecules.

Another function of lysosomes is in the segregation and digestion of components from the surrounding cytoplasm, a process known as *cellular autophagy*. Bodies which resemble lysosomes have been observed under the electron microscope. They consist of a single unit membrane surrounding mitochondria, portions of endoplasmic reticulum, glycogen granules and other cytoplasmic particles in various stages of digestion. Cellular autophagy appears to be enhanced in cells undergoing reorganization during differentiation, although it is not clear how carefully this process is controlled and directed. Thus lysosomes may be involved in the turnover of cellular organelles and the clearance of dead cells, and may form a method of feeding during periods of starvation which avoids permanent damage to the cell.

Lysosomes are concerned in a wide variety of pathological processes, either because they perform their digestive functions improperly or because they perform them in a detrimental way. For example, in the inherited metabolic disease of humans known as metachromatic leukodystrophy, sulphated mucopolysaccharides accumulate in nervous tissue because of a defect in lysosomal aryl sulphatase activity. Again, animals with acute potassium deficiency exhibit increased lysosomal activity in their kidney cells which is believed to result in tissue damage. Lysosomes are also invariably involved in response mechanisms to invading agents of various types. If their defensive function fails, an invading organism may successfully multiply. Similarly the lysosomes may fail to inactivate engulfed toxins, or to destroy, and so regulate, endogenous substances that could build up to toxic levels.

Lysosomes in plant cells

The importance of lysosomal activity in animal cell metabolism stimulated research into the localization of hydrolytic enzymes in plant cells; present evidence suggests that these cells exhibit a similar compartmentation of much of their acid hydrolase activity. There are a number of situations in which lysosomal activity could have an important function in plant metabolism. As in all living cells, plant cells show metabolic turnover of their constituent molecules and organelles which presumably must be degraded at sites distinct from their sites of synthesis. In addition, there are a number of more specific stages in plant cell growth and development which require the participation of hydrolytic activity, including the mobilization of reserves in storage cells, the removal of unwanted materials during differentiation of cells such as the xylem elements and the digestion of damaged cells or cell parts. Evidence for the existence of plant lysosomes has come from two approaches: attempts to isolate plant lysosomes by cell fractionation and the demonstration of sites of hydrolytic activity by the use of microscopic cytochemical techniques with tissue sections.

Biochemical evidence of lysosomes in plant cells

Several early biochemical studies which attempted to isolate lysosomes from plant tissues obtained no positive evidence, perhaps because the methods employed by animal biochemists were applied too rigidly to the very different problems presented by plant tissues. However, encouraging results were reported by Matile and co-workers in Zurich in a number of papers published from 1965 (see Matile 1975, 1976, 1978). For example, Matile, Balz, Semadeni and Jost (1965) subjected extracts from seedlings of corn and tobacco to density gradient centrifugation and examined the various fractions obtained for four hydrolytic enzymes: protease, phosphatase, esterase and ribonuclease. They found that these enzymes were concentrated in two fractions, one of which contained small, single membrane-bound particles as the prominent structure, while the lighter fraction contained larger particles and large digestion vacuoles containing other membrane-bound structures. Not all of the hydrolytic activity was bound to particles since 35–80% of the hydrolase activity was recovered in the soluble fraction of the preparations. This activity may represent free enzymes or may be a result of particle disruption during homogenization. These particles were considered to represent the plant equivalent of animal lysosomes. Both particles have a single bounding membrane, contain acid hydrolase activity and occur in a variety of polymorphic forms.

In a later report, Matile (1968a) found nine hydrolytic enzymes in a 'mitochondrial' fraction isolated from maize root tips; these activities were concentrated largely into two major sub-fractions, called light and heavy lysosomes. The former consisted of vacuoles ranging from 0.3 to 1.5 μm in diameter while the latter were much smaller particles with diameters from 0.1 to 0.3 μm. Further centrifugation separated the heavy lysosome fraction into at least three sub-populations of particles, differing not only in their relative density in sucrose but also in their enzyme content. These different types of particle probably represent stages of differentiation of the lysosomal apparatus, perhaps analogous to the polymorphic forms found with animal lysosomes. The fine structure of the membranes of the lysosomes was similar to that of the endoplasmic reticulum. In addition, the smaller, heavy lysosomes contained certain oxidoreductases which are characteristic of the membranes of the endoplasmic reticulum. These properties were considered by Matile to suggest that the heavy lysosomes are primary derivatives of the endoplasmic reticulum and that synthesis of enzymes continues as the organelles develop and differentiate, a process similar to that postulated for lysosomes in certain animal cells.

The rather drastic nature of cell fractionation methods makes it difficult to identify these hydrolase-containing particles with structures in the intact cell. However, certain more gentle approaches have suggested that the main cell vacuole may be a major site of hydrolase activity in many cells. For example, Doi, Ohtsuru and Matoba (1975) separated the contents of the central vacuole of the giant algae *Nitella* and *Chara* by perfusion after excision of both ends of the large internodal cells. The sap was found to contain a large proportion of the hydrolytic enzymes, acid phosphatase and carboxypeptidase. A very different approach was used by Nishimura and Beevers (1978) who separated and purified vacuoles by gentle treatment of protoplasts (see Ch. 13) isolated from castor bean endosperm tissue. These vacuoles, which arise from the aleurone grains (see below and Fig. 12.10), were the primary sites of enzymes which hydrolyse protein, carbohydrate and phytin and so presumably play a dynamic role in the breakdown of protein which is a major feature of endosperm metabolism. Using a similar technique, Boller and Kende (1979) demonstrated that vacuoles isolated from tobacco, tulip and pineapple tissues were the primary sites of localization of six acid hydrolases; they considered that this supported the concept that the central plant vacuole is analogous to the animal lysosome.

Another potential type of hydrolase-containing particle is the *spherosome*. Spherosomes have been recognized in plant tissues for a long time, being first described as highly refractive bodies by Hanstein in 1880. They were initially called microsomes, but the name was later changed to avoid confusion with similarly named but structurally different particles described in certain animals. Spherosomes are easily recognized as spherical particles, usually between 0·5 and 1·0 μm in diameter, although they may be as large as 2·5 μm in diameter, which occur in abundance in many plant cells (see Fig. 9.7). They appear black under phase contrast microscopy and brightly shining when examined under dark field illumination. They stain with fat-soluble dyes such as Sudan black, Sudan III and Nile blue sulphate and usually have a high oil content. Examination under an electron microscope shows that they are surrounded by a half unit membrane and have a fine granular structure internally, indicative of proteinaceous stroma. The spherosome membrane has been examined in detail by Yatsu and Jacks (1972). It appears as a single line of material which measures some 2 to 3·5 nm wide and thus is about half the width of a normal unit membrane (see p. 138). Membrane structural protein appears to be present and the single line is interpreted as a half unit membrane with the non-polar surface of the lipid layer in contact with the internal storage lipid.

It seems that spherosomes probably originate from the endoplasmic reticulum. Oil accumulates at the end of a strand of endoplasmic reticulum and a small vesicle is then cut off by constriction to form particles which have been called prospherosomes. These grow in size to form spherosomes, although no basic structural changes take place during this stage of development. The function of spherosomes is still somewhat obscure although they clearly may be involved in lipid synthesis and storage (see Ch. 9).

Spherosomes have been isolated from tobacco endosperm tissue (Matile and Spichiger, 1968; Spichiger, 1969) and shown to contain over 90% of the lipid content of the endosperm and a large fraction of the total activities of a number of acid hydrolases (Table 12.2). During germination of the seeds, 90% of the reserve lipids are mobilized and this is accompanied by a temporary rise in the activities of some of the hydrolytic enzymes. These spherosomes are considered to be lysosomes, responsible not only for the accumulation and mobilization of reserve lipids, but also for the digestion of other cytoplasmic components incorporated by phagocytosis. Spherosomes isolated from *Sorghum* seeds have also been shown to contain considerable amounts of soluble hydrolytic enzymes, including ribonuclease, phytase and several glycosidases (Adams and Novellie, 1975). However, this interpretation of spherosomes as lysosomes has been questioned by Pitt (1975), who considers that spherosomes are quite distinct from lysosomes and that the associated hydrolytic activity might be an artefact caused by adsorption of enzymes during isolation.

Table 12.2 Subcellular distribution of hydrolases, lipid and protein in the endosperm cells of dry tobacco seeds. After separation of the oil droplets by flotation the extract was submitted to differential centrifugation yielding a mitochondrial (15 min 20 000 g), a microsomal (30 min 150 000 g), and a soluble fraction. The amounts and activities recovered in the individual fractions are referred to those present in the cell free extract (100%).

	Oil droplets	Mito-chondrial fraction	Micro-somal fraction	Soluble fraction	Extract	Recovery %
Lipid	93·4	1·0	0·67	0	100	95
Protein	23·4	34·0	37·8	6·4	100	102
Phosphatase	67·5	7·2	9·5	22·8	100	107
Esterase	70·0	9·5	11·1	28·8	100	119
Protease	66·0	11·3	3·5	18·8	100	98

Reproduced from Matile and Spichiger (1968)

Lysosomal properties have also been assigned to other storage particles, the *aleurone grains* or *protein bodies* which are approximately spherical in shape, surrounded by a single membrane and of wide occurrence in the endosperm and cotyledons of seeds. They are formed during the later stages of seed ripening and disappear in the early stages of germination. They store protein, chiefly in the form of globulins, and phosphate in the form of phytin, which consists of the insoluble salts of phytic acid or inositol hexaphosphoric acid (Fig. 12.5). It has been demonstrated that aleurone grains contain a wide range of hydrolytic enzymes, including protease and phosphatase which are required for the mobilization of stored protein and phosphate, although the presence of other enzymes such as β-amylase and RNAase suggest that other cell constituents may also be digested (Matile, 1968b, 1975; Pernollet, 1978). The presence of acid phosphatase activity in isolated aleurone grains confirms the observations of Poux (1965), who localized this activity in these particles using cytochemical procedures with the electron microscope. Aleurone vacuoles may therefore have similar functions to the lipid-containing

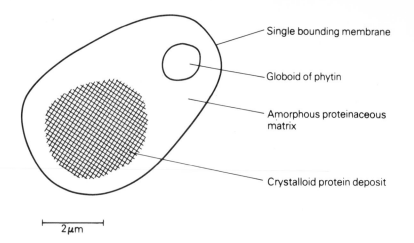

Fig. 12.5   Diagrammatic representation of a typical protein body (aleurone grain). They may be as large as 20 μm in diameter. Protein bodies from some species contain no inclusions or only one type of inclusion.

Single bounding membrane

Globoid of phytin

Amorphous proteinaceous matrix

Crystalloid protein deposit

2 μm

spherosomes isolated from tobacco seeds. They store reserve materials, mobilize them during germination and, in addition, form a compartment for the digestion of other cell components (see Fig. 12.10).

These biochemical studies have clearly demonstrated that plant cells contain a wide range of particles, varying in size, enzyme activity and internal contents, to which the name of lysosome or, at least, lysosome-like body may be applied. The particles are all similar in that they are surrounded by a single membrane, contain acid hydrolytic activity and are probably derived originally from the endoplasmic reticulum although they differ in their subsequent differentiation process.

Cytochemical evidence for lysosomes in plant cells

Cytochemical techniques at the resolution of the light microscope have enabled prominent spherical particles, staining for a number of hydrolytic enzymes, to be observed in sections cut from a wide range of plant materials. Their typical appearance is shown in Fig. 12.6. The particles range in size from 0·5 to 3 μm in diameter, frequently stain for lipids in addition to hydrolases, and are usually identified with spherosomes or lysosome-like bodies. In overall size range, lipid content and enzyme activity, they resemble the particles isolated by cell fractionation techniques from corn and tobacco by Matile and co-workers. In most of the cytochemical studies, latency properties have not been assigned to these bodies, although Gahan (1965) has reported that acid phosphatase-containing particles observed in bean roots showed latency properties when sections were treated with formaldehyde or subjected to repeated freezing and thawing. These treatments reduced the time required to stain these particles for acid phosphatase activity from 20 to 1 min which was interpreted as a result of increased access of substrate to enzyme due to disruption of the bounding membranes. Further cytochemical studies prompted Gahan and Maple (1966) to suggest that lysosomes may play an important rôle in the differentiation of xylem elements. Thus in protoxylem cells of

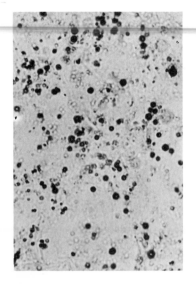

Fig. 12.6   Cortex cells of maize roots stained for acid phosphatase showing intense particulate activity. The technique is described in Ch. 1. × 1 100. Reproduced from Hall (1969).

510

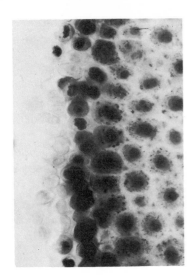

Fig. 12.7   Pea root cap cells stained for acid phosphatase activity showing high particulate activity in the young, inner cells and diffuse staining in the outer cells. ×306. Reproduced from Sexton, Cronshaw and Hall (1971).

bean roots at an early stage of differentiation, acid phosphatase activity was observed to be largely particulate and required a minimum staining time of 14 min. At a later stage of differentiation when further cellulose had been laid down in the cell walls, the minimum staining time was reduced to 4 min which suggested an increased membrane permeability. At an even later stage when some lignification had occurred and the cytoplasm reduced, the stain was observed to be diffuse with no sign of particulate activity. Finally, the fully lignified cells with no cytoplasmic contents showed no phosphatase activity. These observations led to the suggestion that lysosomes are involved in autodigestion of cell contents during stages of differentiation and that this is brought about by the release of hydrolytic enzymes into the cytoplasm. However, this interpretation has been questioned. Phosphatase activity appears to be associated with a variety of structures within these cells, and may not be representative of other hydrolase activity. It is possible that the release of acid phosphatase occurs as a result, rather than a cause, of cell death (Gahan, 1978).

A similar change in phosphatase staining has been observed in the developing root cap cells of pea (Sexton, Cronshaw and Hall, 1971; Fig. 12.7). In the young root cap cells and outer cells of the cortex, large particles staining for acid phosphatase activity were observed. As the cap cells mature, the staining becomes diffuse while the dying cells at the periphery of the root which will eventually be sloughed off show no appreciable phosphatase staining. These changes are presumably associated with the autolysis of cellular materials that accompanies senescence and death.

Further evidence for the presence of lysosomes has come from cytochemical studies with the electron microscope. In many plant cells the major sites of cytoplasmic acid hydrolase activity, which correspond with the particulate activity seen by light microscopy, are the membrane-bound vacuoles (Fig. 12.8). Furthermore, these vacuoles

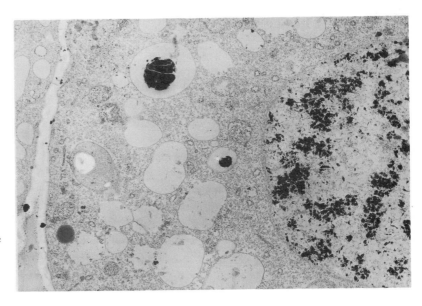

Fig. 12.8   Electron micrograph of part of cortical cell in root tip stained for acid phosphatase and showing large deposits of the reaction product, lead phosphate, in the vacuoles. ×9 400. Reproduced from Sexton, Cronshaw and Hall (1971).

are frequently observed to contain ribosomes, mitochondria and various membrane fragments and, in some instances, appear to be engulfing various cellular components by a process of phagocytosis (Fig. 12.9). These observations support the proposal of Matile (1968a, 1975, 1978) that the plant vacuole represents one type of plant lysosome which is similar to the autophagic vacuoles of animal cells (see Fig. 12.4) and shows a developmental sequence ending in the single, large vacuole of parenchymatous cells.

An interesting example of the lysosomal activity of plant vacuoles has been described by Villiers (1971) in a cytochemical study of *Fraxinus* seeds recovering from a period of enforced dormancy. During this period of dormancy, the membranes of various cell organelles suffer considerable damage and germination of the embryos

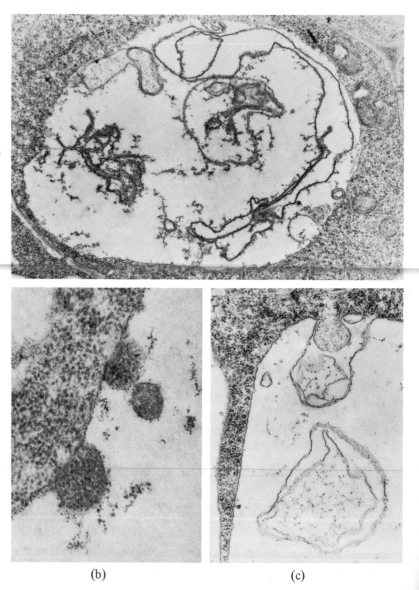

(a)

(b)

(c)

Fig. 12.9   (a) Electron micrograph of part of cortical cell in pea root tip showing vacuole with membrane invaginations and containing various cellular fragments. ×9 700.
(b-c) Electron micrographs of plant cell vacuoles showing the engulfing of cellular components by a process of phagocytosis. (b) ×45 000 (c) ×26 000. (b) reproduced from Hall and Davie (1971).

is delayed by comparison with the germination of normal embryos. However, during this delay, the vacuoles show heavy staining for acid phosphatase activity and are very active in engulfing and digesting various cell organelles. The delay in germination caused by the dormancy period was thought to result from the need to remove damaged organelles and to regenerate active ones.

Cell vacuoles also appear to be involved in the senescence of some tissues. A good example is provided by the short-lived corolla of morning glory, *Ipomoea purpurea* (Matile and Winkenbach, 1971). The rapid wilting of the corolla is accompanied by the breakdown of protein and nucleic acids and a sharp increase in the levels of deoxyribonuclease, ribonuclease and β-glucosidase activities; other hydrolases did not show a marked change in activity. Electron microscopy revealed prominent autophagic activity of the large central vacuole which contained ribosomes, mitochondria and other membranes in various stages of breakdown. These structures appear to be transported into the vacuole by invagination of the tonoplast. The cytoplasm gradually takes on a diluted appearance and eventually complete autolysis of the cellular contents occurs with the disruption of the tonoplast. The breakdown products of this autolysis are presumably exported to growing regions since the phloem of the vascular traces remains intact up to the final stages of the process.

The variety and origin of lysosomes in plant cells

Biochemical and cytochemical observations show that many plant cells contain particles that are approximately spherical, are bounded by a single membrane, usually show little internal structure, contain a variety of hydrolytic enzymes and may show some degree of polymorphism. In these respects they resemble the lysosomes of animal cells, although the majority of these reports have made no mention of latency properties, a characteristic feature of animal lysosomes. Matile (1969) has suggested that plant lysosomes should be defined simply as membrane-bounded cell compartments containing digestive enzymes. This definition therefore includes the spherosomes, aleurone grains and the vacuoles, and these bodies clearly have important rôles in the storage and mobilization of reserve materials and in intracellular digestive processes as discussed above. Future research may well reveal additional types of lysosomes in plant cells.

As in the case of animal cells, the origin of plant lysosomes is not at all clear although it has been suggested that vacuoles, aleurone grains and spherosomes are derived from the endoplasmic reticulum (Matile, 1975). This suggestion is based on ultrastructural studies of the development of these organelles, and on the presence of characteristic enzymes of the endoplasmic reticulum in preparations of isolated vacuoles. The provacuoles derived from the endoplasmic reticulum are analogous to the primary lysosomes of animal cells and develop by fusion to form larger vacuoles. Prospherosomes, on the other hand, develop into spherosomes by the accumulation of lipids which are then

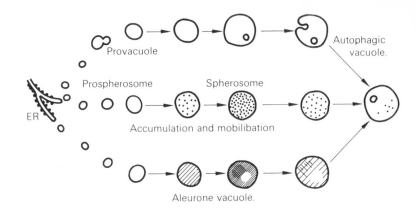

Fig. 12.10 Possible ontogeny and function of lysosome-like organelles in plant cells. Redrawn from Spichiger (1969).

Provacuole

Autophagic vacuole.

ER

Prospherosome

Spherosome

Accumulation and mobilibation

Aleurone vacuole.

Fig. 12.11 Electron micrographs of portions of meristematic cells from the root tip of *Euphorbia characias*. (A), (C), and (D) incubated for acid phosphatase activity showing intense staining in the GERL and provacuole channels. (B) stained with zinc oxide/osmium tetroxide showing stained of the GERL and a nascent provacuole. (C) and (D) are of thick sections examined by high voltage electron microscopy. (A) × 40 000; (B) 32 500; (C) 29 000; (D) × 16 000. gb, Golgi body; ge GERL; pv, provacuole. Reproduced from Marty (1978).

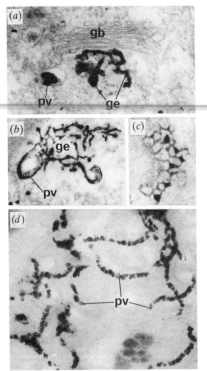

mobilized at a later stage. These ideas have been summarized by Spichiger (1969) in the scheme presented in Fig. 12.10.

The precise pathway of generation of vacuoles from the ER in meristematic cells has been the subject of a detailed study by Marty (1978) using various cytochemical methods with *Euphorbia* roots. These included techniques for the localization of acid phosphatase and esterase activities (see p. 27), which are believed to be marker enzymes for lysosomes, and incubation in a mixture of zinc oxide and osmium tetroxide which specifically stains the endomembrane system (see p. 497) in many cells. The latter was also used in conjunction with high voltage electron microscopy which allows thicker sections to be examined and so gives an insight into the three-dimensional arrangement of cell structures. These studies suggest that the provacuoles are produced from the GERL (Golgi-associated ER from which lysosomes form) as postulated for some animal cells (see Fig. 12.4). This consists of a network of anastomizing tubules situated at the maturing face of a Golgi stack; the provacuoles appear to be budded from the junctions of the anastomizing tubules (Fig. 12.11). In more differentiated cells, provacuoles are elongated into tubes (Fig. 12.11) which may become wrapped around portions of the cytoplasm to produce autophagic vacuoles. Lysosomal enzymes appear to be concentrated and packaged in the provacuoles which swell and fuse to give larger vacuoles; these continue to collect the GERL-derived vesicles throughout the life of the cell.

Other functions of vacuoles

The concept that plant vacuoles function as lysosome-like organelles is a relatively recent one. In many mature cells vacuoles form the largest distinct compartment (see p. 11) and they have traditionally been reviewed as structureless spaces which may act as storage sites for a wide range of substances. These include inorganic ions, sugars, organic acids, amides and amino acids, gums, tannins, lipids, pigments and proteins. The latter include both storage proteins and hydrolytic enzymes and their functions have been discussed in detail above. The turgor pressure resulting from the accumulation of osmotically active solutes pushes the cytoplasmic layer against the cell wall and gives rise to the general rigidity of plant tissues and to the major force driving the extension growth of young cells.

The accumulation of various phenolic compounds in the vacuole is often easily detected, either because they are coloured (e.g. beta-cyanin in red beet storage tissue) or because they can be made electron dense (e.g. iron complexes of certain tannins) (Fig. 12.12). However, the reasons for the synthesis and accumulation of many of these secondary products of metabolism is not clear, although their sequestration in the vacuole may safely remove certain harmful compounds such as alkaloids.

Other accumulated compounds are important metabolites and the vacuoles are assumed to form important storage pools. For example, vacuoles of red beet contain up to 90% of the total sucrose while those of *Bryophyllum* (a CAM plant, see p. 274) contain almost all of the malic acid (see Matile, 1978). The sap which may be squeezed from lemon fruit may contain 0·3 M citric acid with a pH of 2·5; again this is considered to be stored in the vacuole since otherwise it would denature many important enzymes. Isotope labelling experiments provide clear evidence for the presence of discrete pools of certain metabolic intermediates which are not in free equilibrium with the major metabolic pathways. Such pools are probably an important means of maintaining homeostasis within the cytoplasm (Matile 1976,

Fig. 12.12   Electron micrograph of part of a white spruce cell showing tannin materials (TA) in the large central vacuole (V) and in smaller vacuoles which appear to be fusing with the large vacuole. Also present are small tannin-containing vacuoles (arrows) which may have arisen from the endoplasmic reticulum (ER). ×22 000. Reproduced from Chafe and Durzan (1973).

(a)

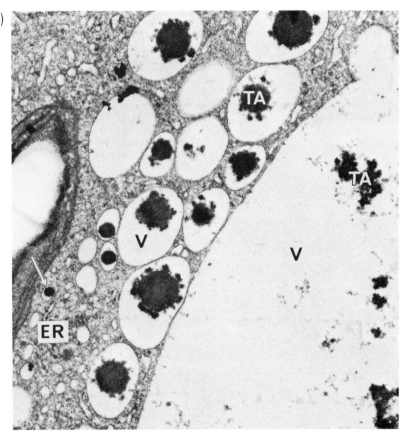

1978): plant cells may experience large changes in environmental conditions and the contents of the vacuolar sap provide a large internal environment which buffers the cytoplasm against these changes. The tonoplast is presumably equipped with a variety of specific permeases since the movement of substances between cytoplasm and vacuole appears to be carefully controlled (Matile, 1978). Although little is known of characteristics of such carriers in higher plant cells at the present time, recently developed techniques for the isolation of intact vacuoles (for example, see p. 508) should lead to rapid advances in this field.

Further Reading

Dean, R. J. (1977) *Lysosomes*. Edward Arnold, London.

Dean, R. J. (1978) *Cellular Degradative Processes*. Chapman and Hall. London.

De Duve, C. (1959) Lysosomes. In *Subcellular Particles*. Ed. T. Hayashi. Ronald Press, New York.

Matile, Ph. (1975) *The Lytic Compartment of Plant Cells*. Springer-Verlag Cell Biology Monographs, Vienna and New York.

Matile, Ph. (1976) Vacuoles. In *Plant Biochemistry*, 2nd edn. Eds. J. Bonner and J. E. Varner. Academic Press, New York.

Matile, Ph. (1978) Biochemistry and function of vacuoles. *Ann. Rev. Plant Physiol.*, **29**, 193.

Pernollet, J-C. (1978) Protein bodies of seeds: ultrastructure, biochemistry, biosynthesis and degradation. *Phytochemistry*. **17**, 1473.

Pitt, D. (1975) *Lysosomes and Cell Function*. Longman, London and New York.

Straus, W. (1967) Lysosomes, phagosomes and related particles. In *Enzyme Cytology*. Ed. D. B. Roodyn. Academic Press, London and New York.

Wanner, G., Formanek, H. and Theimer, R. R. (1981) The ontogeny of lipid bodies (spherosomes) in plants cells. Ultrastructural evidence. *Planta*, **151**, 109.

Literature cited

Adams, C. A. and Novellie, L. (1975) Acid hydrolases and autolytic properties of protein bodies and spherosomes isolated from ungerminated seeds of *Sorghum bicolor* (Linn.). Moench. *Plant Physiol.*, **55**, 7.

Boller, T. and Kende, H. (1979) Hydrolytic enzymes in the central vacuole of plant cells. *Plant Physiol.*, **63**, 1123.

Chafe, S. C and Duran, D. J. (1973) Tannin inclusions in cell suspension cultures of white spruce. *Planta*, **113**, 251.

de Duve, C. and Wattiaux, R. (1968) Functions of lysosomes. *Ann. Rev. Physiol.*, **28**, 435.

Doi, E., Ohtsuru, C. and Matola, T. (1975) Lysosomal nature of plant vacuoles. II. Acid hydrolases in the central vacuole of internodal cells of Charophyta. *Plant Cell Physiol.*, **16**, 581.

Gahan, P. B. (1965) Histochemical evidence for the presence of lysosome-like particles in root meristem cells of *Vicia faba*. *J. exp. Bot.*, **16**, 350.

Gahan, P. B. (1978) A reinterpretation of the cytochemical evidence for acid phosphatase activity during cell death in xylem differentiation. *Ann. Bot.*, **42**, 755.

Gahan, P. B. and Maple, A. J. (1966) The behaviour of lysosome-like particles during cell differentiation. *J. exp. Bot.*, **17**, 151.

Hall, J. L. (1969) Histochemical localization of β-glycerophosphatase activity in young root tips. *Ann. Bot.*, **33**, 399.

Hall, J. L. and Davie, C. A. M. (1971) Localization of acid hydrolase activity in *Zea mays* L. root tips. *Ann. Bot.*, **35**, 849.

Lane, N. J. (1968) Distribution of phosphatases in the Golgi region and associated structures of the thoracic ganglionic neurons in the grasshopper, *Malanoplus differentialis*. *J. Cell Biol.*, **37, 89.**

Marty, F. (1978) Cytochemical studies on GERL, provacuoles and vacuoles in root meristematic cells of *Euphorbia*. *Proc. natl. Acad. Sci. U.S.A.*, **75**, 852.

Matile, Ph. (1968a) Lysosomes of root tip cells in corn seedlings. *Planta*, **79**, 181.

Matile, Ph. (1968b) Aleurone vacuoles as lysosomes. *Z. Pflanzenphysiol.*, **58**, 365.

Matile, Ph., Balz, J. P., Semadeni. E. and Jost, M. (1965) Isolation of spherosomes with lysosomes characteristics from seedlings. *Z. Naturforschg.*, **20**, 693.

Matile, Ph. and Spichiger, J. (1968) Lysosomal enzymes in spherosomes (oil droplets) of tobacco endosperm. *Z. Pflanzenphysiol.*, **58**, 277.

Matile, Ph. and Winkenbach, F. (1971) Function of lysosomes and lysosomal enzymes in the senescing corolla of the morning glory (*Ipomoea purpurea*). *J. exp. Bot.*, **22**, 759.

Nishimura, M. and Beevers, H. (1978) Hydrolases in vacuoles from castor bean endosperm. *Plant Physiol.*, **62**, 44.

Poux, N. (1965) Localization de l'activite phosphatasique acide et des phosphates dan les grains d'aleurone. *J. Microscopie*, **4**, 771.

Sexton, R., Cronshaw, J. and Hall, J. L. (1971) A study of the biochemistry and cytochemical localization of β-glycerophosphatase activity in root tips of maize and pea. *Protoplasma*, **73**, 417.

Spichiger, J. U. (1969) Isolation and charakterterisierung von Spharosomen und Glyoxisomen aus Tabakendosperm. *Planta*, **89**, 56.

Villers, T. A. (1971) Lysosomal activities of the vacuoles in damaged and recovering plant cells. *Nature New Biol.*, **233**, 57.

Yatsu, L. Y. and Jacks, T. J. (1972) Spherosome membranes. Half unit-membranes. *Plant Physiol.*, **49**, 937.

13 Protoplasts

So far in this text we have dealt with the individual organelles and other components which together make up the plant cell. This chapter is concerned with the properties of intact, isolated cells which lack a cell wall. These protoplasts are usually obtained by digesting away the cell wall and are now used extensively in botanical research. The term protoplast is thus widely used today to describe isolated, wall-less cells which are bounded by the plasmalemma and contain all the normal cell constituents. This term differs somewhat from its historical usage in which it described the total living constitutents of the cell regardless of whether a wall was present or not. It must be pointed out however that isolated protoplasts should not simply be considered as cells which lack a wall. The methods used to remove the wall, whether enzymic or physical, and the response of the cells to the absence of a wall may produce significant physiological and structural changes in the protoplasts. Furthermore, the protoplast condition is not a static one since most of these cells will begin the regeneration of a new cell wall within a few hours of isolation and, under appropriate conditions, can be induced to divide, produce a callus, and even regenerate a whole plant. Thus the assumption that protoplasts behave as plant cells without walls should be treated with caution. Alternatively protoplasts may be regarded as stressed cells which require a recovery period and the results obtained with these cells, particularly in relation to their metabolism, may not be directly applicable to intact cells within whole plants. Some of the structural and metabolic changes which result from isolation are discussed in a later section. Nevertheless it must be remembered that protoplasts usually maintain active photosynthesis and other metabolic processes, and regenerate into perfectly normal cells or whole plants.

The use of isolated protoplasts opens up new approaches to a variety of problems in plant biology. They may be treated as cultured animal cells and induced to fuse. This introduces the potential to produce somatic hybrids and is particularly interesting in relation to plants that are normally sexually incompatible. The absence of the cell wall conveys further advantages in relation to the properties of the surface membrane. Although its properties may be altered during protoplast isolation, the exposed plasmalemma is more accessible to study, while particles are more readily taken up into the cell. This latter property is particularly useful to plant pathologists who can achieve synchronous, high infection rates by viruses using protoplasts.

The lack of a cell wall also means that protoplasts can be disrupted without the use of high shear forces needed with intact cells, and this property is proving to be valuable in the isolation of intact organelles. Finally, the stages and controlling factors in cell wall formation can be studied in protoplasts which are regenerating walls, although again the relevance of the findings to the processes that occur in the intact plant must always be questioned.

Isolation of protoplasts

Protoplasts can be isolated from plant tissues by both mechanical and enzymic means. The historical development of these techniques is fully described by Cocking (1972). If certain tissues are plasmolysed and then randomly sliced, some of the retracted protoplasts will be released undamaged from the cut cells. This method was first used at the end of the last century and is particularly applicable to storage tissues. However, the numbers of protoplasts produced are low and the method can only be applied to tissues that may be extensively plasmolysed. Thus protoplasts cannot be obtained from meristematic cells by this means.

These limitations in the mechanical method meant that most of the research effort in the preparation of protoplasts has been concerned with the development of methods which use cell wall-degrading enzymes to isolate protoplasts. Earlier studies attempted to use the gastric juice from snails but, from 1960 onwards, the enzymic approach has relied largely on preparations of pectinase (often called macerozyme) and cellulase extracted from fungi. This procedure has a number of advantages over the mechanical method (Cocking, 1972; Ruesink, 1971). The most important is that large numbers of protoplasts can now be produced from a wide variety of plant tissues including meristematic regions. In addition, less osmotic shrinkage is required and no cutting damage occurs. On the other hand, the enzymes, and perhaps impurities contained in the preparations, may have harmful effects on the structure and metabolism of the cells.

The isolation and regeneration of protoplasts by enzymic wall digestion has now become a routine research technqiue; the most important steps in this procedure are illustrated in Fig. 13.1. This is a simplified scheme and does not show the wide variation in conditions employed by different workers with different tissues (for details, see Bajaj, 1977; Cocking, 1972, 1974; Evans and Cocking, 1977). However, there are certain basic steps common to all of these procedures and these will be described here.

Leaf material is frequently used because the large surface: volume ratio allows rapid penetration of the enzyme mixture into the tissue. Sometimes the lower epidermis is peeled off or the tissue sliced to facilitate penetration. Pectinase and cellulase are normally required to digest the cell wall (see Ch. 10 for cell wall composition) and these may be applied together (the mixed method) or separately (the sequential method). However, the enzyme solution must contain a

Fig. 13.1 Diagrammatic representation of the stages involved in the isolation, culture and regeneration of plants from leaf protoplasts. Redrawn after Bajaj (1977).

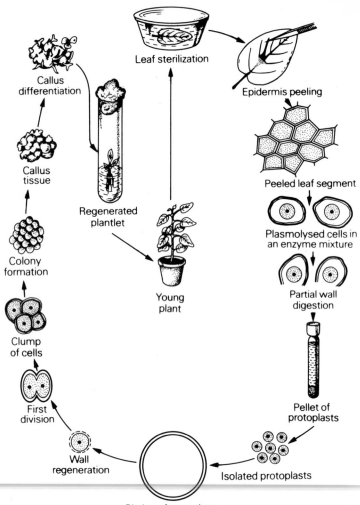

hypertonic osmoticum to stabilize the protoplasts after removal of the cell wall; if not present, the protoplasts simply swell with the uptake of water, eventually rupturing the plasmalemma. This osmotic stabilization is provided by salts, sugars or more commonly, sugar alcohols (e.g. 0·5–0·8 M mannitol or sorbitol).

After incubation in the enzymes, the protoplasts may be separated from the cell debris by a number of methods. These include passage through a nylon sieve, flotation on a hypertonic sucrose solution, or repeated resuspension and recentrifugation. Without a restricting cell wall, isolated protoplasts assume a spherical shape (Fig. 13.2); they are very delicate and must be handled with care. The absence of a cell wall is normally confirmed in two ways: by a negative response to treatment with fluorescent brighteners such as calcofluor which bind to cellulose and, more reliably, by the absence of surface microfibrils when examined by freeze-etching.

Fig. 13.2   Protoplasts isolated from tobacco leaves, clearly showing the peripherally situated chloroplasts. ×850.

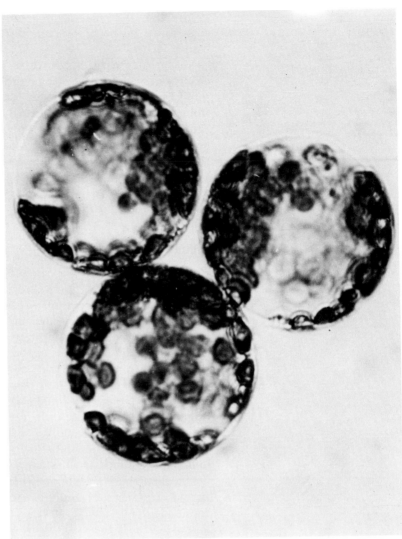

Fig. 13.3   Electron micrography of part of a isolated protoplast from tobacco leaves. Note the general negative-staining appearance of the chloroplasts (chl) and the presence of inclusion bodies (ib). pm, plasmalemma; v, vacuole. ×5 500.

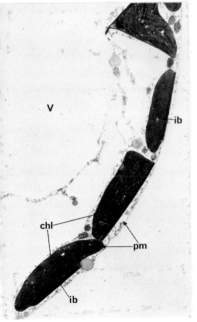

Properties of isolated protoplasts

As mentioned earlier, the extreme conditions required for the isolation of protoplasts may well affect their properties when compared to the intact cell. For example, the fine structure of protoplasts may show important differences from that of the host tissue. The most common difference is the appearance of so-called inclusion bodies in the chloroplasts of protoplasts (Takebe *et al.*, 1973; Taylor and Hall, 1978) (Fig. 13.3). These are semi-crystalline bodies which occur in the stroma and are believed to be a crystalline form of Fraction I protein (see Chs. 5 and 8). They are presumed to be a response to water stress produced by the plasmolysing isolation medium since they have also been reported in intact leaves submitted to various forms of stress. Other ultrastructural changes which have been observed in protoplasts include an increase in the number of osmiophilic bodies and a general negative–staining appearance of the organelles (Taylor and Hall,

1978). The latter may again result from water loss since a similar appearance has been observed in freeze-dried and freeze-substituted tissues.

Since water stress is known to induce a variety of physiological changes in plant tissues, isolated protoplasts might also be expected to show a response to the isolation treatment. Thus protoplasts isolated from maize leaves show a normal respiration rate and response to respiratory inhibitors although they show a marked increase in membrane permeability as judged by leakage of phosphate and rubidium (Taylor and Hall, 1976). Protoplasts may also show a reduced ability to take up amino acids (Ruesink, 1973), although spinach leaf protoplasts are as active as the intact tissues in their photosynthetic activities (Nishimura and Akazawa, 1975).

Culture of protoplasts

If isolated protoplasts are cultured in an appropriate liquid or agar medium, a high proportion will regenerate a cell wall, divide, produce a callus and, in some cases, eventually regenerate a whole plant (Figs. 13.1, 13.4). The culture media used are similar to those employed for the culture of intact cells except, of course, for the additional requirement of osmotic stability in the early stages.

It is interesting to note that cell wall regeneration was first demonstrated by Townsend in 1897 (see Cocking, 1972). He showed that plasmolysed protoplasts could regenerate a new cell wall *within* the original cell wall; in cases where the protoplasm separated into two halves, only the nucleated subprotoplast produced a new wall. The period required for the initiation of wall synthesis varies considerably in different systems. In some cases, cell wall fibres have been observed within 10 minutes of isolation, whereas others show a considerable lag period of perhaps 24 hours or more. However, once initiated, wall formation can be very rapid with a dense mat of fibres produced within a few hours. If the developing wall is then removed by a further treatment with dilute cellulase, new fibres will be produced within a few minutes of return to the culture medium (Burgess, Linstead and Bonsall, 1978). The first lag period is perhaps due to the initial trauma of isolation or to the need to stimulate the wall synthesizing systems in cells which were not primarily engaged in wall synthesis. In protoplasts isolated from *Skimmia* callus tissue, the cultured protoplasts show a marked increase in cell organelles, particularly the endoplasmic reticulum (ER), when compared with the fresh tissue (Robenek and Peveling, 1977). The ER often runs in parallel, and sometimes in contact, with the plasmalemma, and it appears that in this system the ER may be responsible for cell wall regeneration rather than the Golgi apparatus (see Ch. 11) which is only present in sparse amounts.

Cell wall regeneration is normally accompanied by nuclear division and cytokinesis; occasionally cytokinesis does not occur and binucleate cells are formed. Although there were many problems initially, the culture of these dividing protoplasts to produce cell colonies and eventually whole plants has now become a fairly routine procedure

Fig. 13.4   Plant production from leaf protoplasts of *Petunia parodii*.
1. Freshly isolated protoplasts; 2. First division of protoplast after 7 days;
3. Second division to produce four daughter cells (10 days); 4. Shoot differentiation on callus derived from protoplasts (3 months); 5. Adventitious root formation giving rise to plantlet; 6. Normal, fertile plant regenerated from mesophyll protoplast (6 months). Reproduced from Hayward and Power (1975).

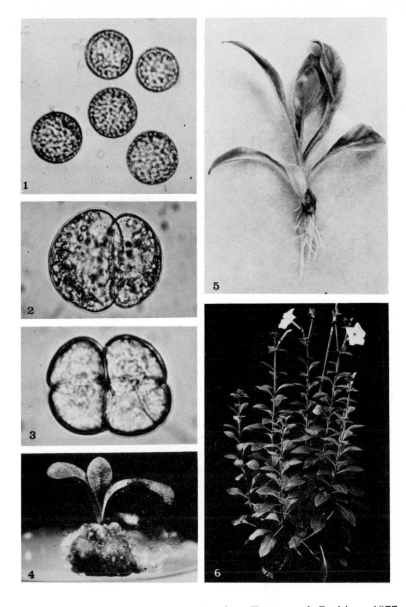

with an increasing range of species (see Evans and Cocking, 1977; Power, 1977). The protoplasts require a basic nutrient medium and a supply of growth regulators, although in some cases the latter requirement may be lost after cell wall regeneration. As cell division proceeds, the concentration of the osmotic stabilizer in the medium can be steadily reduced. Although whole plant regeneration from protoplasts has not been achieved with species from a wide range of taxonomic groups, this has not proved to be possible in all cases. Particular difficulty has been encountered with protoplasts isolated from cereal leaves which will regenerate a new cell wall, but will normally not divide. It is possible to generate whole plants from cultured cells of cereals but the reasons why isolated protoplasts fail to grow are not fully understood. A clue to the solution of this problem has come from studies on the forage crop plant *Sorghum bicolor* (Wernicke and Brettell, 1980). It was shown that only the young parts of the leaf cut from near the growing base have the capacity to regenerate into whole plants. The ability to express totipotency is rapidly lost during leaf maturation which may explain previous failures in the culture of cereal leaf protoplasts.

Use of protoplasts in physiological studies

The production of plant cells without cell walls introduces a new approach to many physiological and developmental problems. Potentially the most important is the use of those cells in somatic hybridization for the improved breeding of plants, and the advances that have been made in this direction are described in the next section. However, the development of methods for the routine isolation of large numbers of protoplasts has led to their use in a variety of other ways. For example, the delicate nature of protoplasts means that they may be disrupted by very mild procedures and this has led to an increasing use as starting material for the isolation of intact and functional cell organelles. Another use is in the study of cell wall synthesis although this possibility has yet to be fully exploited.

One of the most interesting applications involves the increased ability of these cells to take up large molecules (see Bajaj, 1977; Power, 1977). For some time after isolation, protoplasts can show quite extensive endocytotic activity and large molecules, such as ferritin, DNA and polystyrene beads, may be taken up (Fig. 13.5). It has also been shown that bacteria and whole organelles, such as chloroplasts and nuclei, can be introduced into protoplasts. In these cases, membrane fusion or lesions rather than endocytosis may be involved. The uptake of exogenous DNA is particularly interesting since this raises the possibility that the DNA may become stabilized and expressed, leading to genetic transformation of the host cells. Although it is now clear that DNA can be absorbed by protoplasts, unequivocal evidence for the expression and integration of this DNA is lacking. In most cases, it appears to be broken down quite rapidly.

Another important application is in the study of the mechanisms of infection and replication of viruses. With the mechanical barrier of the cell wall removed, protoplasts allow high synchronous infection rates to be achieved. This is normally not possible with intact tissue where infection depends on cut surfaces or damaged cells. The uptake of viruses is enhanced in the presence of polycations, such as poly-L-ornithine, which presumably neutralize the net negative charge at virus and protoplast surfaces and so allow binding to occur. However, it is not clear whether the major effect of the polycations is to enhance pinocytosis or to produce membrane damage which allows virus uptake.

The loss of the cell wall also exposes the plasmalemma and makes it more available for examination. For example, the carbohydrates present at the cell surface may be studied using a class of proteins known as *lectins* or *plant agglutinins* (see p. 141). Membrane carbohy-

Fig. 13.5   Endocytosis of polystyrene spheres by tobacco leaf protoplasts. The electron micrographs show various stages in the uptake of 0·234 μm diameter spheres (S) into vesicles. Note the absence of a cell wall. ER, endoplasmic reticulum; P, plasmalemma. (a); ×89 000. (b), ×120 000. Reproduced from Suzuki *et al.* (1975).

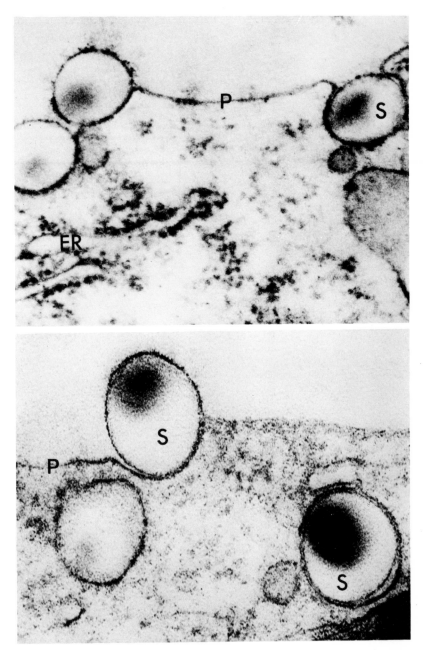

drates are present as glycoproteins or glycolipids. These have important functions in animal plasma membranes although their role in plants is unknown. The lectins have specific recognition properties with binding sites which react with specific sugars. They cause certain animal cells to agglutinate and it has recently been shown that they can have similar effects on plant protoplasts. This response can be used as a probe for the carbohydrates present at the plasmalemma. For example, concanavalin A (con A) causes the agglutination of protoplasts which implies that glucose or mannose are located at the cell surface. Lectin binding at the cell surface can also be detected by electron microscopy if the lectin is attached to an electron-dense label such as haemocyanin, ferritin or gold. Thus the binding of con A to the surface of protoplasts has been demonstrated using con A with colloidal gold or haemocyanin (Williamson *et al.*, 1976; Burgess and Linstead, 1976) (Fig. 13.6). Soybean protoplasts fixed before treat-

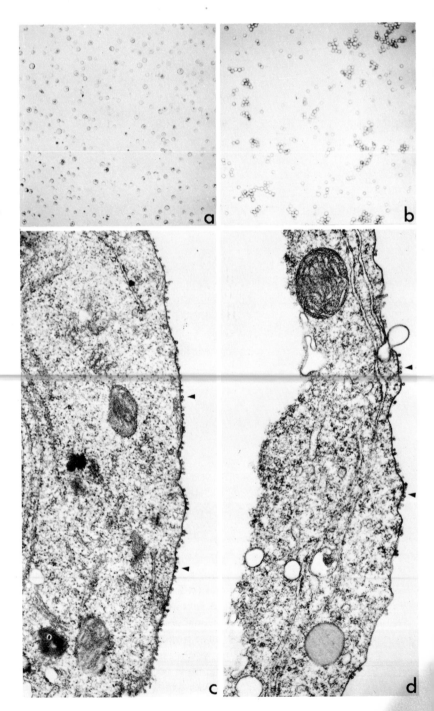

Fig. 13.6   Labelling of concanavalin A sites on the plasmalemma of soybean protoplasts. Figs. (*a*) and (*b*) are light micrographs showing protoplasts dispersed in a sorbitol medium in the absence (*a*) and presence (*b*) of con A. Note the agglutination of the protoplasts in (*a*). ×26. Figs. (*c*) and (*d*) are thin section micrographs showing haemocyanin molecules at the plasmalemma (arrows).
(*c*) Protoplasts treated with con A/haemocyanin after fixation showing an even distribution of the label. (*d*) Protoplasts treated with con A/haemocyanin without prior fixation showing the haemocyanin in tight clusters. Both ×30 000. Reproduced from Williamson *et al.* (1976).

ment with con A showed an even distribution of binding sites (Williamson *et al.*, 1976). In contrast, cells fixed after con A treatment showed an uneven distribution, presumed to be due to cross-linking of binding sites by the tetravalent con A. This means that the sites must be mobile and so is consistent with the fluid mosaic model of membrane structure (see Ch. 3). Similar observations have been made with a variety of animal cells.

Thus we have seen that the production of isolated protoplasts removes an important restriction at the cell surface and increases the experimental scope of the plant physiologist, particularly in relation to the properties of the plasmalemma. Perhaps the most important of these is that the barrier to cell fusion has been removed and so the opportunity of fusing protoplasts from different species to produce somatic hybrids is a real possibility.

Somatic hybridization

Somatic hybridization requires the fusion of somatic cells of different species followed by fusion of the nuclei in the resulting heterokaryon. This may allow a true somatic hybrid to be produced by regeneration of this fusion product. Hybrid cells have been produced by fusion of animal cells but plants have the advantage, as we have seen above, that they can be induced to regenerate a whole plant. The production of somatic hybrids is potentially of very great importance for plant breeding and crop improvement since hybrids may be produced between sexually incompatible plants. It may prove possible to fuse cells with very different biochemical properties. For example, nitrogen-fixing plants, such as blue-green algae, with certain crop plants which do not have this ability, to produce crops which use atmospheric nitrogen with less reliance on fertilizers.

Freely isolated protoplasts do not readily fuse together although spontaneous fusion sometimes occurs during isolation. This appears to take place only between adjacent cells and to be facilitated by the plasmodesmatal connections between them (Withers and Cocking, 1972; Evans and Cocking, 1977); removal of the cell walls allows the plasmodesmata to expand and mixing of the two cytoplasms to occur.

However, reproducible fusion of freely isolated protoplasts may be achieved even between species, by the inclusion of various fusion–inducing agents (Fig. 13.7). Sodium salts, particularly sodium nitrate, were the first to be used for this purpose and their effectiveness is thought to result from a lowering of the net surface negative charge which normally causes protoplasts to repel each other. Protoplasts can then adhere together, small localized regions of the membrane fuse, and eventually complete coalescence occurs. It seems that in addition to surface charge changes, some destabilization of the membrane must also be induced before fusion occurs.

Polyethylene glycol (PEG) in the presence of calcium ions is now the most widely used fusion-inducing agent. When protoplasts are incubated in the presence of PEG and calcium, considerable aggregation

Fig. 13.7 Light micrographs (Nomarski optics) showing various stages in the fusion of two maize root protoplasts. The micrographs (*a–f*) were taken 11, 17, 24, 25, 27 and 28 minutes respectively from the application of 0·3 M NaNO$_3$. ×823. Reproduced from Bright and Northcote (1974).

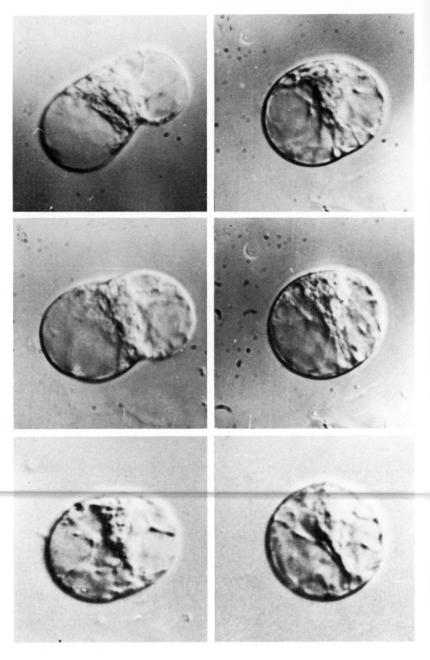

is observed and closely packed clusters are produced (Bajaj; 1977; Evans and Cocking, 1977). This adherence is not fully understood but is presumed to be due to the formation of cross-bridges between proteins in the plasmalemma of adjacent protoplasts. If the system is then disrupted by slowly diluting the PEG with nutrient medium, fusion occurs between a proportion of the protoplasts while others separate and return to their normal shape. Using this method, fusion has been induced both within and between species and even between hen erythrocytes and yeast protoplasts.

The formation of fusion products between different species is only the first, relatively simple, step in the achievement of somatic hybridization. However, nuclear fusion does not necessarily follow and chromosomes may be lost due to differences in the cell cycle or chromosome replication times between the two species (Evans and Cocking, 1977). Stable hybrid cells are therefore formed very infrequently and very careful selection procedures are needed to select the colonies produced by these cells from the numerous colonies produced by the parental protoplasts. For example, in the work of Melchers and Labib (1974), which is discussed below, only 12 hybrids were recovered in $2·2 \times 10^6$ calli which developed from protoplasts. The selection methods used to date are described below in relation to specific reports of somatic hybridization.

The first report that somatic hybridization had been achieved was published by Carlson, Smith and Dearing (1972) using two sexually compatible species of tobacco (*Nicotiana glauca* and *N. langsdorffii*). Selection of the hybrid was based on a previous knowledge of the nutrient requirements of the two species and of the sexual hybrid and so has limited application. Cells were fused in the presence of sodium nitrate and plated out on a medium which would not support the two species but would allow growth of the sexual (and presumably the somatic) hybrid. A few colonies were produced and, after further selection, developed shoots which were grafted on to stocks of *N. glauca* to produce whole plants. The shoots and flowers produced were characteristic of the sexual hybrid. The chromosome number (42) was a summation of the diploid somatic numbers of the parental species (24 + 18), and the electrophoretic pattern of peroxidase isoenzymes was identical to that of the sexual hybrid. Thus this variety of evidence demonstrated that the plants recovered from the fused cells resulted from somatic hybridization.

This report was followed by that of Melchers and Labib (1974) who obtained hybrids from the fusion of protoplasts derived from two chlorophyll-deficient, light sensitive strains of tobacco. The recessive, light sensitive mutants were fused and plated out under high light intensity. Several hybrid plants were recovered since the recessive mutants are presumably complemented in the hybrid to restore the normal green colour and allow growth under high light intensity.

A third approach to the selection of somatic hybrids, which makes use of the naturally occurring differences in the sensitivity of cultured plant protoplasts to growth media and drugs, has been described by Power *et al.* (1976) using two species of *Petunia* (*P. hybrida* and *P. parodii*). Protoplasts of *P. parodii* do not grow beyond small colonies on a medium devised by Murashige and Skoog (M/S medium) while those of *P. hybrida* produces a callus. However, the latter are more sensitive to the drug actinomycin D than protoplasts of *P. parodii*. Thus neither should grow on an M/S medium containing actinomycin D at a concentration which inhibits *P. hybrida*. The selection scheme for the production of somatic hybrids between the two species is shown in Fig. 13.8. If complementation occurs in the somatic hybrid, the *P. hybrida* genome would allow a callus to be produced while the

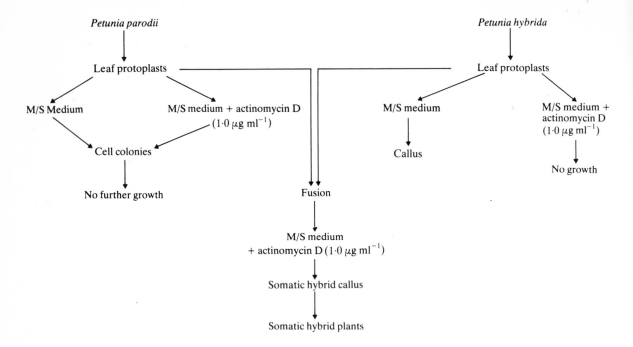

Petunia parodii

Leaf protoplasts

M/S Medium   M/S medium + actinomycin D
                         $(1{\cdot}0\,\mu g\,ml^{-1})$

Cell colonies

No further growth

Fusion

M/S medium
+ actinomycin D $(1{\cdot}0\,\mu g\,ml^{-1})$

Somatic hybrid callus

Somatic hybrid plants

Petunia hybrida

Leaf protoplasts

M/S medium   M/S medium +
                        actinomycin D
                        $(1{\cdot}0\,\mu g\,ml^{-1})$

Callus

No growth

Fig. 13.8 Selection scheme for the production of somatic hybrids of *Petunia parodii* and *P. hybrida* based on differential parental protoplast growth responses to M/S medium and actinomycin D. Divisions of *P. hybrida* protoplasts was limited by actinomycin D at 0·75–1·0 $\mu g\,ml^{-1}$ in M/S medium, yet protoplasts of *P. parodii* were not affected by up to 5·0 $\mu g\,ml^{-1}$. In animal cells, actinomycin D resistance is dominant and likely also to be so in *Petunia*. In M/S medium plus actinomycin D, protoplasts of *P. hybrida* would be expected to survive but not divide, while protoplasts of *P. parodii* would divide to the small colony stage. If complementation occurs in the somatic hybrid, the *P. hybrida* genome would permit division beyond the colony stage, while the *P. parodii* genome would confer resistance to actinomycin D. Reproduced from Power *et al.* (1976).

P. parodii genome would confer resistance to actinomycin D. Calli were selected by this procedure which regenerated into whole plants. These were similar to the sexual hybrid between these species with respect to flower colour and size (Fig. 13.9), and to their peroxidase isoenzyme pattern, both features showing important differences from the parents. Making use of such naturally occurring differences as drug sensitivity should eliminate the need for mutants and may be applicable to a wider range of species.

Thus the research described above clearly demonstrated that somatic hybridization is possible. The next step was to produce somatic hybrids between sexually incompatible species, and this now appears to have been achieved in a few cases. For example, Melchers, Scaristán and Holder (1978) fused mesophyll protoplasts from a tomato (*Lycopersicon esculentum*) mutant with protoplasts isolated from potato (*Solanum tuberosum*) callus; sexual hybrids between these plants do not appear to have been described. Some of the calli produced after culture of the fused protoplasts on a rich medium regenerated normal green shoots which were either transferred to soil or grafted on to a tomato stock. The hybrid nature of these plants was demonstrated by an analysis of the polypeptide pattern of the enzyme ribulose bisphosphate carboxylase prepared from leaf material. The small sub-unit of this chloroplast enzyme is synthesized in the cytoplasm under the control of the nuclear genes (see p. 219). Distinct polypeptide patterns were obtained after isoelectric focusing for the small sub-units isolated from tomato and potato. The plants produced after protoplast fusion as described above contained small sub-unit polypeptides which were characteristic of both tomato and potato, showing the plants to be somatic hybrids of the these species. The authors point out that it may be possible to produce sexual hybrids

Fig. 13.9 Flowers of (from top to bottom) tetraploid *Petunia parodii* (white), tetraploid F_1 hybrid (*P. hybrida* × *P. parodii*) (purple), somatic hybrid (purple), and tetraploid *P. hybrida* (red). The somatic hybrid was also distinguishable from either parent on the basis of corolla tube and peduncle length. Pollen of both parents was yellow, whereas that of the F_1 and somatic hybrids was occasionally purple. Reproduced from Power *et al.* (1976).

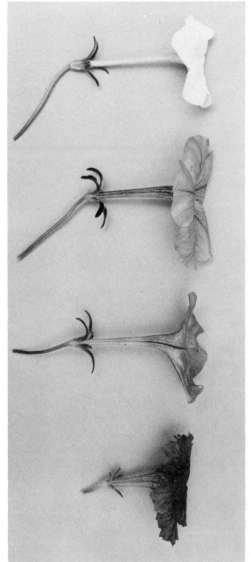

between these plants if techniques are improved. This has been the case in other examples where sexual hybridization was not thought to be possible. Nevertheless the research described above has shown the potential of this approach for plant breeding and crop improvement.

Further reading

Bajaj, Y. P. S. (1977) Protoplast isolation, culture and somatic hybridization. In *Plant Cell Tissue and Organ Culture*. Eds J. Reinert and Y. P. S. Bajaj. Springer-Verlag, Berlin, p. 467.

Butenko, R. G. (1979) Cultivation of isolated protoplasts and hybridization of somatic plant cells. *Int. Rev. Cytol.*, **59**, 323.

Carlson, P. S. (1973) The use of protoplasts for genetic research. *Proc. natl. Acad. Sci.*, *U.S.A.*, **70**, 592.

Cocking, E. C. (1972) Plant cell protoplasts-isolation and development. *Ann. Rev. Plant Physiol.*, **23**, 29.

Evans, P. K. and Cocking, E. C. (1977) Isolated plant protoplasts. In *Plant Tissue and Cell Culture.* Ed. H. E. Street, Blackwell, Oxford, p. 103.

Galun, E. (1981) Plant protoplasts as physiological tools. *Ann. Rev. Plant Physiol.*, **32**, 237.

Power, J. B. (1977) The physiology of isolated plant protoplasts. In *The Molecular Biology of Plant Cells.* Ed. H. Smith. Blackwell, Oxford, p. 418.

Literature cited

Bright, S. W. J., and Northcote, D. H. (1974) Protoplast regeneration from normal and bromodeoxyuridine-resistant sycamore callus. *J. Cell Sci.*, **16**, 445.

Burgess, J. and Linstead, P. J. (1976) Ultrastructural studies on the binding of concanavalin A to the plasmalemma of higher plant protoplasts. *Planta*, **130**, 73.

Burgess, J., Linstead, P. J. and Bonsall, V. E. (1978) Observations of the time course of wall development at the surface of isolated protoplasts. *Planta*, **139**, 85.

Carlson, P. S., Smith, H. H. and Deering, R. D. (1972) Parasexual interspecific plant hybridization. *Proc. natl. Acad. Sci., U.S.A.*, **69**, 2292.

Cocking, E. C. (1974) The isolation of plant protoplasts. In *Methods in Enzymology.* Vol 31. Eds. S. Fleischer and L. Packer. Academic Press. New York, p. 578.

Hayward, C. and Power, J. B. (1975) Plant production from leaf protoplasts of *Petunia parodii. Plant Sci. Lett.*, **4**, 407.

Melchers, G. and Labib, G. (1974) Somatic hybridisation of plants by fusion of protoplasts. *Molec. gen. Genet.*, **135**, 277.

Melchers, G., Sacristán, M. D. and Holder, A. A. (1978) Somatic hybrid plants of potato and tomato regenerated from fused protoplasts. *Carls. Res. Commun.*, **43**, 203.

Nishimura, M. and Akazawa, T. (1975) Photosynthetic activities of spinach, leaf protoplasts. *Plant Physiol.*, **55**, 712.

Power, J. B. Frearson, E. M., Hayward, C., George, D., Evans, P. K., Berry, S. F. and Cocking, E. C. (1976) Somatic hybridisation of *Petunia hybrida* and *P. parodii. Nature, Lond*, **263**, 500.

Robenek, H. and Peveling, E. (1977) Ultrastructure of the cell wall regeneration of isolated protoplasts of *Skimmia japonica* Thumb. *Planta*, **136**, 135.

Ruesink, A. W. (1971) Development of plant cells. In *Methods in Enzymology.* Vol. 23. Ed. A. San Pietro. Academic Press, New York, p. 197.

Ruesink, A. W. (1973) Surface membrane properties of isolated protoplasts. In *Protoplastes et fusion de cellules somatiques vegetales.* CNRS Colloques No. 212, Paris. p. 41.

Suzuki, M., Takebe, I., Kajita, S., Honda, Y. and Matsui, C. (1977) Endocytosis of polystyrene spheres by tobacco leaf protoplasts. *Expl. Cell Res.*, **105**, 127.

Takebe, I., Otsuki, Y., Honda, Y., Nishio, T. and Matsui, C. (1973). Fine structure of isolated mesophyll protoplasts of tobacco. *Planta*, **113**, 21.

Taylor, A. R. D. and Hall, J. L. (1976) Some physiological properties of protoplasts isolated from maize and tobacco tissues. *J. exp. Bot.*, **27**, 383.

Taylor, A. R. D. and Hall, J. L. (1978) Fine structure and cytochemical properties of tobacco leaf protoplasts and comparison with the source tissue. *Protoplasma*, **96**, 113.

Wernicke, W. and Brettell, R. (1980) Somatic embryogenesis from *Sorghum bicolor* leaves. *Nature, Lond.*, **287**, 138.

Williamson, F. A., Fowke, L. C., Constabel, F. C. and Gamborg, O. L. (1976) Labelling of concanavalin A sites of the plasma membrane of soybean protoplasts. *Protoplasma*, **89**, 305.

Withers, L. A. and Cocking, E. C. (1972) Fine-structural studies on spontaneous and induced fusion of higher plant protoplasts. *J. Cell Sci.*, **11**, 59.

Index

Cofactors, 93, 112–14
Colchicine, 175, 230, 466
Coleus, 465
Collenchyma, 432
Companion cells, 7
Compensation point, 336, 423
Competitive inhibition, 99, 100
Conformational coupling hypothesis, 314
Coniferyl alcohol, 442, 469
Co-transport, 150, 152
Coupling factor, 315, 332, 356–8
Coupling site, 295, 307, 353
Crassulacean acid metabolism, 274–6
Cristae, 283–5
Crossover theorem, 305, 307, 353
Crotonase, 411
Crystal-containing body, 404
Cucurbita pepo, 71, 76
Cyanide, 305
Cyclic AMP, 70
Cycloheximide, 217
Cysteine, 80, 211, 385, 386
Cytidine, 69
Cytochemistry, 25–8
Cytochalasin B, 234
Cytochrome, 113, 119, 298–301
    a, 128, 298–301, 305, 306
    b, 128, 298–301, 305, 311, 312, 316
    b_5, 135, 140
    b_6, 351
    c, 128, 298–301, 305, 311, 312, 316
    in chloroplasts, 349, 351
    in mitochondria, 298–301
Cytochrome oxidase, 113, 289, 305, 306, 417
Cytokinin, 204, 214
Cytoplastic streaming, 226, 227
Cytosine, 68, 69, 71, 72, 74, 76
Cytoskeleton, 238
Cytosome, 404

Danthonia bipartita, 377
Dark reactions, 336, 358–75
Daucus carota (carrot), 71, 180, 230
DBMIB, *see* dibromothymoquinone
Dehydroascorbate reductase, 249
Deoxyadenosine, 69
Deoxycytidine, 69
Deoxyribonucleic acid, *see* DNA
Deoxyribose, 47, 49, 68
Deoxythymidine, 69
Desaspidin, 353
3-3'-Diaminobenzidine (DAB), 405–7
Diasterioisomers, 49
Dibromothymoquinone (DBMIB), 353
3-(3, 4 Dichlorophenyl)-1, 1-dimethyl urea (DCMU), 343, 349, 350
Dictyosome, 11, 473–500
Didymium, 228
Difference spectra, 298, 300
Digitaria brownii, 373
Dihydrolipoate, 289, 290
Dihydrolipoate dehydrogenase, 289
Dihydroxyacetone, 46, 48
Dihydroxyacetone phosphate, 246, 363, 364, 393
Dinitrophenol, 309
Disaccharides, 55, 56

DNA, 68, 70–6, 157–9, 161, 164, 165, 171–5
    base composition, 70, 71, 172
    buoyant density, 73, 74, 323, 396
    in chloroplasts, 333, 394, 396–8
    and chromosomes, 164–7
    denaturation, 74–6
    genetic code, 164–7
    polymerase, 173–5, 323
    polymerase II, 174
    polymerase III, 174
    properties, 6, 73, 76
    replication, 172–5
    satellite, 74
    structure, 62, 63
Drosera, 484
Drosophyllum, 478
Dryopteris, 221

Elaioplasts, 333
Electrochemical potential, 148
Electron microscope, 14, 16, 17
    fixation, 18–21, 138, 139
    freeze etching, 22–4, 163
    limitations, 23–5
    negative staining, 22
    sample preparations, 21–3
    scanning, 28, 29
Electromagnetic spectrum, 339
Elodea, 227
Emerson enhancement effect, 343
EMP pathway, 240–51
Enantiomorphs, 46, 49
Endergonic reaction, 16
Endocytosis, 148, 154, 498, 525
Endoglucosidase, 441
Endoglycosidase, 441
Endomembrane system, 497–9
Endoplasmic reticulum, 11, 164, 191, 213, 425, 426
    rough, 191, 213
    smooth, 191
Endosymbiont hypothesis, 376, 397
End-product inhibition, 109
Energy charge, 108, 254
Energy-rich bond, 114
Enolase, 241, 247
cis-2-Enoyl CoA, 412
Enoyl hydratase, 411
Enoyl reductase, 269
Entropy, 125
Enzymes, 85, 92–112
    activators, 112
    active site, 104–6
    allosteric, 109–11, 254
    classification, 102, 103
    co-enzymes, 112–24
    cofactors, 93, 112–24
    constitutive, 112
    cytochemistry, 27, 28, 504, 510, 511
    effects of pH, 97, 98
    effect of temperature, 98, 99
    induced-fit hypothesis, 106–19
    inducible, 112
    inhibitors, 99–102
    kinetics, 94–7
    lock-key hypothesis, 106
    mechanism of action, 104–7

Glycerate, 417, 421
Glycerate kinase, 422
Glycerol, 60
Glycerol ester hydrolase, 409
Glycerol kinase, 271
Glycerol 3-phosphate, 62
Glycine, 80, 83, 124, 211, 317, 367
   in glycolate cycle, 417, 420-2
Glycine max (soybean), 61, 202, 214, 215, 266, 526
Glycolate, 416-23
Glycolate oxidase, 415, 419, 420, 304
Glycolate pathway, 417-23
Glycolate phosphatase, 418
Glycolysis, 240-55, 307
   energy charge, 253, 254
   energy conservation, 246-8
   evidence for, 251
   priming reactions, 244-6
   provision of carbohydrate, 242-4
   regulation, 253-5
   sum, 248, 249
Glycolipids, 65, 66
   in cell membranes, 135
Glycophorin, 140, 145
Glycoproteins, 58, 85, 443, 445, 447, 496, 497
Glycosides, 44, 45, 456-60
   formation of, 456-60
Glycosylation of protein, 496
Glyoxylate cycle, 252, 274, 413-16
Glyoxysome, 252, 403, 408-16
Golgi body, 12, 432, 463, 464, 473-500
   and cellulose synthesis, 492
   and cell wall synthesis, 484-92
   cisternae, 473, 475, 477
   differentiation, 478, 479, 480
   distribution, 474, 476
   enzymes, 494-7
   enzyme markers, 495, 496
   forming face, 478
   function, 483, 494
   glycosylation of protein, 496, 497
   isolation, 479-83
   maturing face, 478, 479
   origin, 497
   secretory function, 492, 493
   structure, 476, 477
   vesicles, 476-8
Gossypium hirsutum (cotton), 71, 265
Grana, 330, 331
Guanine, 68-72, 74, 76
Guanosine, 69

Haem, 119, 120
Haemoglobin, 119, 120
Hatch-Slack pathway, 368-75
Helianthus annuus (sunflower), 61
Helianthus tuberosum, 169, 179
α-Helix, 88
Hemerocallis fulva, 202
Hemicellulase, 441
Hemicellulose, 434, 436, 437, 446, 447, 453, 454
Heptose, 47
Heterochromatin, 158, 164, 170
Hexokinase, 113, 241, 244, 252, 459
Hexose, 47-54
Hill reaction, 337, 341, 343, 396

Histidine, 81, 211, 386
Histones, 159, 165, 186
Homogenization, 30-2
Hordeum vulgare (barley), 29, 183, 214
Hydrogen peroxide, 403, 416-18
Hydrolase, 102
D-3-Hydroxyacyl-CoA, 412
L-3-Hydroxyacyl-CoA dehydrogenase, 411
L-3 Hydroxyacyl-CoA, 410, 411
β-Hydroxyacyl dehydrase, 269
β-Hydroxybenzaldehyde, 442
α-Hydroxy α-carboxyglutarate, 424
p-Hydroxycinnamyl alcohol, 442
Hydroxyphenylpropane, 442
Hydroxyproline, 443, 444, 468, 492
3-Hydroxypropionyl-CoA, 411
Hydroxypyruvate, 420, 421
Hydroxypyruvate reductase, 421, 422
Hypochromism, 74-6, 78

Immunofluorescent labelling, 373
Indole acetic acid, 450-3
Inhibition
   competitive, 99, 100
   end product, 102
   feedback, 102
   irreversible, 99
   non-competitive, 100, 101
   reversible, 99
myo-Inositol, 45-63
   and cell wall metabolism, 459, 466, 467, 494
myo-Inositol oxygenase, 467
myo-Inositol 1-P-phosphatase, 467
Interphase, 157
Intracristal space, 283
Inulin, 57, 58
Invertase, 242, 445
Iodoacetate, 251
Ionophore, 154, 319
Ipomeaea purpurea (morning glory), 513
Iris, 82
Iron sulphur protein, 301, 302, 304
Isocitrate, 104, 289, 413, 414
Isocitrate dehydrogenase, 110, 289, 291, 413
Isocitrate lyase, 427
Isocitric enzyme, 272
Isocitritase, 412, 414
Isoelectric point, 83
Isoleucine, 80, 110, 211, 385
Isomerase, 103, 257
Δ^3-Isopentyl pyrophosphate, 66
Isoprene, 66

Juniperus chinensis, 229
Juglans regia (walnut), 61

Kalanchoe, 273
3-Keto-acyl-CoA, 410, 411
β-Ketoacyl dehydrogenase, 269
α-Ketoglutarate dehydrogenase, 289-92, 424
Ketose, 46
Klebsiella aerogene, 380
Kranz-type anatomy, 368

Lactate, 248-50, 416
Lactate dehydrogenase, 113, 250

Mitochondria (*cont.*)
   transport processes, 318–20
   volume changes, 319, 320
Mitosis, 164–6
   spindle, 166–7
Molecular hybridisation, 65, 66
Monactin, 154
Monosaccharides, 49–55, 57
   heptose, 47
   hexose, 47, 49–55
   mutarotation, 50–2
   pentose, 47–9
   ring structure, 50–4
   triose, 46, 47
Monosomes, 192, 194
Mucoprotein, 85
Multi-enzyme complex, 108
Myoglobin, 90

NAD, *see* Nicotinamide adenine dinucleotide
NADH oxidase, 249
NADH oxidation, 249, 250
NADP, *see* nicotinamide adenine dinucleotide phosphate
NAD pyrophosphorylase, 117
Naphthoquinone, 3
Negative staining, 22, 284, 477
Nernst equation, 149
Neurospora, 110, 194, 325, 407
Nicotinamide adenine dinucleotide (NAD), 116–18, 128, 129
   absorption spectrum, 117
   in β-oxidation, 411
   in glycolysis, 246, 247
   in mitochondria, 290, 296, 297, 303, 305
   oxidation, 248–50, 302, 303, 305, 308
   in pentose phosphate pathway, 263
   structure, 116
Nicotinamide adenine dinucleotide phosphate (NADP), 116–18, 128, 349
   in fatty acid synthesis, 266
   in pentose phosphate pathway, 263, 264
   in photosynthesis, 349–52
   structure, 116
Nicotiana tabacum (tobacco), 194, 216, 217, 404–7, 509, 521
Nitella, 3, 132, 133, 227, 396, 453, 508
Nitrate reductase, 112, 379
Nitrite reductase, 112, 379
Nitrogenase, 375, 376
Nitrogen metabolism, 375–89
Nodules, 377
Non-competitive inhibition, 99, 100, 101
Nonsense triplets, 211
Nostoc glauca, 211, 529
Nucleic acids, 43, 45, 67–78
Nucleoid, 157
Nucleolus, 9, 10, 158, 167–70, 198
Nucleolar organizer, 169, 170
   RNA synthesis, 176–82
Nucleoprotein, 85
Nucleosides, 68–70
Nucleoside diphosphate sugars, 116, 456–60
   formation, 457
   interconversion, 458
Nucleosides, 68–70
Nucleosome, 165, 166
Nucleotide diphosphate kinase, 292
Nucleotides, 68–70

Nucleus, 10, 12, 20, 157–90, 527, 529
   chromatin, 157, 158, 164, 186
   chromosomes, 11, 158, 164–6, 184
   DNA, 157–61, 164, 170–5, 184–7
   envelope, 157, 161–4, 497, 498
   functions, 170–82
   pores, 161–4
   protein, 159, 186
   RNA, 159, 167, 185–8
   sap, 157
   structure, 161–70

Okazaki fragments, 174
Oleic acid, 60, 61, 269, 270
Oligomycin, 309, 315
Oligosaccharides, 55, 56
Oocystis marsonii, 24
Optical isomerism, 45, 46
Organelle
   chloroplast, 11, 372–402
   definition, 8, 10
   dictyosome, 11, 473
   Golgi body, 11, 473–500
   lysosome, 11, 501–17
   microbody, 11, 403–29
   mitochondrion, 11, 281–326
   nucleus, 11, 157–90
   ribosome, 11, 191–223
   vacuole, 11, 506–16
Osmium tetroxide, 19, 138, 227, 282, 514
Oxalate, 420
Oxaloacetate, 117
   and β-carboxylation, 273–7
   in glyoxylate cycle, 413–16
   in photosynthesis, 370–2, 393
   in TCA cycle, 289, 291, 294
Oxidation, 127
β-Oxidation, 265, 317, 409–13
   and gluconeogenesis, 415
   stoichiometry, 413
Oxidation-reduction potential, 127
Oxidative phosphorylation, 281, 295–316
   chemical coupling hypothesis, 308, 309
   chemi-osmotic hypothesis, 309–16
Oxidoreductase, 102, 507
Oxygenase, 270, 366, 418
Oxygen electrode, 295

P700, 346, 348–51, 354
Palmitic acid, 60, 61
   synthesis of, 266–8
Paramural body, 493
Parenchyma, 7, 8
Pasteur effect, 255
Pectic acid, 435
Pectin, 434–6, 440, 446, 453, 454
Pectinase, 441
Pectin methylesterase, 445
Pentose, 47, 49
Pentose phosphate pathway, 255–64
   as a cycle, 260
   control, 263
   evidence, 259
   function, 255, 256
   NADH/NADPH, 109, 263
Peptides, 84, 85, 114

Ultracentrifuge
 analytical, 33–5, 216
 preparative, 33, 224, 408
Ulva taeniata, 340
Uncoupling agent, 295, 309, 314
Unit membrane, 138
Units of measurement, 2
Uracil, 68, 69, 74, 76
Urate oxidase, 405, 415
Uric acid, 415
Uricase, 415
Uridine, 69
 nucleotides, 115, 116
Uronic acids, 58, 59

Vacuole, 11, 511–14
Valine, 80, 211, 385
Valonia, 3, 439, 450
Vanillin, 442
Vanilomycin, 154
van't Hoff rule, 49
Vessels, 7, 8
Vicia faba (broad bean), 170, 213, 389
Vicilin, 214
Vitamin K_1, 350

Waxes, 62
Wobble hypothesis, 204

Xanthophylls, 341
X-ray diffraction, 72, 87–9, 106, 107
Xylan, 436
Xylem, 7, 8, 465, 506, 510
Xyloglucan, 436–8
Xylose, 47–9, 57
 in hemicellulose, 436
Xylulose, 47, 48
 5-phosphate, 122
  in pentose phosphate pathway, 257–9
  in photosynthesis, 363–5, 418

Zea mays (corn, maize), 3, 6, 43, 61, 194, 221, 225, 234, 236, 244, 251, 262, 282, 322, 334, 368, 369, 389, 473, 481, 482, 484, 510, 528
Z scheme, 349
Zwischenferment, 256
Zwitterion, 82
Zymogen granules, 478, 487